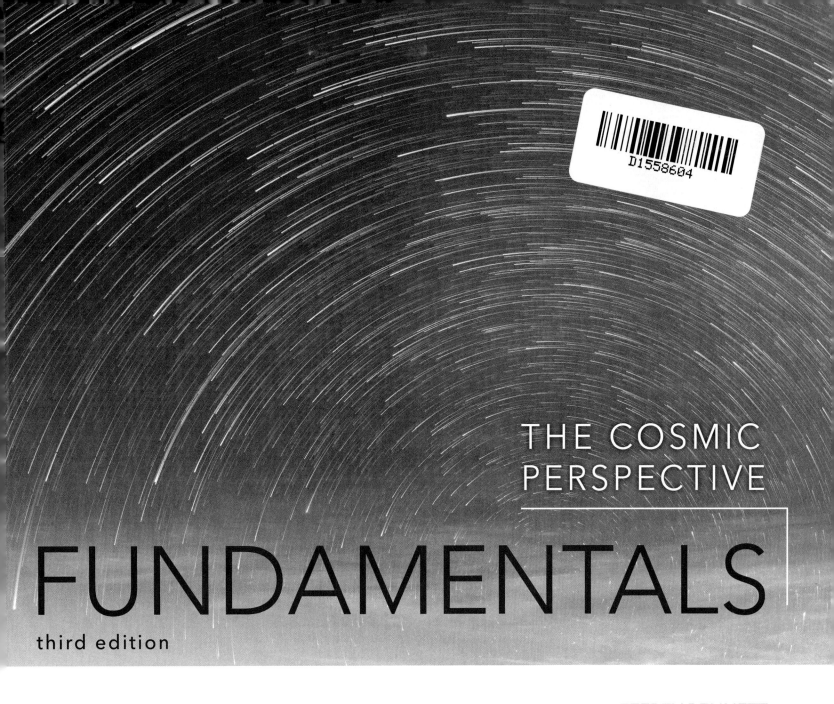

THE COSMIC PERSPECTIVE
FUNDAMENTALS

third edition

JEFFREY BENNETT
University of Colorado at Boulder

MEGAN DONAHUE
Michigan State University

NICHOLAS SCHNEIDER
University of Colorado at Boulder

MARK VOIT
Michigan State University

Director Physical Science Portfolio: Jeanne Zalesky
Physics & Astronomy Portfolio Analyst: Ian Desrosiers
Content Producer: Mary Tindle
Managing Producer: Kristen Flathman
Product Portfolio Editorial Assistant: Frances Lai
Rich Media Content Producer: Ali Candlin
Full-Service Vendor: Lifland et al., Bookmakers
Copyeditor: Lifland et al., Bookmakers
Compositor: SPi Global
Design Manager: Jerilyn Bockorick
Interior and Cover Designer: Alisha Webber

Illustrators: Rolin Graphics; Troutt Visuals
Rights & Permissions Project Manager: Ben Ferrini
Rights & Permissions Management: Matthew Perry
Photo Researcher: Matthew Perry
Manufacturing Buyer: Stacey Weinberger
Director of Field Marketing: Tim Galligan
Director of Product Marketing: Allison Rona
Field Marketing Manager: Yez Alayan
Product Marketing Manager: Elizabeth Bell

Cover Photo Credit: Rawin Cheasagul/Getty Images

Copyright © 2020, 2016, 2010 Pearson Education, Inc. 221 River Street, Hoboken, NJ 07030. All Rights Reserved. Printed in the United States of America. This publication is protected by copyright, and permission should be obtained from the publisher prior to any prohibited reproduction, storage in a retrieval system, or transmission in any form or by any means, electronic, mechanical, photocopying, recording, or otherwise. For information regarding permissions, request forms and the appropriate contacts within the Pearson Education Global Rights & Permissions department.

Attributions of third party content appear on pages 271 and 272, which constitute an extension of this copyright page.

PEARSON, ALWAYS LEARNING and Mastering™ Astronomy are exclusive trademarks in the U.S. and/or other countries, owned by Pearson Education, Inc. or its affiliates.

Unless otherwise indicated herein, any third-party trademarks that may appear in this work are the property of their respective owners and any references to third-party trademarks, logos or other trade dress are for demonstrative or descriptive purposes only. Such references are not intended to imply any sponsorship, endorsement, authorization, or promotion of Pearson's products by the owners of such marks, or any relationship between the owner and Pearson Education, Inc. or its affiliates, authors, licensees or distributors.

Library of Congress Cataloging-in-Publication Data

Names: Bennett, Jeffrey O., author. | Donahue, M. (Megan), 1962- author. |
  Schneider, Nicholas, author. | Voit, Mark, author.
Title: The cosmic perspective fundamentals / Jeffrey Bennett (University of
  Colorado at Boulder), Megan Donahue (Michigan State University), Nicholas Schneider
  (University of Colorado at Boulder), Mark Voit (Michigan State University).
Description: Third edition. | Hoboken, NJ : Pearson, c2020.
Identifiers: LCCN 2018051683 | ISBN 9780134988504
Subjects: LCSH: Astronomy--Textbooks. | Astronomy--Study and teaching (Higher)
Classification: LCC QB43.3 .C683 2020 | DDC 520--dc23
LC record available at https://lccn.loc.gov/2018051683

ISBN 10: 0-134-98850-7;
ISBN 13: 978-0-134-98850-4 (Student edition)

www.pearson.com

Dedication

TO ALL WHO HAVE EVER WONDERED about the mysteries of the universe. We hope this book will answer some of your questions—and that it will also raise new questions in your mind that will keep you curious and interested in the ongoing human adventure of astronomy.

And, especially, to Michaela, Emily, Sebastian, Grant, Nathan, Brooke, and Angela. The study of the universe begins at birth, and we hope that you will grow up in a world with far less poverty, hatred, and war so that all people will have the opportunity to contemplate the mysteries of the universe into which they are born. ✶

Brief Contents

- **1** A Modern View of the Universe   1
- **2** Understanding the Sky   17
- **3** Changes in Our Perspective   35
- **4** Origin of the Solar System   53

- **5** Terrestrial Worlds   74
- **6** The Outer Solar System   94
- **7** Planets Around Other Stars   113
- **8** The Sun and Other Stars   128

- **9** Stellar Lives   146
- **10** The Bizarre Stellar Graveyard   166
- **11** Galaxies   182
- **12** Galaxy Distances and Hubble's Law   200

- **13** The Birth of the Universe   215
- **14** Dark Matter and Dark Energy   232
- **15** Life in the Universe   252

CREDITS   271
APPENDIXES   A-1
GLOSSARY   G-1
INDEX   I-1

Explore Modern Astronomy and Its Connections to Our Lives

The Cosmic Perspective Fundamentals provides a brief, engaging, and up-to-date introduction to astronomy for anyone who is curious about the universe. As respected teachers and active researchers, the authors present astronomy using a coherent narrative and a thematic approach that engages students immediately and guides them through connecting ideas. The **Third Edition** has been fully updated with the latest scientific discoveries, including detection of gravitational waves, results from recent planetary missions, and new insights into extrasolar planets. **Mastering Astronomy** includes a wealth of author-created resources for students to use before, during, and after class.

A Clear Framework Makes Learning Targeted and Expectations Clear for Students . . .

Each chapter begins with an opening page that includes a brief overview of the chapter content and a clear set of **Learning Goals** associated with the chapter. Each Learning Goal is phrased as a question to engage students as they read. Each section is written to address the **Learning Goal questions** from the chapter-opening page.

For more than four centuries after the Copernican revolution taught us that Earth is just one member of our Sun's planetary system, the study of planetary systems remained limited to our own. Then, less than three decades ago, a new scientific revolution began with the first discoveries of planets around other stars. The image above, from the Large Binocular Telescope, shows infrared light from four planets (marked b, c, d, e) orbiting the star HR 8799; light from the star itself (center) was mostly blocked out during the exposure, as indicated by the solid red circle. The discovery that planetary systems are common around other stars has profound implications, making it seem more likely that we might someday find life elsewhere, perhaps even intelligent life. It also allows us to learn more about the general nature of planets and how they form, giving us deeper insights into our cosmic origins. In this chapter, we'll explore the exciting new science of other planetary systems.

LEARNING GOALS

7.1 Detecting Planets Around Other Stars
- How do we detect planets around other stars?
- What properties of extrasolar planets can we measure?

7.2 Characteristics of Extrasolar Planets
- How do extrasolar planets compare with planets in our solar system?
- Are Earth-like planets common?

 THE PROCESS OF SCIENCE IN ACTION

7.3 Extrasolar Planets and the Nebular Theory
- Do we need to modify our theory of solar system formation?

P. 113

Each chapter concludes with a **visual summary** that provides a concise review of the answers to the Learning Goal questions. The summary is followed by a 12-question **Quick Quiz** and a set of **short-answer, essay, and quantitative questions**.

summary of key concepts

7.1 Detecting Planets Around Other Stars

- How do we detect planets around other stars?

 We can look for a planet's gravitational effect on its star through the **astrometric method**, which looks for small shifts in stellar position, or the **Doppler method**, which looks for the back-and-forth motion of stars revealed by Doppler shifts. For the small fraction of planetary systems with orbits aligned edge-on to Earth, we can search for **transits**, in which a planet blocks a little of its star's light as it passes in front of it.

- What properties of extrasolar planets can we measure?

All detection methods allow us to determine a planet's orbital period and distance from its star. The astrometric and Doppler methods can provide masses (or minimum masses), while the transit method can provide sizes. In cases where transit and Doppler methods are used together, we can determine average density. In some cases, transits (and eclipses) can provide other data, including limited data about atmospheric composition and temperature.

7.2 Characteristics of Extrasolar Planets

- How do extrasolar planets compare with planets in our solar system?

The known extrasolar planets have a much wider range of properties than the planets in our solar system. Many orbit much closer to their stars and with more eccentric orbital paths; even some jovian planets, called **hot Jupiters**, are found close to their stars. We also find properties indicating planetary types such as "water worlds" that do not fall neatly into the traditional terrestrial and jovian categories.

- Are Earth-like planets common?

Current data indicate that planetary systems are very common, and that Earth-size planets are also common. We do not yet have enough data to know whether such planets have Earth-like orbits, but we should learn the answer as more data are collected.

 THE PROCESS OF SCIENCE IN ACTION

7.3 Extrasolar Planets and the Nebular Theory

- Do we need to modify our theory of solar system formation?

 Our basic theory of solar system formation seems to be sound, but we have had to modify it to allow for planetary migration and a wider range of planetary types than we find in our solar system. Many mysteries remain, but they are unlikely to require major change to the nebular theory of solar system formation.

P. 126

... with Emphasis on the Process of Science

tools of science — Spectroscopy

Three Types of Spectra

Spectroscopy is the study of light that has been spread out into a spectrum by a device such as a prism. The figure shows the three basic kinds of spectra that we observe: (1) a **continuous spectrum** contains light that is smoothly spread across a broad range of wavelengths; (2) an **emission line spectrum** has bright lines on a dark background; and (3) an **absorption line spectrum** has dark lines on a continuous background.

Spectra provide a wealth of information about the objects we study. We previously discussed how a *thermal radiation* spectrum—which is one type of continuous spectrum—tells us an object's surface temperature (see Tools of Science, page 79) and how *Doppler shifts* in the wavelengths of spectral lines tell us the speed of an object toward or away from us (see Tools of Science, page 117). We learn even more from the spectral lines themselves.

The laws of quantum mechanics (see Tools of Science, page 152) dictate that each different kind of atom, ion, or molecule produces spectral lines at a unique set of wavelengths—in essence, its "chemical fingerprint." We can therefore determine the composition of a distant object by identifying and comparing all the different "fingerprints" in its spectrum. It does not matter whether the lines are absorption lines or emission lines; the wavelengths of the lines are what tell us an object's composition.

We can put these ideas together to illustrate how the spectrum of a star like the Sun reveals its composition. The hot Sun emits thermal radiation, much like the hot light bulb in the figure, producing a continuous spectrum. The low-density, uppermost layers of the Sun act like a cloud of gas that the continuous spectrum must pass through on its way to Earth. Atoms in these layers absorb light at specific wavelengths, producing an absorption line spectrum. The wavelengths of those lines tell us the kinds of atoms that are present in the Sun, and a more detailed analysis tells us the relative proportions of those elements.

Examples of conditions under which we see the three basic types of spectra.

Tools of Science boxes introduce key tools when they are first needed with the subject matter.

The final section of each chapter focuses on a topic that illustrates **the Process of Science in Action.**

The ripples represent gravitational waves carrying energy away from the system.

The loss of orbital energy means the orbit must decay . . .

. . . until the two objects ultimately merge, producing a particularly strong pulse of gravitational waves.

▲ **FIGURE 10.16**
This sequence uses a rubber sheet analogy to illustrate the gravitational waves produced by a binary system with two neutron stars. The gravitational waves carry away orbital energy, causing the orbit to decay until the two neutron stars collide and merge together. The time required for this sequence to play out (from left to right) would typically be a few tens of millions of years.

THE PROCESS OF SCIENCE IN ACTION

10.3 Gravitational Waves

Einstein's general theory of relativity makes many predictions that have been tested and verified, including the effects of gravity on time discussed in the previous section. Another important prediction is the existence of what we call *gravitational waves*. Einstein himself doubted that we would ever detect such waves, but we have, and how we detect them is this chapter's case study in the process of science in action. As you will see, gravitational-wave detectors have given astronomers an entirely new way to observe the universe.

Mastering Astronomy's Study Area Helps Students Come Prepared to Class . . .

The Study Area features a Quick Quiz for each chapter, many videos and interactive figures, a set of self-guided tutorials covering key concepts, a media workbook, World Wide Telescope tours, and much more — PLUS access to a full etext of *The Cosmic Perspective Fundamentals*.

NEW! Dozens of new videos about key concepts and figures in the text, all written and most narrated by the authors to ensure consistency of terminology and pedagogy. Most videos include embedded pause-and-predict questions that allow students to check their understanding as they watch. Students can use these videos to help prepare for lectures, while instructors will find the same videos with assignable tutorials in the instructor-accessible Item Library.

...While Instructors Can Access a Large Library of Homework and Test Questions

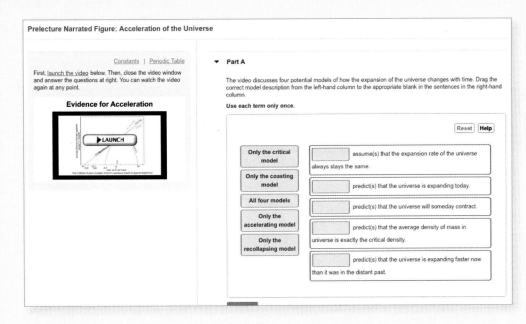

The Item Library consists of more than 250 assignable tutorials—all written or co-written by the textbook authors—including new tutorials based on all of the videos and interactive figures as well as updated tutorials on key concepts, process of science, vocabulary, and much more.

Many of the **assignable tutorials** use ranking or sorting tasks, which research shows to be particularly effective in building conceptual understanding. **The Item Library** also includes all end-of-chapter exercises from the book and a large test bank.

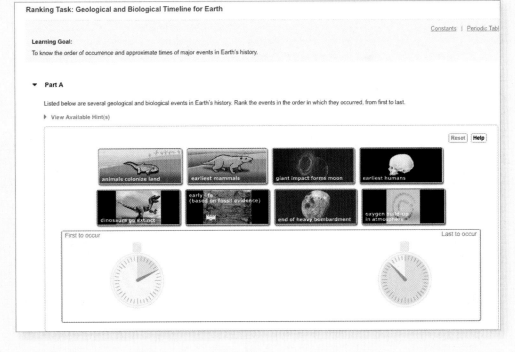

Reach Every Student with Pearson eText

Pearson eText, optimized for mobile, seamlessly integrates videos and other rich media with the text and gives students access to their textbook anytime, anywhere.

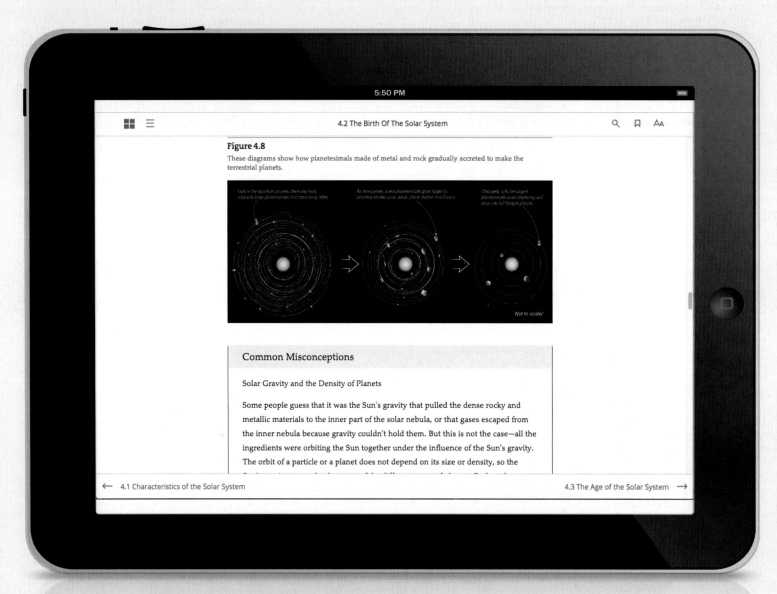

Engage Students Before and During Class with Dynamic Study Modules and Learning Catalytics

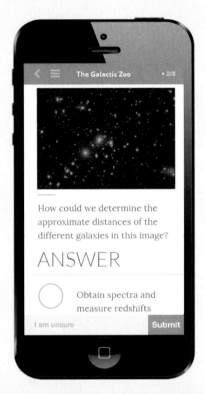

NEW! Dynamic Study Modules in Mastering Astronomy help students study effectively—and at their own pace—by keeping them motivated and engaged. The assignable modules rely on the latest research in cognitive science, using methods—such as adaptivity, gamification, and intermittent rewards—to stimulate learning and improve retention.

With **Learning Catalytics**, you'll hear from every student when it matters most. You pose a variety of questions that help students recall ideas, apply concepts, and develop critical-thinking skills. Your students respond using their own smartphones, tablets, or laptops. You can monitor responses with real-time analytics and find out what your students do—and don't—understand. Then you can adjust your teaching accordingly, and even facilitate peer-to-peer learning, helping students stay motivated and engaged.

Detailed Contents

PREFACE XVII | ABOUT THE AUTHORS XIX | HOW TO SUCCEED IN YOUR ASTRONOMY COURSE XX

1 A Modern View of the Universe 1

1.1 The Scale of the Universe 2
What is our place in the universe? 2
common misconceptions *The Meaning of a Light-Year* 4
How big is the universe? 6

1.2 The History of the Universe 8
How did we come to be? 8
common misconceptions *Confusing Very Different Things* 10
tool of science *Doing the Math* 10
How do our lifetimes compare to the age of the universe? 11

THE PROCESS OF SCIENCE IN ACTION

1.3 Defining Planets 12
What is a planet? 13

2 Understanding the Sky 17

2.1 Understanding the Seasons 18
What causes the seasons? 18
common misconceptions *The Cause of Seasons* 19
cosmic context FIGURE 2.3 *The Cause of Seasons* 20
common misconceptions *High Noon* 22
Why do the constellations we see depend on the time of year? 22

2.2 Understanding the Moon 25
Why do we see phases of the Moon? 25
common misconceptions *Shadows and the Moon* 26
tools of science *Angular Sizes and Distances* 26
common misconceptions *Moon in the Daytime* 27
common misconceptions *The "Dark Side" of the Moon* 27
What causes eclipses? 28

THE PROCESS OF SCIENCE IN ACTION

2.3 The Puzzle of Planetary Motion 31
Why did the ancient Greeks reject the real explanation for planetary motion? 31

3 Changes in Our Perspective 35

3.1 From Earth-Centered to Sun-Centered 36
How did the Greeks explain planetary motion? 36
common misconceptions *Columbus and a Flat Earth* 36
How did the Copernican revolution change our view of the universe? 37
tools of science *Telescopes* 42

3.2 Hallmarks of Science 43
How can we distinguish science from nonscience? 43
common misconceptions *Eggs on the Equinox* 45
What is a scientific theory? 45
cosmic context FIGURE 3.11 *The Copernican Revolution* 46

THE PROCESS OF SCIENCE IN ACTION

3.3 The Fact and Theory of Gravity 49
How does the fact of gravity differ from the theory of gravity? 49

4 Origin of the Solar System 53

4.1 Characteristics of the Solar System 54
What does the solar system look like? 54
cosmic context FIGURE 4.1 *The Solar System* 56
What features of our solar system provide clues to how it formed? 60

4.2 The Birth of the Solar System 61
What theory best explains the orderly patterns of motion in our solar system? 61
tools of science *Conservation Laws* 62
How does our theory account for the features of planets, moons, and small bodies? 64
common misconceptions *Solar Gravity and the Density of Planets* 65

THE PROCESS OF SCIENCE IN ACTION

4.3 The Age of the Solar System 70
How do we determine the age of Earth and the solar system? 70

xiv Detailed Contents

5 Terrestrial Worlds 74

5.1 Terrestrial Surfaces and Atmospheres 75
What determines a world's level of geological activity? 75
common misconceptions *Earth Is Not Full of Molten Lava* 76
How does an atmosphere affect conditions for life? 78
tools of science *Basic Properties of Light* 79

5.2 Histories of the Terrestrial Worlds 81
common misconceptions *Why Is the Sky Blue?* 81
Why did the terrestrial worlds turn out so differently? 81
common misconceptions *The Greenhouse Effect Is Bad* 86
What unique features of Earth are important for life? 87

THE PROCESS OF SCIENCE IN ACTION

5.3 Global Warming 89
What is the evidence for global warming? 90

6 The Outer Solar System 94

6.1 Jovian Planets, Rings, and Moons 95
What are jovian planets like? 95
tools of science *Newton's Version of Kepler's Third Law* 96
Why are jovian moons so geologically active? 99

6.2 Asteroids, Comets, Dwarf Planets, and the Impact Threat 105
Why are asteroids and comets grouped into three distinct regions? 105
common misconceptions *Dodge Those Asteroids!* 105
Do small bodies pose an impact threat to Earth? 108

THE PROCESS OF SCIENCE IN ACTION

6.3 Extinction of the Dinosaurs 109
Did an impact kill the dinosaurs? 109

7 Planets Around Other Stars 113

7.1 Detecting Planets Around Other Stars 114
How do we detect planets around other stars? 114
tools of science *The Doppler Effect* 116
What properties of extrasolar planets can we measure? 117
cosmic context FIGURE 7.6 *Detecting Extrasolar Planets* 118

7.2 Characteristics of Extrasolar Planets 121
How do extrasolar planets compare with planets in our solar system? 121
Are Earth-like planets common? 122

THE PROCESS OF SCIENCE IN ACTION

7.3 Extrasolar Planets and the Nebular Theory 124
Do we need to modify our theory of solar system formation? 124

8 The Sun and Other Stars 128

8.1 Properties of the Sun 129
Why does the Sun shine? 129
How does energy escape from the Sun? 130
common misconceptions *The Sun Is Not on Fire* 130
tools of science *Spectroscopy* 132

8.2 Properties of Other Stars 134
How do we measure the properties of stars? 134
common misconceptions *Photos of Stars* 136
What patterns do we find in the properties of stars? 137

THE PROCESS OF SCIENCE IN ACTION

8.3 Visualizing Patterns Among Stars 140
How did we discover the patterns in stellar properties? 140
cosmic context FIGURE 8.17 *Reading an H-R Diagram* 142

9 Stellar Lives 146

9.1 Lives in Balance 147
Why do stars shine so steadily? 147
Why do a star's properties depend on its mass? 149

9.2 Star Death 151
What will happen when our Sun runs out of fuel? 152
tools of science *Quantum Laws and Astronomy* 152
How do high-mass stars end their lives? 156

THE PROCESS OF SCIENCE IN ACTION

9.3 Testing Stellar Models with Star Clusters 159
What do star clusters reveal about the lives of stars? 160
cosmic context FIGURE 9.25 *Stellar Lives* 162

10 The Bizarre Stellar Graveyard 166

10.1 White Dwarfs and Neutron Stars 167
What are white dwarfs? 167
What are neutron stars? 169

10.2 Black Holes 171
What are black holes? 171
tools of science Einstein's Theories of Relativity 172
common misconceptions Black Holes Don't Suck 174
What is the evidence for black holes? 175
cosmic context FIGURE 10.15 *Balancing Pressure and Gravity* 176

THE PROCESS OF SCIENCE IN ACTION
10.3 Gravitational Waves 178
What are gravitational waves and how have astronomers detected them? 178

11 Galaxies 182

11.1 Our Galaxy: The Milky Way 183
What does our galaxy look like? 183
tools of science Observing Different Kinds of Light 184
common misconceptions The Sound of Space 187
How did the Milky Way form? 188

11.2 Galaxies Beyond the Milky Way 189
What are the major types of galaxies? 190
Why do galaxies differ? 192

THE PROCESS OF SCIENCE IN ACTION
11.3 Seeking Supermassive Black Holes 195
What is the evidence for supermassive black holes at the centers of galaxies? 195

12 Galaxy Distances and Hubble's Law 200

12.1 Measuring Cosmic Distances 201
How do we measure the distances to galaxies? 201
tools of science Measuring Distances with Standard Candles 203
What is Hubble's law? 204

12.2 The Implications of Hubble's Law 206
In what sense is the universe expanding? 206
How do distance measurements tell us the age of the universe? 208

common misconceptions What Is the Universe Expanding Into? 208
common misconceptions Beyond the Horizon 210

THE PROCESS OF SCIENCE IN ACTION
12.3 Observing Galaxy Evolution 210
What do we see when we look back through time? 210

13 The Birth of the Universe 215

13.1 The Big Bang Theory 216
What were conditions like in the early universe? 216
tools of science Particle Accelerators 217
How did the early universe change with time? 218

13.2 Evidence for the Big Bang 221
How do observations of the cosmic microwave background support the Big Bang theory? 221
cosmic context FIGURE 13.5 *The Early Universe* 222
How do the abundances of elements support the Big Bang theory? 225

THE PROCESS OF SCIENCE IN ACTION
13.3 Inflation 226
Did the universe undergo an early episode of inflation? 227

14 Dark Matter and Dark Energy 232

14.1 Evidence for Dark Matter 233
What is the evidence for dark matter? 233
tools of science The Orbital Velocity Formula 235
What might dark matter be made of? 237

14.2 Gravity versus Expansion 239
How did structures like galaxies form? 239
cosmic context FIGURE 14.13 *Galaxy Evolution* 242
Will the universe continue expanding forever? 244

THE PROCESS OF SCIENCE IN ACTION
14.3 Evidence for Dark Energy 245
What is the evidence for dark energy? 245
cosmic context FIGURE 14.17 *Dark Matter and Dark Energy* 248

15 Life in the Universe 252

15.1 The Search for Life in the Solar System 253
What are the necessities of life? 253
Could there be life elsewhere in our solar system? 255

15.2 The Search for Life Among the Stars 257
tools of science *Planetary Spacecraft* 257
How can we identify potentially habitable planets? 258
Is there intelligent life beyond Earth? 259

THE PROCESS OF SCIENCE IN ACTION

15.3 Evolution on Earth and Beyond 263
What is the evidence for evolution? 264
cosmic context FIGURE 15.14 *A Universe of Life* 266

CREDITS 271

APPENDIXES A-1
A Useful Numbers A-2
B Useful Formulas A-3
C A Few Mathematical Skills A-4

GLOSSARY G-1
INDEX I-1

Preface

We humans have gazed into the sky for countless generations. We have wondered how our lives are connected to the Sun, Moon, planets, and stars that adorn the heavens. Today, through the science of astronomy, we know that these connections go far deeper than our ancestors ever imagined. This book focuses on the story of modern astronomy and the new perspective—*the cosmic perspective*—that astronomy gives us on ourselves and our planet.

Who Is This Book For?

The Cosmic Perspective Fundamentals is designed to support one-term college courses in introductory astronomy—particularly those in which instructors couple the text with Mastering™ Astronomy to create an active or collaborative learning environment. No prior background in astronomy or physics is assumed, making *The Cosmic Perspective Fundamentals* suitable for both high school courses and college courses for nonscience majors. *The Cosmic Perspective Fundamentals* differs from our more comprehensive texts (*The Cosmic Perspective* and *The Essential Cosmic Perspective*) in covering a smaller set of topics and therefore being much shorter in length, but it is built upon the same "big picture" approach to astronomy and uses the same pedagogical principles.

New to This Edition

Many new discoveries have been made in astronomy during the four years since publication of the second edition of *The Cosmic Perspective Fundamentals,* leading to many changes in this third edition. Indeed, the changes are too many to list here, but those who used the second edition will notice significant updates to almost every chapter in the book, primarily as a result of the many new astronomical discoveries that have occurred in recent years, including results from missions such as *New Horizons, Rosetta, Dawn, Curiosity, MAVEN, Cassini, Juno,* and more; the first direct detections of gravitational waves; and major advances in our understanding of extrasolar planets.

Topical Selection

A briefer, focused text must necessarily cover fewer topics. We have carefully selected those topics using the following four criteria:

- *Importance.* We surveyed a large number of professors to identify the topics considered of greatest importance in a college-level astronomy course, in order to ensure that the most fundamental concepts are covered in this text. Most astronomy courses begin with topics such as the scale of the universe, seasons, and phases of the Moon and then progress to study of the planets, stars, galaxies, and cosmology. Our selected topics have been organized in a similar fashion. The fifteen chapters are designed so that they can be covered in a typical semester at a rate of approximately one chapter per week.

- *Active learning.* Educational research has shown that students learn scientific concepts best by actively solving conceptual problems, both individually and in collaboration with other students. We have emphasized topics that are well suited to active learning, and each chapter includes **Think About It** critical thinking questions for in-class discussion and **See It for Yourself** hands-on activities to further promote active learning. These in-text features are reinforced by a variety of active learning resources on the Mastering Astronomy website.

- *Engagement.* Most students in a college astronomy course are there to satisfy a general education requirement, but the subject is sufficiently interesting that it should be possible to choose topics that students will find highly engaging—and that they will therefore be willing to work hard to learn.

- *Process of science.* We believe that the primary purpose of a general education requirement in science is to ensure that students learn about science itself. Throughout the book, we have chosen topics that illustrate important aspects of the process of science, and each chapter concludes with a section called **The Process of Science in Action,** which presents a case study of how the process of science has helped (or is currently helping) to provide greater insight into key topics in astronomy.

Book Structure

To facilitate student learning, we have created a simple pedagogical structure used in each of the book's fifteen chapters:

- Each chapter begins with an opening page that includes a brief overview of the chapter content and a clear set of **Learning Goals** associated with the chapter. Each Learning Goal is phrased as a question to engage students as they read.

- Each chapter consists of three sections. The first two sections focus on the key topics of the chapter; the third section builds on the ideas from the first two sections, but focuses on **The Process of Science in Action.**

- Each section is written to address the Learning Goal questions from the chapter-opening page.

- Each chapter concludes with a **visual summary** that provides a concise review of the answers to the Learning Goal questions.

- The summary is followed by a 12-question **Quick Quiz** and a set of **short-answer, essay, and quantitative questions.**

  Additional features of the book include the following:

- **Tools of Science** boxes, which present a brief overview of key tools that astronomers use, including theories, equations, observational techniques, and technology. Each chapter includes one Tools of Science box related to the chapter content.

- **Common Misconceptions** boxes, which address popularly held but incorrect ideas about topics in the text
- **Annotated Figures and Photos**, which act like the voice of an instructor, walking students through the key ideas presented in complex figures, photos, and graphs
- **Cosmic Context Figures**, which combine text and illustrations into accessible and coherent two-page visual summaries that will help improve student understanding of essential topics

Mastering Astronomy—A New Paradigm in Astronomy Teaching

What is the single most important factor in student success in astronomy? Both research and common sense reveal the same answer: *study time.* No matter how good the teacher or how good the textbook, students learn only when they spend adequate time studying. Unfortunately, limitations on resources for grading have prevented most instructors from assigning much homework despite its obvious benefits to student learning. And limitations on help and office hours have made it difficult for students to make sure they use self-study time effectively. That, in a nutshell, is why we created Mastering Astronomy. For students, it provides personalized learning designed to coach them *individually*—responding to their errors with specific, targeted feedback and providing optional hints for those who need additional guidance. For professors, Mastering Astronomy provides the unprecedented ability to automatically monitor and record students' step-by-step work and evaluate the effectiveness of assignments and exams.

All students registered for Mastering Astronomy receive full access to the Study Area. Key resources available in the Study Area include the following:

- A large set of prelecture videos, narrated figures, and interactive figures that will help students understand key concepts from the textbook
- A set of self-study tools, including a Quick Quiz for each chapter and interactive self-guided tutorials that go into depth on topics that some students find particularly challenging
- A downloadable set of group activities
- Additional videos covering basic math skills, as well as selected videos of the authors speaking to the public
- And much more, including a media workbook, Starry Night activities, World Wide Telescope tours, and even access to a full etext of *The Cosmic Perspective Fundamentals*

Instructors have access to many additional resources, including a large Item Library featuring more than 250 assignable tutorials, organized by chapter, that include guidance for understanding key concepts, assessments based on the large set of prelecture videos, ranking tasks, sorting tasks, and more. There is also a set of Math Review tutorials to help students who need work on topics including scientific notation, working with units, metric units, and problem-solving skills.

Finally, please note that nearly all the content available at the Mastering Astronomy site for *The Cosmic Perspective Fundamentals* has been written or co-written by the textbook authors. This means that you can count on consistency between the textbook and web resources, with both emphasizing the same concepts and using the same terminology and the same pedagogical approaches. This type of consistency ensures that students will be able to study in the most efficient way possible.

Acknowledgments

A textbook may carry author names, but it is the result of hard work by a long list of committed individuals, as well as many reviewers. We could not possibly list everyone who has helped, but we would especially like to thank our editorial team at Pearson, our production team at Lifland et al., and the more than 100 professors who have reviewed our texts in depth, providing valuable feedback; a list of these professors can be found in *The Cosmic Perspective*, ninth edition.

Jeff Bennett
Megan Donahue
Nick Schneider
Mark Voit

About the Authors

Jeffrey Bennett, a recipient of the American Institute of Physics Science Communication Award, holds a B.A. in biophysics (UC San Diego) and an M.S. and Ph.D. in astrophysics (University of Colorado). He specializes in science and math education and has taught at every level from preschool through graduate school. Career highlights include serving 2 years as a visiting senior scientist at NASA headquarters, where he developed programs to build stronger links between research and education, proposing and helping to develop the Voyage scale model solar system on the National Mall (Washington, DC) and developing the free app *Totality by Big Kid Science* to help people learn about total solar eclipses. He is the lead author of textbooks in astronomy, astrobiology, mathematics, and statistics and of critically acclaimed books for the public including *Beyond UFOs* (Princeton University Press), *Math for Life* (Big Kid Science), *What Is Relativity?* (Columbia University Press), *On Teaching Science* (Big Kid Science), and *A Global Warming Primer* (Big Kid Science). He is also the author of six science picture books for children, titled *Max Goes to the Moon*, *Max Goes to Mars*, *Max Goes to Jupiter*, *Max Goes to the Space Station*, *The Wizard Who Saved the World*, and *I, Humanity*; all six have been launched to the International Space Station and read aloud by astronauts for NASA's Story Time From Space program. His personal website is www.jeffreybennett.com.

Megan Donahue is a full professor in the Department of Physics and Astronomy at Michigan State University (MSU), a Fellow of the American Physical Society and of the American Association for the Advancement of Science, and President of the American Astronomical Society (2018–2020). Her research focuses on using x-ray, UV, infrared, and visible light to study galaxies and clusters of galaxies: their contents—dark matter, hot gas, galaxies, active galactic nuclei—and what they reveal about the contents of the universe and how galaxies form and evolve. She grew up on a farm in Nebraska and received an S.B. in physics from MIT, where she began her research career as an x-ray astronomer. She has a Ph.D. in astrophysics from the University of Colorado. Her Ph.D. thesis on theory and optical observations of intergalactic and intracluster gas won the 1993 Robert Trumpler Award from the Astronomical Society for the Pacific for an outstanding astrophysics doctoral dissertation in North America. She continued postdoctoral research as a Carnegie Fellow at Carnegie Observatories in Pasadena, California, and later as an STScI Institute Fellow at Space Telescope. Megan was a staff astronomer at the Space Telescope Science Institute until 2003, when she joined the MSU faculty. She is also actively involved in advising national and international astronomical facilities and NASA, including planning future NASA missions. Megan is married to Mark Voit, and they collaborate on many projects, including this textbook, over 70 peer-reviewed astrophysics papers, and the nurturing of their children, Michaela, Sebastian, and Angela. Megan has run three full marathons, including Boston. These days she runs trails with friends, orienteers, and plays piano and bass guitar for fun and no profit.

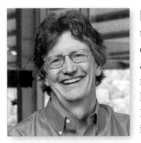

Nicholas Schneider is a full professor in the Department of Astrophysical and Planetary Sciences at the University of Colorado and a researcher in the Laboratory for Atmospheric and Space Physics. He received his B.A. in physics and astronomy from Dartmouth College in 1979 and his Ph.D. in planetary science from the University of Arizona in 1988. His research interests include planetary atmospheres and planetary astronomy. One research focus is the odd case of Jupiter's moon Io. Another is the mystery of Mars's lost atmosphere, which he is helping to answer by leading the Imaging UV Spectrograph team on NASA's *MAVEN* mission now orbiting Mars. Nick enjoys teaching at all levels and is active in efforts to improve undergraduate astronomy education. Over his career he has received the National Science Foundation's Presidential Young Investigator Award, the Boulder Faculty Assembly's Teaching Excellence Award, and NASA's Exceptional Scientific Achievement Medal. Off the job, Nick enjoys exploring the outdoors with his family and figuring out how things work.

Mark Voit is a full professor in the Department of Physics and Astronomy and Associate Dean for Undergraduate Studies at Michigan State University. He earned his A.B. in astrophysical sciences at Princeton University and his Ph.D. in astrophysics at the University of Colorado in 1990. He continued his studies at the California Institute of Technology, where he was a research fellow in theoretical astrophysics, and then moved on to Johns Hopkins University as a Hubble Fellow. Before going to Michigan State, Mark worked in the Office of Public Outreach at the Space Telescope, where he developed museum exhibitions about the Hubble Space Telescope and helped design NASA's award-winning HubbleSite. His research interests range from interstellar processes in our own galaxy to the clustering of galaxies in the early universe, and he is a Fellow of the American Association for the Advancement of Science. He is married to coauthor Megan Donahue and cooks terrific meals for her and their three children. Mark likes getting outdoors whenever possible and particularly enjoys running, mountain biking, canoeing, orienteering, and adventure racing. He is also author of the popular book *Hubble Space Telescope: New Views of the Universe*.

How to Succeed in Your Astronomy Course

If Your Course Is	Times for Reading the Assigned Text (per week)	Times for Homework Assignments (per week)	Times for Review and Test Preparation (average per week)	Total Study Time (per week)
3 credits	2 to 4 hours	2 to 3 hours	2 hours	6 to 9 hours
4 credits	3 to 5 hours	2 to 4 hours	3 hours	8 to 12 hours
5 credits	3 to 5 hours	3 to 6 hours	4 hours	10 to 15 hours

The Key to Success: Study Time

The single most important key to success in any college course is to spend enough time studying. A general rule of thumb for college classes is that you should expect to study about 2 to 3 hours per week *outside* of class for each unit of credit. For example, based on this rule of thumb, a student taking 15 credit hours should expect to spend 30 to 45 hours each week studying outside of class. Combined with time in class, this works out to a total of 45 to 60 hours spent on academic work—not much more than the time a typical job requires, and you get to choose your own hours. Of course, if you are working or have family obligations while you attend school, you will need to budget your time carefully.

The table above gives rough guidelines for how you might divide your study time. If you find that you are spending fewer hours than these guidelines suggest, you can probably improve your grade by studying longer. If you are spending more hours than these guidelines suggest, you may be studying inefficiently; in that case, you should talk to your instructor about how to study more effectively.

Using This Book

Each chapter in this book is designed to make it easy for you to study effectively and efficiently. To get the most out of each chapter, you might wish to use the following study plan.

- A textbook is not a novel, and you'll learn best by reading the elements of this text in the following order:

  1. Start by reading the Learning Goals and the introductory paragraph at the beginning of the chapter so that you'll know what you are trying to learn.
  2. Get an overview of key concepts by studying the illustrations and their captions and annotations. The illustrations highlight most major concepts, so this "illustrations first" strategy gives you an opportunity to survey the concepts before you read about them in depth. You will find the two-page Cosmic Context figures especially useful.
  3. Read the chapter narrative, trying the Think About It questions and the See It for Yourself activities as you go along, but save the boxed features (Common Misconceptions, Tools of Science) to read later. As you read, make notes on the pages to remind yourself of ideas you'll want to review later. Take notes as you read, but avoid using a highlight pen (or a highlighting tool if you are using an e-book), which makes it too easy to highlight mindlessly.
  4. After reading the chapter once, go back through and read the boxed features.
  5. Review the Summary of Key Concepts, ideally by trying to answer the Learning Goal questions for yourself before reading the given answers.

- After completing the reading as outlined above, test your understanding with the end-of-chapter exercises. A good way to begin is to make sure you can answer all of the Quick Quiz questions; if you don't know an answer, look back through the chapter until you figure it out.

- Further build your understanding by making use of the videos, quizzes, and other resources available at Mastering Astronomy. These resources have been developed specifically to help you learn the most important ideas in your course, and they have been extensively tested to make sure they are effective. They really do work, and the only way you'll gain their benefits is by going to the website and using them.

General Strategies for Studying

- Budget your time effectively. Studying 1 or 2 hours each day is more effective, and far less painful, than studying all night before homework is due or before exams. *Note*: Research shows that it can be helpful to create a "personal contract" for your study time (or for any other personal commitment), in which you specify rewards you'll give yourself for success and penalties you'll assess for failings.

- Engage your brain. Learning is an active process, not a passive experience. Whether you are reading, listening to a lecture, or working on assignments, always make sure that your mind is actively engaged. If you find your mind drifting or find yourself falling asleep, make a conscious effort to revive yourself, or take a break if necessary.

- Don't miss class, and come prepared. Listening to lectures and participating in class activities and discussions is much more effective than reading someone else's notes or watching a video later. Active participation will help you retain what you are learning. Also, be sure to complete any assigned reading *before* the class in which it will be discussed. This is crucial, since class lectures and discussions are designed to reinforce key ideas from the reading.

- Take advantage of resources offered by your professor, whether it be email, office hours, review sessions, online chats, or other opportunities to talk to and get to know your professor. Most professors will go out of their way to help you learn in any way that they can.
- Start your homework early. The more time you allow yourself, the easier it is to get help if you need it. If a concept gives you trouble, do additional reading or studying beyond what has been assigned. And if you still have trouble, ask for help: You surely can find friends, peers, or teachers who will be glad to help you learn.
- Working together with friends can be valuable in helping you understand difficult concepts. However, be sure that you learn *with* your friends and do not become dependent on them.
- Don't try to multitask. Research shows that human beings simply are not good at multitasking: When we attempt it, we do more poorly at all of the individual tasks. And in case you think you are an exception, research has also shown that those people who believe they are best at multitasking are often the worst! So when it is time to study, turn off your electronic devices, find a quiet spot, and concentrate on your work. (If you *must* use a device to study, as with an e-book or online homework, turn off email, text, and other alerts so that they will not interrupt your concentration; some apps will do this for you.)

Preparing for Exams

- Rework problems and other assignments; try additional questions, including the online quizzes available at Mastering Astronomy, to be sure you understand the concepts. Study your performance on assignments, quizzes, or exams from earlier in the term.
- Study your notes from classes, and reread relevant sections in your textbook. Pay attention to what your instructor expects you to know for an exam.
- Study individually *before* joining a study group with friends. Study groups are effective only if every individual comes prepared to contribute.
- Don't stay up too late before an exam. Don't eat a big meal within an hour of the exam (thinking is more difficult when blood is being diverted to the digestive system).
- Try to relax before and during the exam. If you have studied effectively, you are capable of doing well. Staying relaxed will help you think clearly.

Presenting Homework and Writing Assignments

All work that you turn in should be of *collegiate quality:* neat and easy to read, well organized, and demonstrating mastery of the subject matter. Future employers and teachers will expect this quality of work. Moreover, although submitting homework of collegiate quality requires "extra" effort, it serves two important purposes directly related to learning:

1. The effort you expend in clearly explaining your work solidifies your learning. Writing (or typing) triggers different areas of your brain than reading, listening, or speaking. As a result, writing something down will reinforce your learning of a concept, even when you think you already understand it.
2. By making your work clear and self-contained (that is, making it a document that you can read without referring to the questions in the text), you will have a much more useful study guide when you review for a quiz or exam.

The following guidelines will help ensure that your assignments meet the standards of collegiate quality:

- Always use proper grammar, proper sentence and paragraph structure, and proper spelling. Do not use texting shorthand, and don't become over-reliant on spell checkers, which may miss "too two three mistakes, to."
- All answers and other writing should be fully self-contained. A good test is to imagine that a friend is reading your work and to ask yourself whether the friend would understand exactly what you are trying to say. It is also helpful to read your work out loud to yourself, making sure that it sounds clear and coherent.
- In problems that require calculation:
  1. Be sure to *show your work* clearly so that both you and your instructor can follow the process you used to obtain an answer. Also, use standard mathematical symbols, rather than "calculator-ese." For example, show multiplication with the $\times$ symbol (not with an asterisk), and write 10^5, not 10^5 or 10E5.
  2. *Word problems should have word answers*. That is, after you have completed any necessary calculations, make sure that any problem stated in words is answered with one or more *complete sentences* that describe the point of the problem and the meaning of your solution.
  3. *Units are crucial*. If your answer has units, be sure they are stated clearly. For example, if you are asked to calculate a distance, be sure you state whether your answer is in miles, kilometers, or some other distance unit.
  4. Express your word answers in a way that would be *meaningful* to most people. For example, most people would find it more meaningful if you expressed a result of 720 hours as 1 month. Similarly, if a precise calculation yields an answer of 9,745,600 years, it may be more meaningfully expressed in words as "nearly 10 million years."
- Include illustrations whenever they help explain your answer, and make sure your illustrations are neat and clear. For example, if you graph by hand, use a ruler to make straight lines. If you use software to make illustrations, be careful not to make them overly cluttered with unnecessary features.
- If you study with friends, be sure that you turn in your own work stated in your own words—you should avoid anything that might give even the *appearance* of possible academic dishonesty.

A Modern View of the Universe

1

This Hubble Space Telescope photo shows a piece of the sky so small that you could block your view of it with a grain of sand held at arm's length. Yet the photo shows an almost unimaginable expanse of both space and time: Nearly every object within it is a galaxy containing billions of stars, most likely orbited by planets, and some of the smaller smudges are galaxies so far away that their light has taken more than 12 billion years to reach us. A major goal of this book is to help you understand what you see in this photograph. We'll begin with a brief survey of our modern, scientific view of the universe.

LEARNING GOALS

1.1 The Scale of the Universe
- What is our place in the universe?
- How big is the universe?

1.2 The History of the Universe
- How did we come to be?
- How do our lifetimes compare to the age of the universe?

 THE PROCESS OF SCIENCE IN ACTION

1.3 Defining Planets
- What is a planet?

A Modern View of the Universe

1.1 The Scale of the Universe

For most of human history, our ancestors imagined Earth to be stationary at the center of a relatively small universe. This idea made sense at a time when understanding was built upon everyday experience. After all, we cannot feel the constant motion of Earth as it rotates on its axis and orbits the Sun, and if you observe the sky you'll see that the Sun, Moon, planets, and stars all appear to revolve around us each day. Nevertheless, we now know that Earth is a planet orbiting a rather average star in a rather typical galaxy, and that our universe is filled with far greater wonders than our ancestors ever imagined.

◆ What is our place in the universe?

Before we can discuss the universe and its great wonders, we first need to develop a general sense of our place within it. We can do this by thinking about what we might call our "cosmic address," illustrated in Figure 1.1.

Our Cosmic Address Earth is a *planet* in our **solar system**, which consists of the Sun, the planets and their moons, and countless smaller objects that include rocky *asteroids* and icy *comets*. Keep in mind that our Sun is a *star*, just like the stars we see in our night sky.

Our solar system belongs to the huge, disk-shaped collection of stars called the **Milky Way Galaxy**. A **galaxy** is a great island of stars in space, all held together by gravity and orbiting a common center. The Milky Way is a relatively large galaxy, containing more than 100 billion stars, and we think that most of these stars are orbited by planets. Our solar system is located a little over halfway from the galactic center to the edge of the galactic disk.

Billions of other galaxies are scattered throughout space. Some galaxies are fairly isolated, but most are found in groups. Our Milky Way, for example, is one of the two largest among more than 70 galaxies (most relatively small) in the **Local Group**. Groups of galaxies with many more large members are often called **galaxy clusters**.

On a very large scale, galaxies and galaxy clusters appear to be arranged in giant chains and sheets with huge voids between them; the background of Figure 1.1 represents this large-scale structure. The regions in which galaxies and galaxy clusters are most tightly packed are called **superclusters**, which are essentially clusters of galaxy clusters. Our Local Group is located in the outskirts of the Local Supercluster (also called *Laniakea*, Hawaiian for "immense heaven").

think about it Some people think that our tiny physical size in the vast universe makes us insignificant. Others think that our ability to learn about the wonders of the universe gives us significance despite our small size. What do *you* think?

Together, all these structures make up our **universe**. In other words, the universe is the sum total of all matter and energy, encompassing the superclusters and voids and everything within them.

Astronomical Distance Measurements The labels in Figure 1.1 give approximate sizes for the various structures in kilometers (recall that 1 kilometer ≈ 0.6 mile), but many distances in astronomy are so large that kilometers are not the most convenient unit. Instead, we often use two other units:

- One **astronomical unit (AU)** is Earth's average distance from the Sun, which is about 150 million kilometers (93 million miles). We commonly describe distances within our solar system in AU.

Our Cosmic Address

Universe
approx. size: 10^{21} km ≈ 100 million ly

Local Supercluster
approx. size: 3×10^{19} km ≈ 3 million ly

Local Group
approx. size: 10^{18} km ≈ 100,000 ly

Milky Way Galaxy

Solar System (not to scale)
approx. size: 10^{10} km ≈ 60 AU

Earth
approx. size: 10^4 km

▼ **FIGURE 1.1**
These paintings show key levels of structure in our universe; for a more detailed view, see the "You Are Here in Space" foldout diagram in the front of the book.

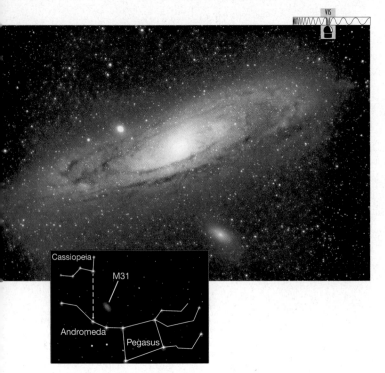

▲ **FIGURE 1.2**
The Andromeda Galaxy (also known as M31). When we look at this galaxy, we see light that traveled through space for 2.5 million years. The inset shows the galaxy's location in the constellation Andromeda.

- One **light-year (ly)** is the *distance* that light can travel in 1 year, which is about 10 trillion kilometers (see Tools of Science, page 10). We generally use light-years to describe the distances of stars and galaxies.

Looking Back in Time Light-years are a unit of distance, but they are related to the time it takes light to travel through space. Consider Sirius, the brightest star in the night sky, which is located about 8 light-years away. Because it takes light 8 years to travel this distance, we see Sirius not as it is today, but rather as it was 8 years ago. The star Betelgeuse, a bright red star in the constellation Orion, is about 600 light-years away, which means we see it as it was about 600 years ago. If Betelgeuse exploded in the past 600 years or so (a possibility we'll discuss in Chapter 9), we would not yet know it, because the light from the explosion would not yet have reached us.

The general idea that light takes time to travel through space leads to a remarkable fact: **The farther away we look in distance, the further back we look in time**. The effect is dramatic for large distances. The Andromeda Galaxy (Figure 1.2) is about 2.5 million light-years away, which means we see it as it looked about 2.5 million years ago. We see more distant galaxies as they were even further in the past.

It's also amazing to realize that any "snapshot" of a distant galaxy is a picture of both space and time. For example, because the Andromeda Galaxy is about 100,000 light-years in diameter, the light we see from the far side of the galaxy must have left on its journey to us some 100,000 years before the light from the near side. Figure 1.2 therefore shows different parts of the galaxy spread over a time period of 100,000 years. When we study the universe, it is impossible to separate space and time.

see it for yourself The glow from the central region of the Andromeda Galaxy is faintly visible to the naked eye and easy to see with binoculars. Use a star chart to find it in the night sky, and remember that you are seeing light that spent 2.5 million years in space before reaching your eyes. If students on a planet in the Andromeda Galaxy were looking at the Milky Way right now, what would they see? Could they know that we exist here on Earth?

The Observable Universe As we'll discuss in Section 1.2, the measured age of the universe is about 14 billion years. This fact, combined with the fact that looking deep into space means looking far back in time, places a limit on the portion of the universe that we can see, even in principle.

Figure 1.3 shows the idea. If we look at a galaxy that is 7 billion light-years away, we see it as it looked 7 billion years ago*—which means we see it as it was when the universe was half its current age. If we look at a galaxy that is 12 billion light-years away (like the most distant ones in the Hubble Space Telescope photo on page 1), we see it as it was 12 billion years ago, when the universe was only 2 billion years old. And if we tried to look beyond 14 billion light-years, we'd be looking to a time more than 14 billion years ago—which is before the universe existed and therefore means that there is nothing to see. This distance of 14 billion light-years therefore marks the boundary (or *horizon*) of our **observable universe**—the portion of the entire universe that we can potentially observe. Note that this fact does not put any limit on the size of the *entire* universe, which may be far larger than our observable universe. We simply have no hope of seeing or studying anything beyond the bounds of our observable universe.

common misconceptions

The Meaning of a Light-Year

Maybe you've heard people say things like "It will take me light-years to finish this homework!" But that statement doesn't make sense, because light-years are a unit of *distance*, not time. If you are unsure whether the term *light-year* is being used correctly, try testing the statement by using the fact that 1 light-year is about 10 trillion kilometers, or 6 trillion miles. The statement then becomes "It will take me 6 trillion miles to finish this homework," which clearly does not make sense.

---

*As we'll discuss in Chapter 12, distances to faraway galaxies in an expanding universe can be described in more than one way; distances like those given here are based on the time it has taken a galaxy's light to reach us (called the *lookback time*).

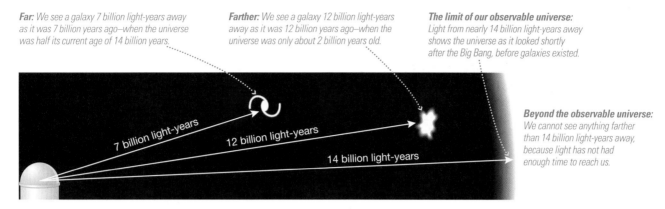

▲ FIGURE 1.3
The farther away we look in space, the further back we look in time. The age of the universe therefore puts a limit on the size of the *observable* universe—the portion of the entire universe that we could observe in principle.

basic astronomical objects, units, and motions

This box summarizes key definitions used throughout this book.

Basic Astronomical Objects

star A large, glowing ball of gas that generates heat and light through nuclear fusion in its core. Our Sun is a star.

planet A moderately large object that orbits a star and shines primarily by reflecting light from its star. According to the current definition, an object can be considered a planet only if it (1) orbits a star, (2) is large enough for its own gravity to make it round, and (3) has cleared most other objects from its orbital path. An object that meets the first two criteria but not the third, like Pluto, is designated a **dwarf planet**.

moon (or **satellite**) An object that orbits a planet. The term *satellite* is also used more generally to refer to any object orbiting another object.

asteroid A relatively small and rocky object that orbits a star.

comet A relatively small and ice-rich object that orbits a star.

small solar system body An asteroid, comet, or other object that orbits a star but is too small to qualify as a planet or dwarf planet.

Collections of Astronomical Objects

solar system The Sun and all the material that orbits it, including the planets, dwarf planets, and small solar system bodies. Although the term *solar system* technically refers only to our own star system (*solar* means "of the Sun"), it is often applied to other star systems as well.

star system A star (sometimes more than one star) and any planets and other materials that orbit it.

galaxy A great island of stars in space, all held together by gravity and orbiting a common center, with a total mass equivalent to millions, billions, or even trillions of stars.

cluster of galaxies (or **group of galaxies**) A collection of galaxies bound together by gravity. Small collections (up to a few dozen galaxies) are generally called *groups*, while larger collections are called *clusters*.

supercluster A gigantic region of space where many individual galaxies and many groups and clusters of galaxies are packed more closely together than elsewhere in the universe.

universe (or **cosmos**) The sum total of all matter and energy—that is, all galaxies and everything between them.

observable universe The portion of the entire universe that can be seen from Earth, at least in principle. The observable universe is probably only a tiny portion of the entire universe.

Astronomical Distance Units

astronomical unit (AU) The average distance between Earth and the Sun, which is about 150 million kilometers. More technically, 1 AU is the length of the semimajor axis of Earth's orbit.

light-year (ly) The distance that light can travel in 1 year, which is about 9.46 trillion kilometers.

Terms Relating to Motion

rotation The spinning of an object around its axis. For example, Earth rotates once each day around its axis, which is an imaginary line connecting the North and South Poles.

orbit (or **revolution**) The orbital motion of one object around another due to gravity. For example, Earth orbits around the Sun once each year.

expansion (of the universe) The increase in the average distance between galaxies as time progresses.

6 CHAPTER 1 • A Modern View of the Universe

▲ **FIGURE 1.4**
This photo shows the pedestals housing the Sun (the gold sphere on the nearest pedestal) and the inner planets in the Voyage scale model solar system (Washington, D.C.). The model planets are encased in the sidewalk-facing disks visible at about eye level on the planet pedestals. To the left is the National Air and Space Museum.

◆ How big is the universe?

Figure 1.1 put numbers on the sizes of different structures in the universe, but these numbers have little meaning for most people—after all, they are literally astronomical. Let's try to put these numbers into perspective.

The Scale of the Solar System One of the best ways to develop perspective on cosmic sizes and distances is to imagine our solar system shrunk down to a scale that would allow you to walk through it. The Voyage scale model solar system (Figure 1.4) makes such a walk possible by showing sizes and distances in the solar system at *one ten-billionth* of their actual values.

Figure 1.5a shows the Sun and planets at their actual sizes (but not distances) on the Voyage scale: The model Sun is about the size of a large grapefruit, Jupiter is about the size of a marble, and Earth is about the size of the ball point in a pen. You can immediately see some key facts about our solar system. For example, the Sun is far larger than any of the planets; in mass, the Sun outweighs all the planets combined by a factor of nearly 1000. The planets also vary considerably in size: The storm on Jupiter known as the Great Red Spot (visible near Jupiter's lower left in the painting) could swallow up the entire Earth.

The scale of the solar system is even more remarkable when you combine the sizes shown in Figure 1.5a with the distances illustrated by the map of the Voyage model in Figure 1.5b. For example, the ball-point-sized Earth is located

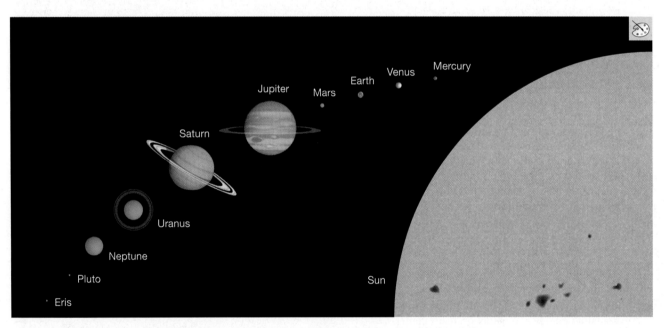

a The scaled sizes (but not distances) of the Sun, the planets, and the two largest known dwarf planets.

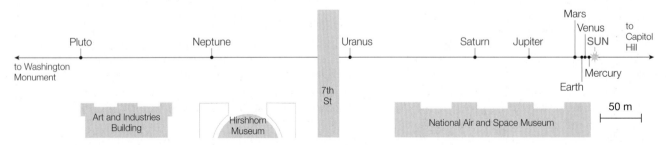

b Locations of major objects in the Voyage model (Washington, D.C.).

▲ **FIGURE 1.5**
The Voyage scale model represents sizes and distances in the solar system at one *ten-billionth* of their actual values. Planets are lined up in the model, but in reality each planet orbits the Sun independently and a perfect alignment never occurs.

about 15 meters (16.5 yards) from the grapefruit-sized Sun, which means you can picture Earth's orbit as a circle of radius 15 meters around a grapefruit.

Perhaps the most striking feature of our solar system when we view it to scale is its emptiness. The Voyage model shows the planets along a straight path, so we'd need to draw each planet's orbit around the model Sun to show the full extent of our planetary system. Fitting all these orbits would require an area measuring more than a kilometer on a side—an area equivalent to more than 300 football fields arranged in a grid. Spread over this large area, only the grapefruit-size Sun, the planets, and a few moons would be big enough to see. The rest of the area would look virtually empty (that's why we call it *space*!).

Seeing our solar system to scale also helps put space exploration into perspective. The Moon, the only other world on which humans have ever stepped (Figure 1.6), lies only about 4 centimeters ($1\frac{1}{2}$ inches) from Earth in the Voyage model. On this scale, the palm of your hand can cover the entire region of the universe in which humans have so far traveled. The trip to Mars is more than 150 times as far as the trip to the Moon, even when Mars is on the same side of its orbit as Earth. And while you can walk from Earth to Pluto in just a few minutes on the Voyage scale, the *New Horizons* spacecraft, which flew past Pluto in 2015, took more than 9 years to make the real journey, despite traveling at a speed nearly 100 times that of a commercial jet.

Distance to Stars If you visit the Voyage model in Washington, D.C., you need to walk only about 600 meters to go from the Sun to Pluto. How much farther would you have to walk to reach the next star on this scale?

Amazingly, you would need to walk to California. If this answer seems hard to believe, you can check it for yourself. A light-year is about 10 trillion kilometers, which becomes 1000 kilometers on the 1-to-10-billion scale (because 10 trillion ÷ 10 billion = 1000). The nearest star system to our own, a three-star system called Alpha Centauri (Figure 1.7), is about 4.4 light-years away. That distance is about 4400 kilometers (2700 miles) on the 1-to-10-billion scale, roughly equivalent to the distance across the United States.

The tremendous distances to the stars give us some perspective on the technological challenge of astronomy. For example, because the largest star of the Alpha Centauri system is roughly the same size and brightness as our Sun, viewing it in the night sky is somewhat like being in Washington, D.C., and seeing a very bright grapefruit in San Francisco (neglecting the problems introduced by the curvature of the Earth). It may seem remarkable that we can see this star at all, but the blackness of the night sky allows the naked eye to see it as a faint dot of light. It looks much brighter through powerful telescopes, but we still cannot see features of the star's surface.

Now, consider the difficulty of detecting *planets* orbiting nearby stars, which is equivalent to looking from Washington, D.C., and trying to find ball points or marbles orbiting grapefruits in California or beyond. When you consider this challenge, it is all the more amazing to realize that we now have technology capable of finding such planets [Section 7.1].

The vast distances to the stars also offer a sobering lesson about interstellar travel. Although science fiction shows like *Star Trek* and *Star Wars* make such travel look easy, the reality is far different. Consider the *Voyager 2* spacecraft. Launched in 1977, *Voyager 2* flew by Jupiter in 1979, Saturn in 1981, Uranus in 1986, and Neptune in 1989. It is now bound for the stars at a speed of close to 50,000 kilometers per hour—about 100 times as fast as a speeding bullet. But even at this speed, *Voyager 2* would take about 100,000 years to reach Alpha Centauri if it were headed in that direction (which it's not). Convenient interstellar travel remains well beyond our present technology.

The Size of the Milky Way Galaxy We must change our scale to visualize the galaxy, because very few stars would even fit on Earth with the 1-to-10-billion

▲ **FIGURE 1.6**
This famous photograph from the first Moon landing (*Apollo 11* in July 1969) shows astronaut Buzz Aldrin, with Neil Armstrong reflected in his visor. Armstrong was the first to step onto the Moon's surface, saying, "That's one small step for a man, one giant leap for mankind." (When asked why this photo became so iconic, Aldrin replied, "Location, location, location!")

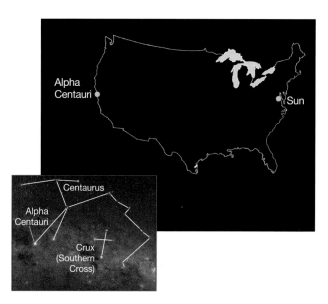

▲ **FIGURE 1.7**
On the same 1-to-10-billion scale on which you can walk from the Sun to Pluto in just a few minutes, you'd need to cross the United States to reach Alpha Centauri, the nearest other star system. The inset shows the location and appearance of Alpha Centauri among the constellations.

▲ FIGURE 1.8
This painting shows the Milky Way Galaxy on a scale where its diameter is the length of a football field. On this scale, stars are microscopic and the distance between our solar system and Alpha Centauri is only 4.4 millimeters. There are so many stars in our galaxy that it would take thousands of years just to count them out loud.

▲ FIGURE 1.9
The number of stars in the observable universe is comparable to the number of grains of dry sand on all the beaches on Earth.

scale we used to visualize the solar system. Let's therefore reduce our scale by another factor of 1 billion (making it a scale of 1 to 10^{19}).

On this new scale, each light-year becomes 1 millimeter, and the 100,000-light-year diameter of the Milky Way Galaxy becomes 100 meters, or about the length of a football field. Visualize a football field with a scale model of our galaxy centered over midfield (Figure 1.8). Our entire solar system is a microscopic dot located around the 20-yard line. The 4.4-light-year separation between our solar system and Alpha Centauri becomes just 4.4 millimeters on this scale—smaller than the width of your little finger. If you stood at the position of our solar system in this model, millions of star systems would lie within reach of your arms.

Another way to put the galaxy into perspective is to consider its number of stars—more than 100 billion. Imagine that tonight you are having difficulty falling asleep (perhaps because you are contemplating the scale of the universe). Instead of counting sheep, you decide to count stars. If you are able to count about one star each second, how long would it take you to count 100 billion stars in the Milky Way? Clearly, the answer is 100 billion (10^{11}) seconds, but how long is that? Amazingly, 100 billion seconds is more than 3000 years. (You can confirm this by dividing 100 billion by the number of seconds in 1 year.) You would need thousands of years just to *count* the stars in the Milky Way Galaxy, and this assumes you never take a break—no sleeping, no eating, and absolutely no dying!

The Observable Universe As incredible as the scale of our galaxy may seem, the Milky Way is only one of more than 100 billion large galaxies in the observable universe. Just as it would take thousands of years to count the stars in the Milky Way, it would take thousands of years to count all these galaxies.

Think for a moment about the total number of stars in all these galaxies. If we assume 100 billion galaxies and 100 billion stars per galaxy, the total number of stars in the observable universe is roughly 100 billion × 100 billion, or 10,000,000,000,000,000,000,000 (10^{22}). How big is this number? Visit a beach. Run your hands through the fine-grained sand. Imagine counting each tiny grain of sand as it slips through your fingers. Then imagine counting every grain of sand on the beach and continuing to count *every* grain of dry sand on *every* beach on Earth. If you could actually complete this task, you would find that the number of grains of sand was comparable to the number of stars in the observable universe (Figure 1.9).

think about it Contemplate the incredible numbers of stars in our galaxy and in the universe, and the fact that each star is a potential sun for a system of planets. How does this perspective affect your thoughts about the possibilities for finding life—or intelligent life—beyond Earth? Explain.

1.2 The History of the Universe

Our universe is vast not only in space, but also in time. In this section, we will briefly discuss the history of the universe as we understand it today.

◆ How did we come to be?

Figure 1.10 summarizes the history of the universe according to modern science. Let's start at the upper left of the figure, and discuss the key events and what they mean.

Our Cosmic Origins

▼ FIGURE 1.10

Birth of the Universe: The expansion of the universe began with the hot and dense Big Bang. The cubes show how one region of the universe has expanded with time. The universe continues to expand, but on smaller scales gravity has pulled matter together to make galaxies.

Galaxies as Cosmic Recycling Plants: The early universe contained only two chemical elements: hydrogen and helium. All other elements were made by stars and recycled from one stellar generation to the next within galaxies like our Milky Way.

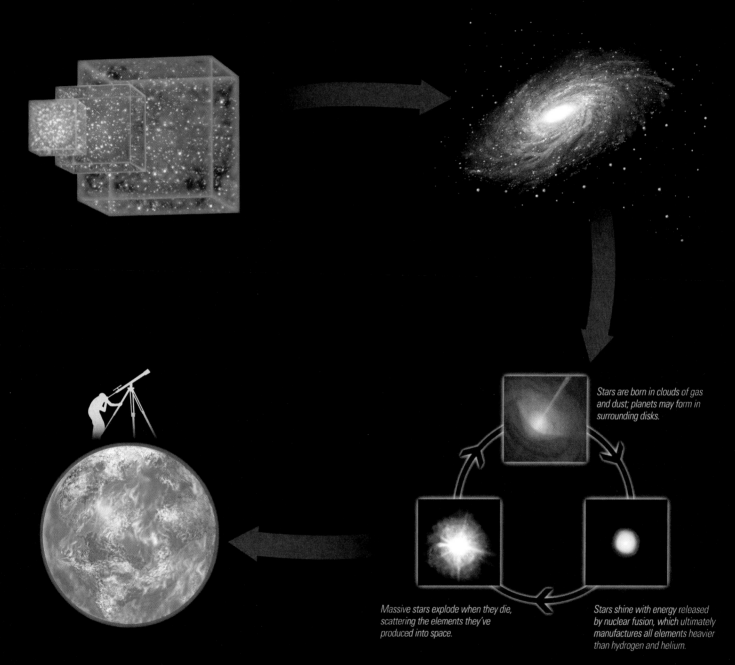

Stars are born in clouds of gas and dust; planets may form in surrounding disks.

Massive stars explode when they die, scattering the elements they've produced into space.

Stars shine with energy released by nuclear fusion, which ultimately manufactures all elements heavier than hydrogen and helium.

Earth and Life: By the time our solar system was born, 4½ billion years ago, about 2% of the original hydrogen and helium had been converted into heavier elements. We are therefore "star stuff," because we and our planet are made from elements manufactured in stars that lived and died long ago.

Life Cycles of Stars: Many generations of stars have lived and died in the Milky Way.

common misconceptions

Confusing Very Different Things

Most people are familiar with the terms *solar system* and *galaxy*, but few realize how incredibly different they are. Our solar system is a single star system, while our galaxy is a collection of more than 100 billion star systems—so many that it would take thousands of years just to count them. Moreover, if you look at the sizes in Figure 1.1, you'll see that our galaxy is about 100 million times larger in diameter than our solar system. So be careful; numerically speaking, mixing up *solar system* and *galaxy* is a gigantic mistake!

The Big Bang, Expansion, and the Age of the Universe Telescopic observations of distant galaxies show that the entire universe is *expanding*, meaning that average distances between galaxies are increasing with time. This fact implies that galaxies must have been closer together in the past, and if we go back far enough, we must reach the point at which the expansion began. We call this beginning the **Big Bang**, and scientists use the observed rate of expansion to calculate that it occurred about 14 billion years ago. The three cubes in the upper left corner of Figure 1.10 represent the expansion of a small piece of the entire universe through time.

The universe as a whole has continued to expand ever since the Big Bang, but on smaller scales the force of gravity has drawn matter together. Structures such as galaxies and galaxy clusters occupy regions where gravity has won out against the overall expansion. That is, while the universe as a whole continues to expand, individual galaxies and galaxy clusters (and objects within them such as stars and planets) do *not* expand. This idea is also illustrated by the three cubes in Figure 1.10. Notice that as the cube as a whole grew larger, the matter within it clumped into galaxies and galaxy clusters. Most galaxies, including our own Milky Way, formed within a few billion years after the Big Bang.

tools of science — Doing the Math

Mathematics is one of the most important tools of science, because it allows scientists to make precise, numerical predictions that can be tested through observations or experiments. These types of tests make it possible for us to gain confidence in scientific ideas. That is why the development of science and mathematics has often gone hand in hand. For example, Sir Isaac Newton developed the mathematics of calculus so that he could do the calculations necessary to test his theory of gravity, and Einstein used new mathematical ideas to work out the details of his general theory of relativity. Fortunately, you don't have to be a Newton or an Einstein to benefit from mathematics in science. Calculations using only multiplication and division can still provide important insights into scientific ideas. Let's look at a few examples.

EXAMPLE 1: How far is a light-year?

Solution: A light-year (ly) is the distance that light can travel in one year; recall that light travels at the *speed of light*, which is 300,000 km/s. Just as we can find the distance that a car travels in 2 hours by multiplying the car's speed by 2 hours, we can find a light-year by multiplying the speed of light by 1 year. Because we are given the speed of light in kilometers per second, we must carry out the multiplication while converting 1 year into seconds (see Appendix C for a review of unit conversions). The result is

$$1 \text{ ly} = \left(300{,}000 \frac{\text{km}}{\text{s}}\right) \times (1 \text{ yr})$$

$$= \left(300{,}000 \frac{\text{km}}{\text{s}}\right) \times \left(1 \text{ yr} \times 365 \frac{\text{day}}{\text{yr}} \times 24 \frac{\text{hr}}{\text{day}} \times 60 \frac{\text{min}}{\text{hr}} \times 60 \frac{\text{s}}{\text{min}}\right)$$

$$= 9{,}460{,}000{,}000{,}000 \text{ km}$$

That is, 1 light-year is equivalent to 9.46 trillion kilometers, which is easier to remember as almost 10 trillion kilometers.

EXAMPLE 2: How big is the Sun on the 1-to-10-billion scale?

Solution: The Sun's actual radius is 695,000 km, which we express in scientific notation as 6.95×10^5 km. (See Appendix C to review powers of 10 and scientific notation.) To find the Sun's radius on the 1-to-10-billion scale, we divide its actual radius by 10 billion, or 10^{10}:

$$\text{scaled radius} = \frac{\text{actual radius}}{10^{10}}$$

$$= \frac{6.95 \times 10^5 \text{ km}}{10^{10}}$$

$$= 6.95 \times 10^{(5-10)} \text{ km}$$

$$= 6.95 \times 10^{-5} \text{ km}$$

This answer is easier to interpret if we convert it to centimeters, which we can do by recalling that there are 1000 ($= 10^3$) meters in a kilometer and 100 ($= 10^2$) centimeters in a meter:

$$6.95 \times 10^{-5} \text{ km} \times \frac{10^3 \text{ m}}{1 \text{ km}} \times \frac{10^2 \text{ cm}}{1 \text{ m}} = 6.95 \text{ cm}$$

On the 1-to-10-billion scale, the Sun is just under 7 centimeters in radius, or 14 centimeters in diameter.

EXAMPLE 3: How fast is Earth orbiting the Sun?

Solution: Earth completes one orbit in 1 year, so we can find its average orbital speed by dividing the circumference of its orbit by 1 year. Earth's orbit is nearly circular with a radius of 1 AU ($= 1.5 \times 10^8$ km); the circumference of a circle is given by the formula $2\pi \times$ radius. If we want the speed to come out in units of km/hr, we divide this circumference by 1 year converted to hours, as follows:

$$\text{orbital speed} = \frac{\text{orbital circumference}}{1 \text{ yr}}$$

$$= \frac{2 \times \pi \times (1.5 \times 10^8 \text{ km})}{1 \text{ yr} \times \frac{365 \text{ day}}{\text{yr}} \times \frac{24 \text{ hr}}{\text{day}}}$$

$$\approx 107{,}000 \text{ km/hr}$$

Earth's average speed as it orbits the Sun is more than 100,000 km/hr.

Stellar Lives and Galactic Recycling Within galaxies like the Milky Way, gravity drives the collapse of clouds of gas and dust to form stars and planets. Stars are not living organisms, but they nonetheless go through "life cycles." A star is born when gravity compresses the material in a cloud to the point at which the center becomes dense enough and hot enough to generate energy by **nuclear fusion**, the process in which lightweight atomic nuclei smash together and stick (or fuse) to make heavier nuclei. The star "lives" as long as it can shine with energy from fusion, and "dies" when it exhausts its usable fuel.

In its final death throes, a star blows much of its content back out into space. The most massive stars die in titanic explosions called *supernovae*. The returned matter mixes with other matter floating between the stars in the galaxy, eventually becoming part of new clouds of gas and dust from which new generations of stars can be born. Galaxies therefore function as cosmic recycling plants, recycling material expelled from dying stars into new generations of stars and planets. This cycle is illustrated in the lower right of Figure 1.10. Our own solar system is a product of many generations of such recycling.

Star Stuff The recycling of stellar material is connected to our existence in an even deeper way. By studying stars of different ages, we have learned that the early universe contained only the simplest chemical elements: hydrogen and helium (and a trace of lithium). We and Earth are made primarily of other elements, such as carbon, nitrogen, oxygen, and iron. Where did these other elements come from? Evidence shows that they were manufactured by stars, some through the nuclear fusion that makes stars shine and most others through nuclear reactions accompanying the explosions that end stellar lives.

By the time our solar system formed, about $4\frac{1}{2}$ billion years ago, earlier generations of stars had already converted up to 2% of our galaxy's original hydrogen and helium into heavier elements. Therefore, the cloud that gave birth to our solar system was made of roughly 98% hydrogen and helium and 2% other elements. This 2% may sound small, but it was more than enough to make the small rocky planets of our solar system, including Earth. On Earth, some of these elements became the raw ingredients of life, which ultimately blossomed into the great diversity of life on Earth today.

In summary, most of the material from which we and our planet are made was created inside stars that lived and died before the birth of our Sun. As astronomer Carl Sagan (1934–1996) said, we are "star stuff."

◆ How do our lifetimes compare to the age of the universe?

We can put the 14-billion-year age of the universe into perspective by imagining this time compressed into a single year, so each month represents a little more than 1 billion years. On this *cosmic calendar*, the Big Bang occurs at the first instant of January 1 and the present is the stroke of midnight on December 31 (Figure 1.11).

On this time scale, the Milky Way Galaxy probably formed in February. Many generations of stars lived and died in the subsequent cosmic months, enriching the galaxy with the "star stuff" from which we and our planet are made.

Our solar system and our planet did not form until early September on this scale, or $4\frac{1}{2}$ billion years ago in real time. By late September, life on Earth was flourishing. However, for most of Earth's history, living organisms remained relatively primitive and microscopic. On the scale of the cosmic calendar, recognizable animals became prominent only in mid-December. Early dinosaurs appeared on the day after Christmas. Then, in a cosmic instant, the dinosaurs disappeared forever—probably because of the impact of an asteroid or a comet [Section 6.3]. In real time the death of the dinosaurs occurred some 65 million years ago, but on the cosmic calendar it was only yesterday. With the dinosaurs gone, small furry mammals

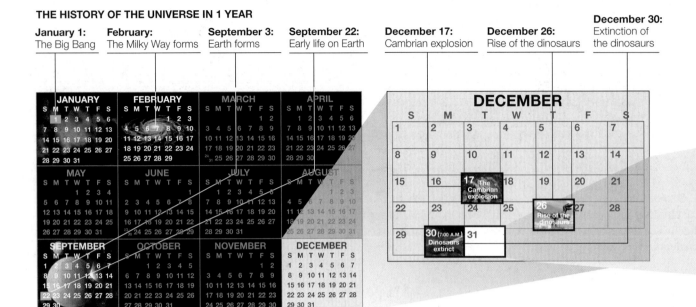

▲ **FIGURE 1.11**
The cosmic calendar compresses the 14-billion-year history of the universe into 1 year, so that each month represents a little more than 1 billion years. (Adapted from the cosmic calendar created by Carl Sagan. For a more detailed version, see the "You Are Here in Time" foldout diagram in the front of the book.)

inherited Earth. Some 60 million years later, or at around 9 P.M. on December 31 of the cosmic calendar, early hominids (human ancestors) began to walk upright.

Perhaps the most astonishing thing about the cosmic calendar is that the entire history of human civilization falls into just the last half-minute. The ancient Egyptians built the pyramids only about 11 seconds ago on this scale. About 1 second ago, Kepler and Galileo proved that Earth orbits the Sun rather than vice versa. The average college student was born about 0.05 second ago, around 11:59:59.95 P.M. on the cosmic calendar. On the scale of cosmic time, the human species is the youngest of infants, and a human lifetime is a mere blink of an eye.

think about it Study the more detailed cosmic calendar found on the foldout in the front of this book. How does an understanding of the scale of time affect your view of human civilization? Explain.

1.3 Defining Planets

One of the goals of this book is to help you learn more about science in general as you study the science of astronomy. We will therefore conclude each chapter with a case study that illustrates the process of science in action. Here, we look at the process of scientific classification, focusing on the challenge of defining the seemingly simple term *planet*.

Science begins with observations of the world around us, and after observing we often try to classify the objects we find. Scientific classification helps us organize our thinking and provides a common language for discussion. In this chapter, we have already classified objects into categories such as planets, stars, and galaxies. The box on page 5 provides basic definitions for these categories but does not explain *how* we came to classify objects in this way. The story behind the definition of *planet* provides an excellent example of how scientists classify objects and how scientific classification must adapt to new discoveries.

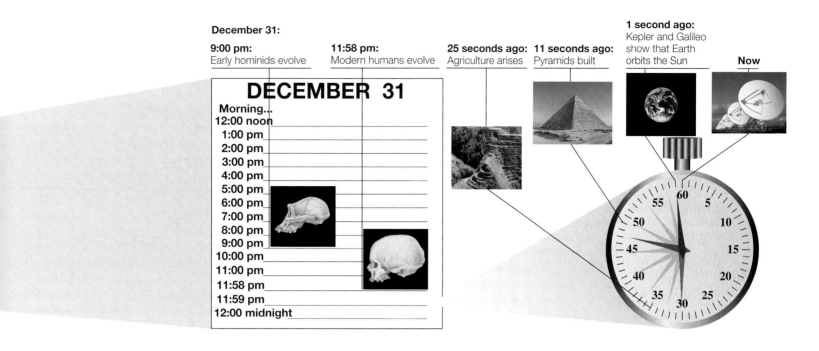

◆ What is a planet?

The difference between a star and a planet is not obvious from a casual glance at the night sky. In fact, the term *star* historically applied to almost any shining object in the night sky, including the planets and even the brief flashes of light known as "shooting stars" (or *meteors*), which we now know to be caused by comet dust entering Earth's atmosphere. To the naked eye, the difference between stars and planets becomes clear only if you observe the sky over a period of many days or weeks: Stars remain fixed in the patterns of the constellations, while planets appear to move slowly among the constellations of stars [Section 2.3].

Planets as Wanderers The word *planet* comes from the Greek for "wanderer," and in ancient times it applied to all objects that appear to move, or wander, among the constellations. The Sun and the Moon were counted as planets, because they move steadily through the constellations. Earth did *not* count as a planet, since it is not something we see in the sky and it was presumed to be stationary at the center of the universe. Ancient observers therefore recognized seven objects as planets: the Sun, the Moon, and the five planets that are easily visible to the naked eye (Mercury, Venus, Mars, Jupiter, and Saturn). The special status of these seven objects is still enshrined in the names of the seven days of the week. In English, only Sunday, "Moonday," and "Saturnday" are obvious, but if you know a romance language like Spanish you'll be able to figure out the rest: Tuesday is Mars day (martes), Wednesday is Mercury day (miércoles), Thursday is Jupiter day (jueves), and Friday is Venus day (viernes).

This original definition of *planet* began to change about 400 years ago, when we learned that Earth is *not* the center of the universe but rather an object that orbits the Sun. The term *planet* then came to mean any object that orbits the Sun, which added Earth to the list of planets and removed the Sun and Moon (because the Moon orbits Earth). This definition successfully accommodated the planets Uranus and Neptune after their discoveries in 1781 and 1846, respectively, and it can also be easily adapted for planets around other stars.

A weakness of this definition became apparent as scientists began to discover asteroids, starting with the discovery of Ceres in 1801. Ceres was initially hailed

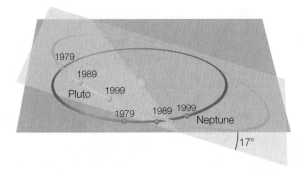

▲ FIGURE 1.12
Pluto's orbit is significantly elongated and tilted with respect to those of the other planets. Pluto even comes closer than Neptune to the Sun for 20 years in each 248-year orbit, as was the case between 1979 and 1999. There's no danger of a collision, however, because Neptune completes exactly three orbits for every two of Pluto's orbits.

as a new "planet," but as the number of known asteroids grew—and as we realized that asteroids were all much smaller than the traditional planets—scientists decided that these relatively small worlds should count only as "minor planets." Today we refer to "minor planets" as asteroids or comets, depending on their composition, though larger ones, such as Ceres, also qualify as *dwarf planets*.

The Case of Pluto The most recent change in the definition of *planet* comes from the story of Pluto. Pluto was quickly given planetary status upon its discovery in 1930, partly because astronomers were actively searching for a planet but also because they initially overestimated its mass. Still, it was clear from the start that Pluto was a misfit among the known planets. Its 248-year orbit around the Sun is more elongated in shape than that of any other planet. Its orbit is also significantly tilted relative to the orbits of the other planets (Figure 1.12). It became even more of a misfit after astronomers pinned down its mass and composition: Pluto is only about $\frac{1}{25}$ as massive as Mercury, smallest of the first eight planets, and its ice-rich composition is more similar to that of a comet than to that of any of the other planets.

Pluto's low mass, unusual orbit, and comet-like composition eventually began to cause controversy over its status as a planet. Scientists had known since the 1950s that many of the comets that we see in the inner solar system come from the same region of the solar system in which Pluto orbits (called the *Kuiper belt* [Section 4.1]). By the 1990s, advancing telescope technology had made it possible for astronomers to discover vast numbers of objects orbiting within the Kuiper belt, including some that were not much smaller than Pluto. Then, in 2005, astronomer Mike Brown announced the discovery of Eris, which is slightly more massive than Pluto. This discovery forced astronomers to consider the question of whether Eris and other objects similar in size to Pluto should count as planets.

Decisions on astronomical names and definitions are officially made by the International Astronomical Union (IAU), which is made up of astronomers from nations around the world. Members of the IAU considered numerous possible definitions of the term *planet*. One proposed definition would have applied the term to any object large enough for its own gravity to make it round. Proponents of this definition argued that it would make planetary status depend only on an object's intrinsic characteristics (as opposed to also including orbital properties), and that roundness is a good indicator of whether an object is large enough to have had any "planet-like" geological activity during its history; this latter idea was confirmed by the dramatic images of Pluto and its moon Charon obtained by the *New Horizons* spacecraft during its 2015 flyby (Figure 1.13). Those arguing against this definition pointed out that, in addition to significantly raising the official number of planets in our solar system, it would also have meant counting numerous large moons — including Earth's Moon — as planets.

The definition that ultimately won out (in an IAU vote held in 2006) focused on both size and orbits, defining a planet as an object that (1) orbits a star (but is not itself a star),* (2) is massive enough for its own gravity to make it round, and (3) dominates its orbital region. This definition leaves out moons because they orbit planets rather than stars, and it leaves out Pluto and Eris because, although they orbit the Sun and are round, they share their orbital region with many similarly sized objects.

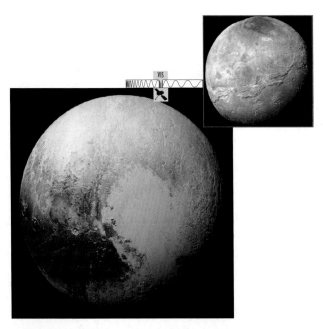

▲ FIGURE 1.13
Pluto and its largest moon, Charon (shown to scale), as seen by the *New Horizons* spacecraft during its 2015 flyby of them. As we'll discuss further in Chapter 6, the visible surface features offer clear evidence that both worlds have had interesting geological histories, and Pluto may have ongoing geological activity.

---

*The officially adopted definition refers only to objects orbiting *our Sun*, but astronomers have generally applied the definition in a way that also allows us to refer to "planets" orbiting other stars.

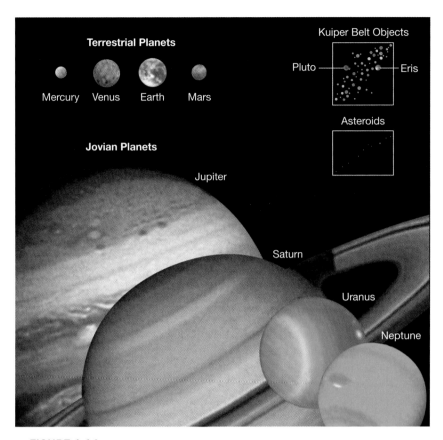

▲ **FIGURE 1.14**
Relative sizes of various objects in the solar system. Notice that the eight planets divide clearly into two groups, known as the *jovian planets* (Jupiter, Saturn, Uranus, and Neptune) and the *terrestrial planets* (Mercury, Venus, Earth, and Mars). Pluto and Eris clearly belong to a group of much smaller but more numerous objects.

Future Challenges The adopted definition of *planet* still stirs some controversy, but it works well in dividing the objects of our solar system by size: Figure 1.14 shows that the eight planets clearly divide into two groups, while Pluto and Eris clearly belong to a different group. Nevertheless, the current definition is likely to face future challenges, including one that may arise fairly soon. Over the past few years, astronomers have discovered that many small objects in the outer solar system appear to be following an orbital pattern suggesting that they are being tugged on by the gravity of an undiscovered object at least several times as massive as Earth. As this book goes to press, astronomers are actively searching for such an object, which has been called "planet nine." But note that, if it really exists, this object would likely be sharing the same general region of the solar system with many other ice-rich objects, which would mean it would not count as a planet under the current definition, despite its large size. And even if this object turns out not to exist in our solar system, we will probably find similar objects in other solar systems, forcing us once again to reconsider the definition of a planet.

Perhaps the most important scientific idea to take away from this debate is that nature need not obey the classification systems we propose. Whether we call an object like Pluto a planet, a dwarf planet, or a large comet may affect the way we think about it, but it does not change what it actually is or our scientific interest in it. A key part of science is learning to adapt our own notions of organization to the underlying reality of nature. As we discover new things, we must sometimes change our definitions.

think about it How would *you* define "planet," and why? Defend your choice.

summary of key concepts

1.1 The Scale of the Universe

◆ **What is our place in the universe?**

Earth is a planet orbiting the Sun. Our Sun is one of more than 100 billion stars in the **Milky Way Galaxy**. Our galaxy is one of more than 70 galaxies in the **Local Group**. The Local Group is one small part of the **Local Supercluster**, which is one small part of the **universe**.

◆ **How big is the universe?**

On a 1-to-10-billion scale, the Sun is the size of a grapefruit, Earth is a ball point about 15 meters away, and the nearest stars are thousands of kilometers away. Our galaxy has so many stars that it would take thousands of years just to count them out loud. The **observable universe** contains more than 100 billion galaxies, and the total number of stars is comparable to the number of grains of dry sand on all the beaches on Earth.

1.2 The History of the Universe

◆ **How did we come to be?**

The universe began in the **Big Bang** and has been expanding ever since, except in localized regions where gravity has caused matter to collapse into galaxies and stars. The Big Bang essentially produced only two chemical elements: hydrogen and helium. The rest have been produced by stars and recycled within galaxies from one generation of stars to the next, which is why we are "star stuff."

◆ **How do our lifetimes compare to the age of the universe?**

On a cosmic calendar that compresses the history of the universe into 1 year, human civilization is just a few seconds old, and a human lifetime lasts only a fraction of a second.

 THE PROCESS OF SCIENCE IN ACTION

1.3 Defining Planets

◆ **What is a planet?**

The definition of the term *planet* has changed over time. It originally referred to objects that wandered among the constellations, but now applies to Earth and seven other objects in our solar system, while Pluto, Eris, and similar objects are classified as *dwarf planets*. The definition may yet change again, showing how scientific classification must adapt to new discoveries.

Investigations

For instructor-assigned homework and other learning materials, go to www.MasteringAstronomy.com.

Quick Quiz

Choose the best answer to each of the following. Explain your reasoning.

1. Which of the following correctly lists our "cosmic address" from smallest to largest? (a) Earth, solar system, Milky Way Galaxy, Local Group, Local Supercluster, universe (b) Earth, solar system, Local Group, Local Supercluster, Milky Way Galaxy, universe (c) Earth, Milky Way Galaxy, solar system, Local Group, Local Supercluster, universe
2. An *astronomical unit* is (a) any planet's average distance from the Sun. (b) Earth's average distance from the Sun. (c) any large astronomical distance.
3. A *light-year* is (a) about 10 trillion kilometers. (b) the time it takes light to reach the nearest star. (c) the time it takes light to travel around the Sun.
4. The star Betelgeuse is about 600 light-years away. If it explodes tonight, (a) we'll know because it will be brighter than the full Moon. (b) we'll know because debris from the explosion will rain down on us from space. (c) we won't know about it until about 600 years from now.
5. Could we see a galaxy that is 50 billion light-years away? (a) Yes, if we had a big enough telescope. (b) No, because it would be beyond the bounds of our observable universe. (c) No, because a galaxy could not possibly be that far away.
6. If we represent the solar system on a scale that allows us to walk from the Sun to Pluto in a few minutes, then (a) the planets are the size of basketballs and the nearest stars are a few miles away. (b) the planets are marble size or smaller and the nearest stars are thousands of miles away. (c) the planets are microscopic and the stars are millions of miles away.
7. The number of stars in the Milky Way Galaxy is roughly (a) 100,000. (b) 100 million. (c) 100 billion.
8. When we say the universe is *expanding*, we mean that (a) everything in the universe is growing in size. (b) the average distance between galaxies is growing with time. (c) the universe is getting older.
9. The *Big Bang* is the name astronomers give to (a) the explosion that occurs when a star dies. (b) the largest explosion ever observed. (c) the birth of the universe.
10. We are "star stuff" in the sense that (a) we are made of elements that were produced in stars. (b) our bodies have the same chemical composition as stars. (c) we are born, live, and die, just like stars.
11. The age of our solar system is about (a) $\frac{1}{3}$ of the age of the universe. (b) $\frac{3}{4}$ of the age of the universe. (c) the same as the age of the universe.
12. The event that triggered the change in Pluto's status from planet to dwarf planet was the discovery that (a) it is smaller than the planet Mercury. (b) it has a comet-like composition of ice and rock. (c) it is not the most massive object in its region of the solar system.

Short-Answer/Essay Questions

Explain all answers clearly, using complete sentences and proper essay structure if needed. An asterisk (*) designates a quantitative problem, for which you should show all your work.

13. *Our Cosmic Origins.* Write one to three paragraphs summarizing why we could not be here if the universe did not contain both stars and galaxies.
14. *Alien Technology.* Some people believe that Earth is regularly visited by aliens who travel here from other star systems. For this to be true, how much more advanced than our own technology would the alien space travel technology have to be? Write one to two paragraphs to give a sense of the technological difference. (*Hint:* The ideas of scale in this chapter can help you contrast the distance the aliens would have to travel with the distances we currently are capable of traveling.)
15. *Looking for Evidence.* In this first chapter, we have discussed the scientific story of the universe but have not yet discussed most of the evidence that backs it up. Choose one idea from this chapter—such as the idea that there are billions of galaxies, that the universe was born in the Big Bang, or that we are "star stuff"—and briefly discuss the type of evidence you would like to see before accepting the idea. (*Hint:* It's okay to look ahead in the book to see the evidence presented in later chapters.)
16. *The Value of Classification.* Section 1.3 discussed difficulties that can arise with attempts to define scientific classifications precisely, such as those that occur with the term *planet*. Make a bullet list of pros and cons (at least three of each) of having classification schemes. Then write a one-paragraph summary stating your opinion of the value (or lack of value) of scientific classification.
17. *The Cosmic Perspective.* Write a one-page essay describing how the ideas presented in this chapter affect your perspectives on your own life and on human civilization.
*18. *Light-Minute.* Just as a light-year is the distance that light can travel in 1 year, a light-minute is the distance that light can travel in 1 minute. What is a light-minute in kilometers?
*19. *Sunlight.* Use the speed of light and the Earth–Sun distance of 1 AU to calculate how long it takes light to travel from the Sun to Earth.
*20. *Cosmic Calendar.* The cosmic calendar condenses the 14-billion-year history of the universe into 1 year. How long does 1 second represent on the cosmic calendar?
*21. *Saturn vs. the Milky Way.* Photos of Saturn with its rings can look so similar to photos of galaxies that children often think they are similar objects, but of course galaxies are far larger. About how many times larger in diameter is the Milky Way Galaxy than Saturn's rings? (*Data:* Saturn's rings are 270,000 km in diameter; the Milky Way is 100,000 light-years in diameter.)
*22. *Galactic Rotation.* Our solar system is located about 27,000 light-years from the galactic center and orbits the center once every 230 million years. How fast are we traveling around the galaxy, in km/hr?

Understanding the Sky

2

This time-exposure photo, taken at Arches National Park in Utah, shows how the entire sky seems to circle daily around a point above Earth's North Pole (or South Pole, for the Southern Hemisphere). Stars far to the north, like those visible within the arch, complete their entire circles above the northern horizon. Other stars—along with the Sun, Moon, and planets—follow circles that cross the horizon, which is why they rise in the east and set in the west each day. This daily circling of the sky explains why most of our ancestors assumed that the universe revolved around Earth. Today, we know that it is actually Earth's daily rotation that makes the sky appear to turn. How did we learn this fact? Through careful study of other, more subtle patterns of change in the sky, including the patterns we will study in this chapter.

LEARNING GOALS

2.1 Understanding the Seasons
- What causes the seasons?
- Why do the constellations we see depend on the time of year?

2.2 Understanding the Moon
- Why do we see phases of the Moon?
- What causes eclipses?

 THE PROCESS OF SCIENCE IN ACTION

2.3 The Puzzle of Planetary Motion
- Why did the ancient Greeks reject the real explanation for planetary motion?

▶ Discovering the Universe for Yourself

2.1 Understanding the Seasons

We're all familiar with seasonal changes, such as longer and warmer days in summer and shorter and cooler days in winter. But *why* do the seasons occur? We'll explore the answer in this section.

◆ What causes the seasons?

Seasons are a result of the way the tilt of Earth's axis causes sunlight to fall differently on Earth at different times of year. To understand exactly how this works, we'll first look at how Earth's rotation produces the apparent daily path of the Sun through the sky, then examine how Earth's axis tilt and orbit cause this path to change over the course of each year.

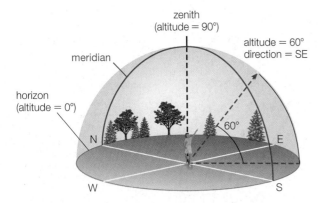

▲ **FIGURE 2.1**
From any place on Earth, the local sky looks like a dome (hemisphere). This diagram shows key reference points in the local sky. It also shows how we can describe any position in the local sky by its altitude and direction.

The Sun's Daily Path The general pattern of the Sun's daily path through the sky is fairly simple: Unless you live within the Arctic or Antarctic circle, the Sun always rises somewhere in the east, reaches its highest point around noon, then sets somewhere in the west. Ancient people assumed that the Sun actually circled Earth each day, but we now know that its daily motion arises from Earth's rotation. Just as the world seems to circle around you if you spin in place, the Sun seems to go around us on our rotating planet. However, while the general pattern is always the same, the Sun's precise path through the sky varies with the seasons.

To describe the Sun's daily path more clearly, we need to define a few key reference points in the sky (Figure 2.1). The boundary between Earth and sky defines the **horizon**. The point directly overhead is the **zenith**. The **meridian** is an imaginary half-circle stretching from the horizon due south, through the zenith, to the horizon due north. We can pinpoint the position of any object in the sky by stating its *direction* along the horizon (sometimes expressed as *azimuth*) and its *altitude* above the horizon.

Figure 2.2 shows how the Sun's daily path differs at different times of year for a typical Northern Hemisphere location. Notice that the Sun's path is long and high in summer, with the Sun rising well north of due east and setting well north of due west. In winter, the Sun's path is short and low, as it rises well south of due east and sets well south of due west. Notice also that the Sun *never* passes directly overhead at this latitude. The Sun can reach the zenith only for locations within the *tropics*, meaning locations on Earth that are between latitudes of $23\frac{1}{2}°$N and $23\frac{1}{2}°$S.

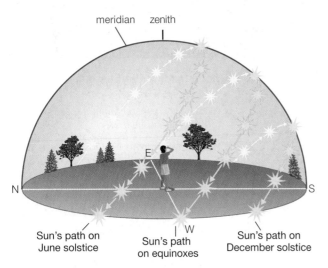

▲ **FIGURE 2.2**
This diagram shows the Sun's path in different seasons for a Northern Hemisphere location (latitude 40°N). The precise paths are different for other latitudes.

▶ The Reason for Seasons

The Reason for Seasons The causes of these seasonal changes in the Sun's path are summarized in the four steps of Figure 2.3 (pp. 20–21). Step 1 illustrates how Earth's rotation axis is tilted with respect to its orbit. Notice that the axis remains pointed in the same direction in space (toward the North Star, Polaris) throughout the year. As a result, the orientation of the axis *relative to the Sun* changes over the course of each orbit: The Northern Hemisphere is tipped toward the Sun in June and away from the Sun in December, while the reverse is true for the Southern Hemisphere. That is why the two hemispheres experience opposite seasons.

Step 2 shows Earth in June, when the axis tilt causes sunlight to strike the Northern Hemisphere at a steeper angle and the Southern Hemisphere at a shallower angle. The steeper sunlight angle makes it summer in the Northern Hemisphere, for two reasons. First, as shown in the zoom-out, the steeper angle means more concentrated sunlight, which tends to make it warmer. Second, if you visualize what happens as Earth rotates each day, you'll see that the steeper angle also means the Sun follows a longer and higher path through the sky (the path shown for June in Figure 2.2), giving the Northern Hemisphere more hours of daylight during which it is warmed by the Sun. The opposite is true for the Southern Hemisphere at this time: The shallower sunlight angle makes it winter there, because sunlight is less concentrated and the Sun follows a shorter, lower path through the sky.

The sunlight angle gradually changes as Earth orbits the Sun. At the opposite side of Earth's orbit, Step 4 shows that it has become winter for the Northern Hemisphere and summer for the Southern Hemisphere. In between these two extremes, Step 3 shows that both hemispheres are illuminated equally in March and September. It is therefore spring for the hemisphere that is on the way from winter to summer, and fall for the hemisphere on the way from summer to winter.

The key point is this: Seasons occur because of the combination of Earth's axis tilt and its orbit around the Sun. If Earth did not have an axis tilt, we would not have seasons.

think about it Jupiter has an axis tilt of about 3°, small enough to be insignificant. Saturn has an axis tilt of about 27°, slightly greater than that of Earth. Both planets have nearly circular orbits around the Sun. Do you expect Jupiter to have seasons? Do you expect Saturn to have seasons? Explain.

Solstices and Equinoxes To help us mark the changing seasons, we define four special moments in the year, each of which corresponds to one of the four special positions in Earth's orbit shown in Figure 2.3.

- The **June solstice**, called the *summer solstice* in the Northern Hemisphere, occurs around June 21 and is the moment when the Northern Hemisphere is tipped most directly toward the Sun and receives the most direct sunlight.
- The **December solstice**, called the *winter solstice* in the Northern Hemisphere, occurs around December 21 and is the moment when the Northern Hemisphere receives the least direct sunlight.
- The **March equinox**, called the *spring* (or *vernal*) *equinox* in the Northern Hemisphere, occurs around March 21 and is the moment when the Northern Hemisphere goes from being tipped slightly away from the Sun to being tipped slightly toward the Sun.
- The **September equinox**, called the *fall* (or *autumnal*) *equinox* in the Northern Hemisphere, occurs around September 22 and is the moment when the Northern Hemisphere first starts to be tipped away from the Sun.

The exact dates and times of the solstices and equinoxes can vary by up to a couple days from the dates given above, depending on where we are in the leap year cycle. Leap years add a day—February 29—to the calendar every fourth year, except century years that are *not* divisible by 400. (So the years 1700, 1800, and 1900 were not leap years, but the year 2000 was.) Leap days keep the calendar aligned with the seasons by making the average length of the calendar year match the true length of the year, which is very close to $365\frac{1}{4}$ days.

First Days of Seasons We usually say that each equinox or solstice marks the first day of a season. For example, the day of the June solstice is usually called the "first day of summer" in the Northern Hemisphere. Notice, however, that the Northern Hemisphere has its *maximum* tilt toward the Sun at this time. You might then wonder why we consider the summer solstice to be the beginning rather than the midpoint of summer.

The choice is somewhat arbitrary, but it makes sense in at least two ways. First, it was much easier for ancient people to identify the days on which the Sun reached extreme positions in the sky—such as when it reached its highest point

---

common misconceptions

The Cause of Seasons

Many people guess that seasons are caused by variations in Earth's distance from the Sun. But if this were true, the whole Earth would have summer or winter at the same time, and it doesn't: The seasons are opposite in the Northern and Southern Hemispheres. In fact, Earth's slightly varying orbital distance has virtually no effect on the weather. The real cause of seasons is Earth's axis tilt, which causes the two hemispheres to take turns being tipped toward the Sun over the course of each year.

cosmic context — FIGURE 2.3 The Cause of Seasons

Earth's seasons are caused by the tilt of its rotation axis, which is why the seasons are opposite in the two hemispheres. The seasons do *not* depend on Earth's distance from the Sun, which varies only slightly throughout the year.

① Axis Tilt: Earth's axis points in the same direction throughout the year, which causes changes in Earth's orientation *relative to the Sun*.

perpendicular to ecliptic plane — 23½° — N

S

② Northern Summer/Southern Winter: In June, sunlight falls more directly on the Northern Hemisphere, which makes it summer there because solar energy is more concentrated and the Sun follows a longer and higher path through the sky. The Southern Hemisphere receives less direct sunlight, making it winter.

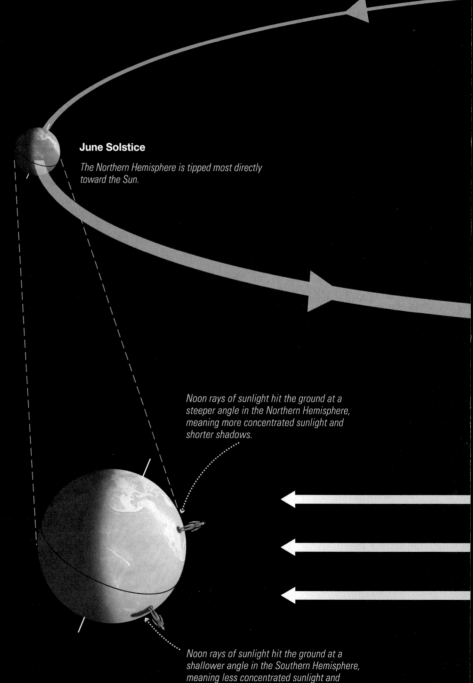

June Solstice
The Northern Hemisphere is tipped most directly toward the Sun.

Noon rays of sunlight hit the ground at a steeper angle in the Northern Hemisphere, meaning more concentrated sunlight and shorter shadows.

Noon rays of sunlight hit the ground at a shallower angle in the Southern Hemisphere, meaning less concentrated sunlight and

Interpreting the Diagram

To interpret the seasons diagram properly, keep in mind:

1. Earth's size relative to its orbit would be microscopic on this scale, meaning that both hemispheres are at essentially the same distance from the Sun.

2. The diagram is a side view of Earth's orbit. A top-down view (below) shows that Earth orbits in a nearly perfect circle and comes closest to the Sun in January.

March Equinox
147.1 million km — January 3
152.1 million km
July 4

③ **Spring/Fall:** Spring and fall begin when sunlight falls equally on both hemispheres, which happens twice a year: In March, when spring begins in the Northern Hemisphere and fall in the Southern Hemisphere; and in September, when fall begins in the Northern Hemisphere and spring in the Southern Hemisphere.

④ **Northern Winter/Southern Summer:** In December, sunlight falls less directly on the Northern Hemisphere, which makes it winter because solar energy is less concentrated and the Sun follows a shorter and lower path through the sky. The Southern Hemisphere receives more direct sunlight, making it summer.

March Equinox
The Sun shines equally on both hemispheres.

The variation in Earth's orientation relative to the Sun means that the seasons are linked to four special points in Earth's orbit:

Solstices *are the two points at which sunlight becomes most extreme for the two hemispheres.*

Equinoxes *are the two points at which the hemispheres are equally illuminated.*

December Solstice
The Southern Hemisphere is tipped most directly toward the Sun.

September Equinox
The Sun shines equally on both hemispheres.

Noon rays of sunlight hit the ground at a shallower angle in the Northern Hemisphere, meaning less concentrated sunlight and longer shadows.

Noon rays of sunlight hit the ground at a steeper angle in the Southern Hemisphere, meaning more concentrated sunlight and shorter shadows.

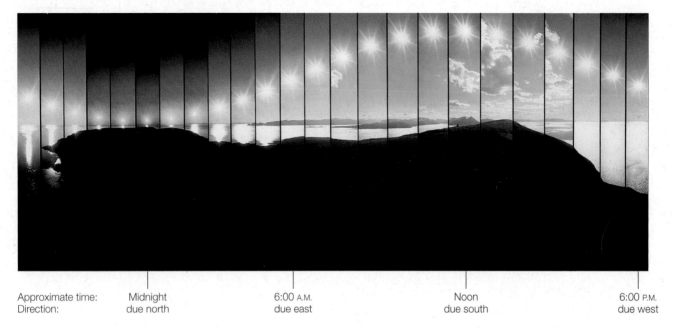

▲ FIGURE 2.4
This sequence of photos shows the progression of the Sun all the way around the horizon on the June solstice from a location near the Arctic Circle. Notice that the Sun skims the northern horizon at midnight, then gradually rises higher, reaching its highest point when it is due south at noon.

common misconceptions

High Noon

When is the Sun directly overhead in your sky? Many people answer "at noon." It's true that the Sun reaches its *highest* point each day when it crosses the meridian, giving us the term "high noon" (though the meridian crossing is rarely at precisely 12:00). However, unless you live in the tropics (between latitudes 23.5°S and 23.5°N), the Sun is *never* directly overhead. In fact, anytime you can see the Sun as you walk around, you can be sure it is *not* at your zenith. Unless you are lying down, seeing an object at the zenith requires tilting your head back into a very uncomfortable position.

on the summer solstice—than other days in between. Second, we usually think of the seasons in terms of weather, and the warmest summer weather tends to come 1 to 2 months after the solstice. To understand why, think about what happens when you heat a pot of cold soup. Even though you may have the stove turned on high from the start, it takes a while for the soup to warm up. In the same way, it takes some time for sunlight to heat the ground and oceans from the cold of winter to the warmth of summer. "Midsummer" in terms of weather therefore comes in late July and early August for the Northern Hemisphere, which makes the June solstice a pretty good choice for the "first day of summer." Similar logic applies to the starting times for spring, fall, and winter.

Seasons Around the World The seasons have different characteristics in different parts of the world. High latitudes have more extreme seasons. For example, Vermont has much longer summer days and much longer winter nights than Florida. At the Arctic Circle (latitude $66\frac{1}{2}°$), the Sun remains above the horizon all day long on the June solstice (Figure 2.4), and never rises on the December solstice (although bending of light by the atmosphere makes the Sun *appear* to be about a half-degree higher than it really is). The most extreme cases occur at the North and South Poles, where the Sun remains above the horizon for 6 months in summer and below the horizon for 6 months in winter.

Seasons also differ in equatorial regions, because the equator gets its most direct sunlight on the two equinoxes and its least direct sunlight on the solstices. As a result, instead of the four seasons experienced at higher latitudes, equatorial regions generally have rainy and dry seasons, with the rainy seasons coming when the Sun is higher in the sky.

◆ Why do the constellations we see depend on the time of year?

We have seen how and why we have seasonal changes in the weather and in the Sun's daily path through the sky. You can also track the seasons and the time of year by observing the nighttime sky. For example, the stars of Orion shine prominently on clear evenings in January (Figure 2.5), but are not visible at all on evenings in July. To understand why, we must first understand the general appearance of the night sky.

Constellations People of nearly every culture gave names to patterns they saw in the sky. We usually refer to such patterns as *constellations*, but to astronomers the term has a more precise meaning: A **constellation** is a *region* of the sky with well-defined borders, and just as every spot of land in the continental United States is part of some state, every spot in the sky belongs to some constellation. The familiar patterns of stars merely help us locate these constellations. For example, while we identify Orion by recognizing the pattern of stars outlined in Figure 2.5, even the seemingly empty spots in that region of the sky are part of the constellation Orion.

The patterns of stars in the constellations have seemed to remain fixed on the time scale of human civilization, but stars actually move at fairly high speeds relative to one another. Their positions seem to stay fixed only because they are so far away that their movements are not noticeable to the naked eye, even over many centuries. However, if you could watch a time-lapse movie made over tens of thousands of years, you would see dramatic changes in the patterns of the constellations.

The Celestial Sphere Together, all the constellations appear to fill a great **celestial sphere** surrounding Earth (Figure 2.6). Of course, the entire celestial sphere is an illusion: The sky looks like a great sphere only because the stars are so far away that we have no depth perception when we look into space, and Earth seems to be in the center of the celestial sphere only because it is where we are located as we look into space. Nevertheless, the celestial sphere is a useful illusion, because it allows us to map the sky as seen from Earth. For reference, we identify four special points and circles on the celestial sphere:

- The **north celestial pole** is the point directly over Earth's North Pole.
- The **south celestial pole** is the point directly over Earth's South Pole.
- The **celestial equator**, which is a projection of Earth's equator into space, makes a complete circle around the celestial sphere.
- The **ecliptic** is the path the Sun follows as it appears to circle around the celestial sphere once each year. It crosses the celestial equator at a $23\frac{1}{2}°$ angle, because that is the tilt of Earth's axis.

Daily Paths of Stars Through the Sky Stars appear to move through the sky during the night for the same reason the Sun appears to move through the sky during the day: Earth's daily west-to-east rotation causes everything on the celestial sphere to appear to move around us from east to west (Figure 2.7). If we could stop Earth from rotating, the stars would appear motionless.

The daily motion of the sky looks very simple if you visualize it on the celestial sphere, with each object appearing to make a daily circle around Earth (Figure 2.7). However, the motion looks a little more complex in the local sky, because your horizon cuts the celestial sphere in half. Figure 2.8 shows the idea for a typical Northern Hemisphere location (latitude 40°N). Because your location means you are standing at an angle to Earth's equator, your horizon appears to slice through the celestial sphere at the same angle to the celestial equator. This angle causes the daily circles of the stars on the celestial sphere to appear tilted in your local sky. Stars that are near the north celestial pole make daily circles above your horizon, like those within the arch in the photo on page 17. Stars that are near the south celestial pole are never visible in your sky. In between, all other stars—along with the Sun, Moon, and planets—appear to rise up from your eastern horizon and set into your western horizon.

think about it Do distant galaxies also rise and set like the stars in our sky? Why or why not?

The same general ideas apply to all locations on Earth, but the details vary with latitude. Because latitude describes the angle that your zenith makes with Earth's equator, it also determines the angle at which your horizon slices through

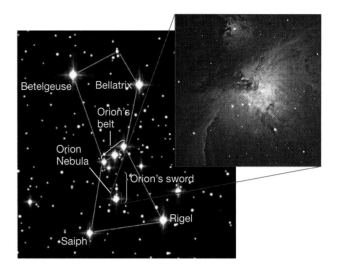

▲ **FIGURE 2.5**
The constellation Orion is prominent in the evening sky during Northern Hemisphere winter (Southern Hemisphere summer). With binoculars or a telescope, you can see that one of its fainter "stars" is actually an interstellar cloud, called the Orion Nebula, in which many stars are being born. (Note: Viewed from the Southern Hemisphere, the constellation Orion would appear rotated compared to the photo shown here, and you would see it in the northern half of the sky rather than the southern half.)

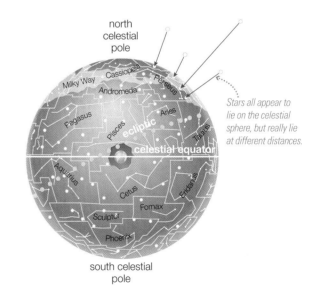

▲ **FIGURE 2.6**
The constellations appear to fill a great celestial sphere that surrounds Earth.

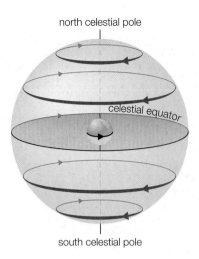

▲ **FIGURE 2.7**
Earth rotates from west to east (black arrow), making the celestial sphere *appear* to rotate around us from east to west (red arrows).

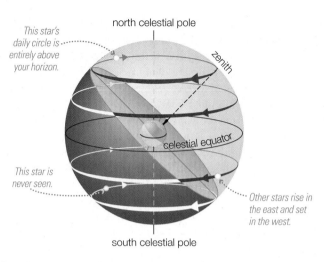

▲ **FIGURE 2.8**
The local sky for a location at latitude 40°N. The horizon slices through the celestial sphere at an angle to the celestial equator, causing the daily circles of stars to appear tilted in the local sky. Note: It is easier to follow the star paths in the local sky if you rotate the page so that the zenith points up.

the celestial sphere. The portion of the celestial sphere that you can see and the daily paths of stars through the sky therefore both depend on your latitude; that is why the night sky looks quite different at the southern latitudes of Australia than it does at the northern latitudes of the United States.

Seasonal Changes in the Night Sky Look back at Figure 2.6 and notice the yellow dots marking the path we call the *ecliptic*. These dots show that the Sun appears to move gradually around the celestial sphere, completing one full circuit in one year. This is a direct consequence of Earth's yearly orbit around the Sun.

Figure 2.9 shows how this works. From our vantage point on Earth, the annual orbit of Earth around the Sun makes the Sun *appear* to move steadily

▼ **FIGURE 2.9**
The Sun appears to move steadily eastward along the ecliptic as Earth orbits the Sun, so we see the Sun against the background of different zodiac constellations at different times of year. For example, on August 21 the Sun appears to be in Leo, because it is between us and the much more distant stars that make up Leo.

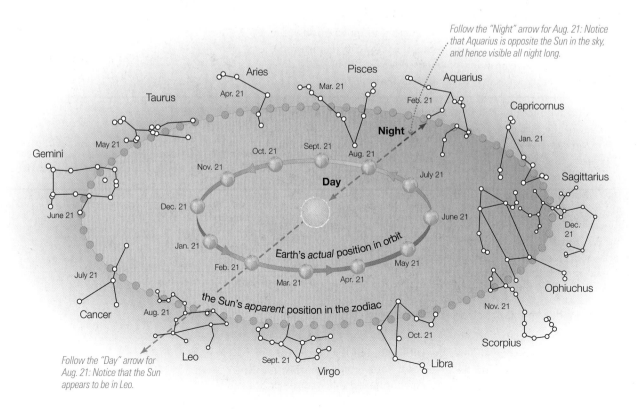

eastward along the ecliptic, with the stars of different constellations in the background at different times of year. The constellations along the ecliptic make up what we call the **zodiac**; tradition places 12 constellations along the zodiac, but the official borders include a 13th constellation, Ophiuchus.

The Sun's apparent location along the ecliptic determines which constellations we see at night. For example, Figure 2.9 shows that the Sun appears to be in Leo in late August. We therefore cannot see Leo at this time (because it is in our daytime sky), but we can see Aquarius all night long because of its location opposite Leo on the celestial sphere. Six months later, in February, we see Leo at night while Aquarius is above the horizon only in the daytime.

2.2 Understanding the Moon

Aside from the seasons and the daily circling of the sky, the most familiar pattern of change in the sky is that of the phases of the Moon. We will explore these changes in this section—along with the rarer changes that occur with eclipses—and see that they are consequences of the Moon's orbit around Earth.

◆ Why do we see phases of the Moon?

The easiest way to understand lunar phases is with the simple demonstration illustrated in Figure 2.10. Take a ball outside on a sunny day. (If it's dark or cloudy, you can use a flashlight instead of the Sun; put the flashlight on a table a

▼ **FIGURE 2.10**
In this demonstration, your head represents Earth and the ball represents the Moon as it orbits Earth. The half of the ball (Moon) facing the Sun is always illuminated while the other half is dark, but from your perspective in the middle, you see the ball (Moon) go through the phases shown in the photos. The labels indicate the approximate times of day at which each phase rises, reaches its highest point in the sky, and sets (exact times vary with location, time of year, and orbital details).

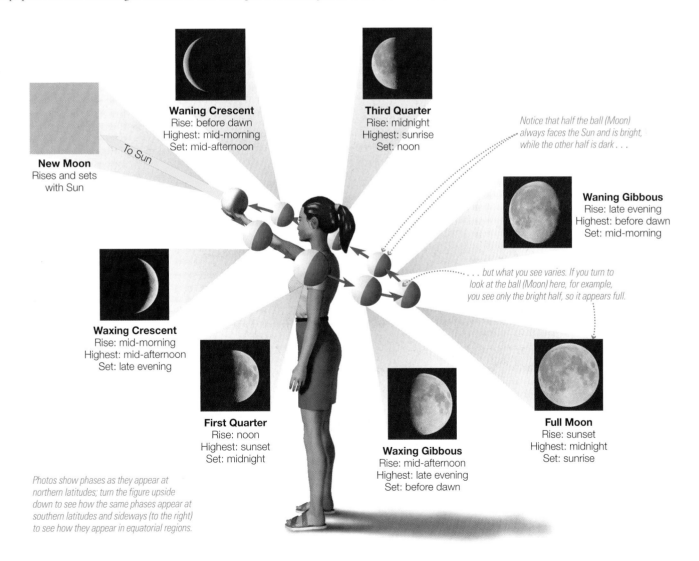

common misconceptions

Shadows and the Moon

Many people guess that the Moon's phases are caused by Earth's shadow falling on its surface, but this is not the case. Instead, the Moon's phases are caused by the fact that we see different portions of its daylight and night sides at different times as it orbits around Earth. The only time Earth's shadow falls on the Moon is during the relatively rare event of a lunar eclipse.

few meters away and shine it toward you.) Hold the ball at arm's length to represent the Moon, while your head represents Earth. Slowly spin counterclockwise so that the ball goes around you the way the Moon orbits Earth. (If you live in the Southern Hemisphere, spin clockwise because you view the sky "upside down" compared to someone in the Northern Hemisphere.) As you turn, you'll see the ball go through phases just like the Moon. If you think about what's happening, you'll realize that the phases of the ball result from just two basic facts:

1. Half the ball always faces the Sun (or flashlight) and therefore is bright, while the other half faces away from the Sun and is dark.
2. As you look at the ball at different positions in its "orbit" around your head, you see different combinations of its bright and dark faces.

For example, when you hold the ball directly opposite the Sun, you see only the bright portion of the ball, which represents the "full" phase. When you hold the ball at its "first-quarter" position, half the face you see is dark and the other half is bright.

tools of science — Angular Sizes and Distances

Our lack of depth perception on the celestial sphere means we have no way to judge the true sizes or separations of the objects we see in the sky. We can directly measure only *angular* sizes or separations.

Figure 1 shows the basic idea behind angular measurement. The **angular size** of an object is the angle it appears to span in your field of view. For example, the angular sizes of the Sun and Moon are each about $\frac{1}{2}°$ (Figure 1a). The **angular distance** between a pair of objects in the sky is the angle that appears to separate them. For example, the angular distance between the "pointer stars" at the end of the Big Dipper's bowl is about 5° (Figure 1b). You can use your outstretched hand to estimate angles in the sky (Figure 1c).

For greater precision, we subdivide each degree into 60 **arcminutes** (symbolized by ′) and subdivide each arcminute into 60 **arcseconds** (symbolized by ″), as shown in Figure 2. For example, we read 35°27′15″ as "35 degrees, 27 arcminutes, 15 arcseconds." Note that one arcminute is approximately the thickness of a fingernail at arm's length, and one arcsecond is approximately the thickness of someone else's fingernail viewed from across a football field.

Angular measurements are very useful in astronomy. We use angles to describe the locations of objects in the sky or on the celestial sphere, much as we describe the location of an object on Earth with latitude and longitude. Even more important, angular sizes can help us to calculate physical sizes if we also know an object's true distance, or vice versa. For example, we can use the Moon's distance and angular size to calculate its physical size, or we can use a planet's angular size and physical size to calculate its distance.

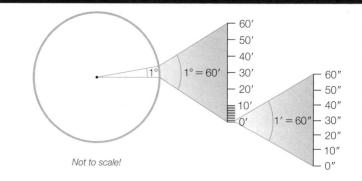

Not to scale!

▲ **FIGURE 2**

a The angular sizes of the Sun and the Moon are about 1/2°.

b The angular distance between the "pointer stars" of the Big Dipper is about 5°, and the angular length of the Southern Cross is about 6°.

Stretch out your arm as shown here.

c You can estimate angular sizes or distances with your outstretched hand.

▲ **FIGURE 1**

We see lunar phases for the same reason. Half the Moon is always illuminated by the Sun, but the amount of this illuminated half that we see from Earth depends on the Moon's position in its orbit. The photographs in Figure 2.10 show how the phases look. The new moon photo shows blue sky, because a new moon is nearly in line with the Sun and therefore hidden from view in the bright daytime sky.

Each complete cycle of phases, from one new moon to the next, takes about $29\frac{1}{2}$ days—hence the origin of the word *month* (think "moonth"). The phase determines the time the Moon rises, reaches its highest point in the sky, and sets. For example, the full moon must rise around sunset, because it occurs when the Moon is opposite the Sun in the sky. It therefore reaches its highest point in the sky at midnight and sets around sunrise. Similarly, a first-quarter moon must rise around noon, reach its highest point around sunset, and set around midnight, because it occurs when the Moon is about 90° east of the Sun in our sky.

think about it Suppose you go outside in the morning and notice that the visible face of the Moon is half light and half dark. Is this a first-quarter or third-quarter moon? How do you know?

Notice that the phases from new to full are said to be *waxing*, which means "increasing." Phases from full to new are *waning*, or "decreasing." Also notice that no phase is called a "half moon." Instead, we see half the Moon's face at first-quarter and third-quarter phases; these phases mark the times when the Moon is one-quarter or three-quarters of the way through its monthly cycle (which begins at new moon). The phases just before and after new moon are called *crescent*, while those just before and after full moon are called *gibbous* (pronounced with a hard *g* as in "gift").

The Moon's Synchronous Rotation Although we see many *phases* of the Moon, we do not see many *faces*. From Earth, we always see (nearly) the same face of the Moon. This happens because the Moon rotates on its axis in the same amount of time that it takes to orbit Earth, a trait called **synchronous rotation**. A simple demonstration shows this idea (Figure 2.11). Place a ball on a table to represent Earth, while you represent the Moon. The only way you can face the ball at all times is by completing exactly one rotation while you complete one orbit. Note that the Moon's synchronous rotation is *not* a coincidence; it is a consequence of Earth's gravity affecting the Moon in much the same way that the Moon's gravity causes tides on Earth.

The View from the Moon A good way to solidify your understanding of the lunar phases is to imagine that you live on the side of the Moon that faces Earth. For example, what would you see if you looked at Earth when people on Earth saw a new moon? By remembering that a new moon occurs when the Moon is

common misconceptions

Moon in the Daytime

Night is so closely associated with the Moon in traditions and stories that many people mistakenly believe that the Moon is visible only in the nighttime sky. In fact, the Moon is above the horizon as often in the daytime as at night, though it is easily visible only when its light is not drowned out by sunlight. For example, a first-quarter moon is easy to spot in the late afternoon as it rises through the eastern sky, and a third-quarter moon is visible in the morning as it heads toward the western horizon.

common misconceptions

The "Dark Side" of the Moon

Some people refer to the far side of the Moon—meaning the side that we never see from Earth—as the *dark side*. But this is not correct, because the far side is not always dark. For example, during new moon the far side faces the Sun and hence is completely sunlit. In fact, because the Moon rotates with a period of approximately one month (the same time that it takes to orbit Earth), points on both the near and the far side have two weeks of daylight alternating with two weeks of darkness. The only time the far side is completely dark is at full moon, when it faces away from both the Sun and Earth.

a If you do not rotate while walking around the model, you will not always face it.

b You will face the model at all times only if you rotate exactly once during each orbit.

◀ **FIGURE 2.11**
The fact that we always see the same face of the Moon means that the Moon must rotate once in the same amount of time that it takes to orbit Earth once. You can see why by walking around a model of Earth while imagining that you are the Moon.

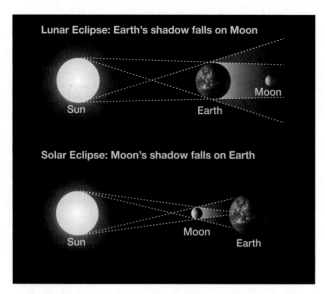

▲ FIGURE 2.12
These illustrations show the two basic types of eclipse. The relative sizes of the Moon and Earth are shown to scale, but the relative distance between them is about 50 times that shown here, the Sun is actually about 100 times larger in diameter than Earth, and the Earth-Sun distance is about 400 times the Earth-Moon distance.

between the Sun and Earth, you'll realize that from the Moon you'd be looking at Earth's daytime side and hence would see a *full Earth*. Similarly, at full moon you would be facing the night side of Earth and would see a *new Earth*. In general, you'd always see Earth in a phase opposite the phase of the Moon seen by people on Earth at the same time.

think about it About how long would each day and night last if you lived on the Moon? Explain. (*Hint*: Recall that a full cycle of phases lasts about a month.)

◆ **What causes eclipses?**

▶ **Eclipses Part 1**

The lunar phases repeat "moonthly," but sometimes the Moon gets involved in more dramatic events: eclipses. There are two basic types of eclipse (Figure 2.12).

- A **lunar eclipse** occurs when Earth comes directly between the Sun and Moon, so that Earth's shadow falls on the Moon.
- A **solar eclipse** occurs when the Moon comes directly between the Sun and Earth, so that the Moon's shadow falls on Earth.

Variations on Eclipses There are variations on the two basic types of eclipse, because the three-dimensional geometry of an eclipse shadow creates two distinct shadow regions (Figure 2.13): a central **full shadow**, or *umbra*, in which sunlight is fully blocked, and a surrounding **partial shadow**, or *penumbra*, in which light from only part of the Sun is blocked.

The shadow geometry leads to three types of lunar eclipse (Figure 2.14). If the Sun, Earth, and Moon are nearly perfectly aligned, the Moon passes through Earth's full shadow to create a **total lunar eclipse**. Earth's shadow gradually progresses across the face of the Moon, with the clear curvature of the full shadow demonstrating that our world is round. **Totality** begins when the Moon is entirely engulfed in the full shadow and typically lasts about an hour, after which we see the shadow gradually move off the Moon. The Moon becomes dark and eerily red during totality, for reasons you can understand by considering the view of an observer on the eclipsed Moon. This observer would see Earth's night side surrounded by the reddish glow of all the sunrises and sunsets occurring on Earth at that moment, which means that this reddish light illuminates the Moon during totality. If the alignment is somewhat less perfect, only part of the full moon passes through the full shadow (with the rest in the partial shadow) and we see a **partial lunar eclipse**. If the Moon passes *only* through Earth's partial shadow (penumbra), we see a **penumbral lunar eclipse**. Penumbral eclipses are the most common, but they are the least visually impressive because the full moon darkens only slightly.

There are also three types of solar eclipse (Figure 2.15), in part because the Moon's distance from Earth varies somewhat over the course of each orbit. If a solar eclipse occurs when the Moon is relatively close to Earth, the Moon's full shadow can cover a small area of Earth's surface (up to about 270 kilometers in diameter). Within this area you will see a **total solar eclipse**. If the eclipse occurs when the Moon is relatively far from Earth, the full shadow may not reach Earth's surface, leading to an **annular eclipse**—a ring of sunlight surrounding the Moon—in the small region of Earth directly behind the full shadow. In either case, the region of totality or annularity will be surrounded by a much larger region (typically about 7000 kilometers in diameter) that falls within the Moon's partial shadow. Here you will see a **partial solar eclipse**, in which only part of the Sun is blocked from view. (Some solar eclipses are only partial, meaning that no locations on Earth see a total or annular eclipse, because the full shadow passes above or below our planet.) The combination of Earth's rotation and the Moon's orbital motion causes the Moon's shadow to race across the face of Earth at a typical speed of about 1700 kilometers

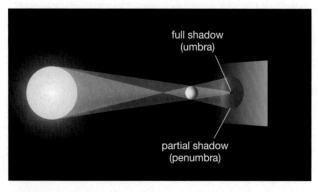

▲ FIGURE 2.13
This diagram shows the two distinct, cone-shaped regions of the shadow cast on an imaginary screen by a world like Earth or the Moon in sunlight. If you looked back from within the full shadow, you would see the Sun fully blocked from view; if you looked from within the partial shadow, the Sun would be only partly blocked from view.

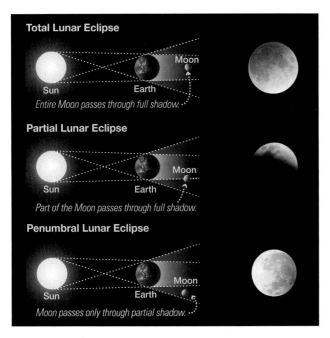

▲ FIGURE 2.14
The three types of lunar eclipse. The Moon is actually much dimmer during totality than during the partial eclipse, but it is shown here as the eye tends to notice it because of its coloration.

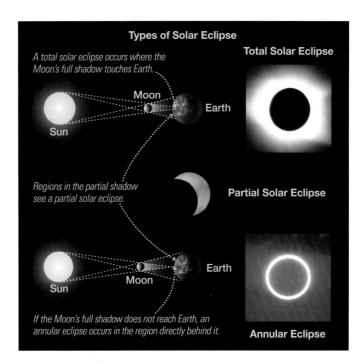

▲ FIGURE 2.15
The three types of solar eclipse. (A partial solar eclipse may also be seen in cases when the Moon's full shadow passes above or below Earth.)

per hour. As a result, the full (or annular) shadow traces a narrow path across Earth, and totality can never last more than a few minutes in any particular place.

A total solar eclipse is a spectacular sight (Figure 2.16). It begins when the disk of the Moon first appears to touch the Sun. Over the next couple of hours, the Moon appears to take a larger and larger "bite" out of the Sun. As totality approaches, the sky darkens and temperatures fall. Birds head back to their nests, and crickets begin their nighttime chirping. During the few minutes of totality, the Moon completely blocks the visible disk of the Sun, allowing the faint *corona* to be seen. The surrounding sky takes on a twilight glow, and planets and bright stars become visible in the daytime. As totality ends, the Sun slowly emerges from behind the Moon. However, because your eyes have adapted to the darkness, totality appears to end far more abruptly than it began.

▶ Eclipses Part 2

Conditions for Eclipses How often do eclipses occur? If you look at a simple Moon phase diagram (such as Figure 2.10), it might seem as if the Sun, Earth, and Moon would line up with every new and full moon, in which case we'd have both a lunar and a solar eclipse every month. But we don't, and the reason is that the Moon's orbit is slightly inclined (by about 5°) to the ecliptic plane (the plane of Earth's orbit around the Sun).

You can visualize this inclination by imagining the ecliptic plane as the surface of a pond (Figure 2.17). The Moon's orbital inclination means that the Moon spends most of its time either above or below the pond surface (representing the ecliptic plane). The Moon crosses *through* this surface only twice during each orbit—once coming out and once going back in—at the two points called the **nodes** of the Moon's orbit.

Notice that the nodes are aligned approximately the same way (diagonally in Figure 2.17) throughout the year, which means they lie along a nearly straight line with the Sun and Earth about twice each year. Eclipses can occur only during these periods, called **eclipse seasons**, which each last about five weeks (on

▲ FIGURE 2.16
This multiple-exposure photograph shows the progression of the 2017 total solar eclipse over Green River Lake, Wyoming. Totality (central image) lasted about 2 minutes. Note: Download the free app "Totality by Big Kid Science" (created by one of the textbook authors) to learn about upcoming total solar eclipses, including the next one in the United States on April 8, 2024.

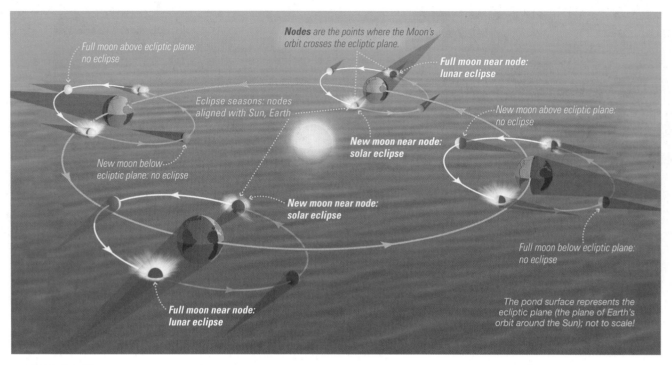

▲ **FIGURE 2.17**
This illustration represents the ecliptic plane as the surface of a pond. The Moon's orbit is tilted by about 5° to the ecliptic plane, so the Moon spends half of each orbit above the plane (the pond surface) and half below it. Eclipses occur only during *eclipse seasons*, when the nodes are closely aligned with the Sun and Earth *and* when the lunar phase is either new (for a solar eclipse) or full (for a lunar eclipse).

average). In other words, eclipses occur only during eclipse seasons, with a lunar eclipse occurring at full moon and a solar eclipse at new moon.

think about it Suppose that, today, some parts of the world saw a partial solar eclipse. Is it possible that you will see a lunar eclipse tonight? in two weeks? in two months? Explain.

Predicting Eclipses Few phenomena have so inspired and humbled humans throughout the ages as eclipses. For many cultures, eclipses were mystical events associated with fate or the gods, and countless stories and legends surround them.

Much of the mystery of eclipses probably stems from the relative difficulty of predicting them. Look again at Figure 2.17, focusing on the two eclipse seasons. If this figure told the whole story, eclipse seasons would always occur 6 months apart and predicting eclipses would be easy. For example, if eclipse seasons always occurred in January and July, eclipses would always occur on the dates of new and full moons in those months. Actual eclipse prediction is more difficult than this because of something the figure does not show: The nodes slowly move around the Moon's orbit, causing the eclipse seasons to occur slightly less than 6 months apart (about 173 days apart). As a result, instead of recurring with an annual pattern, eclipses recur in a pattern that repeats about every 18 years $11\frac{1}{3}$ days, which is called the **saros cycle**.

Some ancient civilizations learned to recognize the saros cycle and therefore were able to predict when eclipses would occur. Even then, however, they could not predict all aspects of particular eclipses, including where they would be visible on Earth. Today, we can predict eclipses with great precision because we know the details of the orbits of Earth and the Moon.

THE PROCESS OF SCIENCE IN ACTION

2.3 The Puzzle of Planetary Motion

We have already discussed how Earth's rotation leads to the daily rise and set of the Sun, Moon, and stars, and how Earth's orbit makes the Sun appear to move through the constellations. Ancient people did not find it difficult to explain these phenomena, though their explanations differed from the modern ones because they believed in an Earth-centered universe.

Planetary motion posed a much bigger puzzle in ancient times, and we will see in Chapter 3 that the quest to understand this motion ultimately led to the realization about four centuries ago that Earth is a planet in orbit around the Sun. But it was not the first time this idea had been raised: The ancient Greeks also considered the possibility that Earth moves around the Sun. In fact, the Greeks took the idea quite seriously, but most of them rejected it because they could not find sufficient evidence to support it with the observations possible at the time. The story of how the Greeks considered but rejected the correct explanation for planetary motion provides an early example of scientific thinking in action, and demonstrates that ancient civilizations thought more deeply about nature than many people realize.

◆ Why did the ancient Greeks reject the real explanation for planetary motion?

Over the course of a single night, planets behave like all other objects in the sky: Earth's rotation makes them appear to rise in the east and set in the west. But if you continue to watch the planets night after night, you will notice that their movements among the constellations are quite complex. Instead of moving steadily eastward relative to the stars, like the Sun and Moon, the planets vary substantially in both speed and brightness. Moreover, while the planets *usually* move eastward through the constellations, they occasionally reverse course, moving westward through the zodiac (Figure 2.18). These periods of **apparent retrograde motion** (*retrograde* means "backward") last from a few weeks to a few months, depending on the planet.

For ancient people who believed in an Earth-centered universe, apparent retrograde motion was very difficult to explain. After all, what could make planets sometimes turn around and go backward? The ancient Greeks came up with some clever ways to explain it, but their explanations (which we'll study in Chapter 3) were quite complex. In contrast, apparent retrograde motion has a simple explanation in a Sun-centered solar system. You can demonstrate it for yourself with the help of a friend (Figure 2.19a). Pick a spot in an open area to represent the Sun. You can represent Earth by walking counterclockwise around the Sun, while your friend represents a more distant planet (such as Mars or Jupiter) by walking in the same direction around the Sun at a greater distance. Your friend should walk more slowly than you, because more distant planets orbit the Sun more slowly. As you walk, watch how your friend appears to move relative to buildings or trees in the distance. Although both of you always walk the same way around the Sun, your friend will appear to move backward against the background during the part of your "orbit" in which you catch up to and pass him or her. Figure 2.19b shows how the same idea applies to Mars. (To understand the apparent retrograde motions of Mercury and Venus, which are closer to the Sun than Earth, simply switch places with your friend and repeat the demonstration.)

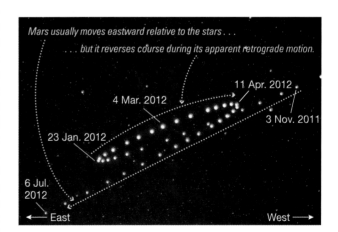

▲ FIGURE 2.18
This composite of images (taken at 5- to 7-day intervals in 2011–2012) shows a retrograde loop of Mars. Note that Mars is biggest and brightest in the middle of the retrograde loop, because that is where it is closest to Earth in its orbit.

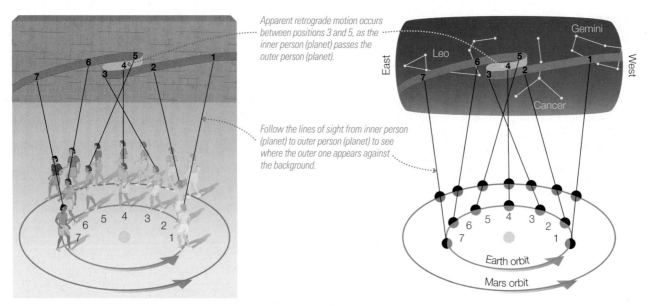

a The retrograde motion demonstration: Watch how your friend (in red) usually appears to move forward against the background of the building in the distance but appears to move backward as you (in blue) catch up to and pass her in your "orbit."

b This diagram shows the same idea applied to a planet. Follow the lines of sight from Earth to Mars in numerical order. Notice that Mars appears to move westward relative to the distant stars (from point 3 to point 5) as Earth passes it by in its orbit.

▲ FIGURE 2.19
Apparent retrograde motion—the occasional "backward" motion of the planets relative to the stars—has a simple explanation in a Sun-centered solar system.

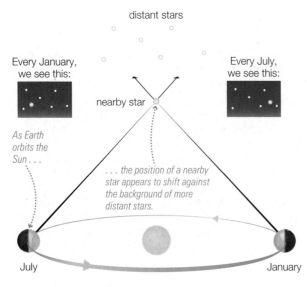

▲ FIGURE 2.20
Stellar parallax is an apparent shift in the position of a nearby star as we look at it from different places in Earth's orbit. This figure is greatly exaggerated; in reality, the amount of shift is far too small to detect with the naked eye.

The ancient Greeks were aware of how simply a Sun-centered system could explain apparent retrograde motion, and this may have been one reason the Greek astronomer Aristarchus proposed in about 260 B.C. that Earth goes around the Sun. However, Aristarchus's contemporaries rejected his idea, and the Sun-centered solar system did not gain wide acceptance until almost 2000 years later.

Although there were many reasons the Greeks were reluctant to abandon the idea of an Earth-centered universe, one of the most important was their inability to detect what we call **stellar parallax**. Extend your arm and hold up one finger. If you keep your finger still and alternately close your left eye and right eye, your finger will appear to jump back and forth against the background. This apparent shifting, called *parallax,* occurs because your two eyes view your finger from opposite sides of your nose. If you move your finger closer to your face, the parallax increases. If you look at a distant tree or flagpole instead of your finger, you may not notice any parallax at all. In other words, parallax depends on distance, with nearer objects exhibiting greater parallax than more distant objects.

If you now imagine that your two eyes represent Earth at opposite sides of its orbit around the Sun and that the tip of your finger represents a relatively nearby star, you have the idea of stellar parallax. Because we view the stars from different places in our orbit at different times of year, nearby stars should *appear* to shift back and forth against the background of more distant stars (Figure 2.20).

Because the Greeks believed that all stars lay on the same celestial sphere, they expected to see stellar parallax in a slightly different way. If Earth orbited the Sun, they reasoned, at different times of year we would be closer to different parts of the celestial sphere and would notice changes in the angular separations of stars. However, no matter how hard they searched, they could find no sign of stellar parallax. They concluded that one of the following must be true:

1. Earth orbits the Sun but the stars are so far away that stellar parallax is undetectable to the naked eye.
2. There is no stellar parallax because Earth remains stationary at the center of the universe.

Aside from a few notable exceptions such as Aristarchus, the Greeks rejected the correct answer (the first one) because they could not imagine that the stars could be *that* far away. Today, we can detect stellar parallax with the aid of telescopes, providing direct proof that Earth really does orbit the Sun.

think about it How far apart are opposite sides of Earth's orbit? How far away are the nearest stars? Using the 1-to-10-billion scale from Chapter 1, describe the challenge of detecting stellar parallax.

The moral of this story is that our ability to do science is inextricably linked with our ability to make careful observations and measurements. The ancient Greeks knew that planetary motion was more easily explained by a Sun-centered system, but they also knew that such a system implied the existence of stellar parallax. Because their measurements were not good enough to detect parallax, they saw no clear reason to abandon their deeply held belief in an Earth-centered universe. As we'll see in the next chapter, evidence sufficient to change this belief came only with greatly improved measurements of planetary positions in the sky.

summary of key concepts

2.1 Understanding the Seasons

◆ **What causes the seasons?**

The tilt of Earth's axis causes the seasons. The axis points in the same direction throughout the year, so as Earth orbits the Sun, sunlight hits different parts of Earth more directly at different times of year.

◆ **Why do the constellations we see depend on the time of year?**
The visible **constellations** vary with the time of year because our night sky lies in different directions in space as we orbit the Sun.

2.2 Understanding the Moon

◆ **Why do we see phases of the Moon?**

The **phase** of the Moon depends on its position relative to the Sun as it orbits Earth. The half of the Moon facing the Sun is always illuminated while the other half is dark, but from Earth we see varying combinations of the illuminated and dark faces.

◆ **What causes eclipses?**

We see a **lunar eclipse** when Earth's shadow falls on the Moon and a **solar eclipse** when the Moon blocks our view of the Sun. We do not see an eclipse at every new and full moon because the Moon's orbit is slightly inclined to the ecliptic plane.

THE PROCESS OF SCIENCE IN ACTION

2.3 The Puzzle of Planetary Motion

◆ **Why did the ancient Greeks reject the real explanation for planetary motion?**

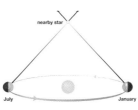

Planets generally move eastward relative to the stars from night to night, but sometimes reverse course for weeks to months of **apparent retrograde motion**. This motion is simply explained in a Sun-centered system, but the Greeks rejected this model in part because they could not detect **stellar parallax**—slight apparent shifts in stellar positions over the course of the year that must occur if Earth orbits the Sun.

Investigations

For instructor-assigned homework and other learning materials, go to www.MasteringAstronomy.com.

Quick Quiz

Choose the best answer to each of the following. Explain your reasoning.

1. When it is summer in Australia, it is (a) winter in the United States. (b) summer in the United States. (c) spring in the United States.
2. In winter, Earth's axis points toward the star Polaris. In spring, (a) the axis also points toward Polaris. (b) the axis points toward Vega. (c) the axis points toward the Sun.
3. Which of the following is *not* true during Northern Hemisphere summer? (a) Sunlight strikes the ground at a steeper angle in the Northern Hemisphere than it does in the Southern Hemisphere. (b) The Sun follows a longer and higher path through the Northern Hemisphere sky than it does through the Southern Hemisphere sky. (c) Noontime shadows are longer in the Northern Hemisphere than in the Southern Hemisphere.
4. We can be sure that variations in Earth's distance from the Sun are *not* the cause of the seasons, because (a) Earth is always exactly the same distance from the Sun. (b) if distance were responsible for the seasons, both the Northern and the Southern Hemisphere would have summer at the same time. (c) if distance were responsible for the seasons, seasonal temperature changes would be much more extreme.
5. Two stars that are in the same constellation (a) must both be part of the same cluster of stars in space. (b) must both have been discovered at about the same time. (c) may actually be very far away from each other.
6. Suppose you see a crescent moon; how much of the Moon's entire surface (the full globe of the Moon) is in daylight? (a) half (b) one-quarter (c) one-eighth
7. One week after full moon, the Moon's phase is (a) first quarter. (b) third quarter. (c) new.
8. The fact that we always see the same face of the Moon tells us that (a) the Moon does not rotate. (b) the Moon's rotation period is the same as its orbital period. (c) the Moon looks the same on both sides.
9. The reason we do *not* have a solar eclipse at every new moon is that (a) a solar eclipse can occur only at full moon. (b) the Moon's orbit is inclined relative to Earth's orbit around the Sun. (c) the nodes of the Moon's orbit gradually move around the orbit.
10. If there is going to be a total lunar eclipse tonight, then you know that the Moon's phase is (a) full. (b) new. (c) first or third quarter.
11. When we see Saturn going through a period of apparent retrograde motion, we know that (a) Saturn is temporarily moving backward in its orbit around the Sun. (b) Earth is passing Saturn in its orbit, with both planets on the same side of the Sun. (c) Saturn and Earth must be on opposite sides of the Sun.
12. Which of the following best describes why the Greeks rejected the correct explanation for apparent retrograde motion? (a) They did not realize that planets sometimes appear to move backward relative to the constellations. (b) They predicted that if Earth orbited the Sun, then stellar parallax should occur, but they could not detect this parallax. (c) They never even considered the possibility that Earth could orbit the Sun.

Short-Answer/Essay Questions

Explain all answers clearly, using complete sentences and proper essay structure if needed. An asterisk (*) designates a quantitative problem, for which you should show all your work.

13. *Earth-Centered or Sun-Centered?* Decide whether each of the following phenomena is consistent or inconsistent with a belief in an Earth-centered system. If consistent, describe how. If inconsistent, explain why, and also explain why the inconsistency did not immediately lead people to abandon the Earth-centered belief.
    a. The daily paths of stars through the sky
    b. Seasons
    c. Phases of the Moon
    d. Eclipses
    e. Apparent retrograde motion of the planets
    f. Stellar parallax
14. *Cause of the Seasons.* Suppose a friend is trying to claim that it is summer when Earth is closest to the Sun and winter when Earth is farthest away. Refute this claim in at least two different ways: (1) with an argument that invokes seasonal differences between the Northern and Southern Hemispheres and (2) with an argument based on the times of year at which Earth actually is closest to and farthest from the Sun. Write out your arguments clearly.
15. *New Planet.* A planet in another solar system has a circular orbit and an axis tilt of 35°. Would you expect this planet to have seasons? If so, would you expect them to be more extreme than the seasons on Earth? If not, why not?
16. *View from the Sun.* Suppose you lived on the Sun (and could ignore the heat). Would you still see the Moon go through phases as it orbits Earth? Why or why not?
17. *View from the Moon.* Assume you live on the Moon, near the center of the face that looks toward Earth.
    a. If you saw a full Earth in your sky, what phase of the Moon would people on Earth see? Explain.
    b. If people on Earth saw a full moon, what phase would you see for Earth? Explain.
    c. If people on Earth saw a waxing gibbous moon, what phase would you see for Earth? Explain.
    d. If people on Earth were viewing a total lunar eclipse, what would you see from your home on the Moon? Explain.
18. *A Farther Moon.* Suppose the distance to the Moon were twice its actual value. Would it still be possible to have a total solar eclipse? Why or why not?
19. *A Smaller Earth.* Suppose Earth were smaller. Would solar eclipses be any different? If so, how? What about lunar eclipses? Explain.
20. *Group Activity: Shadow Phases.* Many people incorrectly guess that the phases of the Moon are caused by Earth's shadow falling on the Moon. Work in small groups to come up with a demonstration that would show that Earth's shadow can't be the cause of phases. Can you expand on your demonstration to explain the conditions necessary for a lunar eclipse, and how the shape of Earth's shadow shows that our world is round?
21. *Project: Observing Planetary Motion.* Find out which planets are currently visible in your evening sky. At least once a week, observe the planets and draw a diagram showing the position of each visible planet relative to stars in a zodiac constellation. From week to week, note how the planets are moving relative to the stars. Can you see any of the wandering features of planetary motion? Explain.
*22. *Seasons Scale.* Figure 2.3 does not show Earth and its orbit to scale. Let's examine how the figure would look if it did.
    a. Use a ruler to measure the diameter of Earth's orbit as shown in the figure (measure in the longest direction, since the perspective distorts the diameter in other directions). Then use the fact that the actual diameter of Earth's orbit around the Sun is about 300,000,000 km to determine the scale on which the orbit is drawn.
    b. How large would Earth itself appear if drawn to the scale you found in part a? (*Hint:* Earth's actual diameter is about 12,800 km.)
    c. Suppose someone is trying to claim that it is summer whenever your hemisphere (either Northern or Southern) is closer to the Sun than the other hemisphere. Based on your answers to parts a and b, how can you refute their argument?

Changes in Our Perspective

3

LEARNING GOALS

3.1 From Earth-Centered to Sun-Centered
- How did the Greeks explain planetary motion?
- How did the Copernican revolution change our view of the universe?

3.2 Hallmarks of Science
- How can we distinguish science from nonscience?
- What is a scientific theory?

 THE PROCESS OF SCIENCE IN ACTION

3.3 The Fact and Theory of Gravity
- How does the fact of gravity differ from the theory of gravity?

Imagine yourself as a satellite, orbiting Earth like the astronaut in the photo and looking down at the thin layer of atmosphere that separates our homes from the blackness of space. You'd see Earth in a way that most of our ancestors could hardly have imagined. Before the advent of science, people assumed Earth to be the center of the universe, not a planet orbiting the Sun. Our perspective began to change as we gathered and sought to interpret increasingly precise observations of the sky. This chapter tells the story of that shift in perspective and discusses how it exemplifies the kind of thinking we now call *science*. We'll examine the key features of scientific thinking and see how science produces reliable knowledge about how the universe works—so reliable, in fact, that we can now launch astronauts into space and return them safely to Earth.

The Science of Astronomy

3.1 From Earth-Centered to Sun-Centered

As we discussed in Chapter 2, the idea that Earth might orbit the Sun was proposed by the ancient Greek philosopher Aristarchus around 260 B.C. Nevertheless, the incorrect Earth-centered view continued to hold sway until about 400 years ago. To understand why the Earth-centered view remained popular for so long—and why it was finally replaced—we must investigate how the ancient Greeks managed to explain the puzzle of planetary motion, especially the "backward" loops of apparent retrograde motion [Section 2.3].

◆ How did the Greeks explain planetary motion?

The key to understanding both the Greek and modern explanations of planetary motion lies in an idea first developed by the Greeks that remains central to science today—the idea of creating *models* of nature.

Scientific Models Scientific models differ somewhat from the models you may be familiar with in everyday life. In our daily lives, we tend to think of models as miniature physical representations, such as model cars or airplanes. In contrast, a scientific **model** is a conceptual representation created to explain and predict observed phenomena. For example, a model of Earth's climate uses logic and mathematics to represent what we know about how the climate works. Its purpose is to explain and predict climate changes, such as the changes that may occur with global warming. Just as a model airplane does not faithfully represent every aspect of a real airplane, a scientific model may not fully explain all our observations of nature. Nevertheless, even the failings of a scientific model can be useful, because they often point the way toward building a better model.

The Greek Geocentric Model The Greeks constructed conceptual models of the universe in an attempt to explain what they observed in the sky. We do not know precisely when the Greeks first began to think that Earth is round, but this idea was being taught as early as about 500 B.C. by the famous mathematician Pythagoras (c. 560–480 B.C.). More than a century later, Aristotle (384–322 B.C.) recognized Earth's curved shadow on the Moon during lunar eclipses as evidence for a spherical Earth. The Greek model was therefore a **geocentric model** of the universe (recall that *geocentric* means "Earth-centered"), with a spherical Earth at the center of a great celestial sphere.

Greek philosophers quickly realized that their model needed more than a single sphere surrounding Earth. To account for the gradual eastward motions of the Sun and Moon through the constellations, they placed the Sun and Moon on separate spheres that turned at different rates from the sphere of the stars. They also added a sphere for each planet, because each planet moves differently as it wanders among the stars (Figure 3.1).

The difficulty with this geocentric model was that it did not easily account for the apparent retrograde motion of the planets. You might guess that the Greeks would simply have allowed the planetary spheres to sometimes turn forward and sometimes turn backward relative to the sphere of the stars, but they did not because doing so would have violated their deeply held belief in "heavenly perfection." According to this idea, enunciated most clearly by Plato, heavenly objects could move only in perfect circles.

The Greeks came up with a number of ingenious ideas for explaining planetary motion while preserving Earth's central position and motion in perfect circles. These ideas were refined for centuries and reached their apex in the work of Claudius Ptolemy (c. A.D. 100–170; pronounced "tol-e-mee"). We refer to Ptolemy's model as the **Ptolemaic model**, to distinguish it from earlier geocentric models.

▲ **FIGURE 3.1**
This model represents the Greek idea of the heavenly spheres (c. 400 B.C.). Earth is a sphere that rests in the center. The Moon, the Sun, and the planets all have their own spheres. The outermost sphere holds the stars.

common misconceptions

Columbus and a Flat Earth

A widespread myth gives credit to Columbus for learning that Earth is round, but knowledge of Earth's shape predated Columbus by nearly 2000 years. Not only were scholars of Columbus's time well aware that Earth is round, but they even knew its approximate size: Earth's circumference was first measured in about 240 B.C. by the Greek scientist Eratosthenes. In fact, a likely reason Columbus had so much difficulty finding a sponsor for his voyages was that he tried to argue a point on which he was wrong: He claimed the distance by sea from western Europe to eastern Asia to be much less than the scholars knew it to be. When he finally found a patron in Spain and left on his journey, he was so woefully underprepared that the voyage would almost certainly have ended in disaster if the Americas hadn't stood in his way.

Circles upon Circles The essence of the Ptolemaic model was that each planet moves on a small circle whose center moves around Earth on a larger circle (Figure 3.2). (The small circle is called an *epicycle,* and the larger circle is called a *deferent.*) A planet following this circle-upon-circle motion traces a loop as seen from Earth, with the backward portion of the loop mimicking apparent retrograde motion.

think about it Review the real cause for apparent retrograde motion in Section 2.3. Why does Mars appear biggest and brightest when it is in the middle of its retrograde loop? Does the Ptolemaic model also account for planets appearing brighter in the middle of their retrograde loops? Explain.

The circle-upon-circle loops successfully explained the general idea of apparent retrograde motion. However, to make his model agree well with observations, Ptolemy had to include a number of other complexities, such as positioning some of the large circles slightly off-center from Earth and adding even smaller circles that moved upon the original set of small circles. As a result, the full Ptolemaic model was mathematically complex and tedious. Many centuries later, while supervising computations based on the Ptolemaic model, the Spanish monarch Alphonso X (1221–1284) is said to have complained, "If I had been present at the creation, I would have recommended a simpler design for the universe."

Despite its complexity, the Ptolemaic model worked remarkably well: It could correctly forecast future planetary positions to within a few degrees of arc, which is about the angular size of your hand held at arm's length against the sky. This was sufficiently accurate to keep the model in use for the next 1500 years. When Ptolemy's book describing the model was translated by Arabic scholars around A.D. 800, they gave it the title *Almagest,* derived from words meaning "the greatest work."

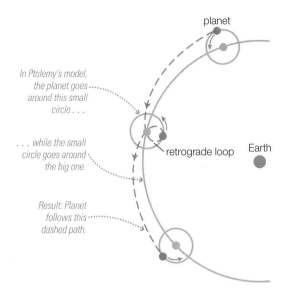

▲ **FIGURE 3.2**
This diagram shows how the Ptolemaic model accounted for apparent retrograde motion. Each planet is assumed to move around a small circle that turns upon a larger circle. The resulting path (dashed) includes a loop in which the planet goes backward as seen from Earth.

◆ How did the Copernican revolution change our view of the universe?

The events that ultimately led to the demise of the Ptolemaic model began in earnest with the work of the Polish scientist Nicholas Copernicus (1473–1543), and hence are generally referred to as events of the **Copernican revolution**. As we will see, the Copernican revolution is in many ways the story of the origin of modern science. It is also the story of several key personalities, beginning with Copernicus himself.

Copernicus Copernicus was born in Torún, Poland, on February 19, 1473. He began studying astronomy in his late teens, and learned that tables of planetary motion based on the Ptolemaic model had been growing increasingly inaccurate with time. He began a quest to find a better way to predict planetary positions.

Copernicus was aware of and adopted Aristarchus's ancient Sun-centered idea [Section 2.3], perhaps because it offered such a simple explanation for the apparent retrograde motion of the planets. But he went beyond Aristarchus in working out mathematical details of the model. In the process, Copernicus discovered simple geometric relationships that strengthened his belief in the Sun-centered idea, because they allowed him to calculate each planet's orbital period and relative distance (compared to Earth's distance) from the Sun.

Copernicus was hesitant to publish his work, fearing that the idea of a moving Earth would be considered absurd. However, he discussed his system with other scholars, including high-ranking officials of the Church, who urged him to publish a book. Copernicus saw the first printed copy of his book, *De Revolutionibus Orbium Coelestium* (On the Revolutions of the Heavenly Spheres), on the day he died—May 24, 1543.

Copernicus (1473–1543)

Publication of the book spread the Sun-centered idea widely, and many scholars were drawn to its aesthetic advantages. However, the Copernican model gained relatively few converts over the next 50 years, for a good reason: It didn't work all that well. The primary problem was that while Copernicus had been willing to overturn Earth's central place in the cosmos, he held fast to the ancient belief that heavenly motion must occur in perfect circles. This incorrect assumption forced him to add numerous complexities to his system (including circles on circles much like those used by Ptolemy) to get it to make decent predictions. In the end, his complete model was no more accurate and no less complex than the Ptolemaic model, and few people were willing to throw out thousands of years of tradition for a new model that worked just as poorly as the old one.

Tycho Part of the difficulty faced by astronomers who sought to improve either the Ptolemaic or the Copernican model was a lack of quality data. The telescope had not yet been invented, and existing naked-eye observations were not very accurate. Better data were needed, and they were provided by the Danish nobleman Tycho Brahe (1546–1601), usually known simply as Tycho (pronounced "tie-koe").

In 1572, Tycho observed a "new star" in the sky and showed that it was much farther away than the Moon, calling into question the ancient Greek belief in unchanging heavens. (Today, we know that Tycho actually saw a *supernova*—the explosion of a distant star [Section 9.2].) In 1577, he observed a comet and showed that it too lay in the realm of the heavens. Others, including Aristotle, had argued that comets were phenomena of Earth's atmosphere. These successes led to royal sponsorship for Tycho's work, which allowed him to build a giant observatory for naked-eye observations. Over a period of three decades, Tycho and his assistants compiled naked-eye observations of planetary positions accurate to within 1 arcminute (1/60 of 1°)—less than the thickness of a fingernail viewed at arm's length.

Despite the quality of his observations, Tycho did not come up with a satisfying explanation for planetary motion. He was convinced that the *planets* must orbit the Sun, but his inability to detect stellar parallax [Section 2.3] led him, like the ancient Greeks, to conclude that Earth must remain stationary. He therefore advocated a model in which the Sun orbited Earth while all other planets orbited the Sun. Few people took this model seriously.

Johannes Kepler (1571–1630)

Kepler—A Successful Model of Planetary Motion Tycho failed to explain the motions of the planets satisfactorily, but he succeeded in finding someone who could: In 1600, he hired a young German astronomer, Johannes Kepler (1571–1630).

Like Copernicus, Kepler believed that planetary orbits should be perfect circles. He therefore worked diligently to match circular motions to Tycho's observations. After years of effort, he found a set of circular orbits that matched most of Tycho's observations quite well. Even in the worst cases, which were for the planet Mars, Kepler's predicted positions differed from Tycho's observations by only about 8 arcminutes.

Kepler surely was tempted to attribute these discrepancies to errors by Tycho. After all, 8 arcminutes is barely one-fourth the angular diameter of the full moon. But Kepler trusted Tycho's work. The small discrepancies finally led Kepler to abandon the idea of circular orbits—and to find the correct solution to the ancient riddle of planetary motion. About this event, Kepler wrote:

> *If I had believed that we could ignore these eight minutes [of arc], I would have patched up my hypothesis accordingly. But, since it was not permissible to ignore, those eight minutes pointed the road to a complete reformation in astronomy.*

Kepler's decision to trust the data over his preconceived beliefs marked an important transition point in the history of science. Once he abandoned perfect circles, he was free to try other ideas, and he ultimately found the correct one: Planetary orbits take the shapes of the special ovals known as *ellipses*. Much as you can draw a circle by pulling a string around one point (the center), you can draw an ellipse by stretching a string around *two* points; each of these two points is called a *focus* of the ellipse. Kepler then used his knowledge of mathematics to put his new model of planetary motion on a firm footing, expressing the key features of the model with what we now call **Kepler's laws of planetary motion**.

- **Kepler's first law**: *The orbit of each planet about the Sun is an ellipse with the Sun at one focus* (Figure 3.3). This law tells us that a planet's distance from the Sun varies during its orbit. Its closest point is called **perihelion** (from the Greek for "near the Sun") and its farthest point is called **aphelion** ("away from the Sun"). The *average* of a planet's perihelion and aphelion distances is the length of its **semimajor axis**.

- **Kepler's second law**: *A planet moves faster in the part of its orbit nearer the Sun and slower when farther from the Sun, sweeping out equal areas in equal times.* As shown in Figure 3.4, the "sweeping" refers to an imaginary line connecting the planet to the Sun, and keeping the areas equal means that the planet moves a greater distance (and hence is moving faster) when it is near perihelion than it does in the same amount of time near aphelion.

- **Kepler's third law**: *More distant planets orbit the Sun at slower average speeds, obeying the precise mathematical relationship $p^2 = a^3$*; p is the planet's orbital period in years and a is its average distance from the Sun in astronomical units. The mathematical statement of Kepler's third law allows us to calculate the average orbital speed of each planet (Figure 3.5).

think about it Suppose a comet has an orbit that brings it quite close to the Sun at perihelion and beyond Mars at aphelion, but with an average distance (semimajor axis) of 1 AU. How long would the comet take to complete each orbit of the Sun? Would it spend most of its time close to the Sun, far from the Sun, or somewhere in between? Explain.

Kepler published his first two laws in 1609 and his third in 1619. Together, they made a model that could predict planetary positions with far greater accuracy than the models of either Ptolemy or Copernicus. Indeed, Kepler's model works so well that we now see it as more than just an abstract model, and instead as revealing a deep underlying truth about planetary motion.

Galileo—Answering the Remaining Objections The success of Kepler's laws in matching Tycho's data provided strong evidence in favor of Copernicus's placement of the Sun, rather than Earth, at the center of the solar system. Nevertheless, many scientists still voiced reasonable objections to the Copernican view. There were three basic objections, all rooted in the 2000-year-old beliefs of Aristotle and other ancient Greeks.

1. Aristotle had held that Earth could not be moving because, if it were, objects such as birds, falling stones, and clouds would be left behind as Earth moved along.

2. The idea of noncircular orbits contradicted Aristotle's claim that the heavens—the realm of the Sun, Moon, planets, and stars—must be perfect and unchanging.

3. Despite many efforts, no one had detected the stellar parallax that should occur if Earth orbits the Sun.

▲ **FIGURE 3.3**
Kepler's first law: The orbit of each planet about the Sun is an ellipse with the Sun at one focus. (The ellipse shown here is more *eccentric*, or stretched out, than any of the actual planetary orbits in our solar system.)

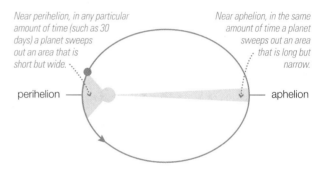

▲ **FIGURE 3.4**
Kepler's second law: As a planet moves around its orbit, an imaginary line connecting it to the Sun sweeps out equal areas (the shaded regions) in equal times.

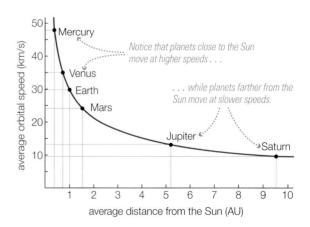

▲ **FIGURE 3.5**
This graph, based on Kepler's third law ($p^2 = a^3$) and modern values of planetary distances, shows that more distant planets orbit the Sun more slowly.

Galileo (1564–1642)

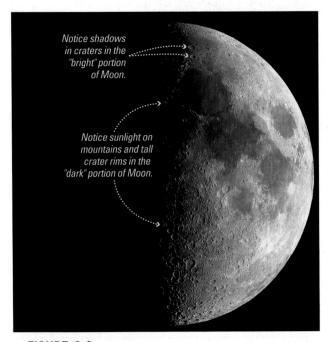

Notice shadows in craters in the "bright" portion of Moon.

Notice sunlight on mountains and tall crater rims in the "dark" portion of Moon.

▲ **FIGURE 3.6**
Shadows visible near the dividing line between the light and dark portions of the lunar face show that the Moon's surface is not perfectly smooth.

Galileo Galilei (1564–1642), usually known by his first name, answered all three objections.

Galileo defused the first objection with experiments that almost single-handedly overturned the Aristotelian view of physics. In particular, he used experiments with rolling balls to demonstrate that a moving object remains in motion *unless* a force acts to stop it (an idea now codified in Newton's first law of motion). This insight explained why objects that share Earth's motion through space—such as birds, falling stones, and clouds—should *stay* with Earth rather than falling behind as Aristotle had argued. This same idea explains why passengers stay with a moving airplane even when they leave their seats.

The second objection had already been challenged by Tycho's supernova and comet observations, which demonstrated that the heavens could change. Galileo shattered the idea of heavenly perfection after he built a telescope in 1609. (Galileo did not invent the telescope, but his innovations made it much more powerful.) Through his telescope, Galileo saw sunspots on the Sun, which were considered "imperfections" at the time. He also used his telescope to observe that the Moon has mountains and valleys like the "imperfect" Earth by noticing the shadows cast near the dividing line between the light and dark portions of the lunar face (Figure 3.6). If the heavens were not perfect, then the idea of elliptical orbits (as opposed to "perfect" circles) was not so objectionable.

The third objection—the absence of observable stellar parallax—had been of particular concern to Tycho. Based on his estimates of the distances of stars, Tycho believed that his naked-eye observations were sufficiently precise to detect stellar parallax if Earth did in fact orbit the Sun. Refuting Tycho's argument required showing that the stars were more distant than Tycho had thought and therefore too distant for him to have observed stellar parallax. Although Galileo didn't actually prove this fact, he provided strong evidence in its favor. For example, he saw with his telescope that the Milky Way resolved into countless individual stars. This discovery helped him argue that the stars were far more numerous and more distant than Tycho had believed.

In hindsight, the final nails in the coffin of the Earth-centered universe came with two of Galileo's earliest discoveries through the telescope. First, he observed four moons clearly orbiting Jupiter, *not* Earth. Soon thereafter, he observed that Venus goes through phases in a way that meant that it must orbit the Sun and not Earth (Figure 3.7).

Although we now recognize that Galileo was correct, the story was more complex in his own time, when Catholic Church doctrine still held Earth to be the center of the universe. On June 22, 1633, Galileo was brought before a Church inquisition in Rome and ordered to recant his claim that Earth orbits the Sun. Nearly 70 years old and fearing for his life, Galileo did as ordered. His life was spared. However, legend has it that as he rose from his knees he whispered under his breath, *Eppur si muove*—Italian for "And yet it moves." The Church did not formally vindicate Galileo until 1992, but Church officials had given up the argument long before that. Moreover, Catholic scientists have long worked at the forefront of much astronomical research, and today's official Church teachings are compatible not only with Earth's planetary status but also with the theories of the Big Bang and the subsequent evolution of the cosmos and of life.

Newton—Culmination of the Revolution Kepler's model worked so well and Galileo so successfully countered the remaining objections to the Sun-centered model that, by about the 1630s, scientists were nearly unanimous in accepting the validity of Kepler's laws of planetary motion. However, no one yet knew *why* the planets should move in elliptical orbits with varying speeds. The question became a topic of great debate, and a few scientists even guessed the correct answer—but they could not prove it, largely because the necessary understanding of physics

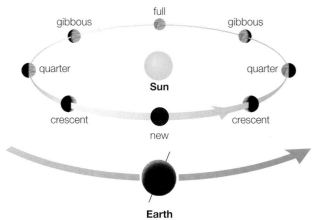

a In the Ptolemaic model, Venus orbits Earth, moving around a smaller circle on its larger orbital circle; the center of the smaller circle lies on the Earth–Sun line. If this view were correct, Venus's phases would range only from new to crescent.

b In reality, Venus orbits the Sun, so from Earth we can see it in many different phases. This is just what Galileo observed, allowing him to prove that Venus orbits the Sun.

▲ **FIGURE 3.7**
Galileo's telescopic observations of Venus showed that it orbits the Sun rather than Earth.

and mathematics didn't exist yet. This understanding finally came through the remarkable work of Sir Isaac Newton (1642–1727), who invented the mathematics of calculus and used it to explain and discover many fundamental principles of physics.

In 1687, Newton published a famous book usually called *Principia,* short for *Philosophiae Naturalis Principia Mathematica* (Mathematical Principles of Natural Philosophy). In it, he laid out precise mathematical descriptions of how motion works in general, ideas that we now describe as **Newton's laws of motion**. For reference, Figure 3.8 illustrates the three laws of motion. Be careful not to confuse *Newton's* three laws, which apply to all motion, with *Kepler's* three laws, which describe only the motion of planets moving about the Sun.

Newton continued on in *Principia* to describe his universal law of gravitation (see Section 3.3), and then used mathematics to prove that Kepler's laws are natural consequences of the laws of motion and gravity. In essence, Newton had created a new model for the inner workings of the universe in which motion is governed by clear laws and the force of gravity. Scientists have continued to build on this model ever since, giving us an ever deeper understanding of nature.

Sir Isaac Newton (1642–1727)

Newton's first law of motion:
An object moves at constant velocity unless a net force acts to change its speed or direction.

Example: A spaceship needs no fuel to keep moving in space.

Newton's second law of motion:
Force = mass × acceleration

Example: A baseball accelerates as the pitcher applies a force by moving his arm. (Once the ball is released, the force from the pitcher's arm ceases, and the ball's path changes only because of the forces of gravity and air resistance.)

Newton's third law of motion:
For any force, there is always an equal and opposite reaction force.

Example: A rocket is propelled upward by a force equal and opposite to the force with which gas is expelled out its back.

◀ **FIGURE 3.8**
Newton's three laws of motion.

tools of science — Telescopes

▶ Telescopes (Learning Outcomes 1, 2)

The Copernican revolution illustrates the importance of precise observations in science. Without Tycho's unprecedented accuracy in naked-eye observations, Kepler could not have come up with his laws of planetary motion. And the case for a Sun-centered solar system was sealed with the help of Galileo's telescopic observations. The need for ever better observations continues today, and in astronomy this means the development of more powerful telescopes.

We characterize the power of a telescope in terms of two fundamental properties: (1) The **light-collecting area** tells us how much total light a telescope can collect at one time. We usually characterize a telescope's "size" as the *diameter* of its light-collecting area; for example, a "10-meter telescope" has a light-collecting area that is 10 meters in diameter. (2) The **angular resolution** of a telescope is the *smallest* angle over which it allows us to tell that two dots—or two stars—are distinct. The human eye has an angular resolution of about 1 arcminute ($\frac{1}{60}°$), meaning that if two stars are separated by less than 1 arcminute, they will appear blended together as a single point of light. Telescopes have much better angular resolution. For example, the Hubble Space Telescope has an angular resolution of about 0.05 arcsecond (for visible light), which would allow you to read this book from a distance of almost 1 kilometer.

Telescopes come in two basic designs. A **refracting telescope** uses transparent glass lenses to collect and focus light (Figure 1). A **reflecting telescope** uses a precisely curved *primary mirror* to gather light (Figure 2); this mirror reflects the gathered light to a *secondary mirror* that lies in front of it, which then reflects the light to a focus at a place where the eye or instruments can observe it. Most research telescopes today are reflecting telescopes. The largest are just over 10 meters in diameter, but astronomers are working on plans for telescopes with diameters as large as 30 to 50 meters.

Many telescopes are now launched into space. One reason for going into space is that Earth's atmosphere can blur light; the familiar twinkling of stars is caused by winds and air movements in our atmosphere. In addition, telescopes in space can observe forms of light that do not reach the ground, such as ultraviolet light and x-rays. As telescope technology continues to improve, we can expect many new discoveries about the universe.

▲ **FIGURE 1**
A refracting telescope collects light with a large transparent lens; the photo shows the 1-meter refractor at the University of Chicago's Yerkes Observatory (the world's largest refractor).

▲ **FIGURE 2**
A reflecting telescope collects light with a precisely curved primary mirror; the photo shows the 8-meter Gemini North telescope (Mauna Kea, Hawaii).

3.2 Hallmarks of Science

The story of how our ancestors gradually figured out the basic architecture of the cosmos exhibits many features of what we now consider "good science." We have seen how models were formulated and tested against observations, and modified or replaced when they failed those tests. The story also illustrates some classic mistakes, such as the apparent failure of anyone before Kepler to question the belief that orbits must be circles. The ultimate success of the Copernican revolution led scientists, philosophers, and theologians to reassess the various modes of thinking used in the 2000-year process of discovering Earth's place in the universe. Let's examine how the principles of modern science emerged from the lessons learned in the Copernican revolution.

◆ How can we distinguish science from nonscience?

It's surprisingly difficult to define the term *science* precisely. The word comes from the Latin *scientia*, meaning "knowledge," but not all knowledge is science. For example, you may know what music you like best, but your musical taste is not a result of scientific study.

Approaches to Science One reason science is difficult to define is that not all science works in the same way. For example, you've probably heard that science is supposed to proceed according to something called the "scientific method." As an idealized illustration of this method, consider what you would do if your flashlight suddenly stopped working. You might suspect that the flashlight's batteries had died. This type of tentative explanation, or **hypothesis**, is sometimes called an *educated guess*—in this case, it is "educated" because you already know that flashlights need batteries. Your hypothesis allows you to make a simple prediction: If you replace the batteries with new ones, the flashlight should work. You can test this prediction by replacing the batteries. If the flashlight now works, you've confirmed your hypothesis. If it doesn't, you must revise or discard your hypothesis, perhaps in favor of another one that you can also test (such as that the bulb no longer works). Figure 3.9 illustrates the basic flow of this process.

The scientific method can be a useful idealization, but real science rarely progresses in such an orderly way. Scientific progress often begins with someone going out and looking at nature in a general way, rather than conducting a careful set of experiments. For example, Galileo wasn't looking for anything in particular when he pointed his telescope at the sky and made his startling discoveries. Furthermore, scientists are human beings, and their intuition and personal beliefs inevitably influence their work. Copernicus, for example, adopted the idea that Earth orbits the Sun not because he had carefully tested it but because he believed it made more sense than the prevailing view of an Earth-centered universe. While his intuition guided him to the right general idea, he erred in the specifics because he still held to the ancient belief that heavenly motion must be in perfect circles.

Given that the idealized scientific method is an overly simplistic characterization of science, how can we tell what is science and what is not? To answer this question, we must look a little deeper at the distinguishing characteristics of scientific thinking.

Hallmarks of Science One way to define scientific thinking is to list the criteria that scientists use when they judge competing models of nature. Historians and philosophers of science have examined (and continue to examine) this issue in great depth, and different experts express different viewpoints on the details. Nevertheless, everything we now consider to be science shares the following three basic characteristics, which we will refer to as the "hallmarks" of science (Figure 3.10):

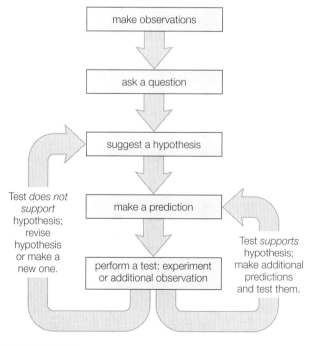

▲ **FIGURE 3.9**
This diagram illustrates what we often call the *scientific method*.

- Modern science seeks explanations for observed phenomena that rely solely on natural causes.
- Science progresses through the creation and testing of models of nature that explain the observations as simply as possible.
- A scientific model must make testable predictions about natural phenomena that would force us to revise or abandon the model if the predictions did not agree with observations.

Each of these hallmarks is evident in the story of the Copernican revolution. The first shows up in the way Tycho's careful measurements of planetary motion motivated Kepler to come up with a better explanation for those motions. The second is evident in the way several competing models were compared and tested, most notably those of Ptolemy, Copernicus, and Kepler. We see the third in the fact that each model could make precise predictions about the future motions of the Sun, Moon, planets, and stars in our sky. Kepler's model gained acceptance because it worked, while the competing models lost favor because their predictions failed to match the observations. Figure 3.11 (pages 46–47) summarizes the Copernican revolution and how it illustrates the hallmarks of science.

Occam's Razor The criterion of simplicity in the second hallmark deserves additional explanation. Remember that Copernicus's original model did *not* match the data noticeably better than Ptolemy's model. If scientists had judged this model solely on the accuracy of its predictions, they might have rejected it immediately. However, many scientists found elements of the Copernican model appealing, such as its simple explanation for apparent retrograde motion. They therefore kept the model alive until Kepler found a way to make it work.

If agreement with data were the sole criterion for judgment, we could imagine a modern-day Ptolemy adding millions or billions of additional circles to the geocentric model in an effort to improve its agreement with observations. A sufficiently complex geocentric model could in principle reproduce the observations with almost perfect accuracy—but it still would not convince us that Earth was the center of the universe. We would still choose the Copernican view over the geocentric view because its predictions would be just as accurate but follow from a much simpler model of nature. The idea that scientists should prefer the simpler of two models that agree equally well with observations is called *Occam's razor*, after the medieval scholar William of Occam (1285–1349).

Verifiable Observations The third hallmark of science forces us to face the question of what counts as an "observation" against which a prediction can be tested. Consider the claim that aliens are visiting Earth in UFOs. Proponents of this claim say that thousands of eyewitness reports of UFO encounters provide evidence that it is true. But do these personal testimonials count as *scientific* evidence? On the surface, the answer isn't obvious, because all scientific studies involve eyewitness accounts on some level. For example, only a handful of scientists have personally made detailed tests of Einstein's theory of relativity, and it is their personal reports of the results that have convinced other scientists of the theory's validity. However, there's an important difference between personal testimony about a scientific test and a UFO: The first can be verified by anyone, at least in principle, while the second cannot.

Understanding this difference is crucial to understanding what counts as science and what does not. Even though you may never have conducted a test of Einstein's theory of relativity yourself, there's nothing stopping you from doing so. It might require several years of study to acquire the necessary background to conduct the test, but you could eventually confirm the results reported by other scientists. In other words, while you may currently be trusting the eyewitness testimony of scientists, you always have the option of verifying their testimony for yourself.

▲ **FIGURE 3.10**
Hallmarks of science.

In contrast, there is no way for you to verify someone's eyewitness account of a UFO. Moreover, scientific studies of eyewitness testimony show it to be notoriously unreliable; different eyewitnesses often disagree on what they saw even immediately after an event has occurred. As time passes, memories of the event may change further. In some cases in which memory has been checked against reality, people have reported vivid memories of events that never happened at all. This explains something that virtually all of us have experienced: disagreements with a friend about who did what and when. Since both people cannot be right in such cases, at least one person must have a memory that differs from reality.

The demonstrated unreliability of eyewitness testimony explains why it is generally considered insufficient for a conviction in criminal court; at least some other evidence is required. For the same reason, we cannot accept eyewitness testimony by itself as sufficient evidence in science, no matter who reports it or how many people offer similar testimony.

Objectivity in Science We generally think of science as being objective, meaning that all people should be able to find the same scientific results. However, there is a difference between the overall objectivity of science and the objectivity of individual scientists who may bring their personal biases and beliefs to their scientific work. For example, most scientists choose their research projects based on personal interests rather than on some objective formula. In extreme cases, scientists have been known to cheat—either deliberately or subconsciously—to obtain a result they desire. For example, in the late 19th century, astronomer Percival Lowell claimed to see a network of artificial canals in blurry telescopic images of Mars, leading him to conclude that there was a great Martian civilization. But no such canals actually exist, so Lowell must have allowed his beliefs about extraterrestrial life to influence the way he interpreted what he saw—in essence a form of cheating, though almost certainly not intentional.

Bias can occasionally show up even in the thinking of the scientific community as a whole. Some valid ideas may not be considered by any scientist because they fall too far outside the general patterns of thought, or **paradigm**, of the time. Einstein's theory of relativity is an example. Many scientists in the decades before Einstein had gleaned hints of the theory but did not investigate them, at least in part because they seemed too outlandish.

The beauty of science is that it encourages continued testing by many people. Even if personal biases affect some results, tests by others should eventually uncover any mistakes. Similarly, if a new idea is correct but falls outside the accepted paradigm, continued testing and verification of the idea will eventually force a paradigm shift. In that sense, *science ultimately provides a means of bringing people to agreement,* at least on topics that can be subjected to scientific study.

◆ What is a scientific theory?

The most successful scientific models explain a wide variety of observations in terms of just a few general principles. When a powerful yet simple model makes predictions that survive repeated and varied testing, scientists elevate its status and call it a **theory**. Some famous examples are Isaac Newton's theory of gravity, Charles Darwin's theory of evolution, and Albert Einstein's theory of relativity.

Note that the scientific meaning of the word *theory* is quite different from its everyday meaning, in which we equate a theory more closely with speculation or hypothesis. For example, someone might say, "I have a new theory about why people enjoy the beach." Without the support of a broad range of evidence that others have tested and confirmed, this "theory" is really only a guess. In contrast, Newton's theory of gravity qualifies as a scientific theory because it uses simple physical principles to explain a great many observations and experiments.

---

common misconceptions

Eggs on the Equinox

One of the hallmarks of science holds that you needn't take scientific claims on faith. In principle, at least, you can always test them for yourself. Consider the claim, repeated in news reports every year, that the spring equinox is the only day on which you can balance an egg on its end. Many people believe this claim, but you'll be immediately skeptical if you think about the nature of the spring equinox. The equinox is merely a point in time at which sunlight strikes both hemispheres equally (see Figure 2.3). It's difficult to see how sunlight could affect an attempt to balance eggs (especially if the eggs are indoors), and there's nothing special about either Earth's gravity or the Sun's gravity on that day.

More important, you can test this claim directly. It's not easy to balance an egg on its end, but with practice you can do it on any day of the year, not just on the spring equinox. Not all scientific claims are so easy to test for yourself, but the basic lesson should be clear: Before you accept any scientific claim, you should demand a reasonable explanation of the evidence that backs it up.

cosmic context FIGURE 3.11 The Copernican Revolution

Ancient Earth-centered models of the universe easily explained the simple motions of the Sun and Moon through our sky, but had difficulty explaining the more complicated motions of the planets. The quest to understand planetary motions ultimately led to a revolution in our thinking about Earth's place in the universe that illustrates the process of science. This figure summarizes the major steps in that process.

① Night by night, planets usually move from west to east relative to the stars. However, during periods of *apparent retrograde motion,* they reverse direction for a few weeks to months. The ancient Greeks knew that any credible model of the solar system had to explain these observations.

② Most ancient Greek thinkers assumed that Earth remained fixed at the center of the solar system. To explain retrograde motion, they therefore added a complicated scheme of circles moving upon circles to their Earth-centered model. However, at least some Greeks, such as Aristarchus, preferred a Sun-centered model, which offered a simpler explanation for retrograde motion.

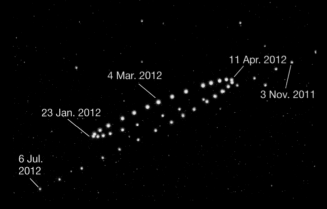

This composite photo shows the apparent retrograde motion of Mars.

The Greek geocentric model explained apparent retrograde motion by having planets move around Earth on small circles that turned on larger circles.

HALLMARK OF SCIENCE **A scientific model must seek explanations for observed phenomena that rely solely on natural causes.** The ancient Greeks used geometry to explain their observations of planetary motion.

(Left page)
A schematic map of the universe from 1539 with Earth at the center and the Sun (Solis) orbiting it between Venus (Veneris) and Mars (Martis).

(Right page)
A page from Copernicus's De Revolutionibus, published in 1543, showing the Sun (Sol) at the center and Earth (Terra) orbiting between Venus and Mars.

③ By the time of Copernicus (1473–1543), predictions based on the Earth-centered model had become noticeably inaccurate. Hoping for improvement, Copernicus revived the Sun-centered idea. He did not succeed in making substantially better predictions because he retained the ancient belief that planets must move in perfect circles, but he inspired a revolution continued over the next century by Tycho, Kepler, and Galileo.

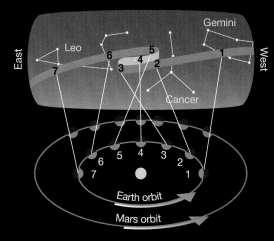

Apparent retrograde motion is simply explained in a Sun-centered system. Notice how Mars appears to change direction as Earth moves past it.

HALLMARK OF SCIENCE Science progresses through creation and testing of models of nature that explain the observations as simply as possible. Copernicus developed a Sun-centered model in hopes of explaining observations better than the more complicated Earth-centered model.

④ Tycho exposed flaws in both the ancient Greek and Copernican models by observing planetary motions with unprecedented accuracy. His observations led to Kepler's breakthrough insight that planetary orbits are elliptical, not circular, and enabled Kepler to develop his three laws of planetary motion.

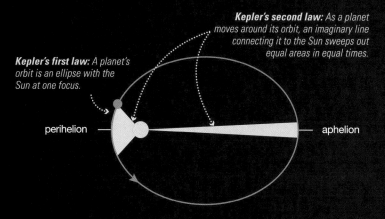

Kepler's first law: *A planet's orbit is an ellipse with the Sun at one focus.*

Kepler's second law: *As a planet moves around its orbit, an imaginary line connecting it to the Sun sweeps out equal areas in equal times.*

Kepler's third law: *More distant planets orbit at slower average speeds, obeying $p^2 = a^3$.*

HALLMARK OF SCIENCE A scientific model makes testable predictions about natural phenomena. If predictions do not agree with observations, the model must be revised or abandoned. Kepler could not make his model agree with observations until he abandoned the belief that planets move in perfect circles.

⑤ Galileo's experiments and telescopic observations overcame remaining scientific objections to the Sun-centered model. Together, Galileo's discoveries and the success of Kepler's laws in predicting planetary motion overthrew the Earth-centered model once and for all.

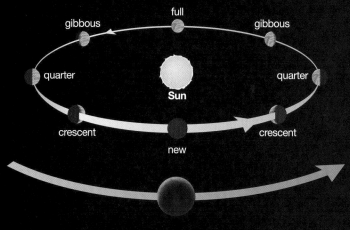

With his telescope, Galileo saw phases of Venus that are consistent only with the idea that Venus orbits the Sun rather than Earth.

"Theory" is just one of many terms that are used with different meaning in science than in everyday life. Table 3.1 summarizes a few of the most common of these terms.

Despite its success in explaining observed phenomena, a scientific theory can never be proved true beyond all doubt, because future observations may disagree with its predictions. However, anything that qualifies as a scientific theory must be supported by a large, compelling body of evidence.

In summary, a scientific theory is not at all like a hypothesis or any other type of guess. We are free to change a hypothesis at any time, because it has not yet been carefully tested. In contrast, the results of the many tests that a scientific theory has already passed cannot simply disappear, so if a theory ever fails a new test, a replacement theory must still explain the past results.

think about it When people claim that something is "only a theory," what do you think they mean? Does this meaning of *theory* agree with the definition of a theory in science? Do scientists always use the word *theory* in its "scientific" sense? Explain.

TABLE 3.1 Scientific Usage Often Differs from Everyday Usage

This table lists some words you will encounter in the media that have a different meaning in science than in everyday life. (Adapted from a table published by Richard Somerville and Susan Joy Hassol in Physics Today.)

Term	Everyday Meaning	Scientific Meaning	Example of Use in Science
model	something you build, like a model airplane	a representation of nature, sometimes using mathematics or computer simulations, that is intended to explain or predict observed phenomena	A model of planetary motion can be used to calculate exactly where planets should appear in our sky.
hypothesis	a guess or assumption of almost any type	a model that has been proposed to explain some observations but that has not yet been rigorously confirmed	Scientists hypothesize that the Moon was formed by a giant impact, but there is not enough evidence to be fully confident in this model.
theory	speculation	a particularly powerful model that has been so extensively tested and verified that we have extremely high confidence in its validity	Einstein's theory of relativity successfully explains a broad range of natural phenomena and has passed a great many tests of its validity.
bias	distortion, political motive	tendency toward a particular result	Current techniques for detecting extrasolar planets are biased toward detecting large planets.
critical	really important; involving criticism, often negative	right on the edge, near a boundary	A boiling point is a "critical value" because above that temperature, a liquid will boil away.
deviation	strangeness or unacceptable behavior	change or difference	The recent deviation in global temperatures from their long-term average implies that something is heating the planet.
enhance/enrich	improve	increase or add more, but not necessarily to make something "better"	"Enhanced colors" means colors that have been brightened. "Enriched with iron" means containing more iron.
error	mistake	range of uncertainty	The "margin of error" tells us how closely measured values are likely to reflect true values.
feedback	a response	a self-regulating (negative feedback) or self-reinforcing (positive feedback) cycle	Gravity can provide positive feedback to a forming planet: Adding mass leads to stronger gravity, which leads to more added mass, and so on.
state (as a noun)	a place or location	a description of current condition	The Sun is in a state of balance, so it shines steadily.
uncertainty	ignorance	a range of possible values around some central value	The measured age of our solar system is 4.55 billion years with an uncertainty of 0.02 billion years.
values	ethics, monetary value	numbers or quantities	The speed of light has a measured value of 300,000 km/s.

THE PROCESS OF SCIENCE IN ACTION

3.3 The Fact and Theory of Gravity

We have discussed the distinction between a hypothesis and a theory in science, but what is the difference between a *fact* and a theory? The basic answer is fairly simple: A fact is something we consider to be demonstrably true, while a theory is a model that explains *why* a body of facts is true and can therefore be modified as we collect more facts. To further illustrate this distinction, let's look at how the theory of gravity has developed over time.

◆ How does the fact of gravity differ from the theory of gravity?

Gravity is clearly a fact: Things really do fall when you drop them, and planets really do orbit the Sun. However, despite our daily experience with gravity, an adequate *theory* of gravity took a long time to develop. In ancient Greece, Aristotle imagined gravity to be an inherent property of heavy objects and claimed that heavier objects would fall to the ground faster than lighter-weight objects. Galileo put this idea to the test in a series of experiments that supposedly included dropping weights from the Leaning Tower of Pisa. His results showed that all objects fall to the ground at the same rate, as long as air resistance is unimportant. Aristotle was therefore wrong about gravity, but Galileo's ideas about gravity still fell short of being a useful theory.

see it for yourself Find a piece of paper and a small rock. Hold both at the same height, one in each hand, and let them go at the same instant. The rock, of course, hits the ground first. Next, crumple the paper into a small ball and repeat the experiment. What happens? Explain how this experiment suggests that, without air resistance, gravity causes all falling objects to fall at the same rate.

▶ **Making Sense of the Universe (Learning Outcome 2)**

Newton's Theory of Gravity The breakthrough in our understanding of gravity came from Isaac Newton. By his own account, Newton experienced a moment of inspiration in 1666 when he saw an apple fall to the ground and suddenly realized that the gravity making the apple fall was the same force that held the Moon in orbit around Earth. With this insight, Newton eliminated the long-held distinction between the realm of the heavens and the realm of Earth. For the first time, the two realms were brought together as one *uni*verse governed by a single set of principles.

Newton worked hard to turn his insight into a theory of gravity, which he published in 1687 (in his book *Principia*). Newton expressed the force of gravity mathematically with his **universal law of gravitation**. Three simple statements summarize this law:

- Every mass attracts every other mass through the force called *gravity*.
- The strength of the gravitational force attracting any two objects is *directly proportional* to the product of their masses. For example, doubling the mass of *one* object doubles the force of gravity between the two objects.
- The strength of gravity between two objects decreases with the *square* of the distance between their centers. We therefore say that the gravitational force follows an **inverse square law** with distance. For example, doubling the distance between two objects weakens the force of gravity by a factor of 2^2, or 4.

▲ FIGURE 3.12
The universal law of gravitation is an *inverse square law*, which means the force of gravity declines with the *square* of the distance *d* between two objects.

These three statements tell us everything we need to know about Newton's universal law of gravitation. Mathematically, all three statements can be combined into a single equation, usually written like this:

$$F_g = G\frac{M_1 M_2}{d^2}$$

where F_g is the force of gravitational attraction, M_1 and M_2 are the masses of the two objects, and d is the distance between their centers (Figure 3.12). The symbol G is a constant called the **gravitational constant**, and its numerical value has been measured to be $G = 6.67 \times 10^{-11} \text{ m}^3/(\text{kg} \times \text{s}^2)$.

think about it How does the gravitational force between two objects change if the distance between them triples? If the distance between them drops by half?

Newton's theory of gravity gained rapid acceptance because it explained a great many facts that other scientists had already discovered. For example, it explained Galileo's observations about falling objects and Kepler's laws of planetary motion. Even more impressively, it led to new predictive successes. Shortly after Newton published his theory, Sir Edmund Halley used it to calculate the orbit of a comet that had been seen in 1682, from which he predicted the comet's return in 1758. Halley's Comet returned on schedule, which is why it now bears his name. In 1846, after carefully examining the orbit of Uranus, the French astronomer Urbain Leverrier used Newton's theory to predict that Uranus's orbit was being affected by a previously undiscovered eighth planet. He predicted the location of the planet and sent a letter suggesting a search to Johann Galle of the Berlin Observatory. On the night of September 23, 1846, Galle discovered Neptune within 1° of the position predicted by Leverrier. It was a stunning triumph for Newton's theory.

A Problem Appears Today, we can apply Newton's theory of gravity to objects throughout the universe, including the orbits of extrasolar planets around their stars, of stars around the Milky Way Galaxy, and of galaxies in orbit of each other. There seems no reason to doubt the universality of the law. However, we also now know that Newton's law does not tell the entire story of gravity.

The first hint of a problem with Newton's theory arose not long after Leverrier's success in predicting the existence of Neptune. Astronomers discovered a slight discrepancy between the observed characteristics of the orbit of Mercury and the characteristics predicted by Newton's theory. The discrepancy was very small, and Mercury was the only planet that showed any problem, but there seemed no way to make it go away: Unless the data were wrong, which seemed highly unlikely, Newton's theory was giving a slightly incorrect prediction for the orbit of Mercury.

Einstein's Solution The solution to the problem of Mercury came in 1915, when Albert Einstein (1879–1955) published his *general theory of relativity* (see Tools of Science, page 172). This theory predicted an orbit for Mercury that matched the observations. Not long after, astronomers put Einstein's new theory to the test during a solar eclipse, finding that it successfully predicted the precise positions of stars visible near the blocked disk of the Sun, while Newton's theory gave a prediction that was slightly off. Scientists have continued to test both Newton's and Einstein's theories ever since. In every case in which the two theories give different answers, Einstein's theory has matched the observations while Newton's theory has not. That is why, today, we consider Einstein's general theory of relativity to have supplanted Newton's theory as our "best" theory of gravity.

Does this mean that Newton's theory of gravity was "wrong"? Remember that Newton's theory successfully explains nearly all observations of gravity in the universe, and it works so well that we can use it to plot the courses of spacecraft to the planets. Moreover, in all cases in which Newton's theory works well, Einstein's theory gives essentially the same answers. The differences in the predictions between the two theories are noticeable only with extremely precise measurements or in cases where gravity is unusually strong. We therefore do not say that Newton's theory was wrong, but rather that it was only an *approximation* to a more exact theory of gravity—Einstein's general theory of relativity. Under most circumstances, the approximation is so good that we can barely tell the difference between the two theories of gravity, but in cases of strong gravity, Einstein's theory works and Newton's fails.

While Einstein's theory of gravity has so far passed every test that it has been subjected to, most scientists suspect that we'll eventually find an even better theory of gravity. The reason is that for the most extreme possible case of gravity—that which occurs at the infinitely small and high-density center of a black hole [Section 10.2]—Einstein's theory of relativity gives a different answer than the equally well tested theory of the very small (known as the theory of quantum mechanics). Because these two theories contradict each other in this special case, scientists know that one or both will ultimately have to be modified.

The Bottom Line Gravity is both a fact and a theory. The *fact* of gravity is obvious in the observations we make of falling objects on Earth and orbiting objects in space. The *theory* of gravity is our best explanation of those observations, and we can use it to make precise predictions of how objects will behave due to gravity. In the future, our theory of gravity may be further improved, but gravity remains a fact regardless of how we revise the theory. Note that gravity is not unique in this way: Scientists make the same type of distinction in many other cases, such as when they talk about the fact of atoms being real and the atomic theory used to explain them, and when they talk about the fact of evolution revealed in the fossil record and the theory used to explain how evolution occurs.

summary of key concepts

3.1 From Earth-Centered to Sun-Centered

- **How did the Greeks explain planetary motion?**

The Greeks developed **models** in an attempt to explain their observations of nature. The Greek Earth-centered model culminated with the model of Ptolemy. The **Ptolemaic model** explained apparent retrograde motion by having each planet move on a small circle whose center moved around Earth on a larger circle.

- **How did the Copernican revolution change our view of the universe?**

The Copernican revolution replaced the ancient Earth-centered view of the universe with a new view in which Earth is just one planet going around the Sun. It also made scientists recognize the importance of precise observations in testing models of nature. The model based on **Kepler's laws of planetary motion** proved much better able to explain the observations than the old Earth-centered model or Copernicus's Sun-centered model in which orbits were perfect circles.

3.2 Hallmarks of Science

- **How can we distinguish science from nonscience?**

Science generally exhibits three hallmarks: (1) Modern science seeks explanations for observed phenomena that rely solely on natural causes. (2) Science progresses through the creation and testing of models of nature that explain the observations as simply as possible. (3) A scientific model must make testable predictions about natural phenomena that would force us to revise or abandon the model if the predictions did not agree with observations.

- **What is a scientific theory?**

A scientific **theory** is a simple yet powerful model that explains a wide variety of observations in terms of just a few general principles and has attained the status of a theory by surviving repeated and varied testing.

THE PROCESS OF SCIENCE IN ACTION

3.3 The Fact and Theory of Gravity

- **How does the fact of gravity differ from the theory of gravity?**

Gravity is a *fact* in that objects really do fall to the ground and planets really do orbit the Sun. The *theory* of gravity is used to explain why gravity acts as it does. While the fact of gravity does not change, the theory can be improved with time, as Einstein's general theory of relativity improved on Newton's theory of gravity.

Investigations

For instructor-assigned homework and other learning materials, go to www.MasteringAstronomy.com.

Quick Quiz

Choose the best answer to each of the following. Explain your reasoning.

1. A *geocentric* model of the universe is (a) any model that places Earth at the center of the universe. (b) any model that has Earth as a planet orbiting the Sun. (c) any model that places the Sun at the center of the universe.
2. In Ptolemy's geocentric model, the retrograde motion of a planet occurs when (a) Earth is about to pass the planet in its orbit around the Sun. (b) the planet actually goes backward in its orbit around Earth. (c) the planet is aligned with the Moon in our sky.
3. Which of the following was *not* a major advantage of Copernicus's Sun-centered model over the Ptolemaic model? (a) It made significantly better predictions of planetary positions in our sky. (b) It offered a more natural explanation for the apparent retrograde motion of planets in our sky. (c) It allowed calculation of the orbital periods and distances of the planets.
4. Tycho Brahe's contributions to astronomy included (a) inventing the telescope. (b) proving that Earth orbits the Sun. (c) collecting data that enabled Kepler to discover the laws of planetary motion.
5. Kepler made a major break from ancient beliefs when he (a) decided to look for a system that could explain actual observations. (b) realized that gravity holds the planets in their orbits. (c) abandoned circular orbits in favor of elliptical orbits.
6. Earth is closer to the Sun in January than in July. Therefore, in accord with Kepler's second law, (a) Earth travels faster in its orbit around the Sun in July than in January. (b) Earth travels faster in its orbit around the Sun in January than in July. (c) it is summer in January and winter in July.
7. According to Kepler's *third* law, (a) Mercury travels fastest in the part of its orbit in which it is closest to the Sun. (b) Jupiter orbits the Sun at a faster speed than Saturn. (c) all the planets have elliptical orbits.
8. Galileo's contributions to astronomy included (a) discovering the laws of planetary motion. (b) discovering the law of gravity. (c) making observations and conducting experiments that dispelled scientific objections to the Sun-centered model.
9. Which of the following is *not* true about scientific progress? (a) Science progresses through the creation and testing of models of nature. (b) Science advances only through strict application of the scientific method. (c) Science avoids explanations that invoke the supernatural.
10. Which of the following is *not* true about a scientific theory? (a) A theory must explain a wide range of observations or experiments. (b) Even the strongest theories can never be proved true beyond all doubt. (c) A theory is essentially an educated guess.
11. If Earth were twice as far as it actually is from the Sun, the force of gravity attracting Earth to the Sun would be (a) twice as strong. (b) half as strong. (c) one-quarter as strong.
12. Acceptance of Einstein's theory of gravity (general relativity) demonstrated that Newton's theory had been (a) wrong. (b) incomplete. (c) really only a guess.

Short-Answer/Essay Questions

Explain all answers clearly, with complete sentences and proper essay structure if needed. An asterisk (*) designates a quantitative problem, for which you should show all your work.

13. State each of Kepler's laws of planetary motion and describe in your own words what it means in a way that a friend could understand.
14. What is the difference between a *hypothesis* and a *theory* in science? Explain clearly.
15. Describe each of the three hallmarks of science and give an example of how we can see each one in the unfolding of the Copernican revolution.
16. *Copernican Players.* Using a bulleted-list format, make a one-page "executive summary" of the major roles that Copernicus, Tycho, Kepler, Galileo, and Newton played in overturning the ancient belief in an Earth-centered universe.
17. *Influence on History.* Based on what you have learned about the Copernican revolution, write a one- to two-page essay about how you believe it altered the course of human history.
18. *What Makes It Science?* Choose one idea from the modern view of the cosmos as discussed in Chapter 1 (such as "The universe is expanding" or "We are made from elements manufactured by stars").
    a. Describe how this idea reflects each of the three hallmarks of science, discussing how it is based on observations, how our understanding of it depends on a model, and how that model is testable.
    b. Describe a hypothetical observation that, if it were actually made, might cause us to call the idea into question. Then briefly discuss whether you think that, overall, the idea is likely or unlikely to hold up to future observations.
19. *Discussion: A Flat Earth?* A few relatively famous people have recently made the claim that Earth is flat. Working in small groups, first find out the basis on which they make these claims, then make a list of key observations that refute the flat Earth claims. Based on what you learn, how can you tell which websites discussing flat Earth claims are scientifically valid and which are not, and what lessons the flat Earth claims may contribute to the more general issue of "fake news" in our modern society?
*20. *The Universal Law of Gravitation.*
    a. How does quadrupling the distance between two objects affect the gravitational force between them?
    b. Suppose the Sun were somehow replaced by a star with twice as much mass. What would happen to the gravitational force between Earth and the Sun?
    c. Suppose Earth were moved to one-third of its current distance from the Sun. What would happen to the gravitational force between Earth and the Sun?
*21. *Eris Orbit.* The recently discovered object Eris, which is slightly larger than Pluto, orbits the Sun every 557 years. What is its average distance (semimajor axis) from the Sun? (*Hint:* Use Kepler's third law.)
*22. *Halley's Orbit.* Halley's Comet orbits the Sun every 76.0 years.
    a. Find its average distance (semimajor axis) from the Sun. (*Hint:* Use Kepler's third law.)
    b. Halley's orbit is a very eccentric (stretched-out) ellipse, so at perihelion Halley's Comet is only about 90 million kilometers from the Sun, compared to more than 5 billion kilometers at aphelion. Does Halley's Comet spend most of its time near its perihelion distance, near its aphelion distance, or halfway in between? Explain.

Origin of the Solar System

4

LEARNING GOALS

4.1 Characteristics of the Solar System
- What does the solar system look like?
- What features of our solar system provide clues to how it formed?

4.2 The Birth of the Solar System
- What theory best explains the orderly patterns of motion in our solar system?
- How does our theory account for the features of planets, moons, and small bodies?

 THE PROCESS OF SCIENCE IN ACTION

4.3 The Age of the Solar System
- How do we determine the age of Earth and the solar system?

From space, we see Earth as a small blue oasis set against a black void. Such images have inspired great works of art, music, and poetry, and have come to symbolize global awareness of the environment. These images also inspire us to ask scientific questions, such as how our planet came to be and why it is such an ideal home for life. As we will see in this and coming chapters, the answers are best understood not just by studying Earth, but also by comparing Earth to other worlds. We'll begin by exploring general features of our solar system and how they are explained by the modern scientific theory of the solar system's formation.

4.1 Characteristics of the Solar System

Our major goal in this chapter is to understand the scientific theory that best explains the origin of Earth and the solar system. To do so, we must first have a general picture of what the solar system looks like and a basic understanding of its individual worlds. We'll then use these ideas to come up with a list of major features that any successful theory of the solar system's birth must explain.

▶ **Our Planetary System**

◆ What does the solar system look like?

It's easy to find data about our solar system, like the planetary data summarized in Table 4.1. But what would we actually see if we could view the solar system from beyond the orbits of the planets? Without a telescope, the answer would be "not much." The Sun and planets are so small compared to the distances between them [Section 1.1] that from the outskirts of our solar system, the planets would be visible only as pinpoints of light and even the Sun would be just a small bright dot in the sky. But if we magnify the sizes of the planets by about a thousand times compared to their distances from the Sun, and show their orbital paths, we get the central picture in Figure 4.1 (pages 56–57).

Even on first glance, the table and figure make it clear that our solar system is *not* a random collection of worlds. Instead, our solar system shows several clear patterns. For example, all the planets orbit the Sun in the same direction and in nearly the same plane, and the four inner planets are much smaller and closer together than the next four planets. We'll discuss these and other patterns in more detail shortly. First, let's build our general picture of the solar system by taking a brief tour of its major regions, starting with the Sun and moving outward.

The Sun The Sun is by far the largest and brightest object in our solar system. It contains more than 99.8% of the solar system's total mass, making it nearly a thousand times as massive as everything else in the solar system combined. The Sun's gravity governs the orbits of the planets, and the Sun is the source of virtually all the visible light in our solar system, since planets and moons shine only by virtue of the sunlight they reflect. It is also the primary influence on the temperatures of planetary surfaces and atmospheres. Charged particles flowing outward from the Sun (which make up the *solar wind*) help shape planetary magnetic fields and can influence planetary atmospheres. However, because we can understand the planets without knowing further details about the Sun, we'll save these details for our study of the Sun as a star in Chapter 8.

think about it Look ahead to Chapter 8. How does the size of Earth compare to the size of a sunspot?

The Inner Planets The four inner planets—Mercury, Venus, Earth, and Mars—are all quite small and close together compared to the outer planets. These four planets also share similar compositions of metal and rock, which is why we often refer to them as the **terrestrial planets** (*terrestrial* means "Earth-like").

Despite their similarities in composition, the four terrestrial planets differ substantially in their details. Mercury is a desolate, cratered world that looks much like our own Moon. Venus, nearly identical in size to Earth, is best known for its extreme temperature and pressure: Its thick, carbon dioxide atmosphere bakes Venus's surface to an incredible 470°C (about 880°F) and creates a surface pressure equivalent to the pressure nearly a kilometer (0.6 mile) beneath

TABLE 4.1 Planetary Data*

Photo	Planet	Relative Size	Average Distance from Sun (AU)	Average Equatorial Radius (km)	Mass (Earth = 1)	Average Density (g/m³)	Orbital Period	Rotation Period	Axis Tilt	Average Surface (or Cloud-Top) Temperature†	Composition	Known Moons (2018)	Rings?
	Mercury	·	0.387	2440	0.055	5.43	87.9 days	58.6 days	0.0°	700 K (day) 100 K (night)	Rocks, metals	0	No
	Venus	•	0.723	6051	0.82	5.24	225 days	243 days	177.3°	740 K	Rocks, metals	0	No
	Earth	•	1.00	6378	1.00	5.52	1.00 year	23.93 hours	23.5°	290 K	Rocks, metals	1	No
	Mars	·	1.52	3397	0.11	3.93	1.88 years	24.6 hours	25.2°	220 K	Rocks, metals	2	No
	Jupiter	●	5.20	71,492	318	1.33	11.9 years	9.93 hours	3.1°	125 K	H, He, hydrogen compounds§	69	Yes
	Saturn	●	9.54	60,268	95.2	0.70	29.4 years	10.6 hours	26.7°	95 K	H, He, hydrogen compounds§	62	Yes
	Uranus	●	19.2	25,559	14.5	1.32	83.8 years	17.2 hours	97.9°	60 K	H, He, hydrogen compounds§	27	Yes
	Neptune	●	30.1	24,764	17.1	1.64	165 years	16.1 hours	29.6°	60 K	H, He, hydrogen compounds§	14	Yes
	Pluto	·	39.5	1185	0.0022	1.9	248 years	6.39 days	112.5°	44 K	Ices, rock	5	No
	Eris	·	67.7	1168	0.0028	2.3	557 years	1.08 days	78°	43 K	Ices, rock	1	No

*Including the dwarf planets Pluto and Eris.
†Surface temperatures for all objects except Jupiter, Saturn, Uranus, and Neptune, for which cloud-top temperatures are listed.
§Includes water (H_2O), methane (CH_4), and ammonia (NH_3).

Cosmic Context   FIGURE 4.1 The Solar System

The solar system's layout and composition offer four major clues to how it formed. The main illustration below shows the orbits of planets in the solar system from a perspective beyond Neptune, with the planets themselves magnified by about a thousand times relative to their orbits. (The Sun is not shown on the same scale as the planets; it would fill the page if it were.)

1 Large bodies in the solar system have orderly motions. All planets have nearly circular orbits going in the same direction in nearly the same plane. Most large moons orbit their planets in this same direction, which is also the direction of the Sun's rotation.

Seen from above, planetary orbits are nearly circular.

White arrows indicate the rotation direction of the planets and Sun.

Red circles indicate the orbital direction of major moons around their planets.

Each planet's axis tilt is shown, with small circling arrows to indicate the direction of the planet's rotation.

Orbits are shown to scale, but planet sizes are exaggerated about 1000 times and the Sun's size is exaggerated about 50 times relative to the orbits.

Orange arrows indicate the direction of orbital motion.

② Planets fall into two major categories: Small, rocky terrestrial planets and large, hydrogen-rich jovian planets.

terrestrial planet   jovian planet

Terrestrial Planets:
- small in mass and size
- close to the Sun
- made of metal and rock
- few moons and no rings

Jovian Planets:
- large mass and size
- far from the Sun
- made of H, He, and hydrogen compounds
- rings and many moons

③ Swarms of asteroids and comets populate the solar system. Vast numbers of rocky asteroids and icy comets are found throughout the solar system, but are concentrated in three distinct regions.

*Asteroids are made of metal and rock, and most orbit in the **asteroid belt** between Mars and Jupiter.*

*Comets are ice-rich, and many are found in the **Kuiper belt** beyond Neptune's orbit.*

*Even more comets orbit the Sun in the distant, spherical region called the **Oort cloud**, and only a rare few ever plunge into the inner solar system.*

Kuiper belt

Saturn

④ Several notable exceptions to these trends stand out. Some planets have unusual axis tilts, unusually large moons, or moons with unusual orbits.

Uranus's odd tilt

Uranus

Uranus rotates nearly on its side compared to its orbit, and its rings and major moons share this "sideways" orientation.

Earth's relatively large moon

Our own Moon is much closer in size to Earth than most other moons in comparison to their planets.

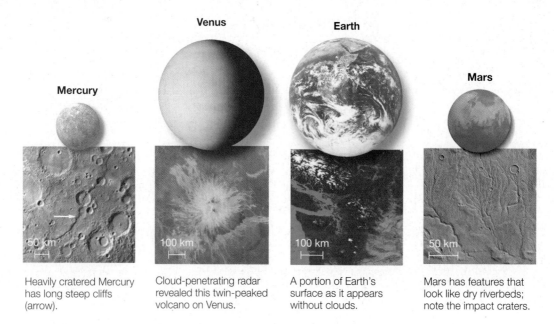

▲ **FIGURE 4.2**
The terrestrial planets, shown to scale, along with sample close-ups viewed from orbit.

the ocean's surface on Earth. Our home planet, Earth, is the only world in our solar system with surface oceans of liquid water. Mars, though it lacks any surface liquid water today, shows clear evidence of having had flowing water in its distant past, which is why scientists are so interested in the search for past or present life on Mars. Figure 4.2 shows the four terrestrial planets to scale, along with sample views of their surfaces.

The Asteroid Belt Notice the donut-shaped region of dots between the orbits of Mars and Jupiter in Figure 4.1. This region represents the **asteroid belt,** which contains millions of asteroids; an **asteroid** is essentially a chunk of metal and rock that orbits the Sun much like a planet, but it is much smaller in size (Figure 4.3). Despite their large numbers, the total mass of all the asteroids combined is much smaller than the mass of our Moon. Moreover, despite what you may have seen in movies, asteroids are spread out over such a large region of space that there would be little danger of hitting one if you flew a spaceship through the asteroid belt.

The Outer Planets Beyond the asteroid belt we enter the realm of the four outer planets: Jupiter, Saturn, Uranus, and Neptune (Figure 4.4). These planets are much larger and much farther apart than the terrestrial planets. They are also very different in composition from the terrestrial planets, because they contain vast amounts of materials that would be gaseous on Earth, including hydrogen, helium, and *hydrogen compounds* such as water (H_2O), methane (CH_4), and ammonia (NH_3). Their compositions mean they have no solid surfaces; instead, if you entered any of their atmospheres, you would simply plunge deeper and deeper into the interior until you were crushed by the growing pressure. Because Jupiter is the largest of this group of planets, we refer to them jointly as **jovian planets.** (*Jovian* means "Jupiter-like.")

a Mathilde, photographed by the *Near-Earth Asteroid Rendezvous* (*NEAR*) spacecraft.

b Itokawa, photographed by the Japanese *Hayabusa* mission, which captured a dust sample from the surface and returned it to Earth.

▲ **FIGURE 4.3**
Images of two small asteroids that have been visited by spacecraft. Like these, most asteroids are not large enough for gravity to have made them round.

The jovian planets also differ from the terrestrial planets in another important way: While the terrestrial planets have only three moons among them (one for Earth and two very small moons for Mars), the jovian planets each have many moons. Many of these moons are amazing worlds in their own right. For example, Jupiter's moon Io is the most volcanically active world in our solar system, and its moon Europa is thought to have a deep subsurface ocean of liquid water. Saturn's moon Titan is the only moon in the solar system with a thick atmosphere, and its surface contains cold lakes of liquid methane and ethane. In addition to moons,

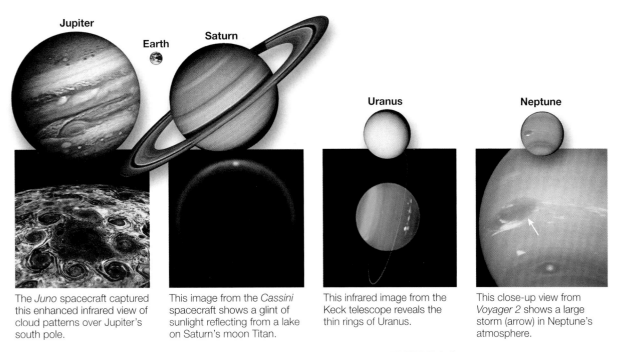

▲ FIGURE 4.4
The jovian planets shown to scale, with Earth for comparison, along with selected features of the planets, rings, or moons.

all four jovian planets are orbited by vast numbers of small particles that make up their **rings,** although only Saturn's rings are easily visible from Earth.

Realm of the Comets People throughout history have been intrigued and inspired by the occasional appearance of a **comet** in our skies (Figure 4.5). Comets are made of rock and ice, so when they enter the inner solar system they can grow spectacular *tails* as ice vaporizes and escapes into space.

The comets that we see in the sky must come from somewhere, and study of their orbital paths has led scientists to conclude that they come from two vast regions shown as Feature 3 in Figure 4.1. The first, known as the **Kuiper belt** (*Kuiper* rhymes with *piper*), is a donut-shaped region beyond the orbit of

▼ FIGURE 4.5
Comet McNaught and the Milky Way over Patagonia, Argentina, in 2007. The fuzzy patches above the comet tail are the Magellanic Clouds, satellite galaxies of the Milky Way.

Neptune that is thought to contain at least 100,000 comets. Some of these "comets," including Pluto and Eris, are large enough for their own gravity to make them round in shape, which qualifies them as *dwarf planets* [Section 1.3].

The second cometary region is called the **Oort cloud** (*Oort* rhymes with *court*). It is much farther from the Sun than the Kuiper belt and may contain as many as a trillion comets. These comets have orbits randomly inclined to the ecliptic plane, giving the Oort cloud a roughly spherical shape.

Four Features of the Solar System

◆ **What features of our solar system provide clues to how it formed?**

We have seen that our solar system is a family of worlds, exhibiting numerous "family traits" that we expect to be explained by any theory of our solar system's formation. We'll focus our attention on four major features of our solar system, each of which is associated with one of the numbered steps in Figure 4.1.

Feature 1: Patterns of Motion Figure 4.1 shows several clear patterns of motion among the large bodies of our solar system. For example:

- All planetary orbits are nearly circular and lie nearly in the same plane.
- All planets orbit the Sun in the same direction: counterclockwise as viewed from high above Earth's North Pole.
- Most planets rotate in the same direction in which they orbit, with fairly small axis tilts. The Sun also rotates in this same direction.
- Most of the solar system's large moons exhibit similar properties in their orbits around their planets, such as orbiting in their planet's equatorial plane in the same direction that the planet rotates.

Together, these orderly patterns represent the first major feature of our solar system. As we'll see shortly, our theory of solar system formation explains these patterns as consequences of processes that occurred during the early stages in the birth of our solar system.

Feature 2: Two Types of Planets Our brief planetary tour showed that the four terrestrial planets are quite different in character from the four jovian planets. Table 4.2 contrasts the general traits of these two types of planets. As we will see, the fact that the terrestrial planets lie closer than the jovian planets to the Sun provides an important clue to the history of the solar system.

Feature 3: Asteroids and Comets The third major feature of the solar system is the existence of vast numbers of small objects orbiting the Sun: rocky *asteroids* and icy *comets*. Moreover, rather than being randomly spread throughout the solar system, we've found that these small bodies are found mainly in three regions: Most asteroids orbit the Sun within the *asteroid belt* between Mars and Jupiter, while comets are found both in the *Kuiper belt* just beyond the orbit of Neptune and in the more distant, spherical *Oort cloud*. Our theory of solar system formation must account both for the vast numbers of small bodies and for their distribution in these three major regions.

Feature 4: Exceptions to the Rules The fourth key feature of our solar system is that there are a few notable exceptions to the general rules. For example, while most of the planets rotate in the same direction as they orbit, Uranus rotates nearly on its side and Venus rotates "backward" (opposite the direction of

TABLE 4.2 Comparison of Terrestrial and Jovian Planets

Terrestrial Planets	Jovian Planets
Smaller size and mass	Larger size and mass
Higher density	Lower density
Made mostly of rock and metal	Made mostly of hydrogen, helium, and hydrogen compounds
Solid surface	No solid surface
Few (if any) moons and no rings	Rings and many moons
Closer to the Sun (and closer together), with warmer surfaces	Farther from the Sun (and farther apart), with cool temperatures at cloud tops

planetary orbits). Similarly, while most large moons orbit in their planet's equatorial plane in the same direction as their planet rotates, many small moons have inclined or backward orbits.

One of the most interesting exceptions is our own Moon. While the other terrestrial planets have either no moons (Mercury and Venus) or very tiny moons (Mars has two small moons), Earth has one of the largest moons in the solar system. Our formation theory must be able to account for such exceptions, even while it explains the general rules.

4.2 The Birth of the Solar System

The four major features of our solar system would be difficult to attribute to coincidence, and in science we assume that these features have a natural explanation. In this section, we'll discuss how our modern theory of the birth of our solar system successfully explains the four major features.

◆ What theory best explains the orderly patterns of motion in our solar system?

As we noted briefly in Chapter 1 (see Figure 1.10), evidence indicates that star systems are born through the gravitational collapse of huge clouds of gas in space. This idea was first proposed in 1755 by the German philosopher Immanuel Kant, but it gained acceptance only much more recently, after detailed models based on this idea proved successful at explaining the major features of the solar system. The success of such models elevated the idea to the status of a scientific *theory* [Section 3.2]. We now call it the **nebular theory** of solar system formation, because an interstellar cloud is usually called a *nebula* (Latin for "cloud"), and its successes have continued with observations of other planetary systems (see Chapter 7), which the theory predicted should exist before they were discovered.

The particular cloud that gave birth to our own solar system about $4\frac{1}{2}$ billion years ago is usually called the **solar nebula**. Recall from Chapter 1 that the universe had already been around for more than 9 billion years by that time, and generations of stars had converted about 2% (by mass) of the original hydrogen and helium from the Big Bang into heavier elements. That is, the solar nebula was made up of about 98% hydrogen and helium, and only 2% everything else combined. The rocky terrestrial planets are therefore made from material that represented only a tiny fraction of the solar nebula.

think about it Could a solar system like ours have formed with the first generation of stars after the Big Bang? Explain.

The solar nebula probably began as a large and roughly spherical cloud of very cold, low-density gas. Initially, this gas was probably so spread out—perhaps over a region a few light-years in diameter—that gravity alone may not have been able to pull it together to start its collapse. Instead, the collapse may have been triggered by a cataclysmic event, such as the impact of a shock wave from the explosion of a nearby star (a *supernova*).

Once the collapse started, the law of gravity enabled it to continue. Recall that the strength of gravity follows an inverse square law with distance [Section 3.3]. The mass of the cloud remained the same as it shrank, so the strength of gravity increased as the diameter of the cloud decreased.

Because gravity pulls inward in all directions, you might at first guess that the solar nebula would have remained spherical as it shrank. Indeed, the idea

tools of science | Conservation Laws

A candy bar looks quite different if you smash it into little pieces, but the total amount of candy remains the same no matter how it breaks apart. In science, we would say that the total amount of candy is *conserved*, and we could say that the process of smashing the candy bar obeys a "law of conservation of candy." While candy conservation is somewhat obvious, scientists have discovered other conservation laws that reveal deep secrets of nature. Here we'll discuss two that are especially important in astronomy.

The **law of conservation of energy** tells us that energy cannot appear out of nowhere or disappear into nothingness. Objects can gain or lose energy only by exchanging energy with other objects. Energy can change forms, however, so we must know some of the different forms that energy can take (Figure 1):

- **Kinetic energy** is energy of motion. Falling rocks, orbiting planets, and cyclists in motion all have kinetic energy in an amount that depends on the object's mass and speed. Note that individual atoms and molecules are always in motion and therefore also have kinetic energy; the collective kinetic energy of many individual particles in a substance like a rock or in the air is called **thermal energy**.

- **Potential energy** is stored energy that might later be converted into some other form. For example, a rock perched on a ledge has **gravitational potential energy** because it will fall if it slips off the edge, and food contains **chemical potential energy** that can be converted into the kinetic energy of the moving cyclist. Einstein discovered that mass itself is a form of potential energy, often called **mass-energy:** The amount of mass-energy in an object of mass m is given by Einstein's famous equation $E = mc^2$.

- **Radiative energy** is energy carried by light. (The word *radiation* is often used as a synonym for *light*.) All light carries energy, which is why light can affect molecules in our eyes—thereby allowing us to see—or warm the surface of a planet.

A couple of examples should help clarify the idea of energy conservation. Consider a rock that falls off a ledge. Before it falls, it has gravitational potential energy by virtue of its height, but no kinetic energy because it is not moving. As it falls, its gravitational potential energy gradually turns into kinetic energy, so the rock accelerates as it falls. Next consider the Sun, which sends radiative energy into space as it shines. This radiative energy is generated by nuclear fusion in the Sun's core, which releases some of the mass-energy stored in the Sun's hydrogen.

A second important conservation law for astronomy is the **law of conservation of angular momentum**. To understand this law, think about a spinning ice skater (Figure 2). We say that he has angular momentum due to the fact that he is spinning, and the amount of angular momentum he has is described by the formula $m \times v \times r$, where m is his mass, v is his velocity as he spins, and r is his "radius," represented (approximately) by how far he extends his arms as he spins. Because there is little friction on the ice to slow him down, the law of conservation of angular momentum tells us that his angular momentum must stay the same no matter what he does with his arms. If he pulls his arms in, thereby decreasing his radius, his velocity of spin must increase to keep the product $m \times v \times r$ constant. For an astronomical example, consider a contracting cloud of interstellar gas: Because its radius decreases as it shrinks in size, its rate of spin must increase in order to keep its angular momentum unchanged.

Energy can be converted from one form to another.

kinetic energy (including thermal energy)

radiative energy (energy of light)

potential energy (chemical potential energy, gravitational potential energy, mass-energy, ...)

▲ **FIGURE 1**
The law of conservation of energy means that energy can be converted from one form to another, but cannot be created or destroyed.

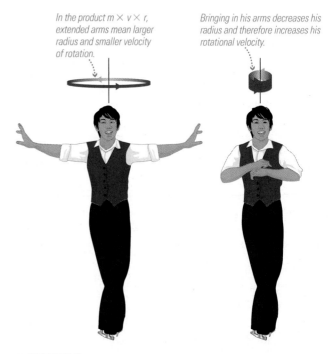

In the product $m \times v \times r$, extended arms mean larger radius and smaller velocity of rotation.

Bringing in his arms decreases his radius and therefore increases his rotational velocity.

▲ **FIGURE 2**
A spinning skater conserves angular momentum.

▲ FIGURE 4.6
This sequence of illustrations shows how the gravitational collapse of a large cloud of gas causes it to become a spinning disk of matter. The hot, dense central bulge becomes a star, while planets can form in the surrounding disk.

that gravity pulls in all directions explains why the Sun and the planets are spherical. However, gravity is not the only physical law that affects a collapsing cloud of gas, and three key processes altered the solar nebula's density, temperature, and shape (Figure 4.6):

- *Heating.* The temperature of the solar nebula increased as it collapsed. Such heating represents energy conservation in action (see Tools of Science, page 62). As the cloud shrank, its gravitational potential energy was converted to the kinetic energy of individual gas particles falling inward. These particles crashed into one another, converting the kinetic energy of their inward fall into the random motions of thermal energy. The Sun formed in the center of the cloud, where temperatures and densities were highest.

- *Spinning.* Like an ice skater pulling in his arms as he spins, the solar nebula rotated faster and faster as it shrank in radius. This increase in rotation rate represents conservation of angular momentum in action (see Tools of Science, page 62). The rotation of the cloud may have been imperceptibly slow before its collapse began, but over time the cloud's shrinkage made fast rotation inevitable. The rapid rotation also ensured that material in the cloud remained spread out rather than all collapsing into the center.

- *Flattening.* The solar nebula flattened into a disk. This flattening is a natural consequence of collisions between particles in a spinning cloud. A cloud may start with any size or shape, and different clumps of gas within the cloud may be moving in random directions at random speeds. These clumps collide and merge as the cloud collapses, and each new clump has the average velocity of the clumps that formed it. The random motions of the cloud therefore become more orderly as it collapses, transforming the cloud into a rotating, flattened disk in which all the particles had nearly circular orbits.

We can now see how the nebular theory successfully explains our first major feature of the solar system, which is its orderly motions. The planets all orbit the Sun in nearly the same plane because they formed in a flat disk. The direction in which the disk was spinning became the direction of the Sun's rotation and the orbits of the planets. Computer models show that planets would have tended to rotate in this same direction as they formed—which is why most planets rotate the same way—though the small sizes of planets compared to

TABLE 4.3 Materials in the Solar Nebula

A summary of the four types of materials present in the solar nebula. The squares represent the relative proportion of each type (by mass).

	Examples	Typical Condensation Temperature	Relative Abundance (by mass)
Hydrogen and Helium Gas	hydrogen, helium	do not condense in nebula	98%
Hydrogen Compounds	water (H_2O), methane (CH_4), ammonia (NH_3)	<150 K	1.4%
Rock	various minerals	500–1300 K	0.4%
Metals	iron, nickel, aluminum	1000–1600 K	0.2%

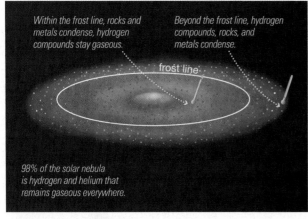

▲ **FIGURE 4.7**
Temperature differences in the solar nebula led to different kinds of condensed materials at different distances from the Sun.

Within the frost line, rocks and metals condense, hydrogen compounds stay gaseous.

Beyond the frost line, hydrogen compounds, rocks, and metals condense.

frost line

98% of the solar nebula is hydrogen and helium that remains gaseous everywhere.

---

the entire disk allowed some exceptions to arise. The fact that collisions in the disk tended to make orbits more circular explains why most planets have nearly circular orbits.

see it for yourself You can demonstrate the development of orderly motion, much as it occurred in the solar system, by sprinkling pepper into a bowl of water and stirring it quickly in random directions. The water molecules continually collide with one another, so the motions of the pepper grains tend to settle down into a slow rotation representing the average of the original random velocities. Try the experiment several times, stirring the water differently each time. Do the random motions ever cancel out exactly, resulting in no rotation at all? Describe what happens and how this demonstration is similar to what took place in the solar nebula.

◆ How does our theory account for the features of planets, moons, and small bodies?

The nebular theory also accounts for the other three major features of the solar system. To see how, we must investigate what happened after the solar nebula took the shape of a spinning, flattened disk.

In the center of the disk, gravity drew together enough material to form the Sun. In the surrounding disk, however, the gaseous material was too spread out for gravity alone to clump it up. Instead, material had to begin clumping in some other way and to grow in size until gravity could start pulling it together into planets. In essence, planet formation required the presence of "seeds"—solid bits of matter around which gravity could ultimately build planets.

The basic process of seed formation was probably much like that of snowflake formation in clouds on Earth: When the temperature is low enough, some atoms or molecules in a gas may bond together and solidify. The general process in which solid (or liquid) particles form in a gas is called **condensation**—we say that the particles *condense* out of the gas. Different materials condense at different temperatures. As summarized in Table 4.3, the ingredients of the solar nebula fell into four major categories:

- *Hydrogen and helium gas (98% of the solar nebula).* These gases never condense under the conditions present in a nebula.
- *Hydrogen compounds (1.4% of the solar nebula).* Materials such as water (H_2O), methane (CH_4), and ammonia (NH_3) can solidify into **ices** at low temperatures (below about 150 K under the low pressure of the solar nebula).
- *Rock (0.4% of the solar nebula).* Rocky material is gaseous at high temperatures, but condenses into solid form at temperatures between about 500 K and 1300 K, depending on the type of rock.
- *Metal (0.2% of the solar nebula).* Metals such as iron, nickel, and aluminum are also gaseous at high temperatures, but condense into solid form at higher temperatures than rock—typically in the range of 1000 K to 1600 K.

Because hydrogen and helium gas made up about 98% of the solar nebula's mass and did not condense, the vast majority of the nebula remained gaseous at all times. However, other materials could condense wherever the temperature allowed (Figure 4.7). Close to the Sun, it was too hot for any material to condense. Farther out, it became cool enough for metal and then various types of rock to condense into tiny solid particles, but still too hot for hydrogen compounds to condense into ices. Ices could form only beyond the **frost line,** which lay between the present-day orbits of Mars and Jupiter. The frost line therefore marked the key transition between the warm inner region of the solar system, in which the solid seeds of planetary formation consisted only of metal and rock, and the cool outer region, in which the seeds also contained ice.

think about it Consider a region of the solar nebula in which the temperature was about 1300 K. What fraction of the material in this region was gaseous? What were the solid particles made of? Answer the same questions for a region with a temperature of 100 K. Would the 100 K region be closer to or farther from the Sun? Explain.

Explaining the Two Types of Planets The process by which small "seeds" grew into planets is called **accretion** (Figure 4.8). Accretion began with the microscopic solid particles that condensed from the gas of the solar nebula and followed the same orderly, circular orbits as the gas from which they condensed. Individual particles therefore moved at nearly the same speed as neighboring particles, so "collisions" were more like gentle touches. Although the particles were far too small to attract each other gravitationally at this point, they were able to move together through electrostatic forces—the same "static electricity" that makes hair stick to a comb. Small particles thereby began to combine into larger ones. As the particles grew in mass, they began to attract each other through gravity, accelerating their growth into boulders large enough to count as **planetesimals,** which means "pieces of planets."

In the inner solar system, where only metal and rock could condense into solid particles, the planetesimals ended up being made of metal and rock. These planetesimals grew rapidly at first, with some probably reaching hundreds of kilometers in size in only a few million years—a long time in human terms, but only about 1/1000 the present age of the solar system. At this point, gravitational encounters between planetesimals began to play an important role. These encounters tended to alter orbits, particularly those of the smaller planetesimals. With different orbits crossing each other, collisions occurred at higher speeds,

common misconceptions

Solar Gravity and the Density of Planets

Some people guess that it was the Sun's gravity that pulled the dense rocky and metallic materials to the inner part of the solar nebula, or that gases escaped from the inner nebula because gravity couldn't hold them. But this is not the case—all the ingredients were orbiting the Sun together under the influence of the Sun's gravity. The orbit of a particle or a planet does not depend on its size or density, so the Sun's gravity was not the cause of the different types of planets. Rather, the different temperatures in the solar nebula were the cause.

▼ **FIGURE 4.8**
These diagrams show how planetesimals made of metal and rock gradually accrete into terrestrial planets.

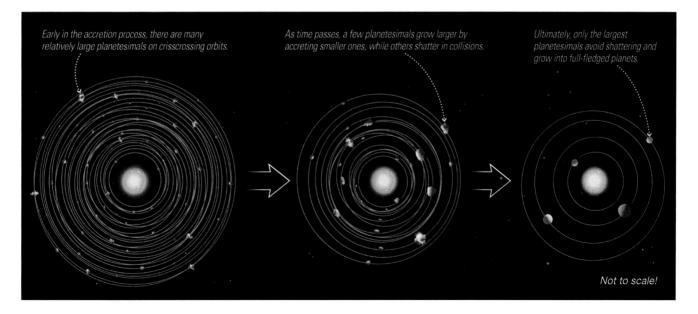

Early in the accretion process, there are many relatively large planetesimals on crisscrossing orbits.

As time passes, a few planetesimals grow larger by accreting smaller ones, while others shatter in collisions.

Ultimately, only the largest planetesimals avoid shattering and grow into full-fledged planets.

Not to scale!

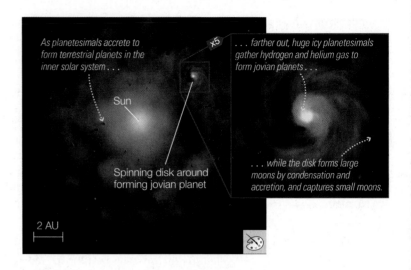

▶ FIGURE 4.9
The jovian planets grew as large icy planetesimals captured hydrogen and helium gas from the solar nebula and became surrounded by spinning disks of gas as they formed, much like the disk of the entire solar nebula but smaller in size. This painting shows the gas and planetesimals surrounding one jovian planet in the larger solar nebula.

often shattering the planetesimals. Only the largest planetesimals could avoid shattering and grow to become planets.

Beyond the frost line, lower temperatures allowed ices to condense along with metal and rock. Because ices were more abundant than rock and metal (see Table 4.3), a few of the icy planetesimals in the outer solar system were able to grow to masses many times that of Earth. With these large masses, their gravity became strong enough to capture and hold some of the hydrogen and helium gas that made up the vast majority of the surrounding solar nebula. The more they grew, the stronger their gravity became. This model explains the existence of the four jovian planets: Each began its formation with the accretion of icy planetesimals, but ultimately collected so much gas that it ended up with a composition rich in hydrogen and helium.

This model also explains most of the large moons of the jovian planets. The same processes of heating, spinning, and flattening that made the disk of the solar nebula should also have affected the gas drawn by gravity to the young jovian planets. Each jovian planet came to be surrounded by its own disk of gas, spinning in the same direction as the planet rotated (Figure 4.9). Moons that accreted from icy planetesimals within these disks ended up with nearly circular orbits going in the same direction as their planet's rotation and lying close to their planet's equatorial plane.

One key question remains: Given that the vast majority of the hydrogen and helium gas in the solar nebula never became part of any planet, what happened to it? Models and observations of other star systems suggest that it was cleared away by a combination of energetic light from the young Sun and the *solar wind*—a stream of charged particles continually blown outward in all directions from the Sun [Section 8.1]. The compositions of the planets have remained nearly the same since this gas cleared. If the gas had remained longer, it might have continued to cool until hydrogen compounds condensed into ices even in the inner solar system. In that case, the terrestrial planets might have accreted abundant ice, and perhaps some hydrogen and helium gas as well, changing their basic nature. At the other extreme, if the gas had been blown out much earlier, the raw materials of the planets might have been swept away before the planets could fully form. Although these extreme scenarios did not occur in our solar system, they may sometimes occur around other stars.

Explaining Asteroids and Comets The planetary formation process also explains the existence of so many asteroids and comets: They are simply "leftover" planetesimals from the era of planet formation. Asteroids are the rocky leftover planetesimals of the inner solar system, while comets are the icy leftover planetesimals of the outer solar system. (We'll discuss why they are found in three distinct regions in Chapter 6.)

The asteroids and comets that exist today probably represent only a small fraction of the leftover planetesimals that roamed the young solar system. Most of the rest must have collided with planets or moons. On worlds with solid surfaces, we see the evidence of past collisions as *impact craters*. Careful study of impact craters on the Moon shows that the vast majority of these collisions occurred in the first few hundred million years of our solar system's history, during the period we call the **heavy bombardment.** Every world in our solar system must have been pelted by impacts during the heavy bombardment (Figure 4.10), and most of the craters on the Moon and other worlds date from this period.

These early impacts, including many that occurred before the planets finished forming, probably played a key role in making our existence possible. The metal and rock planetesimals that built the terrestrial planets should not have contained much water or other hydrogen compounds, because it was too hot for these compounds to condense in our region of the solar nebula. How, then, did Earth come to have the water that makes up our oceans and the gases that first formed our atmosphere? The likely answer is that water, along with other hydrogen compounds, was brought to Earth and other terrestrial planets by the impacts of water-bearing planetesimals that formed farther from the Sun. Remarkably, the water we drink and the air we breathe probably once were part of planetesimals that accreted beyond the orbit of Mars.

Explaining the Exceptions We have seen how the nebular theory accounts for the first three major features of our solar system. For the fourth, the exceptions to the rules, the nebular theory suggests that most of them arose from collisions or close gravitational encounters.

Let's start with unusual moons. We have explained the orbits of most large jovian planet moons by their formation in a disk that swirled around the forming planet. But how do we explain moons with unusual orbits, such as those that go in the "wrong" direction (opposite their planet's rotation) or that have large inclinations to their planet's equator? These moons are probably leftover planetesimals that originally orbited the Sun but were then captured into planetary orbit.

It's not easy for a planet to capture a moon, because the law of gravity dictates that a small object passing a large planet will simply fly on by. An object can be captured only if it loses enough orbital energy to settle into a circular or elliptical orbit. For the jovian planets, this type of capture could have occurred when the planets were very young and still surrounded by extended clouds of relatively dense gas. A passing planetesimal would have been slowed by friction with this gas, and if it was slowed enough, it could have become an orbiting moon. Because of the random nature of the capture process, captured moons would not necessarily orbit in the same direction as their planet or in its equatorial plane. Computer models suggest that this capture process can explain most of the small moons with unusual orbits around the jovian planets. Mars may also have captured its two small moons, Phobos and Deimos, in a similar way at a time when it had a much more extended atmosphere than it does today (Figure 4.11).

Capture processes cannot explain our own Moon, however, because it is much too large to have been captured by a small planet like Earth. We can also rule out the possibility that our Moon formed simultaneously with Earth, because if both had formed together, they would have accreted from planetesimals of the same type and would therefore have approximately the

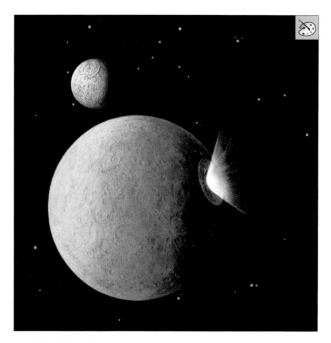

▲ **FIGURE 4.10**
Around 4 billion years ago, Earth, its Moon, and the other planets were heavily bombarded by leftover planetesimals. This painting shows the young Earth and Moon, with an impact in progress on Earth.

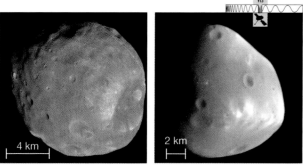

a Phobos   b Deimos

▲ **FIGURE 4.11**
The two moons of Mars may be captured asteroids. Phobos is only about 13 kilometers across and Deimos is only about 8 kilometers across—making each of these two moons small enough to fit within the boundaries of a typical large city. (Images from the *Mars Reconnaissance Orbiter*.)

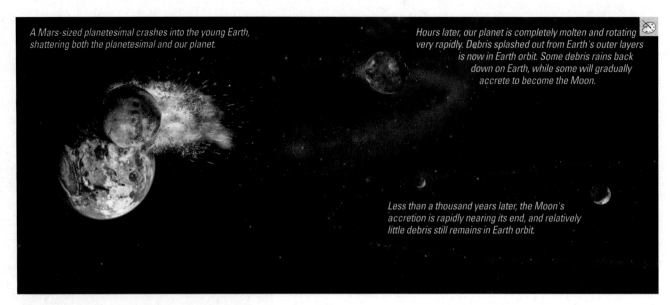

▲ **FIGURE 4.12**
Artist's conception of the giant impact hypothesis for the formation of our Moon. The fact that ejected material came mostly from Earth's outer rocky layers explains why the Moon contains very little metal. The impact must have occurred more than 4.4 billion years ago, since that is the age of the oldest Moon rocks. As shown, the Moon formed quite close to a rapidly rotating Earth, but over billions of years tidal forces have slowed Earth's rotation and moved the Moon's orbit outward.

same composition and density. But this is not the case: The Moon's density is considerably lower than Earth's, indicating that it has a very different average composition. So how did we get our Moon? The leading hypothesis suggests that it formed as the result of a **giant impact** between Earth and a huge planetesimal.

According to models, a few leftover planetesimals may have been as large as Mars. If one of these Mars-size objects struck a young planet, the blow might have tilted the planet's axis, changed the planet's rotation rate, or completely shattered the planet. The giant impact hypothesis holds that a Mars-size object hit Earth at a speed and angle that caused Earth's outer layers to be blasted into space. According to computer simulations, this material could have collected into orbit around our planet, and accretion within this ring of debris could have formed the Moon (Figure 4.12).

Strong support for the giant impact hypothesis comes from two features of the Moon's composition. First, the Moon's overall composition is quite similar to that of Earth's outer layers—just as we should expect if it were made from material blasted away from those layers. Second, the Moon has a much smaller proportion of easily vaporized ingredients (such as water) than Earth. This fact supports the hypothesis because the heat of the impact would have vaporized these ingredients. As gases, they would not have participated in the process of accretion that formed the Moon.

Giant impacts may also explain other exceptions. For example, Mercury's surprisingly high density may be the result of a giant impact that blasted away its outer, lower-density layers. Giant impacts might also have been responsible for tilting the axes of planets (including Earth), for tipping Uranus on its side, and perhaps for Venus's slow and backward rotation. Although we cannot go back in time to see exactly how a particular exception occurred, the main point is clear: The chaotic processes that accompanied planet formation, including the many collisions that surely occurred, are *expected* to have caused at least a few exceptions.

Summary We have found that the nebular theory successfully explains all the major features of our solar system. That is why scientists are confident that the theory captures the essence of what actually happened, even though we may not yet know all the details. Figure 4.13 summarizes the formation of our solar system according to the nebular theory.

FIGURE 4.13 A summary of the process by which our solar system formed, according to the nebular theory.

THE PROCESS OF SCIENCE IN ACTION

4.3 The Age of the Solar System

We've said that the solar system formed about $4\frac{1}{2}$ billion years ago. But since no one was around when the solar system was born, how can we possibly know how old it is? The answer is that careful observations and experiments enable scientists to measure the ages of things much older than we are. For example, we can measure ages of trees up to a few thousand years old by counting tree rings, and we can study Earth's climate over a period of a few hundred thousand years by drilling deep into the ice sheets of Greenland and Antarctica, where we see distinct layers laid down each year. To study things that are millions or billions of years old, we must determine the ages of rocks. The technique scientists use to measure these enormous ages is the focus of this chapter's case study in the process of science in action.

◆ **How do we determine the age of Earth and the solar system?**

The most reliable method for measuring the age of a rock is **radiometric dating**, which is based on careful measurement of the proportions of various atoms in the rock. The method works because some atoms undergo changes with time that allow us to determine how long they have been held in place within the rock's solid structure.

Isotopes and Radioactive Decay You may recall from high school that each chemical element is uniquely characterized by the number of protons in its nucleus. Different **isotopes** of the same element differ only in their number of neutrons. For example, any atom with 6 protons in its nucleus is an atom of carbon, but carbon atoms come in three different isotopes (Figure 4.14): carbon-12, which has 6 neutrons in addition to the 6 protons; carbon-13, which has 7 neutrons; and carbon-14, which has 8 neutrons.

Most of the atoms and isotopes we encounter in daily life are stable, meaning that their nuclei stay the same at all times. For example, most of the carbon in our bodies is carbon-12, which is stable. But some isotopes are unstable, meaning that their nuclei are prone to spontaneous change, or *decay*, such as breaking apart or having a proton turn into a neutron. These unstable nuclei are said to be **radioactive**. Carbon-14 is an example of a radioactive isotope, because it undergoes spontaneous change that turns it into nitrogen-14.

Radioactive decay always occurs at the same rate for any particular radioactive isotope, and scientists can measure these rates in the laboratory. We generally characterize the decay rate of an element by stating its **half-life**—the length of time it would take for half its nuclei to decay. Note that it takes only a few months to years of laboratory measurements to pin down the decay rate and determine the half-life, even if the half-life is billions of years. The half-life of carbon-14 is about 5700 years, which makes it useful to archaeologists trying to determine the dates of ancient human settlements, but which is far too short for dating rocks that are millions or billions of years old. However, many isotopes have much longer half-lives.

Isotopes of Carbon

carbon-12 carbon-13 carbon-14

^{12}C ^{13}C ^{14}C
(6 protons + 6 neutrons)   (6 protons + 7 neutrons)   (6 protons + 8 neutrons)

▲ **FIGURE 4.14**
Different isotopes of a chemical element contain the same number of protons, but different numbers of neutrons.

▶ **Radiometric Dating**

Radiometric Dating To see how dating of rocks works, we'll consider the radioactive isotope potassium-40, which decays into argon-40 with a half-life of 1.25 billion years. (Potassium-40 also decays by other paths, but we focus only on decay into argon-40 to keep the discussion simple.) Imagine a small piece of rock that contained 1 microgram of potassium-40 and no argon-40 when it formed (solidified) long ago. The half-life of 1.25 billion years means that half the original potassium-40 would have decayed into argon-40 by the time the rock was 1.25 billion years old, so at that time the rock would contain $\frac{1}{2}$ microgram of potassium-40 and $\frac{1}{2}$ microgram of argon-40. Half of this remaining potassium-40 would then have decayed by the end of the next 1.25 billion years, so after 2.5 billion years the rock would contain $\frac{1}{4}$ microgram of potassium-40 and $\frac{3}{4}$ microgram of argon-40. After three half-lives, or 3.75 billion years, only $\frac{1}{8}$ microgram of potassium-40 would remain, while $\frac{7}{8}$ microgram would have become argon-40. Figure 4.15 summarizes the gradual decrease in the amount of potassium-40 and corresponding rise in the amount of argon-40.

You can now see the essence of radiometric dating. Suppose you find a rock that contains equal numbers of atoms of potassium-40 and argon-40. If you assume that all the argon came from potassium decay (and if the rock shows no evidence of subsequent heating that could have allowed any argon to escape), then it must have taken precisely one half-life for the rock to end up with equal amounts of the two isotopes. You can therefore conclude that the rock is 1.25 billion years old. The only question is whether you are right in assuming that the rock lacked argon-40 when it formed. In this case, knowing a bit of "rock chemistry" helps. Potassium-40 is a natural ingredient of many minerals in rocks, but argon-40 is a gas that does not combine with other elements and did not condense in the solar nebula. If you find argon-40 gas trapped inside minerals, it must have come from radioactive decay of potassium-40.

Radiometric dating is possible with many other radioactive isotopes as well. In many cases, a rock contains more than one radioactive isotope, and agreement between the ages calculated from the different isotopes gives us confidence that we have dated the rock correctly. We can also check results from radiometric dating against ages measured or estimated with other methods. For example, some archaeological artifacts have original dates printed on them, and these dates agree with ages found by radiometric dating. We can validate the $4\frac{1}{2}$-billion-year radiometric age for the solar system as a whole by comparing it to an age based on detailed study of the Sun. Theoretical models of the Sun, along with observations of other stars, show that stars slowly expand and brighten as they age. The model ages are not nearly as precise as radiometric ages, but they confirm that the Sun is between about 4 and 5 billion years old. Overall, the technique of radiometric dating has been checked in so many ways and relies on such basic scientific principles that there is no serious scientific debate about its validity.

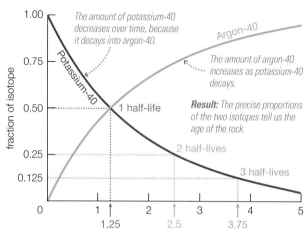

▲ **FIGURE 4.15**
Potassium-40 is radioactive, decaying into argon-40 with a half-life of 1.25 billion years. The red curve shows the decreasing amount of potassium-40, and the blue curve shows the increasing amount of argon-40. The amount of potassium-40 remaining drops in half with each successive half-life.

Ages of Earth Rocks, Moon Rocks, and Meteorites Radiometric dating allows us to determine the length of time that has passed since the atoms in a rock became locked together in their present arrangement, which in most cases means the time *since the rock last solidified*. The ages of rocks therefore vary greatly. Some Earth rocks are quite young because they formed recently from molten lava, and other Earth rocks have a great variety of ages because they melted and then re-solidified at different times in Earth's history. The oldest Earth rocks are over 4 billion years old, and some small mineral grains date to nearly 4.4 billion years ago. Earth as a whole must be even older than these oldest rocks and minerals.

Moon rocks brought back by the *Apollo* astronauts date to as far back as 4.4 billion years ago. Although they are older than Earth rocks, these Moon rocks must still be younger than the Moon itself. The ages of these rocks also tell us that the giant impact thought to have created the Moon must have occurred more than 4.4 billion years ago.

To go all the way back to the origin of the solar system, we must find rocks that have not melted or vaporized since they first condensed in the solar nebula. Meteorites that have fallen to Earth are our source of such rocks. Many meteorites appear to have remained unchanged since they condensed and accreted in the early solar system. Careful analysis of radioactive isotopes in meteorites shows that the oldest ones formed about 4.56 billion years ago, so this time must mark the beginning of accretion in the solar nebula. Because the planets accreted within a few tens of millions of years, Earth and the other planets must have formed by about 4.5 billion years ago.

The Bottom Line In science, we often want to know the ages of things that are far older than we are. Scientists have therefore developed techniques that in many cases can date extremely old objects with amazing precision. Radiometric dating is one of the most important techniques, and it is also one of the best tested and verified. As a result, we have great confidence in ages established by radiometric dating, including the age of Earth and our solar system.

summary of key concepts

4.1 Characteristics of the Solar System

♦ **What does the solar system look like?**

Our solar system consists of the Sun, the planets and their moons, and vast numbers of asteroids and comets. The planets are tiny compared to the distances between them. Each world has its own unique character, but there are many clear patterns among the worlds.

♦ **What features of our solar system provide clues to how it formed?**

Four major features provide clues: (1) The Sun, planets, and large moons generally rotate and orbit in a very organized way. (2) The planets divide clearly into two groups: **terrestrial** and **jovian**. (3) The solar system contains huge numbers of asteroids and comets. (4) There are some notable exceptions to these general patterns.

4.2 The Birth of the Solar System

♦ **What theory best explains the orderly patterns of motion in our solar system?**

The **nebular theory** holds that the solar system formed from the gravitational collapse of a great cloud of gas and dust. As the **solar nebula** shrank in size, it got hotter, spun faster, and flattened out. The orderly motions we observe today came from the orderly motion of this spinning disk.

♦ **How does the nebular theory account for the features of planets, moons, and small bodies?**

Planets formed around solid "seeds" of matter that condensed from gas and then grew through **accretion** into larger **planetesimals**. Terrestrial planets formed in the inner solar system, where high temperatures allowed only metal and rock to condense. Beyond the **frost line**, cold temperatures allowed more abundant ices to condense along with metal and rock, allowing a few planetesimals to grow large enough for their gravity to draw in hydrogen and helium gas, building massive jovian planets. **Asteroids** are the rocky leftover planetesimals of the inner solar system, and **comets** are the icy leftover planetesimals of the outer solar system. Most exceptions to the general trends probably arose from collisions or close encounters with leftover planetesimals, and our Moon probably resulted from a **giant impact** between a Mars-size planetesimal and the young Earth.

THE PROCESS OF SCIENCE IN ACTION

4.3 The Age of the Solar System

♦ **How do we determine the age of Earth and the solar system?**

Radiometric dating allows us to determine the age of a rock by carefully measuring its proportions of a radioactive isotope and the decay product of that isotope. We determine the age of Earth and the solar system by measuring the ages of the oldest meteorites, which are about 4.56 billion years old.

Investigations

For instructor-assigned homework and other learning materials, go to www.MasteringAstronomy.com.

Quick Quiz

Choose the best answer to each of the following. Explain your reasoning.

1. How many of the planets orbit the Sun in the same direction as Earth does? (a) a few (b) most (c) all
2. Planetary orbits are (a) very eccentric (stretched-out) ellipses and in the same plane. (b) fairly circular and in the same plane. (c) fairly circular but oriented in random directions.
3. Which have more moons on average? (a) jovian planets (b) terrestrial planets (c) Terrestrial and jovian planets both have about the same number of moons.
4. Which of the following three kinds of object resides closest to the Sun on average? (a) comets (b) asteroids (c) jovian planets
5. The solar nebula was 98% (a) rock and metal. (b) hydrogen compounds. (c) hydrogen and helium.
6. Which of the following did *not* occur during the collapse of the solar nebula? (a) spinning faster (b) heating up (c) concentrating denser materials nearer the Sun
7. The *frost line* of the solar nebula refers to the location beyond which (a) only metal and rock could condense into solid form. (b) only ice could condense into solid form. (c) metal, rock, and ice could condense into solid form.
8. Which lists the major steps of solar system formation in the correct order? (a) collapse, accretion, condensation (b) collapse, condensation, accretion (c) accretion, condensation, collapse
9. Leftover ice-rich planetesimals are called (a) comets. (b) asteroids. (c) meteors.
10. Which of the following discoveries provides strong evidence in favor of the nebular theory? (a) Planets are common around other stars. (b) The universe is expanding. (c) Mars once had conditions that may have allowed for life.
11. The leading hypothesis for the origin of the Moon is that it (a) formed along with Earth. (b) formed from material ejected from Earth in a giant impact. (c) split out of a rapidly rotating Earth.
12. About how old is the solar system? (a) 4.5 million years (b) 4.5 billion years (c) 4.5 trillion years

Short-Answer/Essay Questions

Explain all answers clearly, with complete sentences and proper essay structure, if needed. An asterisk (*) designates a quantitative problem, for which you should show all your work.

13. *Planetary Tour.* Based on the brief planetary tour in this chapter, which planet besides Earth do you think is the most interesting, and why? Defend your opinion.
14. *Patterns of Motion.* In one or two paragraphs, summarize the orderly patterns of motion in our solar system and explain why they suggest that the Sun and the planets all formed at one time from one cloud of gas, rather than as individual objects at different times.
15. *Solar System Trends.* Answer the following based on the data in Table 4.1.
    a. Describe the relationship between distance from the Sun and surface temperature. Explain the reason behind the general trend, as well as any notable exceptions to the trend.
    b. Describe in general how the columns for density, composition, and distance from the Sun support the classification of planets into the terrestrial and jovian categories, with Pluto fitting neither category.
    c. Describe the trend in orbital periods and explain it in terms of Kepler's third law.
    d. Which data column would you use to find out which planet has the shortest days? Discuss any notable differences in day lengths between the terrestrial and jovian planets.
    e. Which planets would you expect to have seasons? Explain.
16. *Two Kinds of Planets.* In words a friend would understand, explain why the jovian planets differ from the terrestrial planets in each of the following aspects: composition, size, density, distance from the Sun, and number of moons.
17. *Explaining the Past.* Test the nebular theory against each of the three hallmarks of science discussed in Chapter 3. Be as detailed as possible in explaining whether the theory does or does not satisfy these hallmarks. Use your findings to discuss whether it really is possible for science to inform us about how our solar system formed. Defend your opinion.
18. *An Early Solar Wind.* Suppose the solar wind had cleared away the solar nebula before the seeds of the jovian planets could gravitationally draw in hydrogen and helium gas. How would the planets of the outer solar system be different? Would they still have many moons? Explain your answer in a few sentences.
19. *Understanding Radiometric Dating.* Imagine you had the good fortune to find a rocky meteorite in your backyard. How would you expect its ratio of potassium-40 to argon-40 to be different from that of other rocks in your yard? Explain your answer in a few sentences.
20. *Lucky to Be Here?* Considering the overall process of solar system formation, do you think formation of a planet like Earth was likely? Could random events in the early history of the solar system have prevented our being here today? What implications do your answers have for the possibility of finding planets with Earth-like conditions around other stars? Defend your opinions.
*21. *Size Comparisons.* How many Earths could fit inside Jupiter (assuming you could fill up its total volume)? How many Jupiters could fit inside the Sun? (*Hint:* The equation for the volume of a sphere is $V = \frac{4}{3}\pi r^3$.)
*22. *Radiometric Dating.* For each of the following, assume that all the argon-40 comes from radioactive decay of potassium-40, a process with a half-life of 1.25 billion years.
    a. You find a rock that contains equal amounts of potassium-40 and argon-40. How old is it? Explain.
    b. You find a rock that contains three times as much argon-40 as potassium-40. How old is it? Explain.
*23. *What Are the Odds?* The fact that all the planets orbit the Sun in the same direction is cited as support for the nebular hypothesis. Imagine that there's a different hypothesis under which planets can be created orbiting the Sun in either direction. Under this hypothesis, what is the probability that eight planets would end up traveling in the same direction? (*Hint:* It's the same as the probability of flipping a coin eight times and getting all heads.)

5 Terrestrial Worlds

LEARNING GOALS

5.1 Terrestrial Surfaces and Atmospheres
- What determines a world's level of geological activity?
- How does an atmosphere affect conditions for life?

5.2 Histories of the Terrestrial Worlds
- Why did the terrestrial worlds turn out so differently?
- What unique features of Earth are important for life?

THE PROCESS OF SCIENCE IN ACTION

5.3 Global Warming
- What is the evidence for global warming?

The photo at right shows a self-portrait (assembled from dozens of individual images) of NASA's *Curiosity* rover on the surface of Mars. The surface looks almost Earth-like, but the air lacks oxygen and is so thin that you could not survive more than a few minutes outside without a pressurized space suit. Moreover, there's no liquid water anywhere on the surface, the temperature is usually well below freezing, and the lack of an ozone layer would expose you to dangerous ultraviolet light from the Sun. Venus offers an opposite extreme, with a thick atmosphere that makes the surface hotter than a pizza oven. How did it come to be that Earth is so hospitable to life, while our planetary neighbors are so different? In this chapter, we'll investigate the nature of the terrestrial worlds as we seek to understand what makes our planet so suited to diverse life.

5.1 Terrestrial Surfaces and Atmospheres

We surveyed the four terrestrial planets (Mercury, Venus, Earth, and Mars) briefly in Section 4.1. Because our Moon is nearly three-fourths the size of Mercury in diameter, we often think of it as a fifth terrestrial world. Although the five terrestrial worlds look quite different today, they all formed from the accretion of rocky planetesimals, and all have been shaped by the same basic geological and atmospheric processes. In this section, we'll explore these processes, focusing first on geology and then on atmospheres.

▶ **Planetary Geology**

◆ **What determines a world's level of geological activity?**

A quick comparison of the terrestrial worlds reveals that **geological activity**—the degree to which a planet's surface continues to change—depends primarily on a world's size. Earth is the largest and most geologically active of the terrestrial worlds, with a surface continually reshaped by volcanic eruptions, earthquakes and other movements of the crust, and erosion. Venus, the next largest, is probably also quite active, though we know relatively little about its geological activity because its thick atmosphere allows us to "see" the surface only with the aid of radar. Mars, middling in size, appears to have had extensive geological activity in its past, but is far less active today. Mercury and the Moon, the two smallest, undergo so little surface change today that we generally think of them as being geologically dead.

Size, Internal Heat, and Interior Structure Size matters because most geological activity is driven by internal heat. For example, volcanoes erupt when internal heat melts rock and releases gases that drive the molten rock upward. Earthquakes occur when heat-driven internal motions cause rock to slip or shift. All the terrestrial worlds were quite hot when they were young, because accretion of planetesimals converted gravitational potential energy into thermal energy. Decay of radioactive materials inside the terrestrial worlds provided additional heat (and still provides some heating today). Since their youths, the terrestrial worlds have gradually lost heat as it escapes from their surfaces into space. Just as a hot pea cools much more quickly than a hot potato, smaller worlds cool more quickly than larger ones. Mercury and the Moon cooled so quickly that they probably lost the heat needed to power volcanoes before they were 2 billion years old. Mars still retains enough internal heat to drive some geological activity, but much less than it had long ago. Venus and Earth are cooling so slowly that radioactive decay still keeps their interiors warm, which is why they continue to have strong geological activity.

see it for yourself It's easy to demonstrate that larger objects take longer to cool than smaller objects. The next time you eat something large and hot, cut off a small piece; notice how much more quickly the small piece cools than the rest of it. A similar experiment demonstrates the time it takes a cold object to warm up: Find two ice cubes of the same size, crack one into small pieces, and then compare the rates of melting. Explain your observations.

Internal heat also determines what terrestrial worlds are like on the inside. All the terrestrial worlds were hot enough to have had molten interiors when they were very young, allowing materials to separate by density in the process

▲ **FIGURE 5.1**
Interior structures of the terrestrial worlds, shown to scale and in order of decreasing size. The thicknesses of the crust and lithosphere on Venus and Earth are exaggerated to make them visible in this figure.

▲ **FIGURE 5.2**
Earth's hot interior allows the mantle to undergo convection. Arrows indicate the direction of flow in a portion of the mantle.

common misconceptions

Earth Is Not Full of Molten Lava

Many people guess that Earth is full of molten lava (or *magma*). This misconception may arise because we see molten lava emerging from inside Earth when a volcano erupts. However, Earth's mantle and crust are almost entirely solid. The lava that erupts from volcanoes comes only from a narrow region of partially molten material beneath the lithosphere. The only part of Earth's interior that is fully molten is the outer core, which lies so deep within the planet that its material never erupts directly to the surface.

called **differentiation** (because it results in layers made of *different* materials). Differentiation led to three major layers within each world:

- **Core.** The highest-density material, consisting primarily of metals such as nickel and iron, sank to the central core.
- **Mantle.** Rocky material of moderate density—mostly minerals that contain silicon, oxygen, and other elements—formed the thick mantle that surrounds the core.
- **Crust.** The lowest-density rock, such as granite and basalt, formed the thin crust that serves as a world's outer skin.

Figure 5.1 shows these layers for the five terrestrial worlds, as determined through a combination of seismic studies on Earth and modeling of the other worlds. Notice that each world also has a region called a **lithosphere** that extends downward from the surface, and it is thicker for the smaller worlds. The lithosphere encompasses the crust and part of the mantle of each world and represents the region in which rock is relatively cool and rigid. Beneath the lithosphere, warmer temperatures make the rock softer, allowing it to flow slowly over millions and billions of years. The rigid rock of the lithosphere essentially "floats" upon the warmer, softer rock below.

Interior heat drives geological activity by supplying the energy needed to melt or move rock and reshape the surface. If the deep interior of a terrestrial world is hot enough, hot rock can gradually rise within its mantle, while cooler rock at the top of the mantle gradually falls (Figure 5.2). The process by which hot material expands and rises while cooler material contracts and falls is called **convection.** Keep in mind that mantle convection primarily involves *solid* rock, not molten rock. Because solid rock flows quite slowly, mantle convection is a very slow process. At the typical rate of mantle convection on Earth—a few centimeters per year—it would take 100 million years for a piece of rock to be carried from the base of the mantle to the top.

Interior heat plays another important role: It can help create a global **magnetic field** because convection in the core can generate electrical currents. Among the terrestrial planets, only Earth has a strong magnetic field, which is important to life because it helps protect our planet from charged particles flowing outward from the Sun. These particles could strip away atmospheric gas and cause genetic damage to living organisms, but Earth's magnetic field deflects most of them (Figure 5.3).

▶ **Four Geological Processes**

The Four Geological Processes Now that you understand how internal heat drives geological activity, we are ready to examine the specific processes that

◀ **FIGURE 5.3**
This diagram shows how Earth's magnetosphere deflects solar wind particles, with some accumulating in charged particle belts encircling our planet. Charged particles can spiral into Earth's atmosphere in polar regions, causing the aurora.

▲ **FIGURE 5.4**
Impact cratering. Meteor Crater in Arizona is more than a kilometer across and almost 200 meters deep. It was created about 50,000 years ago by the impact of a metallic asteroid about 50 meters across.

shape planetary surfaces. The terrestrial worlds show a huge variety of geological surface features, but almost all of them can be explained by just four major geological processes:

- **Impact cratering:** the creation of bowl-shaped *impact craters* by asteroids or comets striking a planet's surface
- **Volcanism:** the eruption of molten rock, or *lava*, from a planet's interior onto its surface
- **Tectonics:** the disruption of a planet's surface by internal stresses
- **Erosion:** the wearing down or building up of geological features by wind, water, ice, and other phenomena of planetary weather

Let's examine each of these processes in a little more detail. *Impact cratering* is the only process with an external cause: impacts of objects from space. These objects typically hit the surface at a speed between about 40,000 and 250,000 kilometers per hour, releasing enough energy to vaporize solid rock and blast out a large crater (Figure 5.4). As we discussed in Chapter 4, impacts must have battered all the terrestrial worlds during the heavy bombardment, with relatively few impacts occurring after that time. Why, then, do small worlds like the Moon and Mercury have so many impact craters today, while large ones like Venus and Earth have very few? The answer is that geological activity such as volcanic eruptions and erosion gradually erases the impact craters on larger worlds. *We can therefore estimate the age of a surface region from the number of impact craters, with more craters indicating an older surface.* For example, the heavily cratered regions of the Moon must be virtually unchanged since the end of the heavy bombardment some 4 billion years ago, moderately cratered regions of Mars's surface are 2–3 billion years old, and young regions of Earth's surface show only the relative handful of impacts that have occurred during the past few million years.

The second process is *volcanism*, which occurs when underground molten rock finds a path through the lithosphere to the surface (Figure 5.5). Volcanism therefore requires substantial internal heat. Eruptions can create tall, steep volcanoes if the molten rock is thick, or flat lava plains if the molten rock is runny. In addition, volcanic eruptions release gases that were trapped beneath the surface when the planet formed. This process, called **outgassing**, is thought to have been responsible for creating the atmospheres of Venus, Earth, and Mars. Measurements of erupting volcanoes on Earth indicate that the most common gases released by outgassing are water vapor (H_2O), carbon dioxide (CO_2),

▲ **FIGURE 5.5**
Volcanism. This photo shows the eruption of an active volcano on the flanks of Kilauea on the Big Island in Hawaii. The inset shows the underlying process: Molten rock collects in a *magma chamber* and can erupt upward.

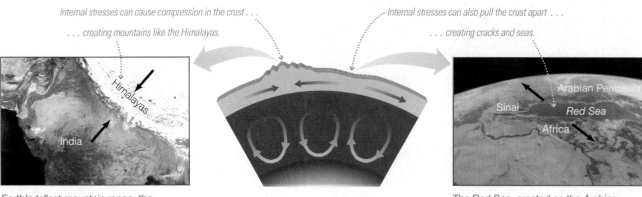

▲ FIGURE 5.6
Tectonics. Tectonic forces can produce a wide variety of features. Mountains created by tectonic compression and valleys or seas created by tectonic stretching are among the most common. Both images are satellite photos.

nitrogen (N_2), and sulfur-bearing gases (H_2S or SO_2). Because Venus, Earth, and Mars have similar compositions, all three planets probably released similar proportions of these gases. As we'll discuss later, the three atmospheres differ today because of differences in the ways that various gases have been absorbed by the surfaces or have escaped into space. Outgassing also occurred on Mercury and the Moon, but gravity was too weak to retain the gases on these small worlds, which is why they have no significant atmospheres today.

The third process, *tectonics*, occurs when stretching, compression, or other forces acting on the lithosphere reshape a world's surface. Figure 5.6 shows two examples of tectonic features on Earth, one created by surface compression (the Himalayas) and one created by surface stretching (the Red Sea). Tectonic activity usually goes hand in hand with volcanism, because both require internal heat, and therefore tectonics occurs today only on larger terrestrial worlds. Much of the tectonic activity on any planet is a direct or indirect result of mantle convection. Tectonics is particularly important on Earth, because the underlying mantle convection fractured Earth's lithosphere into more than a dozen pieces, or *plates*. These plates move over, under, and around each other, leading to a special kind of tectonics that we call **plate tectonics**. While tectonics has affected every terrestrial world, plate tectonics appears to be unique to Earth.

The last of the four major geological processes is *erosion*, which refers to the breakdown or transport of materials through the action of ice, liquid, or gas. The shaping of valleys by glaciers (ice), the carving of canyons by rivers (liquid), and the shifting of sand dunes by wind (gas) are all examples. Erosion plays a far more important role on Earth than on any other terrestrial world, primarily because our planet has both strong winds and plenty of liquid water. Indeed, much of the surface rock on Earth was built by erosion. Over long periods of time, erosion has piled sediments into layers on the floors of oceans and seas, forming **sedimentary rock** (Figure 5.7). Erosion can occur only on worlds with atmospheres, so it does not affect Mercury or the Moon. We see evidence of past water erosion on Mars, but today Mars has only modest wind erosion. Venus has a thick atmosphere but relatively little erosion because it lacks liquid water and has very little wind.

Planetary Atmospheres

◆ How does an atmosphere affect conditions for life?

Venus, Earth, and Mars all have atmospheres that profoundly affect their surfaces. On Earth, the atmosphere creates the conditions that make our existence possible. The atmosphere provides the air we breathe, and its pressure and temperature allow liquid water to rain down and flow on Earth's surface. The atmosphere

▲ FIGURE 5.7
Erosion. The walls of the Grand Canyon consist of layers of sedimentary rock built up by erosion over hundreds of millions of years. These walls are exposed because the Colorado River has carved the deep canyon over the past few million years.

tools of science | Basic Properties of Light

As discussed in this chapter, interactions of light and matter explain much about planetary surfaces. In addition, we learn about planets and virtually all other astronomical objects by studying the light we receive from them. It is therefore important to understand basic properties of light.

Light is a form of energy that travels through space at a speed of about 300,000 kilometers per second. Light is characterized by rapidly changing electric and magnetic fields, which is why we often call light an *electromagnetic wave* (Figure 1). However, light also comes in distinct pieces, called **photons**, each of which has a **wavelength** (the distance between adjacent peaks of the electric or magnetic field) and a **frequency** (the rate at which the electric and magnetic fields change). A simple formula relates the wavelength and frequency of a photon:

$$\text{wavelength} \times \text{frequency} = \text{speed of light}$$

We therefore find that *longer wavelength means lower frequency* and *shorter wavelength means higher frequency*. The energy of a photon is proportional to its frequency, so higher frequency light has higher energy.

Light can have any wavelength, and we refer to the full range of possibilities as the **electromagnetic spectrum** (Figure 2). Note that visible light has only a small range of wavelengths (from about 400 to 700 nanometers). Light with slightly longer wavelength than visible light is called **infrared** (because it is beyond the red end of the visible-light spectrum); **radio waves** represent light with even longer wavelength. Going in the other direction, **ultraviolet light** (beyond the violet end of the visible-light spectrum) has slightly shorter wavelength than visible light, followed by the even shorter wavelengths of **x-rays** and **gamma rays**.

Light can be produced in many ways, but for the moment we'll focus on light emitted by objects such as planets and stars. This light has a spectrum that depends only on the object's surface temperature, so it is called **thermal radiation**. Figure 3 shows thermal radiation spectra for objects of different temperature. These spectra obey two laws:

1. Hotter objects emit more light per unit surface area at all wavelengths. For example, a 15,000 K star emits a lot more light per unit area at every wavelength than a 3000 K star.

2. Hotter objects emit photons with a higher average energy, which means a shorter average wavelength. For example, the peak wavelength of emission for the 15,000 K star is much shorter than that for the 3000 K star.

Notice that a star like the Sun emits more strongly in visible light than at any other wavelength, while a typical planet emits infrared light but no visible light at all. This fact is crucial to understanding the greenhouse effect and planetary atmospheres, as we'll discuss in this chapter.

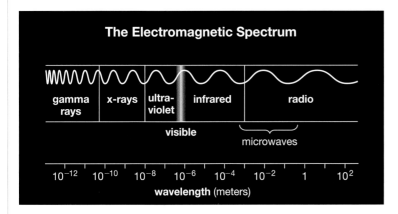

▲ FIGURE 1
Light is an electromagnetic wave, but also comes in individual pieces called photons, each characterized by a wavelength and frequency.

▲ FIGURE 2
The electromagnetic spectrum.

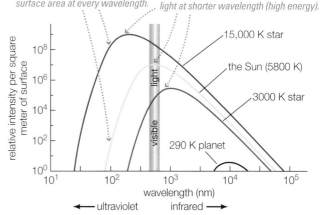

▲ FIGURE 3
Idealized thermal radiation spectra for objects of different temperature.

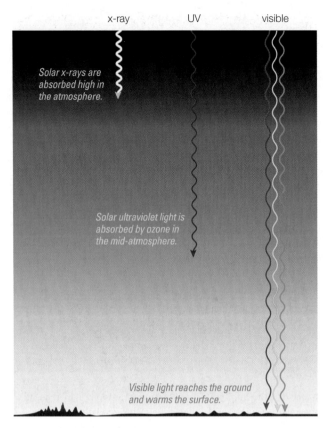

▲ FIGURE 5.8
This diagram summarizes how different forms of light from the Sun are affected by Earth's atmosphere.

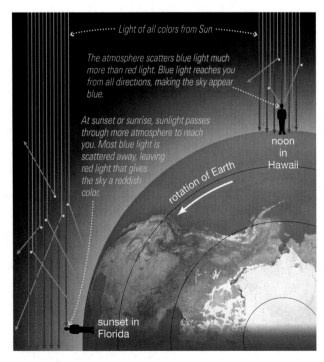

▲ FIGURE 5.9
This diagram summarizes why the sky is blue and sunsets (and sunrises) are red.

is also crucial to life in less obvious ways. Without the atmosphere, dangerous solar radiation would render Earth's surface lifeless, and it would be so cold that most of the water would be perpetually frozen.

Remarkably, our atmosphere plays all these roles despite being very thin. About two-thirds of the air in Earth's atmosphere lies within 10 kilometers of the surface. You could represent this air on a standard globe with a layer only as thick as a sheet of paper. Let's examine how interactions between light and the atmosphere protect and warm Earth's surface. We will then be prepared to understand how atmospheres affect the other terrestrial worlds.

Surface Protection The Sun emits visible light that allows us to see, but it also emits dangerous ultraviolet and x-ray radiation. In space, astronauts need thick space suits to protect them from the hazards of this radiation. On Earth, the atmosphere protects us (Figure 5.8).

X-rays carry enough energy to *ionize*, or knock electrons free from, almost any atom or molecule. That is why they can damage living tissue. Fortunately, atoms and molecules in Earth's atmosphere absorb x-rays so easily that none reach the ground. Ultraviolet light is not so easily absorbed. Most gases are transparent to ultraviolet light, allowing it to pass through unhindered. We owe our protection from ultraviolet light to a relatively rare gas called **ozone** (O_3). Ozone resides primarily in a middle layer of Earth's atmosphere (the *stratosphere*), where it absorbs most of the dangerous ultraviolet radiation from the Sun.

Visible light from the Sun passes easily through Earth's atmosphere and thereby provides light and heat to Earth's surface. However, not all visible-light photons pass straight through Earth's atmosphere. Some are scattered randomly around the sky, which is why the daytime sky is bright; without scattering, the Sun would be just a very bright disk set against a black, starry sky. Scattering also explains why the daytime sky is blue (Figure 5.9). Sunlight consists of all the colors of the rainbow, but the colors are not all scattered equally. Gas molecules scatter blue light much more effectively than red light. When the Sun is overhead, this scattered blue light reaches our eyes from all directions and the sky appears blue. At sunset or sunrise, the sunlight must pass through a greater amount of atmosphere on its way to us. Most of the blue light is scattered away, leaving only red light to color the sky.

The atmospheres of Venus and Mars similarly absorb x-rays from the Sun, but they lack ozone or other gases to absorb ultraviolet light. Both visible and ultraviolet light can therefore penetrate their atmospheres, though Venus's clouds are so thick that they scatter most of this light away before it reaches the ground.

The Greenhouse Effect Visible light warms Earth's surface, but not as much as you might guess. Calculations based on Earth's distance from the Sun and the percentages of visible light absorbed and reflected show that, by itself, visible light would heat Earth to an average surface temperature of only $-16°C$ ($+3°F$) —well below the freezing point of water. Earth's actual global average temperature is about $15°C$ ($59°F$), plenty warm enough for liquid water to flow and life to thrive. Why is Earth so much warmer than it would be from visible-light warming alone? The answer is that our atmosphere traps additional heat through what we call the **greenhouse effect**.

Figure 5.10 shows the basic idea behind the greenhouse effect. Some of the visible light that reaches the ground is reflected and some is absorbed. The absorbed energy must be returned to space—otherwise the ground would rapidly heat up—but planetary surfaces are too cool to emit visible light. Instead, planets emit energy primarily in the form of infrared light.

The greenhouse effect occurs when an atmosphere contains gases that absorb infrared light. Such gases are called **greenhouse gases**, and they include water vapor (H_2O), carbon dioxide (CO_2), and methane (CH_4). These gases absorb infrared light effectively because their molecular structures make them prone to begin rotating or vibrating when struck by an infrared photon. After absorbing the photon, a greenhouse gas molecule reemits a similar photon in some random direction. This photon can then be absorbed by another greenhouse gas molecule, which does the same thing. The net result is that greenhouse gases tend to slow the escape of infrared radiation from the lower atmosphere, while their molecular motions heat the surrounding air. In this way, the greenhouse effect makes the surface and the lower atmosphere warmer than they would be from sunlight alone. The more greenhouse gases present, the more the surface will be warmed.

think about it Molecules that consist of two atoms of the same type—such as the N_2 and O_2 that make up 98% of Earth's atmosphere—are poor infrared absorbers. But imagine if this were not the case: How would Earth be different if nitrogen and oxygen absorbed infrared light effectively?

The greenhouse effect also warms both Venus and Mars, because both have atmospheres made primarily of carbon dioxide. Venus's thick atmosphere creates an extreme greenhouse effect that raises its surface from a temperature that would otherwise be well below freezing to an astounding 470°C. In contrast, Mars's thin atmosphere creates a weak greenhouse effect, leaving Mars with an average surface temperature well below freezing.

5.2 Histories of the Terrestrial Worlds

Given that the same basic geological and atmospheric processes can occur on any terrestrial world, how did the terrestrial worlds end up so different from one another? We can answer this question by briefly studying the histories of the terrestrial worlds.

◆ Why did the terrestrial worlds turn out so differently?

We can trace most of the differences among the terrestrial worlds to differences either in size or in distance from the Sun. Size has the greatest influence on geology, so we'll work our way up from the two smallest worlds.

The Moon and Mercury: Active No More The Moon lacks appreciable geological activity today, which is why much of its surface still bears many impact craters from the period of heavy bombardment that ended nearly 4 billion years ago. However, parts of the lunar surface are not as heavily cratered, indicating that the Moon experienced some geological activity after the heavy bombardment ended. You can see these smoother, less cratered regions with the naked eye (Figure 5.11). They are called **lunar maria**, and they got their name because they look much like oceans seen from afar (*maria* is Latin for "seas").

A combination of models and studies of Moon rocks has allowed scientists to trace the origin of the maria to a period several hundred million years after the end of the heavy bombardment, when the Moon was roughly a billion years old. By that time, the decay of radioactive elements had built up enough interior heat to melt much of the Moon's mantle. Molten rock welled up through surface fractures made by the largest impacts of the heavy bombardment, and lava flooded the huge craters made by those impacts. The relatively few craters within the maria today were made by impacts that occurred after the maria

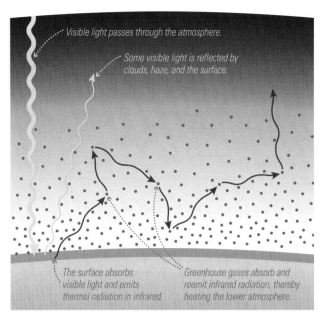

▲ **FIGURE 5.10**
The greenhouse effect. The lower atmosphere becomes warmer than it would be if it had no greenhouse gases such as water vapor, carbon dioxide, and methane.

common misconceptions

Why Is the Sky Blue?

If you ask around, you'll find a wide variety of misconceptions about why the sky is blue. Some people guess that the sky is blue because of light reflecting from the oceans, but that would not explain blue skies over inland areas. Others claim that "air is blue," a vague statement that is clearly wrong: If air molecules emitted blue light, then air would glow blue even in the dark; if they were blue because they reflected blue light and absorbed red light, then no red light could reach us at sunset. The real explanation for our blue sky is light scattering, as shown in Figure 5.9, and this scattering also explains our red sunsets.

▲ **FIGURE 5.11**
The familiar face of the full Moon shows numerous dark, smooth maria.

▲ FIGURE 5.12
Mercury's surface shows many craters, along with smooth lava plains. Bright streaks are caused by material ejected from large craters. Mercury also has many tremendous cliffs (inset), thought to have been created when the entire planet cooled and shrank. (Images from MESSENGER.)

formed, when the heat from radioactive decay was no longer sufficient to produce lava flows.

think about it Did the Moon have more or fewer craters before the maria formed than it does today? Explain.

Mercury is not much bigger than the Moon, and therefore has suffered much the same fate. In fact, Mercury looks so much like the Moon that it can be difficult to tell which world you are looking at in surface photos (Figure 5.12). Impact craters are visible almost everywhere, though they are less crowded together than in the most ancient regions of the Moon. That fact, along with smooth regions within and between craters, suggests that molten lava covered up some of the craters that formed during the heavy bombardment. As on the Moon, these lava flows probably occurred when heat from radioactive decay accumulated enough to melt part of the mantle.

Mercury also differs from the Moon in having tremendous cliffs that run hundreds of kilometers in length, with vertical faces sometimes more than 3 kilometers high. These cliffs probably formed when early tectonic forces compressed the crust, causing the surface to crumple. Scientists suspect that these forces arose from shrinkage of the entire planet, but what could have caused the entire planet to shrink? The answer may lie with Mercury's large, metallic core (see Figure 5.1). Metal expands and contracts more with temperature than rock, and models suggest that as this large core cooled, it would have contracted by as much as 20 kilometers in radius. The mantle and lithosphere then had to contract along with the core, generating tectonic stresses that created the great cliffs.

Mars: Once Warm and Wet? Mars is about half as large in diameter as Earth, which means it is much larger than Mercury and the Moon. We should therefore expect it to have had far more geological activity than the two smaller worlds, and its surface confirms this to be the case. Mars has several towering volcanoes, including Olympus Mons, which rises three times as high as Mount Everest on Earth and has a base that covers an area the size of Arizona (Figure 5.13). Mars also shows evidence of past tectonics, such as the prominent system of valleys called *Valles Marineris,* which is as long as the United States is wide and almost four times as deep as Earth's Grand Canyon (Figure 5.14). However, the most intriguing surface features on Mars are the ones produced by erosion, which suggest that Mars was once relatively warm and wet—conditions that may have been conducive to life.

No liquid water exists on the surface of Mars today. We know this not only because we've studied most of the surface in detail, but also because the surface conditions do not allow it. In most places and at most times, Mars is so cold that any liquid water would immediately freeze into ice. Even when the temperature rises above freezing, as it sometimes does at midday near the equator, the air pressure is so low that liquid water would quickly evaporate. If you put on a space suit and took a cup of water outside your pressurized spaceship, the water would rapidly freeze or boil away (or a combination of both). The only water we find on Mars today is frozen, some at the poles and some underground (Figure 5.15).

Nevertheless, strong evidence indicates that water once flowed on the surface. Figure 5.16 shows some of the evidence from orbital studies. Figure 5.16a shows a broad region of the ancient, heavily cratered southern highlands. Notice the indistinct rims of many large craters and the relative lack of small craters. Both facts argue for ancient rainfall, which would have eroded crater rims and

▲ FIGURE 5.13
Olympus Mons, the tallest mountain in the solar system, photographed from orbit. Note the tall cliff around its rim and the central volcanic crater from which lava erupted. The inset shows a three-dimensional perspective of this immense volcano.

▲ **FIGURE 5.14**
Valles Marineris is a huge system of valleys, created in part by tectonic stresses, that looks like a big "gash" across Mars. The inset shows a perspective view looking north across the center of the canyon. (Main image from *Viking*, inset from *Mars Express*.)

▲ **FIGURE 5.15**
This enhanced color image from the *Mars Reconnaissance Orbiter* shows an approximately 80-meter-thick layer of water ice (blue) exposed on the face of a steep cliff on Mars. As this image suggests, water ice is abundant under the Martian surface, in addition to being present at the poles.

erased small craters altogether. Figure 5.16b shows a three-dimensional perspective of the surface that suggests water once flowed between two ancient crater lakes. Figure 5.16c shows what looks like a river delta where water flowed into an ancient crater; spectra indicate clay minerals on the crater floor, probably deposited by sediments flowing down the river.

Surface studies add further evidence. For example, the *Curiosity* rover (which landed on Mars in 2012) has been exploring Gale Crater, which has a tall central

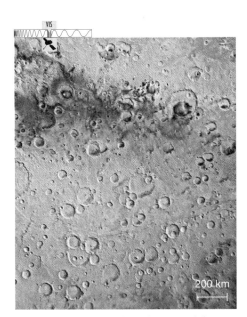

a This photo shows a broad region of the southern highlands on Mars. The eroded rims of large craters and the relative lack of small craters suggest erosion by rainfall.

b This computer-generated perspective view shows how a Martian valley forms a natural passage between two possible ancient lakes (shaded blue). Vertical relief is exaggerated 14 times to reveal the topography.

c Combined visible/infrared image of an ancient river delta that formed where water flowing down a valley emptied into a lake filling a large crater (portions of the crater wall are identified). Clay minerals are identified in green.

▲ **FIGURE 5.16**
Orbital evidence of past water on Mars. The region shown in c, called Jezero Delta, is the planned landing site for NASA's *Mars 2020* rover.

▲ FIGURE 5.17
This image, taken in 2015, captures the *Curiosity* rover's view looking up a river valley in Gale Crater, where it landed in 2012. *Curiosity*'s studies support the idea that, more than 3 billion years ago, a lake filled the crater, allowing layered sedimentary rocks to build up from the crater floor. Later wind erosion exposed much of the crater, but left a tall mountain (Mt. Sharp) at its center.

mountain called Mount Sharp that is composed of layered sedimentary rock (Figure 5.17). The rover's surface studies have revealed sedimentary rocks indicating that a lake once resided in the crater, as well as clumps of pebbles that almost certainly formed in an ancient streambed (Figure 5.18). *Curiosity* also undertook chemical analysis of some of the sedimentary layers, finding that they contain minerals known to form only in relatively pure water, much like that found in most lakes on Earth.

The fact that Mars once had flowing water means that the planet's temperature and atmospheric pressure must once have been much higher than they are today. This idea seems reasonable, because Mars is large enough for gravity to have retained the water vapor and carbon dioxide outgassed by volcanoes in its early history. If Martian volcanoes outgassed carbon dioxide and water in the same proportions as volcanoes on Earth, Mars would have had enough water to fill oceans tens or even hundreds of meters deep, as well as a thick carbon dioxide atmosphere that would have produced a strong greenhouse effect. The big question is what happened to all the atmospheric gas and water.

Counts of craters near riverbeds (such as those visible in the bottom right photo in Figure 4.2) indicate that the warm and wet period ended at least 2–3 billion years ago, and the leading hypothesis ties Mars's global climate change to a change in Mars's magnetic field that may have occurred by that time (Figure 5.19). Early in its history, Mars probably had enough internal heat to have molten metals convecting in its core, much like Earth today, which should have produced a magnetic field that would have protected the atmosphere from the solar wind. However, the magnetic field would have weakened as the small planet cooled and core convection ceased, and at that point particles of the solar wind would have begun to strip gases out of the Martian atmosphere and into space.

Much of the water once present on Mars is also probably gone for good. Like the carbon dioxide gas, some water vapor may have been stripped away by the solar wind. However, Mars also lost water in another way. Because the Martian atmosphere lacks ultraviolet-absorbing gases, atmospheric water molecules would have been easily broken apart by ultraviolet photons. The hydrogen atoms

▼ FIGURE 5.18
Rock formations photographed by *Curiosity* closely resemble those found among sedimentary rocks on Earth formed in the presence of abundant and fairly pure water.

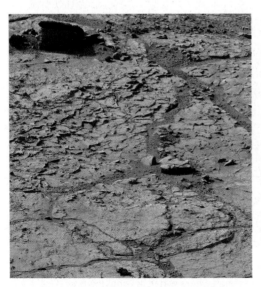

a *Curiosity* drilled into this rock formation of dry clay, finding minerals that should form only in water that was not very acidic or salty.

b These clumps of rounded pebbles show a structure nearly identical to that found in typical streambeds on Earth, providing strong evidence that the pebbles were rounded by flowing water before being cemented together into rock.

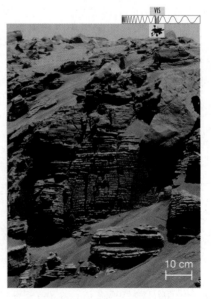

c The even layers of the foreground rocks in this image from *Curiosity* are characteristic of sediment deposited over time in the delta of a river that emptied into a lake.

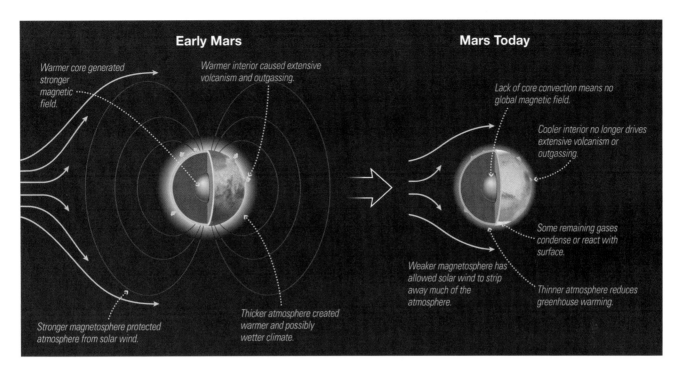

▲ FIGURE 5.19
This figure shows the leading hypothesis explaining the lack of liquid water on Mars today: Mars underwent dramatic climate change as its interior cooled, ensuring that rain could never fall again.

that broke away from the water molecules were then lost to space, along with some of the oxygen from the water molecules. The rest of the oxygen was drawn out of the atmosphere through chemical reactions with surface rock, a process that literally rusted the Martian rocks and gave the "red planet" its distinctive tint. NASA's *MAVEN* mission, which has been orbiting Mars since 2014, has confirmed major elements of the hypothesis illustrated in Figure 5.19 through careful measurements of the rates at which Mars still loses atmospheric gas today.

In summary, Mars's fate was probably sealed by its relatively small size. It was big enough for volcanism and outgassing to release water and atmospheric gas early in its history, but too small to maintain the internal heat and magnetic field needed to keep this water and gas. If Mars had been as large as Earth, so that it could still have outgassing and a global magnetic field, it might have a moderate climate today.

think about it Could Mars still have flowing water today if it had formed closer to the Sun? Use your answer to explain how Mars's fate has been shaped not just by its size, but also by its distance from the Sun.

Venus: Runaway Greenhouse Effect Venus is only about 5% smaller than Earth in radius, and its overall composition is about the same as that of Earth. We therefore expect Venus and Earth to be geologically similar, and in many respects they are. For example, Figure 5.20 shows a surface region of Venus, as revealed by cloud-penetrating radar observations from spacecraft. Much as on Earth, we see volcanoes, tectonic features, and occasional impact craters. However, Venus lacks two major features of Earth's geology. First, as we discussed earlier, Venus has little erosion because of its lack of liquid water and its slow surface winds, which are a consequence of its extremely slow rotation (one Venus rotation takes 243 Earth days). Second, we find no evidence of Earth-like plate tectonics on Venus, suggesting that Venus has a thicker and stronger lithosphere than Earth.

The bigger surprise lies in the differences between the atmospheres and surface conditions of Venus and Earth (Figure 5.21). Given their similar sizes

▲ FIGURE 5.20
This computer-generated image shows typical volcanic and tectonic features on Venus, along with a rare impact crater. The image is based on *Magellan* radar data and exaggerates vertical structure by ten times to reveal features. Bright regions represent rough terrain.

a This color-coded image shows Venus at ultraviolet wavelengths as seen by the Japanese Space Agency's *Akatsuki* spacecraft. At most wavelengths, clouds completely prevent any view of the surface.

b This photo from one of the former Soviet Union's *Venera* landers shows Venus's surface; part of the lander is in the foreground, with sky in the background. The lander survived only a short time before it was essentially fried by the searing 470°C surface temperature and the intense pressure created by Venus's thick, carbon dioxide atmosphere.

▲ FIGURE 5.21
Venus has a thick atmosphere consisting mostly of carbon dioxide.

common misconceptions

The Greenhouse Effect Is Bad

The greenhouse effect is often in the news, usually in discussions about environmental problems, but the greenhouse effect is not a bad thing. In fact, we could not exist without it, since it is responsible for keeping our planet warm enough for liquid water to flow in the oceans and on the surface. Why, then, is the greenhouse effect discussed as an environmental problem? The reason is that human activity is adding more greenhouse gases to the atmosphere—and scientists agree that the additional gases are warming Earth's climate. So while the greenhouse effect makes Earth livable, it is also responsible for the searing 470°C temperature of Venus—proving that it's possible to have too much of a good thing.

and compositions, volcanoes should have outgassed similarly large amounts of water and carbon dioxide on both planets. But today, Venus's thick atmosphere contains nearly 200,000 times as much carbon dioxide as Earth's atmosphere, creating Venus's extreme greenhouse effect, and water fills Earth's oceans but is absent on Venus. Therefore, if the planets really did have similar outgassing, Earth somehow lost its carbon dioxide while Venus lost its water.

We can easily account for the "missing" carbon dioxide on Earth. On Earth, moderate temperatures allowed the outgassed water vapor to condense into rain that ultimately filled the oceans. This, in turn, caused our atmosphere to lose its carbon dioxide, which dissolves in water and then can undergo chemical reactions to make *carbonate rocks* (rocks rich in carbon and oxygen) such as limestone. Earth has about 170,000 times as much carbon dioxide locked up in rocks as in its atmosphere—which means that Earth has almost as much total carbon dioxide as Venus. Of course, the fact that Earth's carbon dioxide is mostly in rocks makes all the difference in the world: If this carbon dioxide were in our atmosphere, our planet would be nearly as hot as Venus and uninhabitable.

We are left with the question of what happened to Venus's water. Venus has essentially no water today, either on its surface (which is too hot for liquid water) or in its atmosphere. This absence of water explains why Venus retains so much atmospheric carbon dioxide: Without oceans, carbon dioxide cannot dissolve or become locked away in carbonate rocks. Assuming that a huge amount of water really was outgassed on Venus, this water somehow disappeared. The leading hypothesis for the disappearance of this water is that, much like on Mars, ultraviolet light from the Sun broke apart molecules of water vapor and the released hydrogen atoms escaped to space. Acting over billions of years, this process can easily explain the loss of an ocean's worth of water from Venus—as long as the water was in the atmosphere as a gas rather than in liquid oceans.

Why didn't Venus develop oceans like Earth? To understand the answer, think about what would happen if we could magically move Earth to the orbit of Venus (Figure 5.22). The greater intensity of sunlight would almost immediately raise Earth's global average temperature by about 30°C, from its current 15°C to about 45°C (113°F). Although this is still well below the boiling point of water, the higher temperature would lead to increased evaporation of water from the oceans and would allow the atmosphere to hold more water vapor before the vapor condensed to make rain. Because water vapor is also a greenhouse gas, the added water vapor would strengthen the greenhouse effect and drive temperatures even higher. The higher temperatures, in turn, would lead to even more ocean evaporation and more water vapor in the atmosphere, strengthening the greenhouse effect even further. In other words, we'd have a positive feedback loop in which each little bit of additional water vapor in the atmosphere would lead to higher temperature and even more water vapor. The process would careen rapidly out of control as a **runaway greenhouse effect**, resulting in the complete vaporization of Earth's oceans.

The runaway greenhouse effect would cause Earth to heat up until the oceans completely evaporated and the carbonate rocks released all their carbon dioxide back into the atmosphere. By the time the process was complete, temperatures on the relocated Earth would be even higher than they are on Venus today, thanks to the combined greenhouse effects of carbon dioxide and water vapor in the atmosphere. The water vapor would then gradually disappear, as ultraviolet light broke water molecules apart and the hydrogen escaped to space. In short, moving Earth to Venus's orbit would essentially turn our planet into another Venus.

We have arrived at a simple explanation of why Venus is so much hotter than Earth. Even though Venus is only about 30% closer to the Sun than Earth is, this difference was enough to be critical. On Earth, it was cool enough for water to rain down and make oceans. The oceans then dissolved carbon dioxide

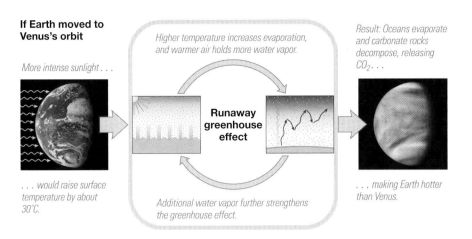

◀ **FIGURE 5.22**
This diagram shows how, if Earth were placed at Venus's distance from the Sun, the runaway greenhouse effect would cause the oceans to evaporate completely.

and chemical reactions locked it away in carbonate rocks, leaving our atmosphere with only enough greenhouse gases to make our planet pleasantly warm. On Venus, the greater intensity of sunlight made it just enough warmer that oceans either never formed or soon evaporated, leaving Venus with a thick atmosphere full of greenhouse gases.

think about it We've seen that moving Earth to Venus's orbit would cause our planet to become Venus-like. If we could somehow move Venus to Earth's orbit, would it become Earth-like? Why or why not?

◆ What unique features of Earth are important for life?

If you think about the other terrestrial worlds, you can probably identify a number of features that are unique to Earth. Four of them turn out to be particularly important to our existence: surface liquid water, atmospheric oxygen, plate tectonics, and a climate that has remained relatively stable throughout Earth's history. Let's explore each of these features and how it is important to life.

All life on Earth requires liquid water, and the abundance of liquid water on Earth's surface makes possible a great abundance of life. As we have discussed, this water originally outgassed from Earth's volcanoes, then rained down on the surface to make the oceans, which neither froze nor evaporated thanks to our moderate greenhouse effect and distance from the Sun.

Oxygen (O_2), which is crucial to animal life, makes up about 21% of Earth's atmosphere. But oxygen is not a product of volcanic outgassing. It is such a highly reactive gas that it would be absorbed by rocks and disappear from the atmosphere in just a few million years if it were not continuously resupplied. The ongoing source of oxygen for our atmosphere is life itself: Plants and many microorganisms release oxygen through photosynthesis. Today, photosynthetic organisms return oxygen to the atmosphere in approximate balance with the rate at which animals and chemical reactions consume it, which is why the oxygen concentration remains relatively steady. Note that oxygen is also what makes possible Earth's protective ozone layer, since ozone (O_3) is produced from ordinary oxygen (O_2).

think about it Suppose that, somehow, all photosynthetic life died out. What would happen to the oxygen in our atmosphere? Could animals, including us, still survive?

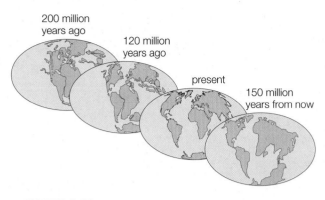

▲ **FIGURE 5.23**
Plate tectonics rearranges the continents over millions of years, as shown in these maps of Earth at different times.

▲ **FIGURE 5.24**
This diagram shows how the CO_2 cycle continually moves carbon dioxide from the atmosphere to the ocean to rock and back to the atmosphere. Note that plate tectonics plays a crucial role in the cycle.

The third and fourth items on our list of Earth's unique features—plate tectonics and climate stability—turn out to be closely linked. Plate tectonics is a slow process; on average, Earth's plates move only a few centimeters per year, which is about the speed at which your fingernails grow. Nevertheless, over millions of years, these motions rearrange the continents, open ocean basins between continents, build mountain ranges, and much more (Figure 5.23). Even more important, plate tectonics turns out to be crucial to climate stability and therefore to the ongoing evolution of life on Earth.

Earth's climate is not perfectly stable—our planet has endured numerous ice ages and warm periods in the past. Nevertheless, even in the deepest ice ages and warmest warm periods, Earth's surface temperature has remained in a range in which some liquid water could still exist and harbor life. This long-term climate stability is even more remarkable in light of models suggesting that the Sun has brightened substantially (about 30%) over the past 4 billion years, yet Earth's temperature has managed to stay in nearly the same range throughout this time. Apparently, the strength of the greenhouse effect self-adjusts to keep the climate stable.

The mechanism by which Earth self-regulates its temperature is called the **carbon dioxide cycle**, or the **CO_2 cycle** for short. Figure 5.24 shows how the cycle works. Atmospheric carbon dioxide dissolves in rainwater, which erodes rocks and leads to minerals being carried to the oceans. In the oceans, these minerals combine with dissolved carbon dioxide and fall to the ocean floor, making carbonate rocks. Over millions of years, motions associated with plate tectonics push the carbonate rocks down into the mantle, where they melt and release their carbon dioxide. This carbon dioxide then returns to the atmosphere by being outgassed from volcanoes.

The CO_2 cycle acts as a long-term thermostat for Earth, because the overall rate at which carbon dioxide is pulled from the atmosphere is very sensitive to temperature: the higher the temperature, the higher the rate at which carbon dioxide is removed. Consider what happens if Earth warms up a bit. Warmer temperature means more evaporation and rainfall, pulling more CO_2 out of the atmosphere. The reduced atmospheric CO_2 concentration leads to a weakened greenhouse effect, which counteracts the initial warming and cools the planet back down. Similarly, if Earth cools a bit, rainfall decreases and less CO_2 is dissolved in rainwater, allowing the CO_2 released by volcanism to build back up in the atmosphere. The increased CO_2 concentration strengthens the greenhouse effect and warms the planet back up.

We can now see why plate tectonics is so intimately connected to our existence. Plate tectonics is a crucial part of the CO_2 cycle; without plate tectonics, CO_2 would have remained locked up in seafloor rocks rather than being recycled through outgassing. Earth's climate might then have undergone changes as dramatic as those that occurred on Venus and Mars. We know that Mars lacks plate tectonics because of its small size. But why doesn't Venus have plate tectonics, given its similar size to Earth? The leading hypothesis is that the lack of water on Venus—which we've traced to the high temperatures due to the runaway greenhouse effect—has given Venus a thicker and stronger lithosphere that has been able to resist fracturing into plates. If this hypothesis is correct, then two primary characteristics of our planet are responsible for allowing life: (1) a size that was large enough to retain internal heat and drive plate tectonics and (2) a distance from the Sun that allowed outgassed water vapor to rain down to form oceans. In that case, life could be common in the universe among planets around other stars that have similar size and orbit to Earth. Figure 5.25 summarizes how a planet's size and distance from the Sun determine its fate.

The Role of Planetary Size

Small Terrestrial Planets

Interior cools rapidly...

...so that tectonic and volcanic activity cease after a billion years or so. Many ancient craters therefore remain.

Lack of volcanism means little outgassing, and low gravity allows gas to escape more easily; no atmosphere means no erosion.

Large Terrestrial Planets

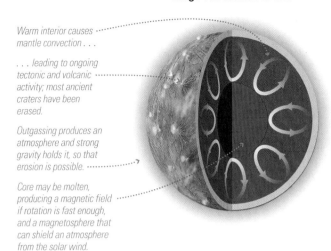

Warm interior causes mantle convection...

...leading to ongoing tectonic and volcanic activity; most ancient craters have been erased.

Outgassing produces an atmosphere and strong gravity holds it, so that erosion is possible.

Core may be molten, producing a magnetic field if rotation is fast enough, and a magnetosphere that can shield an atmosphere from the solar wind.

The Role of Distance from the Sun

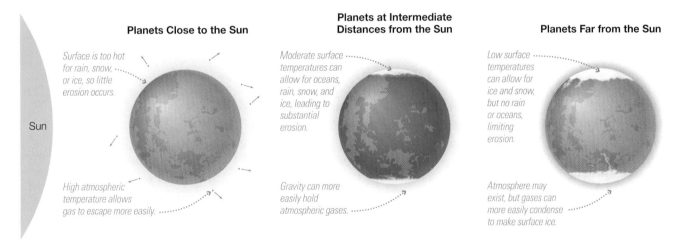

Planets Close to the Sun

Surface is too hot for rain, snow, or ice, so little erosion occurs.

High atmospheric temperature allows gas to escape more easily.

Planets at Intermediate Distances from the Sun

Moderate surface temperatures can allow for oceans, rain, snow, and ice, leading to substantial erosion.

Gravity can more easily hold atmospheric gases.

Planets Far from the Sun

Low surface temperatures can allow for ice and snow, but no rain or oceans, limiting erosion.

Atmosphere may exist, but gases can more easily condense to make surface ice.

▲ **FIGURE 5.25**
This illustration shows how a terrestrial world's size and distance from the Sun determine its geological history. Earth has oceans and life because it is large enough and at a suitably moderate distance from the Sun.

THE PROCESS OF SCIENCE IN ACTION

 Global Warming Evidence

5.3 Global Warming

Our planet seems to regulate its own climate quite effectively over long time scales, but fossil and geological evidence tells us that substantial and rapid swings in global climate can occur on shorter ones. Past climate changes have been due to natural causes. Today, Earth is undergoing climate change for a new reason: human activity that is rapidly increasing the atmospheric concentration of carbon dioxide and other greenhouse gases, leading to **global warming** (Figure 5.26). Global warming is one of the most important issues of our time, so we will devote this chapter's case study of the process of science in action to its study and to the role of models in predicting its implications for the future.

▲ **FIGURE 5.26**
Average global temperatures from 1880 through 2017. Notice the clear global warming trend of the past few decades. (Data from the National Climate Data Center.)

◆ What is the evidence for global warming?

As with most topics in science, our understanding of global warming rests on a combination of observations and modeling. Models are designed to explain the observations as accurately as possible, using the simplest explanation consistent with the observed facts, and we judge the models by how well they predict the results of additional observations [Section 3.2]. The case linking global warming with human activity rests on three basic facts.

1. The greenhouse effect is a simple and well-understood scientific model. We can be confident in our understanding of it because it so successfully explains the observed surface temperatures of other planets. Given this basic model, there is no doubt that a rising concentration of greenhouse gases would make our planet warm up more than it would otherwise; the only debate is about how soon and how much.

2. Human activity such as the burning of fossil fuels is clearly increasing the amounts of greenhouse gases in the atmosphere. Observations show that the atmospheric concentration of carbon dioxide is now about 40% higher than it was before the industrial revolution began or at any other time during the past million years, and it is continuing to rise rapidly (Figure 5.27). We can be confident that the rise is a result of human activity, because the atmosphere is becoming enriched in molecules of CO_2 carrying the distinct ratio of isotopes present in fossil fuels.

3. Climate models that ignore human activity fail to match the observed rise in global temperatures. In contrast, climate models that include the enhanced greenhouse effect from human emission of greenhouse gases match the observed temperature trend quite well (Figure 5.28), providing additional support for the idea that global warming results from human activity.

With scientific evidence in hand that links global warming to human activity, how do we proceed? Again, the comparison between observations and models is crucial. When making predictions for the future, we need to consider the whole range of models that are currently consistent with the data. There are still uncertainties in models of climate change, but they generally predict that if current trends in the greenhouse gas concentration continue—that is, if we do nothing to

▼ **FIGURE 5.27**
This diagram shows the atmospheric concentration of carbon dioxide over the past 800,000 years. Data for the past few decades come from direct measurements; earlier data come from studies of air bubbles trapped in Antarctic ice (ice core samples). The CO_2 concentration is measured in parts per million (ppm), which is the number of CO_2 molecules among every 1 million air molecules. (Ice core data from the European Project for Ice Coring in Antarctica; inset data from NOAA.)

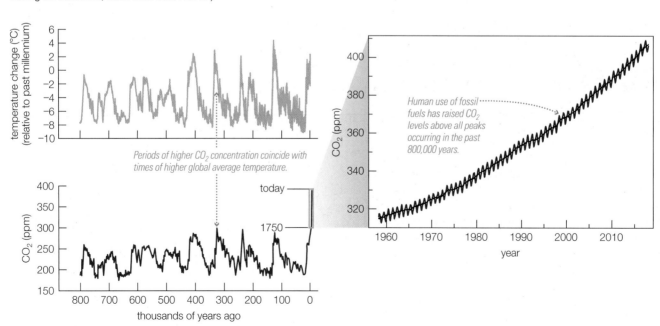

slow our emissions of carbon dioxide and other greenhouse gases—the warming trend will accelerate. By the end of this century, the global average temperature will be 3°C–5°C (6°F–10°F) higher than it is now, giving our children and grandchildren the warmest climate that any generation of *Homo sapiens* has ever experienced.

Although a temperature increase of a few degrees might not sound so bad, small changes in *average* temperature can lead to much more dramatic changes in climate patterns. For example, Figure 5.29 shows that Arctic regions have warmed much more than the average. Moreover, the warming of the atmosphere and oceans means there is more total energy available in the climate system, a fact that may explain why extreme weather events seem to have increased in frequency and severity in recent decades. Other likely consequences of global warming include more severe drought (which can lead to more wildfires) when local conditions are dry, heavier downpours (which can lead to flooding and mudslides) when rains finally come, longer and hotter summers, damage to coral reefs and fish stocks in the ocean, and a rise in sea level. Indeed, recent data suggest a surprisingly rapid change in Greenland's ice sheet, leading some scientists to worry that ice melt could cause sea level to rise by a meter or more—enough to flood coastline around the world—by the end of this century. Longer term, complete melting of the polar ice caps would increase sea level by some 70 meters (more than 200 feet). Although such melting would probably take centuries or millennia, it suggests the disconcerting possibility that future generations will have to send deep-sea divers to explore the underwater ruins of many major cities.

Fortunately, climate models suggest that we still have time to avert the most serious consequences of global warming, provided we dramatically and rapidly curtail our greenhouse gas emissions. That can happen only if humans broadly understand the consequences of their actions. A century ago, as cars were first mass produced and electricity became common, our use of fossil fuels seemed to provide only benefit to humanity; no one at the time foresaw the problems that would later be created by the release of toxic pollutants and carbon dioxide.

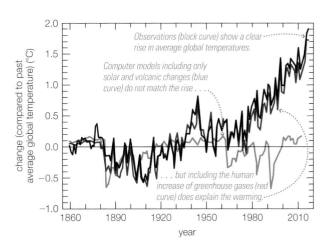

▲ **FIGURE 5.28**
This graph compares observed temperature changes (black curve) with the predictions of climate models that include only natural factors (blue curve), such as changes in the brightness of the Sun and effects of volcanoes, and models that also include the human contribution to increasing greenhouse gas concentrations (red curve). Only the red curve matches the observations well. (The red and blue model curves are averages of many scientists' independent models of global warming, which generally agree with each other within 0.1°C–0.2°C.) Data from the Fourth National Climate Assessment.

▲ **FIGURE 5.29**
This map shows how regional average temperatures for the period 2013–2017 compared to the average from 1951–1980. Notice the large regional differences, and the particularly large warming in the north polar region.

If we hope to avoid other problems in the future, we must continue to study the potential consequences of our actions, through a combination of scientific modeling and observations, and make our choices accordingly.

think about it What do *you* think we should do to alleviate the threat of global warming?

summary of key concepts

5.1 Terrestrial Surfaces and Atmospheres

◆ **What determines a world's level of geological activity?**

A world's size determines its level of geological activity. Larger worlds retain their internal heat longer and therefore experience more **volcanism** and **tectonics**. They also experience more **erosion** because they retain thick atmospheres. **Impact cratering** occurred extensively on all terrestrial worlds, but its effects are visible mainly on worlds where other geological activity has been minimal.

◆ **How does an atmosphere affect conditions for life?**

Two crucial effects are (1) protecting the surface from dangerous solar radiation, such as x-rays and ultraviolet light, and (2) the **greenhouse effect**, which traps heat in a planet's atmosphere and makes it warmer.

5.2 Histories of the Terrestrial Worlds

◆ **Why did the terrestrial worlds turn out so differently?**

A terrestrial world's current state depends on its size and its distance from the Sun. Small worlds like Mercury and the Moon are now geologically dead but experienced volcanism and tectonics when they had warmer interiors. Moderately sized Mars is large enough to have had a thick atmosphere that made it wet and warm in the past, but it became a frozen wasteland after it lost much of its atmosphere. Venus and Earth are almost the same size, but Venus's slightly closer distance from the Sun led to a **runaway greenhouse effect** that has made it much hotter than Earth.

◆ **What unique features of Earth are important for life?**

Unique features of Earth on which we depend for survival are (1) surface liquid water, made possible by Earth's moderate temperature; (2) atmospheric oxygen, a product of photosynthetic life; (3) plate tectonics, driven by internal heat; and (4) climate stability, a result of the **carbon dioxide cycle**, which relies on plate tectonics.

THE PROCESS OF SCIENCE IN ACTION

5.3 Global Warming

◆ **What is the evidence for global warming?**

Scientists are convinced that human activity causes **global warming** because such warming is clearly expected, given our understanding of the greenhouse effect and the fact that we are adding carbon dioxide to the atmosphere. Observations of rising global temperatures agree with this expectation.

Investigations

For instructor-assigned homework and other learning materials, go to www.MasteringAstronomy.com.

Quick Quiz

Choose the best answer to each of the following. Explain your reasoning.

1. What is the name of the rigid outer layer of a planet? (a) crust (b) mantle (c) lithosphere
2. What is the longest-lasting internal heat source responsible for geological activity? (a) accretion (b) radioactive decay (c) sunlight
3. In general, what kind of terrestrial planet would you expect to have the thickest lithosphere? (a) a large planet (b) a small planet (c) a planet located far from the Sun
4. Which of a planet's fundamental properties has the greatest effect on its level of volcanic and tectonic activity? (a) size (b) distance from the Sun (c) rotation rate
5. Which describes our understanding of flowing water on Mars? (a) It was never important. (b) It was important once, but is no longer. (c) It is a major process on the Martian surface today.
6. What do we conclude if a planet has few impact craters of any size? (a) The planet was never bombarded by asteroids or comets. (b) Its atmosphere stopped impactors of all sizes. (c) Other geological processes have erased many craters.
7. Which terrestrial world has the most atmospheric gas? (a) Venus (b) Earth (c) Mars
8. Which of the following is a strong greenhouse gas? (a) nitrogen (b) water vapor (c) oxygen
9. Most of the CO_2 that was outgassed from Earth's volcanoes (a) is still in the atmosphere. (b) escaped to space. (c) is locked up in rocks.
10. What is the leading hypothesis for Venus's lack of water? (a) Venus formed with very little water. (b) Its water is underground. (c) Its water molecules were split by sunlight, allowing the hydrogen to escape to space.
11. The oxygen in Earth's atmosphere was released by (a) volcanic outgassing. (b) the CO_2 cycle. (c) life.
12. What gas is primarily responsible for recent global warming? (a) ozone (b) water vapor (c) CO_2

Short-Answer/Essay Questions

Explain all answers clearly, with complete sentences and proper essay structure if needed. An asterisk (*) designates a quantitative problem, for which you should show all your work.

13. *Miniature Mars.* Suppose Mars had turned out to be significantly smaller than its current size—say, the size of our Moon. How would this have affected the number of Mars's geological features due to each of the four major geological processes? Would this smaller Mars still be a good candidate for harboring extraterrestrial life?
14. *Two Paths Diverged.* Use the fundamental properties of size and distance from the Sun to explain why Earth has oceans and very little atmospheric carbon dioxide, while Venus has a thick carbon dioxide atmosphere.
15. *Change in Formation Properties.* Choose either size or distance from the Sun and suppose it had been different for Earth. (For example, suppose Earth had been smaller in size or at a greater distance from the Sun.) Describe how this change might have affected Earth's subsequent history and the possibility of life on Earth.
16. *Predictive Geology.* Suppose another star system has a rocky terrestrial planet that is twice as large as Earth but at the same distance from its star (which is just like our Sun). Describe the type of geology you would expect the planet to have.
17. *Skeptic Claims on Global Warming.* A small but vocal group of people still dispute that humans are causing global warming. Do some research to find the basis of their claims. Then defend or refute their findings, based on your own studies and your understanding of the hallmarks of science discussed in Chapter 3. Note: You may find it useful to visit globalwarmingprimer.com, which features a free online primer (written by one of the authors of this textbook) about global warming and many of the skeptics' claims.
18. *Experiment: Geological Properties of Silly Putty.* Roll room-temperature Silly Putty into a ball and measure its diameter. Place the ball on a table and gently place one end of a heavy book on it. After 5 seconds, measure the height of the squashed ball. Repeat the experiment two more times, the first time warming the Silly Putty in hot water before you start and the second time cooling it in ice water before you start. How do the different temperatures affect the rate of squashing? How does the experiment relate to planetary geology? Explain.
19. *Experiment: Planetary Cooling in a Freezer.* To simulate the cooling of planetary bodies of different sizes, use a freezer and two small plastic containers of similar shape but different size. Fill each container with cold water and put both in the freezer at the same time. Checking every hour or so, record the time and your estimate of the thickness of the "lithosphere" (the frozen layer) in the two tubs. How long does it take the water in each tub to freeze completely? Describe in a few sentences the relevance of your experiment to planetary geology. Extra credit: Plot your results on a graph, with time on the x-axis and lithospheric thickness on the y-axis. What is the ratio of the two freezing times?
20. *Amateur Astronomy: Observing the Moon.* Any amateur telescope has adequate resolution to identify geological features on the Moon. The light highlands and dark maria should be evident, and shadowing is visible near the line between night and day. Try to observe the Moon near the first- or third-quarter phase. Sketch or photograph the Moon at low magnification, and then zoom in on a region of interest. Again sketch or photograph your field of view, then label the features and identify the geological process that created them. Look for craters, volcanic plains, and tectonic features. Estimate the size of each feature by comparing it to the size of the whole Moon (radius = 1738 km).
*21. *Internal vs. External Heating.* Earth's internal heat leaks out through the surface with a power of about 3 trillion watts. Divide this amount by Earth's total surface area to find the rate of heat loss per square meter. Compare your result to the amount of power contributed by sunlight, which provides an average of about 400 watts per square meter in daylight.
*22. *Plate Tectonics.* Plates typically move relative to each other at a speed of 1 centimeter per year. How long would it take for two continents 3000 kilometers apart to collide?

6 The Outer Solar System

LEARNING GOALS

6.1 Jovian Planets, Rings, and Moons
- What are jovian planets like?
- Why are jovian moons so geologically active?

6.2 Asteroids, Comets, Dwarf Planets, and the Impact Threat
- Why are asteroids and comets grouped into three distinct regions?
- Do small bodies pose an impact threat to Earth?

 THE PROCESS OF SCIENCE IN ACTION

6.3 Extinction of the Dinosaurs
- Did an impact kill the dinosaurs?

Beyond the realm of the terrestrial worlds, we enter the outer solar system, dominated by the four large jovian planets. These planets are not alone out there. Saturn, shown above, is orbited by dozens of moons—including one, Titan, that has a thick atmosphere—and the countless small particles that make up its beautiful rings. [The small blue dot of light just inside Saturn's rings at the left (about the 10:00 position) is Earth, far in the distance.] The other jovian planets also have rings and numerous moons, some with amazing features such as active volcanoes or subsurface oceans. The rest of the outer solar system contains vast numbers of small bodies—asteroids and comets. Although these objects orbit the Sun independently, their orbits are shaped by the gravity of the jovian planets. In this chapter, we'll explore the fascinating objects of the outer solar system and see how, on occasion, they can influence life on Earth.

▶ Jovian Planet Systems

6.1 Jovian Planets, Rings, and Moons

The mythological namesakes of the jovian planets were rulers among gods: Jupiter was the king of the gods, Saturn was Jupiter's father, Uranus was the lord of the sky, and Neptune ruled the sea. The true majesty of the jovian planets may be even greater. The smallest of these four worlds, Neptune, has a volume more than 50 times that of Earth. The largest, Jupiter, could swallow up more than 1000 Earths. These worlds also differ from the terrestrial planets in composition and nature, and have no solid surface on which to stand. In this section, we'll investigate these worlds and their wondrous rings and moons.

◆ **What are jovian planets like?**

Most of what we know about the jovian planets comes from the observations made by spacecraft. We have sent spacecraft on flybys past all four of the jovian planets. Orbiting spacecraft have allowed longer-term observation of Jupiter, orbited by the *Galileo* spacecraft from 1995 to 2003 and by *Juno* since 2016, and Saturn, orbited by *Cassini* from 2004 to 2017.

Size and Composition Figure 6.1 shows the four jovian planets to scale, along with some basic data. Notice that Jupiter and Saturn are quite similar to each other, but fairly different from Uranus and Neptune. The most important differences are in composition. Jupiter and Saturn are both made almost entirely of hydrogen and helium, with just a few percent of their masses in the form of hydrogen compounds and even smaller amounts of rock and metal. In fact, their overall compositions are much more similar to the composition of the Sun than to the compositions of the terrestrial planets. Some people even call Jupiter a "failed star" because it has a starlike composition but lacks the nuclear fusion needed to make it shine. Its lack of fusion is due to its size: Although Jupiter is large for a planet, it is much less massive than any star. As a result, its gravity

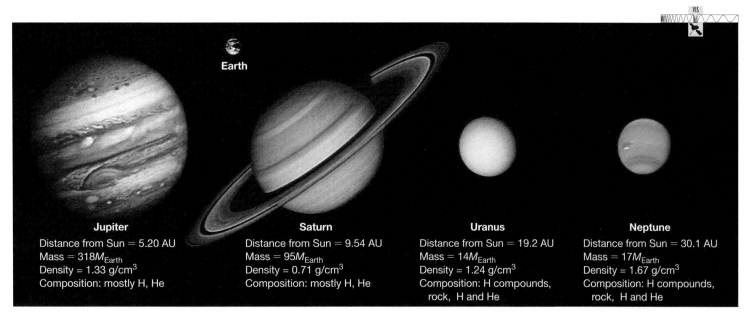

▲ FIGURE 6.1
Jupiter, Saturn, Uranus, and Neptune, shown to scale with Earth for comparison.

is too weak to compress its interior to the extreme temperatures and densities needed for nuclear fusion. (Jupiter would have needed to grow to about 80 times its current mass to have become a star.)

Uranus and Neptune are much smaller than Jupiter and Saturn, and while they also contain substantial amounts of hydrogen and helium, they are made primarily of hydrogen compounds such as water (H_2O), methane (CH_4), and ammonia (NH_3), along with smaller amounts of metal and rock.

We can understand the differences in composition and size among the jovian planets by looking at their formation according to the nebular theory [Section 4.2]. Recall that the jovian planets formed beyond the frost line (see Figure 4.7), where it was cold enough for hydrogen compounds to condense into ices. Because hydrogen compounds were so much more abundant than metal and rock, some of the ice-rich planetesimals of the outer solar system grew large enough for their gravity to draw in the hydrogen and helium gas that surrounded them. All four jovian planets are thought to have grown from ice-rich planetesimals of about the same mass—roughly 10 times the mass of Earth—so their differences in composition must stem from the amounts of hydrogen and helium gas that they captured. Jupiter and Saturn captured so much hydrogen and helium that these gases now make up the vast majority of their masses. Uranus and Neptune pulled in much less hydrogen and helium gas, leaving their bulk compositions similar to the compositions of the ice-rich planetesimals around which they grew. The reason Uranus and Neptune pulled in less gas likely lies in their greater distances from the Sun: The density of the solar nebula was lower at greater distances, so the ice-rich planetesimals that grew into Uranus and Neptune probably took longer to form than those that became Jupiter and Saturn. Uranus and Neptune therefore had less time to capture gas from the solar nebula before the solar wind blew the nebula's remaining gas into interstellar space.

think about it Suppose that solar wind had taken longer to clear the solar nebula. How would you expect the jovian planets to be different in that case? Explain.

tools of science — Newton's Version of Kepler's Third Law

Have you ever wondered how we measure the masses of the jovian planets, the Sun, or other distant objects in the universe? Newton's universal law of gravitation gives us the key tool, as long as we can observe an orbiting object. To see how the idea works, let's consider how the orbits of planets tell us the mass of the Sun.

The Sun's gravity holds the planets in orbit. The stronger the gravity, the stronger the pull on a planet and therefore the faster it travels in its orbit around the Sun. That is why Kepler's third law holds true: Planets closer to the Sun orbit at higher speed than planets farther away, because gravity is stronger closer to the Sun. Now consider what would happen if we replaced the Sun with a more massive star. In that case, the pull of gravity would be even stronger at any particular distance, so all the planets would have to orbit faster if they stayed at their current orbital distances.

Newton made these ideas quantitative by using his universal law of gravitation to determine the relationship between the period and orbital distance of an orbiting object. In doing so, he discovered that Kepler's third law in its original form ($p^2 = a^3$) is actually just a special case of a more general formula that follows directly from the universal law of gravitation. This formula is called **Newton's version of Kepler's third law**:

$$p^2 = \frac{4\pi^2}{G(M_1 + M_2)} a^3$$

where M_1 and M_2 are the masses of two objects that are gravitationally attracting each other, p is the orbital period of one object around the other, a is the distance between the centers of the two objects, and G is the gravitational constant.

Let's see how this formula allows us to calculate the mass of the Sun. Because the Sun is so much more massive than any planet, the sum of the masses, $M_{Sun} + M_{planet}$, is pretty much just M_{Sun}. Therefore, we can use the orbital period p and distance a of any planet to calculate the mass of the Sun; note that we'll get the same answer for the Sun's mass no matter which planet we use.

The same idea works for other distant objects. We can calculate the mass of Jupiter from the orbital period and distance of any of its moons. We can calculate the mass of Earth from the orbital period and distance of our Moon or of an orbiting satellite. We can even calculate the mass of extrasolar planets from their orbits around their stars, or the sum of the masses of two stars in a distant binary star system from their orbits around each other. When Newton extended Kepler's law, he achieved one of a scientist's main goals: to take what's been learned in one situation and apply it more universally. Today, Newton's version of Kepler's third law is the primary means by which we determine masses throughout the universe.

Interior Structure The jovian planets are often called "gas giants," making it sound as if they were entirely gaseous, like air on Earth. The reality is more complex, because the strong gravity of these planets compresses most of the "gas" into forms of matter quite unlike anything we are familiar with in everyday life on Earth. To explore this idea, imagine plunging into Jupiter in a futuristic space suit that could allow you to survive the extreme interior conditions.

The visible clouds of Jupiter extend downward only a few tens of kilometers into Jupiter's atmosphere. Below that, you'd find increasingly higher temperatures and pressures as you descended. These temperatures and pressures would soon destroy any real spacecraft entering Jupiter. When the *Galileo* spacecraft dropped a scientific probe into Jupiter in 1995, the probe provided valuable data about Jupiter's atmosphere, but it survived to a depth of only about 200 kilometers below the cloud tops.

We use computer models to determine the structure of Jupiter's interior at greater depths, and Figure 6.2 shows the layers you'd encounter as you continued into Jupiter. The layers do not differ much in composition—all except the core are mostly hydrogen and helium—but they differ in the phase (such as liquid or gas) of their hydrogen. The uppermost layer, which extends about 10% of the way downward, has conditions in which hydrogen remains in its familiar gaseous form. Below that, the temperatures and pressures become so extreme that hydrogen is forced into a compact form that has the electrical properties of metal but still flows like a liquid. Jupiter's strong magnetic field is generated in this layer of metallic hydrogen. Only when you had plunged to a depth of some 60,000 kilometers would you finally encounter Jupiter's core, which contains a mix of hydrogen compounds, rock, and metal. However, the high temperature and pressure ensure that this mix bears little resemblance to familiar solids or liquids. In fact, while the core contains about 10 times as much mass as the entire Earth, its high density makes it only about the same size as Earth.

Saturn's interior layers are very similar to those of Jupiter, except Saturn's lower mass and weaker gravity means that the pressures necessary for liquid and metallic hydrogen occur at greater depths. Pressures within Uranus and Neptune are not high enough to form liquid or metallic hydrogen at all. Each of these two planets has only a thick layer of gaseous hydrogen surrounding its core of hydrogen compounds, rock, and metal. In fact, the core material in Uranus and Neptune may be liquid, making for very odd "oceans" buried deep inside these planets. Some scientists have speculated that life might exist in these cores, but they are so deeply buried that it would be difficult to explore them.

Jovian Weather Jovian atmospheres have dynamic winds and weather, with colorful clouds and enormous storms. Weather on these planets is driven not only by energy from the Sun (as on the terrestrial planets) but also by heat generated within the planets themselves. While the amount of heat escaping from the interiors of the terrestrial planets is very small compared to the amount of heat provided by sunlight, the heat escaping from within the jovian planets can rival solar heating. All but Uranus generate a great deal of internal heat. No one knows the precise source of this internal heat, but it probably comes from conversion of gravitational potential energy into thermal energy, which is occurring either because these planets continue to shrink slowly but imperceptibly in size or because they have ongoing differentiation as heavier materials continue to sink toward their cores.

The jovian planets all have cloudy atmospheres, but their clouds differ from those on Earth. Clouds form when a gas condenses to make tiny liquid droplets or solid flakes. Earth's atmosphere contains only one ingredient that can condense into clouds: water vapor. The jovian planets have several gases that can condense to form clouds. Because different gases condense at different temperatures, these planets have distinct cloud layers at different altitudes. For example, Jupiter has three primary cloud layers (Figure 6.3). The lowest layer occurs at a depth of about 100 kilometers, where temperatures are Earth-like and water

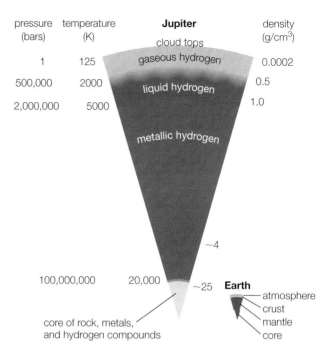

▲ **FIGURE 6.2**
Jupiter's interior structure, labeled with the pressure, temperature, and density at various depths. Earth's interior structure is shown to scale for comparison. (Notes on units: 1 *bar* is approximately the atmospheric pressure at sea level on Earth; the density of liquid water is 1 g/cm³.)

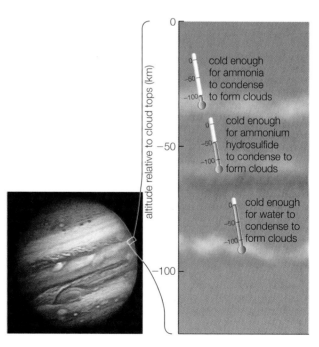

▲ **FIGURE 6.3**
Jupiter has at least three distinct cloud layers because different atmospheric gases condense at different temperatures and hence at different altitudes. (The tops of the ammonia clouds are usually considered the zero altitude for Jupiter, which is why lower altitudes are negative.)

can condense to form clouds. The temperature is lower at higher altitudes; about 50 kilometers above the water clouds, it is cold enough for the gas ammonium hydrosulfide (NH$_4$SH) to condense into clouds. These ammonium hydrosulfide clouds give Jupiter its distinctive tan and red colors. Higher still, the temperature is so cold that ammonia (NH$_3$) condenses to make an upper layer of white clouds. Jupiter's striped appearance occurs because atmospheric flow patterns create cloudy and clear zones at different latitudes, much as Earth's equatorial rainforests are cloudier than the desert regions to the north and south. Saturn has the same set of three cloud layers as Jupiter, but these layers occur deeper in Saturn's cooler atmosphere, making the colors more subdued. Uranus and Neptune contain more methane gas, and they're so cold that some of it condenses into methane clouds. Methane gas absorbs red light from the Sun and therefore gives Uranus and Neptune their distinctive blue colors.

All the jovian planets also have strong winds and powerful storms. The most famous storm is Jupiter's *Great Red Spot*, which is more than twice as wide as Earth. It is somewhat like a giant hurricane, except that its winds circulate around a high-pressure region rather than a low-pressure region (Figure 6.4). It is also extremely long-lived compared to storms on Earth: Astronomers have seen it throughout the two centuries during which telescopes have been powerful enough to detect it. No one knows why the Great Red Spot has lasted so long. However, storms on Earth tend to lose their strength when they pass over land—perhaps Jupiter's biggest storms last for centuries simply because there's no solid surface there to sap their energy.

Rings All of the jovian planets have rings, but only Saturn's rings are easily visible from Earth. They are also the best-studied rings, so we'll use them to discuss general properties of ring systems.

Earth-based views make Saturn's rings appear to be continuous, concentric sheets of material separated by a large gap called the *Cassini division* (Figure 6.5a). Spacecraft images reveal these sheets to be made of many individual rings, each separated from the next by a narrow gap (Figure 6.5b). But even these images are somewhat deceiving. If we could wander into Saturn's rings, we'd find that they are made of countless icy particles, ranging in size from dust grains to large boulders, sometimes clumped together by their mutual gravity (Figure 6.5c). Each individual ring particle orbits Saturn independently in accord with Kepler's laws.

Close-up images show an astonishing number of rings, as well as gaps, ripples, and other features. Scientists are still struggling to explain all these features, but some general ideas are now clear. Rings and gaps are caused by particles

▲ **FIGURE 6.4**
Jupiter's Great Red Spot, viewed here in enhanced color from the *Juno* spacecraft, is a huge high-pressure storm that is large enough to swallow Earth. The overlaid diagram shows a weather map of the region. The global image was taken by the *Cassini* spacecraft when it flew past Jupiter on its way to Saturn.

a This Earth-based telescopic view of Saturn makes the rings look like large, concentric sheets. The dark gap within the rings is called the *Cassini division*.

b This image of Saturn's rings from the *Cassini* spacecraft reveals many individual rings separated by narrow gaps.

c Artist's conception of particles in a ring system. Particles clump together because of gravity, but small random velocities cause collisions that break them up.

▲ **FIGURE 6.5**
Zooming in on Saturn's rings.

bunching up at some orbital distances and being forced out at others. This bunching happens when gravity nudges the orbits of ring particles in some particular way. One source of nudging comes from small moons located within the gaps in the rings themselves, sometimes called *gap moons* (Figure 6.6).

Ring particles also may be nudged by the gravity of larger, more distant moons. For example, a ring particle orbiting about 120,000 kilometers from Saturn's center will circle the planet in exactly half the time it takes Saturn's moon Mimas to orbit. Every time Mimas returns to a certain location, the ring particle will also be at its original location and therefore will experience the same gravitational nudge from Mimas. The periodic nudges reinforce one another and clear a gap in the rings—in this case, the Cassini division. This type of reinforcement due to repeated gravitational nudges is called an **orbital resonance**; the name comes from the idea that the repeated nudges *resonate* with one another, amplifying the effects. Other orbital resonances, caused by moons both within the rings and farther out from Saturn, probably explain most of the intricate structures visible in ring photos.

Where do rings come from? Scientists once guessed that ring particles might be chunks of rock and ice left over from the time when the planets formed, but we now know that particles of the size we find in the rings today could not have survived for billions of years. Ring particles are continually being ground down in size, primarily by the impacts of the countless sand-size particles that orbit the Sun—the same types of particles that become meteors in Earth's atmosphere. Millions of years of such tiny impacts would have ground Saturn's existing ring particles to dust long ago. (In the case of Uranus, ring particles are also lost as a result of drag in the planet's thin upper atmosphere.)

We are left with only one reasonable possibility: New ring particles must be continually replacing those that are destroyed. The most likely source is numerous small *moonlets*—moons the size of gap moons (see Figure 6.6)—that formed in the disks of material orbiting the young jovian planets. Tiny impacts are gradually grinding away these small moonlets, but they are large enough to still exist despite $4\frac{1}{2}$ billion years of such sandblasting. This allows them to contribute ring particles in two ways. First, each tiny impact releases particles from a small moonlet's surface, and these released particles become new dust-size ring particles. Second, occasional larger impacts can shatter a small moonlet completely, creating a supply of boulder-size ring particles. Figure 6.7 summarizes the way small moonlets are thought to supply the particles that make up the rings around the jovian planets.

▲ FIGURE 6.6
Small moons within the rings have important effects on ring structure. Images from the *Cassini* spacecraft.

◆ Why are jovian moons so geologically active?

In addition to their rings, the jovian planets are orbited by more than 170 known moons, including more than 60 each for Jupiter and Saturn. It's helpful to organize these moons into three groups by size: large moons more than 1500 kilometers in diameter, medium moons ranging from about 300 to 1500 kilometers in diameter,

▲ FIGURE 6.7
This illustration summarizes the origin of rings around the jovian planets.

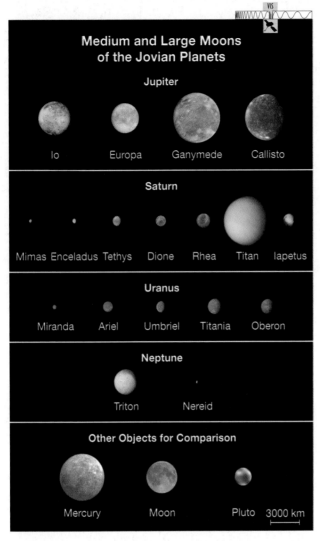

▲ **FIGURE 6.8**
The medium and large moons of the jovian planets, with sizes (but not distances) shown to scale. Mercury, the Moon, and Pluto are included for comparison.

and small moons less than about 300 kilometers in diameter. The vast majority of the jovian moons are small; many are no more than a few kilometers in diameter. Most of these small moons are probably asteroids or comets that were captured into orbit when the jovian planets were still surrounded by swirling clouds of gas [Section 4.2].

Figure 6.8 shows all of the medium and large moons to scale. These worlds are all large enough for gravity to have shaped them into spheres, so they would qualify as dwarf planets or planets if they orbited the Sun independently. Because they have solid surfaces, these moons are shaped by the same four geological processes as the terrestrial worlds—impact cratering, volcanism, tectonics, and erosion. The most surprising aspect of these moons is their level of geological activity. Because even the largest of them are only slightly larger than the Moon and Mercury, we might expect all of these moons to be geologically dead, like those two terrestrial worlds. Instead, we find evidence of tremendous past geological activity on the jovian moons, and some of them remain geologically active today.

How can such small worlds have so much geological activity? The best way to understand the answer is to look at a few of the most interesting cases. As we will see, their geological activity can be traced to two major factors. First, because they formed in the cold outer solar system, the jovian moons have compositions that include substantial amounts of ice in addition to metal and rock, and much less heat is required for "ice geology" than for rock geology. Second, interactions between orbiting moons help to create a source of heating, called *tidal heating*, that is not present among the terrestrial worlds.

The Galilean Moons of Jupiter Jupiter's four largest moons—Io, Europa, Ganymede, and Callisto—are collectively known as the *Galilean moons*, because they were discovered by Galileo [Section 3.1]. Let's start with Io.

For anyone who thinks of moons solely as barren, cratered worlds like our own Moon, Io shatters the stereotype. Active volcanoes pockmark its entire surface, making Io by far the most volcanically active world in our solar system (Figure 6.9). Io's volcanic activity tells us that it must be quite hot inside. However, Io is only about the size of our Moon, so it should have long ago lost any heat from its birth and is too small for radioactivity to provide much ongoing heat. It must therefore have some other source of heat for its interior. Scientists have identified this source as **tidal heating**, which arises from tidal forces exerted by Jupiter.

You probably know that the Moon's gravity creates tides on Earth. Tides arise because the strength of gravity declines with distance, so the gravitational attraction

a Most of the black, brown, and red spots on Io's surface are recently active volcanic features. White and yellow areas are sulfur dioxide (SO_2) and sulfur deposits, respectively, from volcanic gases. (Photographs from the *Galileo* spacecraft; some colors slightly enhanced or altered.)

b Two views of Io's volcanoes taken by *New Horizons* on its way to Pluto.

▲ **FIGURE 6.9**
Io is the most volcanically active body in the solar system.

between the Moon and the side of Earth facing the Moon is stronger than that between the Moon and the side of Earth facing away from the Moon. This difference in attraction creates a "stretching force," or **tidal force**, that stretches the entire Earth to create two tidal bulges: one facing the Moon and one opposite the Moon. Earth's rotation carries every point on the surface through the two bulges each day, which is why there are two high tides each day. Although the tidal rise and fall of the oceans is far more noticeable, tides also cause land to rise and fall twice each day (by about a centimeter). This daily motion generates friction inside Earth, and over millions of years this friction is gradually slowing Earth's rotation.

Just as the Moon's gravity causes it to exert a tidal force on Earth, Earth's gravity exerts a tidal force on the Moon. In fact, because Earth is much more massive than the Moon, Earth's tidal force on the Moon has a much greater effect than the Moon's tidal force on Earth. This strong tidal force explains the Moon's synchronous rotation (see Figure 2.11): The Moon probably once rotated faster, but Earth's tidal force generated friction inside the Moon that slowed its rotation until it kept the same face toward Earth at all times.

Other planets and moons also exert tidal forces on each other. Jupiter's tidal force on the Galilean moons is so strong that it has caused all four moons to keep the same face toward Jupiter at all times. Tides cause heating on Io, Europa, and Ganymede because these three moons have elliptical orbits around Jupiter, which means the strength of Jupiter's gravity and tidal force acting on them varies as these moons orbit. This varying tidal force flexes the interiors of these moons continuously, generating friction and heat. But why do these moons have elliptical orbits, when almost all other large moons have nearly circular orbits? The answer lies in an orbital resonance much like those that affect Saturn's rings: During the time Ganymede takes to complete one orbit of Jupiter, Europa completes exactly two orbits and Io completes exactly four orbits (Figure 6.10). The three moons therefore line up periodically, and over time the repeated gravitational tugs with these alignments have stretched the orbits from circles into ellipses. Tidal heating is strongest for Io because it is closest to Jupiter. Precise calculations show that tidal heating can indeed generate enough heat to explain Io's incredible volcanic activity.

Tidal heating is weaker for Europa but still strong enough to make Europa one of the most interesting places in the solar system. Europa's surface is covered with water ice that shows cracks and other visual evidence suggesting that the ice sometimes melts and refreezes (Figure 6.11). This fact, along with calculations based on tidal heating and studies of Europa's magnetic field, has led scientists to conclude that Europa almost certainly has a vast, liquid water ocean beneath its surface. Detailed models indicate that Europa has a metallic core and rocky mantle, surrounded by a layer of water that is frozen solid near the surface and liquid a few kilometers down (Figure 6.12).

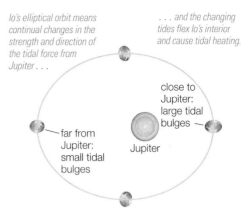

a Tidal heating arises because Io's elliptical orbit (exaggerated in this diagram) causes varying tides.

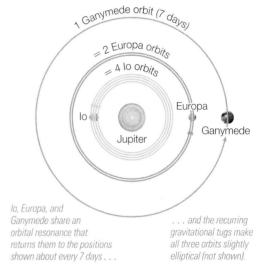

b Io's orbit is elliptical because of the orbital resonance Io shares with Europa and Ganymede.

▲ **FIGURE 6.10**
These diagrams explain the cause of tidal heating on Io. Tidal heating has a weaker effect on Europa and Ganymede, because they are farther from Jupiter and tidal forces weaken with distance.

▲ **FIGURE 6.11**
Europa's icy crust may hide a deep, liquid water ocean beneath it. These photos are from the *Galileo* spacecraft; colors are enhanced in the global view.

Europa may have a 100-km-thick ocean under an icy crust.

Rising plumes of warm water may sometimes create lakes within the ice, causing the crust above to crack.

▲ **FIGURE 6.12**
This diagram shows one model of Europa's interior structure. There is little doubt that the H₂O layer is real, but questions remain about whether the material beneath the icy crust is liquid water, relatively warm convecting ice, or some of each.

If it really exists, Europa's ocean may be more than 100 kilometers deep and contain more than twice as much liquid water as all of Earth's oceans combined. Moreover, the internal heat generated by tidal heating would likely mean that undersea volcanic vents dot Europa's seafloor. Knowing that primitive life thrives near seafloor vents on Earth has made scientists wonder if Europa might also harbor life, a possibility we will explore further in Chapter 15.

Ganymede and Callisto also have surfaces of water ice, and some data suggest that these moons have subsurface oceans. However, tidal heating alone is not strong enough to explain how Ganymede could have an ocean, and tidal heating plays no role at all on Callisto, which does not participate in the orbital resonance that affects the other three Galilean moons. If these moons have subsurface oceans, the heat may be supplied by ongoing radioactive decay. With or without oceans, these worlds show fascinating surface geology. Ganymede has some regions that are dark and densely cratered, suggesting that they look much the same today as they did billions of years ago, while other regions are light-colored with very few craters, suggesting that liquid water has recently flowed and refrozen (Figure 6.13). Callisto, by contrast, looks like a heavily cratered iceball, just as we might expect given its lack of tidal heating (Figure 6.14).

Saturn's Active Moons: Titan and Enceladus We next turn to two of Saturn's moons, Titan and Enceladus. Titan is the second largest moon in the solar system (after Ganymede). It is also unique among the moons of our solar system in having a thick atmosphere—so thick that it hides the surface from view, except at a few specific wavelengths of light (Figure 6.15, left). The atmosphere is more than 95% nitrogen, not that different from the 77% nitrogen content of Earth's atmosphere. However, the rest of Earth's atmosphere is mostly oxygen, while the rest of Titan's atmosphere consists of argon and methane (CH_4), ethane (C_2H_6), and other hydrogen compounds.

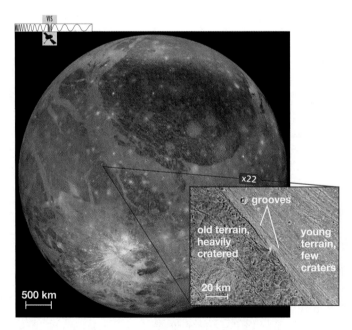

▲ **FIGURE 6.13**
Ganymede, the largest moon in the solar system, has both old and young regions on its surface of water ice. The dark regions are heavily cratered and must be billions of years old, while the light regions are younger landscapes where eruptions of water have presumably erased ancient craters; the long grooves in the light regions were probably formed by water erupting along surface cracks. Notice that the boundary between the two types of terrain can be quite sharp.

Callisto is heavily cratered, indicating an old surface that nonetheless may hide a deeply buried ocean.

Close-up photo shows a dark powder overlaying the low areas of the surface.

▲ **FIGURE 6.14**
Callisto, the outermost of the four Galilean moons, has a heavily cratered icy surface.

▲ FIGURE 6.15
This sequence zooms in on the *Huygens* landing site on Titan. Left: A global view taken by the *Cassini* orbiter using filters designed to peer through the atmosphere at the specific near-infrared wavelengths least affected by the atmosphere. Center: An aerial mosaic of images taken by the probe during descent. Right: A surface view taken by the probe after landing; the "rocks," which are 10 to 20 centimeters across, are presumably made of ices. Keep in mind that you are looking at the surface of a world more than a billion kilometers away.

Methane and ethane are both greenhouse gases and therefore give Titan a greenhouse effect [Section 5.1] that makes it warmer than it would be otherwise. Still, because of its great distance from the Sun, Titan's surface temperature is a frigid 93 K (−180°C). The surface pressure is about 1.5 times the sea level pressure on Earth, which would be fairly comfortable if not for the lack of oxygen and the cold temperatures.

A moon with a thick atmosphere is intriguing enough, but we have at least two other reasons for special interest in Titan. First, its complex atmospheric chemistry probably produces numerous organic chemicals—the chemicals that are the basis of life. Second, although it is far too cold for liquid water to exist on Titan, conditions are right for methane or ethane rain. NASA and the European Space Agency (ESA) joined forces to explore Titan with the *Cassini* spacecraft that orbits Saturn and a European-built probe, called *Huygens* (pronounced "Hoy-guns"), that parachuted to a soft landing on Titan in January 2005. During its descent, the probe photographed river valleys merging together, flowing down to what looks like a shoreline (Figure 6.15, center). On the ground, instruments discovered that the surface has a hard crust but is a bit squishy below, like sand with liquid mixed in. The view from the surface shows "ice boulders" rounded by erosion (Figure 6.15, right). All these results support the idea of a wet climate—but wet with liquid methane or ethane rather than liquid water. *Cassini* radar observations confirmed the presence of liquid methane or ethane lakes on Titan (Figure 6.16).

We can attribute Titan's geological activity to the combination of its relatively large size and its icy composition. Because all of Saturn's other moons are much smaller, scientists had expected them to show very little evidence of active geology. The reality has proven to be quite different. Each of Saturn's six medium-size moons shows evidence of substantial geological activity in its past. Most amazingly, one moon—Enceladus—shows clear evidence of *ongoing* geological activity, despite being barely 500 kilometers across, small enough to fit inside the borders of Colorado. Enceladus features strange grooves near its south pole that vent huge clouds of water vapor and ice crystals (Figure 6.17). These fountains are driven by internal heat, which apparently arises from tidal heating created by an orbital resonance (with Saturn's moon Dione). Moreover, the fountains must have some subsurface source, and other *Cassini* data suggest that Enceladus has a

▲ FIGURE 6.16
Radar image of Ligeia Mare near Titan's north pole showing lakes of liquid methane and ethane at a temperature of −180°C. Most solid surfaces reflect radar well, and these regions are artificially shaded tan to suggest land. The liquid surfaces reflect radar poorly, and these regions are shaded blue and black to suggest lakes.

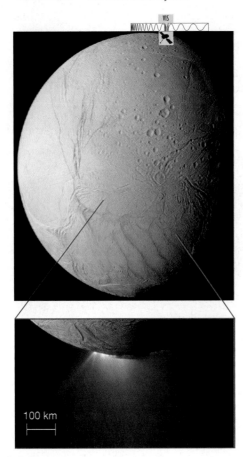

▲ **FIGURE 6.17**
Cassini photo of Enceladus. The blue "tiger stripes" near the bottom of the main photo are regions of fresh ice that must have recently emerged from below. The colors are exaggerated; the image is a composite made at near-ultraviolet, visible, and near-infrared wavelengths. The inset shows Enceladus backlit by the Sun, with fountains of ice particles and water vapor spraying downward.

global, subsurface ocean of liquid water or of a colder water/ammonia mixture, offering at least a slim possibility that Enceladus could harbor life.

think about it The *Cassini* spacecraft continues to study Titan and Enceladus as it orbits Saturn. Do a web search to find recent results. What new things have we learned about these worlds?

Moons of Uranus and Neptune We know far less about the moons of Uranus and Neptune, because they have been photographed close up only once, when *Voyager 2* flew past these worlds in the 1980s. Nevertheless, we again see evidence of surprising geological activity.

Uranus has five medium-size moons (and no large moons), of which at least three show evidence of past volcanism or tectonics. Miranda, the smallest of the five, is the most surprising (Figure 6.18). Despite its small size, it shows tremendous tectonic features and relatively few craters. Apparently, it underwent geological activity well after the heavy bombardment ended [Section 4.2], erasing its early craters.

The surprises continue with Neptune's moon Triton (Figure 6.19). Triton is a strange moon to begin with: It is a large moon, but it orbits Neptune "backward" (opposite to Neptune's rotation) and at a high inclination to Neptune's equator. These are telltale signs of a moon that was captured rather than formed in the disk of gas around its planet. No one knows exactly how a moon as large as Triton could have been captured, but models suggest one possible mechanism: Triton may have once been a member of a binary Kuiper belt object that passed so close to Neptune that Triton lost energy and was captured while its companion gained energy and was flung away at high speed.

Triton's geology is just as surprising as its origin. Triton is smaller than our own Moon, yet its surface shows evidence of relatively recent geological activity. Some regions show evidence of past volcanism, while others show wrinkly ridges (nicknamed "cantaloupe terrain") that appear tectonic in nature. Triton even has a very thin atmosphere that has left some wind streaks on its surface. It's likely that Triton was originally captured into an elliptical orbit, which may have led to enough tidal heating to explain its geological activity.

▲ **FIGURE 6.18**
The surface of Uranus's moon Miranda shows astonishing tectonic activity despite its small size. The cliff walls seen in the inset are higher than those of the Grand Canyon on Earth.

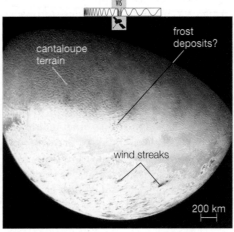

Triton's southern hemisphere as seen by *Voyager 2*.

▲ **FIGURE 6.19**
Neptune's moon Triton shows evidence of a surprising amount of past geological activity.

This close-up shows lava-filled impact basins similar to the lunar maria, but the lava was water or slush rather than molten rock.

Ice Geology The jovian moons have taught us an important lesson about geology. If these moons were similar to the terrestrial worlds, their small sizes would have left them geologically dead long ago. However, there is a crucial difference between the jovian moons and the terrestrial worlds: composition.

Most of the jovian moons contain ices that can melt or deform at far lower temperatures than rock. As a result, they can experience geological activity even when their interiors have cooled to temperatures far below what they were at their births. Indeed, except on Io, most of the volcanism that has occurred in the outer solar system probably did not produce any hot lava at all; it produced icy lava that was essentially liquid water, perhaps mixed with methane and ammonia. The key point is that ice geology is possible at far lower temperatures than rock geology. This fact, combined with the occasional extra heat supplied by tidal heating, explains how the jovian moons have had such interesting geological histories despite their small sizes.

▶ **Asteroids, Comets, and Dwarf Planets**

6.2 Asteroids, Comets, Dwarf Planets, and the Impact Threat

The small bodies of the solar system include asteroids, comets, and dwarf planets. The small sizes of these objects might at first make them seem insignificant, but there is strength in numbers. Moreover, their small sizes allow them to be more easily perturbed by gravitational tugs from other objects, and as a result they sometimes head inward to the inner solar system, where they can appear as spectacular comets or, occasionally, come crashing down to Earth.

◆ **Why are asteroids and comets grouped into three distinct regions?**

As we discussed in Chapter 4, the small bodies of the solar system are essentially leftover planetesimals that now orbit the Sun in three major regions: the *asteroid belt* between Mars and Jupiter, the *Kuiper belt* just beyond the orbit of Neptune, and the spherical *Oort cloud* that extends to great distances from the Sun (see Figure 4.1, Step 3). We can understand why they are grouped in this way by investigating the nature of small bodies in each of these regions.

The Asteroid Belt The asteroid belt probably contains at least a million asteroids with diameters greater than 1 kilometer and many more even smaller in size. Small asteroids are not round (see Figure 4.3) because their gravity is not strong enough to give them that shape. Only the largest, Ceres (diameter close to 1000 kilometers), is round enough to qualify as a dwarf planet (Figure 6.20a), though Vesta (second largest by mass) might also have been round before an impact gave it a huge polar crater (Figure 6.20b). About a dozen other asteroids are large enough that we would call them medium moons if they orbited a planet.

The fact that asteroids are rocky leftover planetesimals similar to those that formed the terrestrial planets suggests that they must once have been spread fairly evenly throughout the inner solar system. Why, then, are they now concentrated in the asteroid belt? The simple answer is that the asteroid belt was the only place where rocky planetesimals could survive. Nearly all of the rocky planetesimals that formed within Mars's orbit ultimately accreted into one of the four inner planets. In contrast, the asteroids in the asteroid belt stay clear of any planet and can therefore survive on their current orbits for billions of years. But

---

common misconceptions

Dodge Those Asteroids!

Science fiction movies often show brave spacecraft pilots navigating through crowded fields of asteroids, dodging this way and that as they heroically pass through with only a few bumps and bruises. It's great drama, but not very realistic. The asteroid belt looks crowded when we draw it on paper as in Figure 4.1, but in reality it is an enormous region of space. Despite their large numbers, asteroids are so far apart on average that it would take incredibly bad luck to crash into one by accident. Indeed, spacecraft must be carefully guided to fly close enough to an asteroid to take a decent photograph. Future space travelers will have plenty of dangers to worry about, but dodging asteroids is not likely to be one of them.

a *Dawn* has been orbiting Ceres since 2015, revealing a world of abundant craters, some with bright spots consisting of salts and other reflective minerals on their floors, along with evidence of past geological activity.

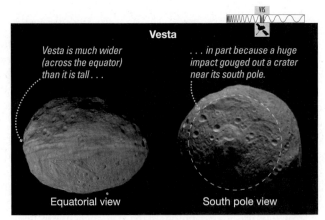

b *Dawn* orbited Vesta in 2011–2012 before heading to Ceres. Vesta is not quite round, probably because of its huge polar crater (dashed circle).

▲ **FIGURE 6.20**
The *Dawn* spacecraft has provided detailed views of the two largest asteroids (by mass), Ceres and Vesta.

this raises a deeper question: Why didn't another planet form in this region and sweep up asteroids, just as the terrestrial worlds did in the inner solar system?

Asteroid orbits provide the key clue: Most asteroids share a few particular orbital periods and distances, leaving gaps in which there are very few asteroids. Careful study shows that these gaps occur at orbital periods that are simple fractions of Jupiter's 12-year orbital period, such as one-half of Jupiter's period or one-third of Jupiter's period. These simple fractions are telltale signs of orbital resonances, much like those in Saturn's rings and among the Galilean moons of Jupiter. We conclude that Jupiter's gravity is the primary influence on the orbits of asteroids in the asteroid belt, and this influence also explains why a planet never formed in there. When the solar system was forming, this region probably contained enough rocky material to form another terrestrial planet. However, resonances with the young planet Jupiter disrupted the orbits of this region's planetesimals, preventing them from accreting into a full-fledged planet. Over the next $4\frac{1}{2}$ billion years, ongoing orbital disruptions gradually kicked pieces of this "unformed planet" out of the asteroid belt altogether. Once booted from the asteroid belt, these objects either crashed into a planet or moon or were flung out of the solar system. The asteroid belt thereby lost most of its original mass, which explains why the total mass of all its asteroids is now much less than that of any terrestrial planet.

The Kuiper Belt The next major region of small bodies is the Kuiper belt, which is similar to the asteroid belt in many ways, except its leftover planetesimals are ice-rich in composition. As a result, when a small Kuiper belt object is gravitationally perturbed and flung into the inner solar system, its ice begins to vaporize and we may see it as a long-tailed comet in our sky (Figure 6.21). (The comets we see in our sky can come from the Oort cloud as well as the Kuiper belt.)

As in the asteroid belt, the orbits of most Kuiper belt objects are shaped by the gravity of a jovian planet—in this case, Neptune. However, the Kuiper belt probably lacks a large planet not because objects were flung off but because the low densities at such great distance from the Sun slowed the rate at which planetesimals grew. As a result, none of them grew large enough to draw in surrounding hydrogen and helium gas before sunlight and the solar wind cleared the solar nebula.

Although no jovian planet formed in the Kuiper belt, some icy planetesimals grew large enough for gravity to make them round, qualifying them as dwarf planets; the two largest known of these are Pluto and Eris. As we discussed briefly in

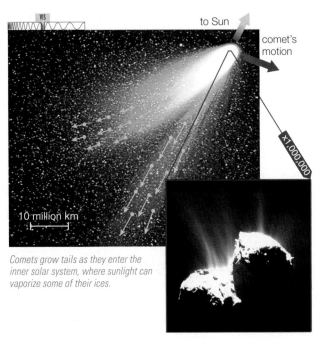

▲ **FIGURE 6.21**
Comets are ice-rich leftover planetesimals that become visible in our sky only if and when they visit the inner solar system; most comets remain perpetually frozen in either the Kuiper belt or the Oort cloud. The main photo shows the tails of Comet Hale-Bopp as seen from Earth, and the inset shows the nucleus of Comet 67P/Churyumov-Gerasimenko, photographed by the *Rosetta* spacecraft.

a Surface features of Pluto, shown with enhanced color. The bright, heart-shaped region in the central image is called Tombaugh Regio after Pluto's discoverer.

Section 1.3, the *New Horizons* flyby of Pluto revealed that dwarf planets can have very interesting geological histories. In fact, Pluto's surface features suggest it may still be geologically active today, and it even has a thin atmosphere (Figure 6.22).

The Oort Cloud The Oort cloud probably extends to a distance nearly 1000 times as far from the Sun as Pluto, which puts its most distant objects nearly one-quarter of the way to the nearest stars. We have never seen a comet in the Oort cloud, because our telescopes are not yet capable of detecting such small objects at such great distances from the Sun. However, based on the numbers of comets that enter the inner solar system, scientists estimate that the Oort cloud contains close to a *trillion* (10^{12}) comets. How did so many small bodies end up orbiting the Sun at such great distances? The only answer that makes scientific sense comes from thinking about what happened to the icy leftover planetesimals that roamed the region in which the jovian planets formed.

The leftover planetesimals that cruised the spaces between Jupiter, Saturn, Uranus, and Neptune were doomed to suffer either a collision or a close gravitational encounter with one of the young jovian planets. The planetesimals that escaped being swallowed up tended to be flung off in all directions. Some may have been cast away at such high speeds that they completely escaped the solar system and now drift through interstellar space. The rest ended up on orbits with very large average distances from the Sun. These became the comets of the Oort cloud. The random directions in which they were flung, along with small gravitational nudges from other stars, explain the roughly spherical shape of the Oort cloud.

b Pluto's jagged mountains, smooth plains, and hazy atmosphere, seen from *New Horizons* as it looked back toward the Sun after the encounter.

▲ **FIGURE 6.22**
New Horizons images of Pluto.

think about it Consider the explanations given for the origins of the Kuiper belt and the Oort cloud. Assuming these explanations are correct, which region contains objects that formed *closer* to the Sun? Explain.

Altering Orbits The existence of impact craters provides proof that small bodies are not always confined to the asteroid belt, Kuiper belt, and Oort cloud. The objects that have crashed into the planets and moons must have had orbits that crossed planetary orbits, and there are still many such objects today. Some of these objects may simply be leftovers from formation that have been lucky enough to avoid a collision to date. More are probably objects that once orbited

▲ **FIGURE 6.26**
Around the world, sedimentary rock layers dating to 65 million years ago share evidence of the impact of a comet or asteroid. Fossils of dinosaurs and many other species appear only in rocks below the iridium-rich layer.

▲ **FIGURE 6.27**
This computer-generated image, based on measurements of small local variations in the strength of gravity, shows an impact crater (dashed circle) that straddles the coast of Mexico's Yucatán Peninsula; the red box in the inset shows the location of the region in the main image.

Impact Evidence In 1978, while analyzing geological samples collected in Italy, a scientific team led by father and son Luis and Walter Alvarez made a startling discovery. They found that a thin layer of dark sediments deposited about 65 million years ago—about the time the dinosaurs went extinct—was unusually rich in the element iridium. Iridium is a metal that is rare on Earth's surface but common in meteorites and therefore presumably common in asteroids and comets. Subsequent studies found the same iridium-rich layer in 65-million-year-old sediments around the world (Figure 6.26). Based on this evidence, the Alvarez team hypothesized that the extinction of the dinosaurs was caused by the impact of an asteroid or comet.

Further evidence for an impact comes from four other features found in the iridium-rich sediment layer: (1) unusually high abundances of several other metals, including osmium, gold, and platinum; (2) grains of *shocked quartz*, quartz crystals with a distinctive structure that indicates they experienced the high-pressure conditions of an impact; (3) spherical *rock droplets* of a type known to form when drops of molten rock cool and solidify in the air; and (4) soot. All these features point to an impact. The metal abundances look more like what we commonly find in meteorites than what we find elsewhere on Earth's surface. Shocked quartz is characteristic of impact sites, including Meteor Crater in Arizona. The rock droplets presumably were made from molten rock splashed into the air by the force and heat of the impact. The soot probably came from vast forest fires ignited by impact debris. Some debris would have been blasted so high that it rose above the atmosphere and spread worldwide. It then would have plunged downward, with atmospheric friction heating it until it became a rain of hot, glowing rock.

More direct evidence comes from the presence of a large impact crater that appears to match the sediment layer in age (Figure 6.27). The size of the crater indicates that it was created by the impact of an asteroid or a comet measuring about 10 kilometers across.

The Extinction We have strong evidence that an impact occurred at the right time, but how did it lead to extinction? Models suggest it could have happened as follows. On that fateful day some 65 million years ago, the asteroid or comet slammed into Mexico with the force of a hundred million hydrogen bombs (Figure 6.28). North America may have been devastated immediately. Hot debris from the impact rained around the rest of the world, igniting fires that killed many more living organisms. Longer-term effects were even more severe. Dust and smoke remained in the atmosphere for weeks or months, blocking sunlight and causing temperatures to fall as if Earth were experiencing a harsh global winter. The reduced sunlight would have stopped photosynthesis for up to a year, killing large numbers of species throughout the food chain. Acid rain may have been another by-product, killing vegetation and acidifying lakes around the world. Chemical reactions in the atmosphere probably produced nitrous oxides and other compounds that dissolved in the oceans and killed marine organisms. Recent evidence also suggests that a period of intense volcanic activity followed the impact—and perhaps was caused by it—which may have further exacerbated the immediate effects of the impact.

If these models are correct, the most astonishing fact may be that some species managed to survive. Among these were a few mammals, which may have survived in part because they lived in underground burrows and managed to store enough food to outlast the global winter that followed the impact. With the dinosaurs gone, these mammals became the new kings of the planet, evolving over the next 65 million years into numerous new species, ultimately including us.

The Bottom Line There is still scientific dispute as to whether an impact was the sole cause of the extinction of the dinosaurs or just one of several causes. But the evidence leaves little doubt that a major impact coincided with the mass extinction of 65 million years ago, and we've discussed a plausible scenario for how the impact might have caused the extinction. This case illustrates a crucial point about how science progresses. The hypothesis of the Alvarez team was quite controversial when they first proposed it, in part because it was the first time anyone had suggested that an astronomical event might change the course of biological evolution. However, the evidence they presented was strong enough that it was published in scientific journals, inspiring other scientists to search for additional geological evidence that could support or refute the impact hypothesis, to learn more about the probability of such a major impact, and to create models of how an impact might affect the environment. These studies showed that large impacts were not only possible but inevitable over geological time scales. Combined with the additional evidence of an impact that was found and a model showing that large impacts could indeed lead to a mass extinction, a hypothesis that was once greeted with great skepticism has since gained widespread acceptance.

▲ FIGURE 6.28
This painting shows an asteroid or comet moments before its impact on Earth, some 65 million years ago. The impact probably caused the extinction of the dinosaurs.

summary of key concepts

6.1 Jovian Planets, Rings, and Moons

◆ **What are jovian planets like?**

The jovian planets are far larger than the terrestrial worlds and are very different in nature. They have no solid surfaces and contain vast amounts of hydrogen, helium, and hydrogen compounds. They have strong weather in their atmospheres and layered interiors. They are each orbited by many moons and by rings that consist of countless individual particles.

◆ **Why are jovian moons so geologically active?**

Even the largest jovian moons are barely larger than the Moon or Mercury, but many of them show evidence of substantial past or present geological activity. The source of this activity can probably be traced to the facts that *ice geology* can occur at lower temperatures than rock geology and that some of these moons have a source of heat— **tidal heating**—that is not important for the terrestrial worlds.

6.2 Asteroids, Comets, Dwarf Planets, and the Impact Threat

◆ **Why are asteroids and comets grouped into three distinct regions?**

Small bodies are found in three major groups: the rocky asteroids of the asteroid belt and the comets of the Kuiper belt and the Oort cloud. In all three cases, the origin and distribution of these small bodies have been governed primarily by gravitational interactions between these bodies and the large jovian planets.

◆ **Do small bodies pose an impact threat to Earth?**

Impacts have clearly occurred in Earth's past and will occur in the future. The level of threat depends on the size of the object: Impacts by smaller bodies cause less serious damage but occur more frequently. Large impacts can cause devastating damage but are quite rare.

 THE PROCESS OF SCIENCE IN ACTION

6.3 Extinction of the Dinosaurs

◆ **Did an impact kill the dinosaurs?**

It may not have been the sole cause, but a major impact clearly coincided with the **mass extinction** that occurred about 65 million years ago, in which the dinosaurs died out. Sediments from this era contain iridium and other evidence of an impact, and an impact crater of the right age lies along the coast of Mexico.

Investigations

For instructor-assigned homework and other learning materials, go to www.MasteringAstronomy.com.

Quick Quiz

Choose the best answer to each of the following. Explain your reasoning.

1. Which lists the jovian planets in order of increasing distance from the Sun? (a) Jupiter, Saturn, Neptune, Uranus (b) Saturn, Jupiter, Uranus, Neptune (c) Jupiter, Saturn, Uranus, Neptune
2. What is Jupiter's Great Red Spot? (a) the largest mountain in the solar system (b) a very large cloud of ammonium hydrosulfide (c) a giant storm
3. The outer layers of most jovian moons are made primarily of (a) rock and metal. (b) ices. (c) hydrogen and helium.
4. Why is Io more volcanically active than our Moon? (a) Io is much larger. (b) Io has a higher concentration of radioactive elements. (c) Io has a different internal heat source.
5. What is unusual about Triton? (a) It orbits its planet backward. (b) It does not keep the same face toward its planet. (c) It is the only moon with rings.
6. Which moon shows evidence of rainfall and erosion by some liquid substance? (a) Europa (b) Titan (c) Ganymede
7. Saturn's many moons affect its rings primarily through (a) tidal forces. (b) orbital resonances. (c) magnetic field interactions.
8. The asteroid belt lies between the orbits of (a) Earth and Mars. (b) Mars and Jupiter. (c) Jupiter and Saturn.
9. Did a large terrestrial planet ever form in the region of the asteroid belt? (a) No, because there was never enough mass there. (b) No, because Jupiter prevented one from accreting. (c) Yes, but it was shattered by a giant impact.
10. Which objects have the most elliptical and tilted orbits? (a) asteroids (b) Kuiper belt comets (c) Oort cloud comets
11. About how often does a 1-kilometer object strike Earth? (a) every year (b) every million years (c) every billion years
12. Which of the following is *not* a key piece of evidence supporting the hypothesis that an impact caused the extinction of the dinosaurs? (a) rare metals found in 65-million-year-old sediments (b) radioactive isotopes found in 65-million-year-old dinosaur bones (c) an impact crater dating to 65 million years ago

Short-Answer/Essay Questions

Explain all answers clearly, with complete sentences and proper essay structure if needed. An asterisk (*) designates a quantitative problem, for which you should show all your work.

13. *Minor Ingredients Matter.* Suppose the jovian planet atmospheres were composed only of hydrogen and helium, with no hydrogen compounds at all. How would their atmospheres be different in terms of clouds, color, and weather? Explain.
14. *The New View of Titan.* What other planet or moon in the solar system does Titan most resemble, in your opinion? Summarize the similarities and differences in a few sentences.
15. *The Role of Jupiter.* Suppose that Jupiter had never existed. Describe at least three ways in which our solar system would be different, and clearly explain why.
16. *Asteroids vs. Comets.* Contrast the compositions and locations of comets and asteroids, and explain in your own words why they turned out differently.
17. *Oort Cloud vs. Kuiper Belt.* Explain in your own words how and why there are two different reservoirs of comets. Be sure to discuss where the two groups of comets formed and what kinds of orbits they travel.
18. *Rise of the Mammals.* Suppose the impact 65 million years ago had not occurred. How do you think our planet would be different? For example, do you think that mammals still would eventually have come to dominate Earth? Would we be here? Defend your opinions.
19. *Observing Project: Jupiter's Moons.* Using binoculars or a small telescope, view the moons of Jupiter. Make a sketch of what you see, or take a photograph. Repeat your observations several times (nightly, if possible) over a period of a couple of weeks. Can you determine which moon is which? Can you measure the moons' orbital periods? Can you determine their approximate distances from Jupiter? Explain.
*20. *Jupiter's Mass.* Io orbits Jupiter at an average distance of 421,600 kilometers and with an orbital period of 1.77 days. Use these facts and Newton's version of Kepler's third law to calculate Jupiter's mass. (*Hint:* Be sure to convert the orbital distance to meters and the period to seconds so that you can use the value of the gravitational constant $G = 6.67 \times 10^{-11} \text{ m}^3/\text{kg} \times \text{s}^2$.)
*21. *Adding Up Asteroids.* It's estimated that there are a million asteroids measuring 1 kilometer across or larger. If a million such asteroids were combined into one object, how big would it be? Compare your answer to the size of Earth. (*Hint:* Assume that each asteroid is a sphere with a radius of 0.5 km; the volume of a sphere is $\frac{4}{3}\pi r^3$.)

Planets Around Other Stars

7

20 AU
0.5"

For more than four centuries after the Copernican revolution taught us that Earth is just one member of our Sun's planetary system, the study of planetary systems remained limited to our own. Then, less than three decades ago, a new scientific revolution began with the first discoveries of planets around other stars. The image above, from the Large Binocular Telescope, shows infrared light from four planets (marked b, c, d, e) orbiting the star HR 8799; light from the star itself (center) was mostly blocked out during the exposure, as indicated by the solid red circle. The discovery that planetary systems are common around other stars has profound implications, making it seem more likely that we might someday find life elsewhere, perhaps even intelligent life. It also allows us to learn more about the general nature of planets and how they form, giving us deeper insights into our cosmic origins. In this chapter, we'll explore the exciting new science of other planetary systems.

LEARNING GOALS

7.1 Detecting Planets Around Other Stars
- How do we detect planets around other stars?
- What properties of extrasolar planets can we measure?

7.2 Characteristics of Extrasolar Planets
- How do extrasolar planets compare with planets in our solar system?
- Are Earth-like planets common?

 THE PROCESS OF SCIENCE IN ACTION

7.3 Extrasolar Planets and the Nebular Theory
- Do we need to modify our theory of solar system formation?

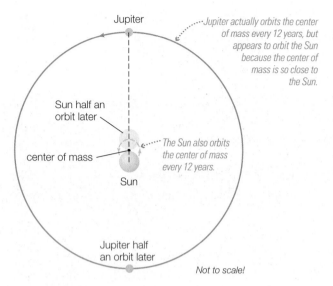

▲ **FIGURE 7.1**
This diagram shows how both the Sun and Jupiter orbit around their mutual center of mass, which lies very close to the Sun. The diagram is not to scale; the sizes of the Sun and its orbit are exaggerated about 100 times compared with the size shown for Jupiter's orbit, and Jupiter's size is exaggerated even more.

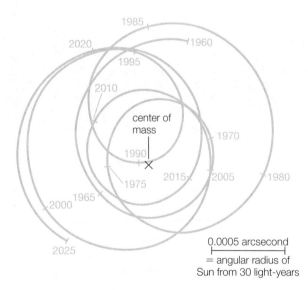

▲ **FIGURE 7.2**
This diagram shows the orbital path of the Sun around the center of mass of our solar system as it would appear over a 65-year period (1960–2025) from a distance of 30 light-years away. Notice that the entire range of motion during this period is only about 0.0015 arcsecond, which is almost 100 times smaller than the angular resolution of the Hubble Space Telescope. Nevertheless, if alien astronomers could measure this motion, they could learn of the existence of planets in our solar system.

▶ **Other Planetary Systems**

7.1 Detecting Planets Around Other Stars

Scientists long suspected that other stars might be orbited by their own planets, known as **extrasolar planets** because they lie beyond our own solar system. However, only in the past three decades has technology reached the point at which we could undertake a search for these planets. Although the search is in its early stages, it has already achieved remarkable success.

◆ How do we detect planets around other stars?

Detecting extrasolar planets poses a huge technological challenge for two basic reasons. First, as we discussed in Chapter 1, planets are extremely tiny compared with the vast distances to other stars. Second, stars are typically *a billion times* brighter than the light reflected by any orbiting planets, so starlight tends to overwhelm any planetary light in photographs. Nevertheless, scientists have begun to meet the challenge.

In a few cases, scientists have been able to obtain low-resolution direct images of extrasolar planets, usually using infrared light; the image that opens this chapter is an example. However, most of our current understanding of extrasolar planets comes from studies in which we *indirectly* infer the existence of planets without actually seeing them. There are two major indirect approaches to finding and studying extrasolar planets:

1. Observing the motion of a star to detect the subtle gravitational tugs of orbiting planets
2. Observing changes in a star's brightness that occur when one of its planets passes in front of the star as viewed from Earth

Gravitational Tugs Although we usually think of a star as remaining still while planets orbit around it, that is only approximately correct. In reality, all objects in a star system, including the star itself, orbit the system's "balance point," or *center of mass*. To understand how this fact allows us to discover extrasolar planets, imagine the viewpoint of extraterrestrial astronomers observing our solar system from afar.

Let's start by considering only the influence of Jupiter (Figure 7.1). The center of mass between Jupiter and the Sun lies just outside the Sun's visible surface (it is close to the Sun because Jupiter is only about 1/1000 the mass of the Sun), so what we usually think of as Jupiter's 12-year orbit around the Sun is really a 12-year orbit around this center of mass. Because the Sun and Jupiter are always on opposite sides of the center of mass (that's what makes it a "center"), the Sun must orbit this point with the same 12-year period. The Sun's orbit traces out only a small ellipse with each 12-year period, because the Sun's average orbital distance is barely larger than its own radius. Nevertheless, with sufficiently precise measurements, extraterrestrial astronomers could detect this orbital movement of the Sun and thereby deduce the existence of Jupiter without having ever seen the planet.

The other planets also exert gravitational tugs on the Sun, each adding a small additional effect to the effect of Jupiter (Figure 7.2). In principle, with sufficiently precise measurements of the Sun's orbital motion made over many decades, an extraterrestrial astronomer could deduce the existence of all the planets of our solar system. This is the essence of the **astrometric method**, in which we make very precise measurements of stellar positions in the sky (*astrometry* means "measurement of the stars"). If a star "wobbles" gradually around its average position (the center of mass), we must be observing the influence of unseen planets. The primary difficulty with the astrometric method is that we are looking for changes in position that are very small even for nearby stars, and smaller still for more distant stars. In addition, the stellar motions are largest for massive planets orbiting far from their star, but the long orbital periods of such planets mean that it can take decades to notice the motion. These difficulties have limited the use of the astrometric method to date. However, the European Space Agency's *GAIA* mission, launched in 2013, is in the process of obtaining astrometric observations of a billion stars to an accuracy that is in some cases better than 10 microarcseconds, equivalent to the angular width of a human hair viewed from a distance of 2000 kilometers. These observations should ultimately enable *GAIA* to detect thousands of extrasolar planets.

A second method for searching for gravitational tugs due to orbiting planets is the **Doppler method**, in which we search for a star's orbital movement around the center of mass by looking for changing *Doppler shifts* in the star's spectrum (see Tools of Science, page 116). As long as a planet's orbit is *not* face-on to us, the planet's gravitational influence will cause its star to move alternately toward and away from us, creating alternating blueshifts and redshifts (relative to the star's average Doppler shift) in the star's spectrum (Figure 7.3). As an example, Figure 7.4 shows data from a star called 51 Pegasi; the 4-day period of the star's motion tells us that the star is orbited by a planet with the same orbital period. This short orbital period implies that the planet lies very close to its star, which means it must have a very high surface temperature. The Doppler data also allow us to determine the planet's approximate mass, because a more massive planet has a greater gravitational effect on the star (for a given orbital distance) and therefore causes the star to move at higher speed around the system's center of mass. In the case of 51 Pegasi, the planet has about half the mass of Jupiter. Scientists therefore refer to this planet as a **hot Jupiter**, because it has a Jupiter-like mass but a much higher surface temperature.

The Doppler method has proven very successful. As of 2018, it had been used to detect more than 700 planets, including more than 100 in multiplanet systems. Keep in mind that because the Doppler method searches for gravitational tugs from orbiting planets, it is much better for finding massive planets like Jupiter than small planets like Earth. It's also best suited to identifying planets that orbit relatively close to their star, because being closer means a stronger gravitational tug and shorter orbital period.

Transits A small fraction (about 1%) of stars with planets should by chance be aligned so that their planets pass in front of their stars once each orbit as seen from Earth. Such events are called **transits**, and they cause small, temporary dips in a system's brightness (Figure 7.5). The larger the planet, the more dimming it will cause. The **transit method** searches for planets by looking for these temporary dimmings of a star. Some transiting planets also undergo a measurable **eclipse** as the planet goes behind the star. Eclipse observations are more easily accomplished in the infrared, because planets contribute a greater proportion of a system's infrared brightness than visible-light brightness.

think about it Suppose you were an alien astronomer searching for planets going around our Sun. Where would your star system have to be located in order for you to observe transits across the Sun? Which planets in our solar system would be the easiest to discover through transits? Explain.

Many modern telescopes can simultaneously monitor large numbers of stars in search of transits. For example, NASA's *Kepler* mission spent about four years (2009 to 2013) measuring the brightnesses of about 150,000 stars about every 30 minutes, in search of dimmings that might indicate transiting planets. Of course, the fact that we see a star dim does not necessarily mean a planet has passed in front of it, because there are many other possible causes for temporary dimming

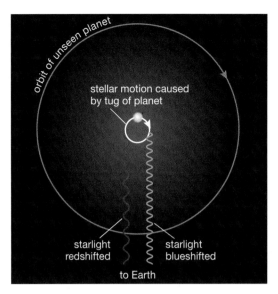

▲ **FIGURE 7.3**
The Doppler method for discovering extrasolar planets: The star's Doppler shift alternates toward the blue and toward the red, allowing us to detect its slight motion—caused by an orbiting planet—around the center of mass.

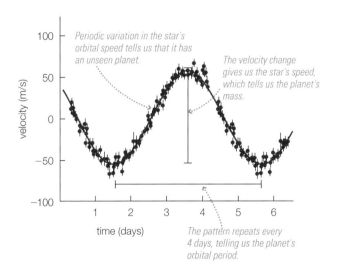

▲ **FIGURE 7.4**
A periodic Doppler shift in the spectrum of the star 51 Pegasi reveals the presence of a large planet with an orbital period of about 4 days. Dots are actual data points; bars through dots represent measurement uncertainty.

tools of science | The Doppler Effect

▶ Light and Matter (Learning Outcome 4)

We can learn about the motions of distant objects relative to us by identifying shifts in their spectra caused by the **Doppler effect**. The Doppler effect is crucial not only to the Doppler method for searching for extrasolar planets but also to many other areas of astronomy that we'll discuss later in the book.

You've probably noticed the Doppler effect on the *sound* of a train whistle near train tracks. If the train is stationary, the pitch of its whistle sounds the same no matter where you stand (Figure 1a). But if the train is moving, the pitch will sound higher when the train is coming toward you and lower when it's moving away from you. Just as the train passes by, you can hear the dramatic change from high to low pitch—a sort of "weeeeeeee–ooooooooh" sound. To understand why, think about what happens to the sound waves coming from the train (Figure 1b). When the train is moving toward you, each pulse of a sound wave is emitted a little closer to you. The bunching up of the waves between you and the train gives them a shorter wavelength and higher frequency (pitch). After the train passes you, each pulse comes from farther away, stretching out the wavelengths and giving the sound a lower frequency.

The Doppler effect causes similar shifts in the wavelengths of light (Figure 1c). If an object is moving toward us, the light waves bunch up between us and the object, so its entire spectrum is shifted to shorter wavelengths. Because shorter wavelengths of visible light are bluer, the Doppler shift of an object coming toward us is called a *blueshift*. If an object is moving away from us, its light is shifted to longer wavelengths. We call this a *redshift*, because longer wavelengths of visible light are redder.

We look for Doppler shifts by studying *spectral lines* in an object's spectrum. (For a discussion of the origin of spectral lines, see Tools of Science, page 132.) For example, suppose we recognize the pattern of hydrogen lines in the spectrum of a distant object (Figure 2). We know the rest wavelengths of the hydrogen lines—their wavelengths in stationary clouds of hydrogen gas—from laboratory experiments in which a tube of hydrogen gas is heated so that the wavelengths of the spectral lines can be measured. If the hydrogen lines from the object appear at longer wavelengths, then we know they are redshifted and the object is moving away from us. If the lines appear at shorter wavelengths, then we know they are blueshifted and the object is moving toward us. The larger the shift, the faster the object is moving. In fact, as long as the object is moving much slower than the speed of light, we can find its speed with a simple formula:

$$\frac{v_{rad}}{c} = \frac{\lambda_{shift} - \lambda_{rest}}{\lambda_{rest}}$$

where v_{rad} is the object's radial velocity, λ_{rest} is the rest wavelength of a particular spectral line, λ_{shift} is the shifted wavelength of the same line, and c is the speed of light. A positive answer means the object has a redshift and is moving away from us; a negative answer means it has a blueshift and is moving toward us.

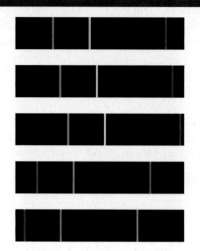

▲ FIGURE 2
Spectral lines provide the crucial reference points for measuring Doppler shifts.

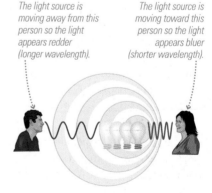

a The whistle sounds the same no matter where you stand near a stationary train.

b For a moving train, the sound you hear depends on whether the train is moving toward you or away from you.

c We get the same basic effect from a moving light source (although the shifts are usually too small to notice by eye).

▲ FIGURE 1
The Doppler effect. Each circle represents the crest of sound (or light) waves going in all directions from the source. For example, the circles from the train might represent waves emitted 0.001 second apart.

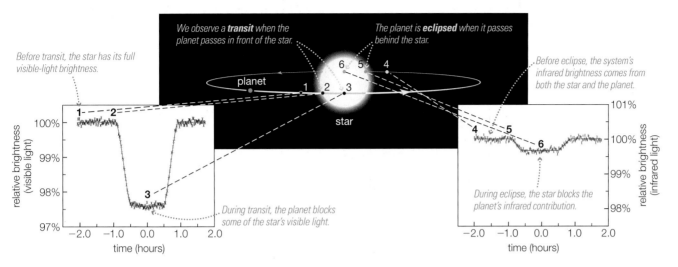

▲ **FIGURE 7.5**
This diagram represents transits and eclipses of the planet orbiting the star HD 189733. The graph shows that each transit lasts about 2 hours, during which the star's visible-light brightness dips by about 2.5%. The eclipses are observable in the infrared, because the star blocks the infrared contribution of the planet. The transits and eclipses each occur once during every 2.2-day orbit of the planet.

of a star, including intrinsic variations in the star's own brightness. Therefore, to be confident that the transit method has really detected a planet, scientists generally have two requirements: (1) The suspected transit event must be observed to repeat at least three times with a regular period, indicating that the same planet is passing in front of the star repeatedly with each orbit; and (2) follow-up observations with another method, such as the Doppler method, should also reveal the same planet.

Transits have been observed with many telescopes, but as of 2018 the vast majority of transit detections had come from the *Kepler* mission, which led to the discovery of more than 2500 confirmed planets, including dozens that are as small as Earth. The transit method is also the cornerstone of NASA's *TESS* mission and the European Space Agency's *CHEOPS* mission.

think about it   *TESS* was launched in 2018 and *CHEOPS* was due for launch in early 2019, shortly after this book went to press. Find the current status of these missions, and briefly describe their important results to date.

Cosmic Context Figure 7.6 summarizes the major planet detection methods that we have discussed.

◆ What properties of extrasolar planets can we measure?

Despite the challenge of detecting extrasolar planets, we can learn a surprising amount about them. Depending on the method or methods, we can determine planetary characteristics including orbital period and distance, orbital eccentricity, mass, size, density, and even a little bit about a planet's atmospheric composition and temperature.

Orbital Characteristics   All three of the major indirect detection methods tell us a planet's orbital period, and once we know the orbital period, we can determine average orbital distance (semimajor axis) with Newton's version of Kepler's third law. Recall that for a small object like a planet orbiting a much more massive object like a star, this law expresses a relationship between the star's mass, the planet's orbital period, and the planet's average distance (see Tools of Science, page 96).

cosmic context   FIGURE 7.6 Detecting Extrasolar Planets

The search for planets around other stars is one of the fastest growing and most exciting areas of astronomy, with known extrasolar planets already numbering well into the thousands. This figure summarizes major techniques that astronomers use to search for and study extrasolar planets.

(1) Gravitational Tugs: We can detect a planet by observing the small orbital motion of its star as both the star and its planet orbit their mutual center of mass. The star's orbital period is the same as that of its planet, and the star's orbital speed depends on the planet's distance and mass. Any additional planets around the star will produce additional features in the star's orbital motion.

Jupiter actually orbits the center of mass every 12 years, but appears to orbit the Sun because the center of mass is so close to the Sun.

The Sun also orbits the center of mass every 12 years.

Not to scale!

(1a) The Doppler Method: As a star moves alternately toward and away from us around the center of mass, we can detect its motion by observing alternating Doppler shifts in the star's spectrum: a blueshift as the star approaches and a redshift as it recedes.

(1b) The Astrometric Method: A star's orbit around the center of mass leads to tiny changes in the star's position in the sky. The GAIA mission is expected to discover many new planets with this method.

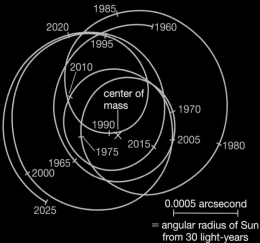

0.0005 arcsecond = angular radius of Sun from 30 light-years

Current Doppler-shift measurements can detect an orbital velocity as small as 1 meter per second—walking speed.

The change in the Sun's apparent position, if seen from a distance of 10 light-years, would be similar to the angular width of a human hair at a distance of 5 kilometers.

Artist's conception of another planetary system, viewed near a ringed jovian planet.

(2) The Transit Method: If a planet's orbital plane happens to lie along our line of sight, the planet will transit in front of its star once each orbit, causing a dip in the star's visible-light brightness. An *eclipse* may occur half an orbit later, during which the system's infrared brightness will decline because the planet's contribution is blocked by the star.

We observe a **transit** when the planet passes in front of the star.

When the planet passes behind the star, we say it is **eclipsed** by the star.

(3) Direct Detection: In principle, the best way to learn about an extrasolar planet is to observe directly either the visible starlight it reflects or the infrared light it emits. Current technology is capable of direct detection in some cases, but only with very low resolution.

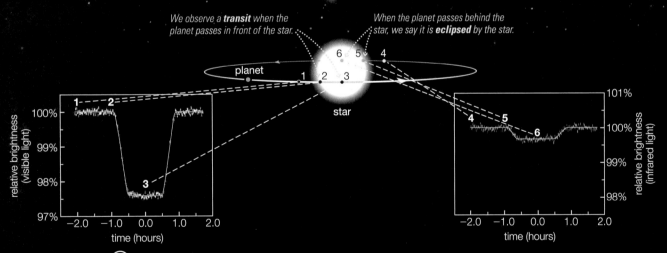

The Large Binocular Telescope imaged the region around the star HR 8799 in infrared light, discovering four planets labeled b through e (a refers to the star itself).

We generally know the masses of the stars with extrasolar planets (through methods we'll discuss in Chapter 8). Therefore, using the star's mass and the planet's orbital period, we can calculate the planet's average orbital distance.

In addition, the astrometric and Doppler methods can tell us the eccentricity of a planet's orbit (recall that eccentricity is a measure of how "stretched out" an ellipse is). A planet with a perfectly circular orbit travels at a constant speed around its star, so the star moves with constant speed around the center of mass. Any variation in the star's speed tells us that the planet is moving with varying speed along its orbit and therefore must have a more eccentric elliptical orbit. Such measurements have revealed that many extrasolar planets have highly eccentric orbits that bring them very close to their star on one side of their orbit and take them much farther from their star on the other side.

Planetary Mass The astrometric and Doppler methods measure stellar motion caused by the gravitational tug of planets, so both methods allow us to estimate planetary masses from the strength of this tug (measured by the star's velocity around the center of mass). The transit method cannot by itself tell us masses for single planets, but it can in some cases reveal planetary masses in multiplanet systems, because the gravitational tug of one planet on another can affect the timing of transits.

There is an important caveat about the Doppler method. Doppler shifts reveal only the part of a star's motion directed toward or away from us, not motion that is across our line-of-sight. Therefore, unless we happen to be viewing an orbit edge-on—in which case it will produce transits—the velocity measured with Doppler shifts will be *less* than the full orbital velocity, which means the mass we calculate will be the planet's *minimum* possible mass (a "lower limit" mass). Statistically, however, the actual planetary mass should be no more than double the minimum mass in at least about 85% of all cases, so masses from the Doppler method provide relatively good estimates for most planets.

Planetary Size and Density We can use transits to measure a planet's size or radius: The larger the planet, the more of a star's light it blocks during a transit. In cases in which we have both sizes from transits and masses from the Doppler or astrometric method, we can then calculate a planet's density (Figure 7.7). These cases are particularly valuable, because density can tell us whether a planet's overall composition is characteristic of a rocky world, a gaseous world, or something in between.

think about it If we discover a planet and measure its size through the transit method, what follow-up observations are needed to learn its mass? If we discover a planet through the Doppler method, can we expect to be able to do follow-up observations to learn its size? Explain.

Atmospheric Composition and Temperature Advancing technology is beginning to allow astronomers to learn about the atmospheres and temperatures of planets. Direct observations have already provided some data and are expected to provide much more with the aid of new instruments that can obtain both images and spectra of extrasolar planets at the same time.

We can also learn about planetary atmospheres from transiting systems by comparing spectra of a star–planet system taken with the planet in front of its star (the transit) or behind it (the eclipse) to spectra taken at other times. Differences between these spectra can reveal spectral lines caused by the planet's atmosphere, which can be used to identify gases in the planet's atmosphere (see Tools of Science, page 132). Infrared observations of eclipses can also provide temperature data, because the system's infrared brightness will drop when the planet is behind its star. The extent of the drop tells us how much infrared the planet emits, which tells us about the planet's temperature (see Tools of Science, page 79). Table 7.1 summarizes how we measure planetary properties.

▲ **FIGURE 7.7**
This diagram summarizes how we can combine data from the Doppler and transit methods to calculate a planet's average density. Data are for the planet Kepler 10b. (The final density calculation requires converting the mass to grams and the volume to units of cubic centimeters.)

Mass = $4.5 M_{Earth}$

For transiting planets, the Doppler method gives an accurate mass.

planet density: $\frac{mass}{volume} = 8.8$ g/cm^3

Size = $1.4 R_{Earth}$

The transit method yields a radius, from which we can calculate the planet's volume.

TABLE 7.1 Major Ways of Measuring Properties of Extrasolar Planets

	Planetary Property	Method(s) Used	Explanation
Orbital Properties	period	Doppler, astrometric, or transit	We directly measure orbital period.
	distance	Doppler, astrometric, or transit	We calculate orbital distance from orbital period using Newton's version of Kepler's third law.
	eccentricity	Doppler or astrometric	Velocity curves and astrometric star positions reveal eccentricity.
Physical Properties	mass	Doppler or astrometric	We calculate mass based on the amount of stellar motion caused by a planet's gravitational tug.
	size (radius)	transit	We calculate size based on the amount of dip in a star's brightness during a transit.
	density	transit plus Doppler	We calculate density by dividing the mass by the volume (using the size from the transit method).
	atmospheric composition, temperature	transit or direct detection	Transits and eclipses provide data on atmospheric composition and temperature; direct spectroscopy is now also possible in some cases.

7.2 Characteristics of Extrasolar Planets

The number of extrasolar planets for which we have measured many key properties is now large enough that we are beginning to gain insight into how these planets compare with the planets of our own solar system. In this section, we'll explore current understanding of extrasolar planets and what they tell us about the likelihood of finding other planetary systems that are similar in nature to our own.

◆ How do extrasolar planets compare with planets in our solar system?

Broadly speaking, we can divide a comparison of extrasolar planets and the planets of our solar system into two categories: a comparison of orbital properties (such as orbital period and semimajor axis) and a comparison of physical properties (such as mass, size, and density). Both types of comparison have revealed a broader range of properties among extrasolar planets than we find in our own solar system.

Orbital Properties of Extrasolar Planets A few extrasolar planets have been found with orbits similar to those of planets in our solar system, but many others exhibit one or both of two orbital properties that seemed quite surprising when they were first discovered. In particular, many planets with Jupiter-like masses or sizes orbit quite close to their stars—in many cases much closer than Mercury orbits the Sun—which is surprising because the jovian planets in our solar system all orbit quite far from the Sun. Moreover, many extrasolar planets have large eccentricities, a clear contrast to the nearly circular orbits of planets in our solar system. We've also found systems in which multiple planets are packed very close together (Figure 7.8), sometimes so close that they tug on each other gravitationally in measurable ways. Some extrasolar planets even orbit within their star's *habitable zone* [Section 15.3]—the range of distances at which it might in principle be possible for them to have surface oceans of liquid water.

Keep in mind that these "surprising" planetary orbits may not be as common as they seem from current data, because these data have all been collected in ways that are biased toward finding planets that orbit close to their stars. Nevertheless, the fact that such surprises exist at all requires explanation, and we'll see in the next section how this led scientists to realize that there is more to the nebular theory than we might have recognized from our solar system alone.

Physical Properties and the Nature of Extrasolar Planets The key physical properties we can measure are mass and size. As we've discussed, the cases for which we have measured both (and hence can calculate density) are

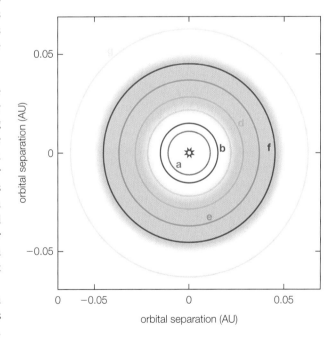

▲ **FIGURE 7.8**
A top-down view of planetary orbits in the TRAPPIST-1 system. All seven planets are similar in size to Earth but orbit much closer to their star than Mercury to the Sun (0.38 AU). Some are nevertheless in the star's habitable zone (green), which is much smaller and closer to the central star than our Sun's habitable zone because the TRAPPIST-1 star is so much lower in mass and fainter than the Sun.

particularly valuable, because they allow us to learn something about the general nature of these planets. This brings us to a key question: Do extrasolar planets fall into the same terrestrial and jovian categories as the planets in our solar system, or do we find additional types of planets?

Scientists address this question by creating models that use our understanding of the behavior of different materials to determine the expected composition of a planet from its mass and radius. The results are shown in Figure 7.9 for a sample of planets for which both mass (usually from the Doppler method) and radius (from transits) are known. Be sure you understand the following key features of the figure:

- The horizontal axis shows planetary mass, in units of Earth masses. (The top of the graph shows the equivalent values in Jupiter masses.) Notice that it uses a scale that rises by powers of 10 because the masses vary over such a wide range.
- The vertical axis shows planetary radius, in units of Earth radii. (The right side of the graph shows the equivalent values in Jupiter radii.)
- Each dot represents one planet for which both mass and radius have been measured. Planets of our solar system are marked in green. The paintings around the graph show artist conceptions of a few selected planets.
- You can estimate a planet's average density from its position relative to three dashed curves showing representative densities.
- The colored regions indicate planets of different general composition, based on models calculated from their masses and radii.

As you study Figure 7.9, notice that extrasolar planets show much more variety than the planets of our own solar system. At the extremes, we've identified planets with average densities as great as that of iron and as low as that of Styrofoam, and we've found planets with almost every average density in between. We see some planets that seem similar to the terrestrial planets in nature, and others that have characteristics very like those of Jupiter and Saturn, but many others appear to fall into other groups. For example, several planets cluster in the region of Uranus and Neptune—and likely share their composition of hydrogen compounds shrouded in an envelope of hydrogen and helium gas—suggesting that we might consider these a category distinct from the more Jupiter-like planets. Two of the most intriguing categories are the "water worlds," which may be made predominantly of water (perhaps in unusual phases) or other hydrogen compounds, and the "super-Earths," which probably have a rock/metal composition similar to that of Earth but are somewhat larger than Earth in size.

The bottom line is that extrasolar planets are abundant and diverse. Unlike our solar system, with only two clear types of "typical planets," these other solar systems have additional types of planets that defy easy categorization.

◆ Are Earth-like planets common?

We have discovered planetary systems around thousands of stars, but these represent only a tiny fraction of the more than 100 billion stars in the Milky Way Galaxy. What about the rest, which we haven't yet studied? Are planetary systems common and, if so, are planets similar to Earth common as well? Remarkably, we can now begin to answer these questions, because the *Kepler* mission observed an essentially random sample of stars in our galaxy. It is possible to use simple geometry to calculate the fraction of these stars that would be expected to have transiting planets if they all had planetary systems. We can then use the statistics of the actual *Kepler* discoveries to estimate the fraction of all stars that have planets of various sizes.

FIGURE 7.9

Masses and sizes for a sample of extrasolar planets for which both have been measured, compared with those of planets in our solar system. Each dot represents one planet. Dashed lines are lines of constant density for planets of different masses. Colored regions indicate expected planet types based on models of their compositions.

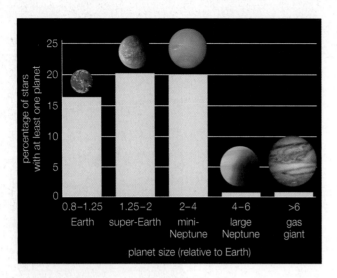

▲ **FIGURE 7.10**
This bar chart shows the estimated proportions of all stars that have planets of different size categories, based on *Kepler* results. Because *Kepler* data have been analyzed only for planets with relatively short orbital periods to date, these estimates are nearly certain to increase as additional data are studied.

Figure 7.10 shows the results of such statistical studies based on *Kepler* data. Two remarkable conclusions are apparent. First, planets are common; by looking across all size categories, astronomers conclude that at least 70% of all stars harbor at least one planet. Second, many of these planets are quite small, suggesting that Earth-size planets are also very common. Again, keep in mind that the statistics remain incomplete: Because planets with larger orbits (which probably means most jovian planets) and smaller sizes are more difficult to detect, the existing data probably underestimate the true numbers of planets in all the size categories. The percentages shown in Figure 7.10 can only increase as we gather more data.

Figure 7.11 shows *Kepler* data in a different way: as a graph of planet size versus orbital period. At first glance, the empty space at the lower right might seem to suggest that there are few Earth-size planets in Earth-size orbits. But the lack of data points in this region of the graph stems from the extreme difficulty of detecting the infrequent, shallow transits such planets make. Accounting statistically for these effects, scientists now estimate that about 20% of stars are likely to have a planet less than twice Earth's size in an Earth-like orbit.

To summarize, the statistics indicate that the vast majority of stars have planetary systems, that many of these systems have multiple planets, and that a good number of these planets are similar in size to Earth. Some of these Earth-size planets also have Earth-like orbits, leaving us with only one major question about them: whether their Earth-like sizes and orbits will also mean that they have Earth-like conditions such as continents and oceans. Answering that question will require a powerful new generation of telescopes that can study these worlds directly, which means the answer may be decades away. Still, we are rapidly closing in on a definitive answer to the question of whether Earth-like planets are common.

think about it Not long before most of today's college students were born, the only known planets were those of our own solar system. Today, the evidence suggests that many or most stars have planets. For the galaxy as a whole, that's a change in the estimated number of planets from fewer than 10 to more than 100 billion. Do you think this change should alter our perspective on our place in the universe? Defend your opinion.

THE PROCESS OF SCIENCE IN ACTION

7.3 Extrasolar Planets and the Nebular Theory

The discovery of extrasolar planets presents us with an opportunity to test our theory of solar system formation. Can our existing theory explain other planetary systems, or will we have to go back to the drawing board?

◆ Do we need to modify our theory of solar system formation?

As we discussed in Chapter 4, the nebular theory holds that our solar system's planets formed as a natural consequence of processes that accompanied the formation of our Sun. If the theory is correct, then the same processes should accompany the births of other stars, so the nebular theory clearly *predicts* the existence of other planetary systems. In that sense, the discovery of extrasolar planets means the theory has passed its most important test, because its most basic prediction has been verified.

▲ **FIGURE 7.11**
This figure shows the orbital periods and sizes of planets identified from *Kepler* data, along with a reference dot for Earth. Note that while many planets are similar in size to Earth, the data do not cover a long enough time period to identify those with orbital periods as long as Earth's. Many of the Earth-size planets may nevertheless be in their stars' habitable zones, because they orbit stars that are lower in mass than our Sun.

However, the characteristics of some known extrasolar planets seem to challenge the details of the nebular theory. Recall that the theory neatly explains the two categories of planet in our solar system by holding that the terrestrial planets formed in the warm region where only rock and metal could condense, while the jovian planets formed beyond the frost line where ices also condensed (see Figure 4.13). How, then, can we explain hot Jupiters that appear to be jovian planets with close-in orbits? And if we have only two categories of planet in our solar system, why do we find more in other planetary systems?

Explaining Planetary Orbits The discovery of hot Jupiters led scientists to reexamine the nebular theory in great detail. For example, might it be possible for jovian planets to form very close to a star? Scientists explored many different potential models for planetary formation, and while we can't completely rule out the possibility that a major flaw has gone undetected, it now seems much more likely that the basic outline of the nebular theory is correct. Scientists therefore suspect that extrasolar jovian planets were born with circular orbits far from their stars, just like the jovian planets of our own solar system, but that those that now have close-in orbits underwent some sort of "planetary migration."

Computer models of solar system formation suggest several ways that migration might occur. For example, in some cases, migration may be caused by waves passing through a gaseous disk (Figure 7.12). The gravity of a planet moving through a disk can create waves that propagate through the disk, causing material to bunch up as the waves pass by. This "bunched up" matter (in the wave peaks) then exerts a gravitational pull on the planet that tends to reduce its orbital energy, causing the planet to migrate inward toward its star. This type of migration is not thought to have played a significant role in our own solar system, because the nebular gas was cleared out before it could have much effect. However, planets may form earlier in some other planetary systems, or the nebular gas may be cleared out later, allowing time for jovian planets to migrate substantially inward. Migration may also occur through close gravitational encounters between planets and large planetesimals. Such encounters may also explain the eccentric orbits of many extrasolar planets.

Explaining Planetary Types Scientists still cannot fully explain the wide range of extrasolar planet properties, but we can envision possible explanations that seem to make sense. For example, the low average densities of some hydrogen-rich extrasolar planets (see Figure 7.9) may simply be a consequence of the fact that these planets orbit close to their stars, giving them high temperatures that puff up their atmospheres. The hydrogen-rich planets with high densities can be explained by planets capturing even more hydrogen and helium gas than Jupiter, which models show would cause their gravity to compress them to smaller sizes. The "water world" planets may also have formed much like the jovian planets, except that their icy planetesimals never got a chance to capture much hydrogen or helium gas; this might occur if an early solar wind blasted the gas away before the "water worlds" had time to capture it. Super-Earths pose a challenge because they have more rocky material than we might expect given that such material makes up only a small fraction of the material in any nebula. Still, their differences from the terrestrial worlds of our own solar system are not great enough to make us question the validity of the entire theory.

An Improved Nebular Theory The bottom line is that discoveries of extrasolar planets have shown us that the nebular theory was incomplete. The original theory explained the formation of planets and the simple layout of a solar system such as ours, but we've needed to add new features—such as planetary migration and variations in the basic planetary types—to explain the differing layouts of other solar systems. We should not be too surprised by this fact, because

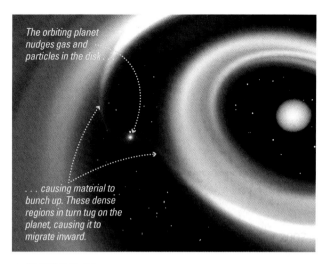

▲ FIGURE 7.12
This figure shows a simulation of waves created by a planet embedded in a disk of material surrounding its star.

scientific theories frequently need modification to accommodate new discoveries. For example, Newton's theory of gravity had to be modified by Einstein to account for effects observed in strong gravitational fields [Section 3.3], and the theory of atoms and subatomic particles has been modified numerous times as new discoveries have been made during the past century. Just as in those cases, modification of the nebular theory has opened new possibilities that we did not previously consider. A much wider variety of solar system arrangements now seems possible than we had guessed before the discovery of extrasolar planets.

summary of key concepts

7.1 Detecting Planets Around Other Stars

◆ **How do we detect planets around other stars?**

We can look for a planet's gravitational effect on its star through the **astrometric method**, which looks for small shifts in stellar position, or the **Doppler method**, which looks for the back-and-forth motion of stars revealed by Doppler shifts. For the small fraction of planetary systems with orbits aligned edge-on to Earth, we can search for **transits**, in which a planet blocks a little of its star's light as it passes in front of it.

◆ **What properties of extrasolar planets can we measure?**

All detection methods allow us to determine a planet's orbital period and distance from its star. The astrometric and Doppler methods can provide masses (or minimum masses), while the transit method can provide sizes. In cases where transit and Doppler methods are used together, we can determine average density. In some cases, transits (and eclipses) can provide other data, including limited data about atmospheric composition and temperature.

7.2 Characteristics of Extrasolar Planets

◆ **How do extrasolar planets compare with planets in our solar system?**

The known extrasolar planets have a much wider range of properties than the planets in our solar system. Many orbit much closer to their stars and with more eccentric orbital paths; even some jovian planets, called **hot Jupiters**, are found close to their stars. We also find properties indicating planetary types such as "water worlds" that do not fall neatly into the traditional terrestrial and jovian categories.

◆ **Are Earth-like planets common?**

Current data indicate that planetary systems are very common, and that Earth-size planets are also common. We do not yet have enough data to know whether such planets have Earth-like orbits, but we should learn the answer as more data are collected.

 THE PROCESS OF SCIENCE IN ACTION

7.3 Extrasolar Planets and the Nebular Theory

◆ **Do we need to modify our theory of solar system formation?**

Our basic theory of solar system formation seems to be sound, but we have had to modify it to allow for planetary migration and a wider range of planetary types than we find in our solar system. Many mysteries remain, but they are unlikely to require major change to the nebular theory of solar system formation.

Investigations

For instructor-assigned homework and other learning materials, go to www.MasteringAstronomy.com.

Quick Quiz

Choose the best answer to each of the following. Explain your reasoning.

1. Which detection method or methods measure the gravitational tug of a planet on its star, allowing us to estimate planetary mass? (a) the transit method only (b) the Doppler method only (c) the astrometric and Doppler methods
2. Which one of the following does the transit method tell us about a planet? (a) its mass (b) its size (c) the eccentricity of its orbit
3. Which method could detect a planet in an orbit that is face-on to Earth? (a) Doppler method (b) transit method (c) astrometric method
4. Which detection method was used by the *Kepler* mission? (a) Doppler method (b) transit method (c) astrometric method
5. How does a hot Jupiter differ from Jupiter? (a) It's much closer to its star. (b) It has a much longer year. (c) It's much more massive.
6. To determine a planet's average density, we can use (a) the transit method alone. (b) the astrometric and Doppler methods together. (c) the transit and Doppler methods together.
7. Based on the model types shown in Figure 7.9, a planet made almost entirely of hydrogen compounds would be considered a (a) terrestrial planet. (b) jovian planet. (c) "water world."
8. Look at the dot for Jupiter in Figure 7.9, then at the red dot directly to Jupiter's left. Compared with the density of Jupiter, the density of the planet represented by that dot is (a) higher. (b) lower. (c) the same.
9. The term "super-Earth" means a planet that is (a) the size of Earth but has more water. (b) larger than Earth but on a close-in orbit that makes it much hotter than Earth. (c) similar in composition to Earth but larger in size.
10. Based on current data, planetary systems appear to be (a) extremely rare. (b) present around about 70% of all stars. (c) present around at least half of all stars.
11. What's the best explanation for the location of hot Jupiters? (a) They formed closer to their stars than Jupiter did. (b) They formed farther out, as Jupiter did, but then migrated inward. (c) The strong gravity of their stars pulled them in close.
12. Based on computer models, when is planetary migration most likely to occur in a planetary system? (a) early in its history, when there is still a gaseous disk around the star (b) shortly after a stellar wind clears the gaseous disk away (c) late in its history, when asteroids and comets occasionally collide with planets

Short-Answer/Essay Questions

Explain all answers clearly, with complete sentences and proper essay structure if needed. An asterisk (*) designates a quantitative problem, for which you should show all your work.

13. *Explaining the Doppler Method.* Explain how the Doppler method works in terms an elementary school child would understand. It may help to use an analogy to explain the difficulty of direct detection and the general phenomenon of the Doppler shift.
14. *Comparing Methods.* What are the strengths and limitations of the Doppler and transit methods? What kinds of planets are easiest to detect with each method? Are there certain planets that each method cannot detect, even if the planets are very large? Explain. What advantages are gained if a planet can be detected by both methods?
15. *No Hot Jupiters Here.* How do we think hot Jupiters formed? Why didn't one form in our solar system?
16. *Low-Density Planets.* Only one planet in our solar system has a density less than 1 g/cm^3, but many extrasolar planets do. Explain why in a few sentences. (*Hint:* Consider the densities of the jovian planets in our solar system, given in Figure 6.1.)
17. *When Is a Theory Wrong?* As discussed in this chapter, in its original form the nebular theory of solar system formation does not explain the orbits of many known extrasolar planets, but it can explain them with modifications such as allowing for planetary migration. Does this mean the theory was "wrong" or only "incomplete" before the modifications were made? Explain. Be sure to look back at the discussion in Chapter 3 of the nature of science and scientific theories.
18. *Unanswered Questions.* As discussed in this chapter, we are only just beginning to learn about extrasolar planets. Briefly describe one important but unanswered question related to the study of planets around other stars. Then write two or three paragraphs in which you discuss how we might answer this question in the future. Be as specific as possible, focusing on the type of evidence necessary to answer the question and how the evidence could be gathered. What are the benefits of finding answers to this question?
*19. *Planet Around 51 Pegasi.* The star 51 Pegasi has about the same mass as our Sun and is orbited by a planet with an orbital period of 4.23 days and estimated mass 0.6 times the mass of Jupiter. Use Kepler's third law to find the planet's average distance (semimajor axis) from its star. (*Hint:* Because the mass of 51 Pegasi is about the same as the mass of our Sun, you can use Kepler's third law in its original form, $p^2 = a^3$. Be sure to convert the period into years before using this equation.)
*20. *Doppler Calculations.* In hydrogen, the transition from level 2 to level 1 has a rest wavelength of 121.6 nm. Suppose you see this line at a wavelength of 120.5 nm in Star A and 121.2 nm in Star B. Calculate each star's speed, and state whether it is moving toward or away from us.

8 The Sun and Other Stars

LEARNING GOALS

8.1 Properties of the Sun
- Why does the Sun shine?
- How does energy escape from the Sun?

8.2 Properties of Other Stars
- How do we measure the properties of stars?
- What patterns do we find in the properties of stars?

 THE PROCESS OF SCIENCE IN ACTION

8.3 Visualizing Patterns Among Stars
- How did we discover the patterns in stellar properties?

Modern astronomy encompasses the study of the entire universe, but the root *astro* comes from the Greek for "star." Stars are therefore the namesakes of astronomy, and in this chapter we will embark on our study of them. We'll begin with our own Sun, which is the only star near enough for us to examine with high resolution. The picture on this page shows the remarkable detail we can see on our star—in this case, a gigantic eruption of hot gas from the solar surface; this eruption led to a "coronal mass ejection" that sent energetic charged particles streaming toward Earth. After briefly studying the Sun, we'll turn our attention to the multitudes of other stars in the sky, comparing their properties with those of our Sun and seeing what these properties reveal about the lives of stars.

Our Star

8.1 Properties of the Sun

Our Sun is the nearest star, and it provides virtually all the light and heat on Earth. In this first section, we'll explore basic properties of the Sun, including how it generates energy and how the flow of that energy determines the Sun's structure.

◆ Why does the Sun shine?

The source of the Sun's great power remained a mystery for most of human history, but today we know that, like all stars, the Sun shines with energy released through nuclear fusion in its core [Section 1.2]. To understand how this occurs, we will examine the Sun's basic properties and then discuss how fusion works.

Basic Properties of the Sun The Sun is immense by any measure. Although it may look relatively small in our sky, we know from its distance and angular size that its radius is about 700,000 kilometers—more than 100 times the radius of Earth. Even **sunspots**, which appear as dark splotches on the Sun's surface, can be larger in size than Earth (Figure 8.1). We can calculate the Sun's mass by applying Newton's version of Kepler's third law (see Tools of Science, page 96): It is about 2×10^{30} kilograms, which is 300,000 times the mass of Earth and nearly 1000 times the combined mass of all the planets in our solar system.

The Sun's immense size is matched by its enormous output of energy. Its total power output, or **luminosity**, is an incredible 3.8×10^{26} watts. If we could somehow capture and store just 1 second's worth of the Sun's luminosity, it would be enough to meet current human energy demands for roughly the next 500,000 years. Of course, only a tiny fraction of the light radiated by the Sun reaches Earth, since it goes out in all directions into space. Much of the Sun's energy is radiated as visible light, but the Sun also radiates light across most of the rest of the electromagnetic spectrum, including ultraviolet rays and x-rays.

In terms of composition, the Sun is essentially a giant ball of hot gas or, more technically, *plasma*—a gas in which many of the atoms are ionized because of the high temperature. Spectroscopy (see Tools of Science, page 132) reveals that this plasma is composed of about 70% hydrogen and 28% helium (by mass), with all other elements making up only about 2%. Careful study of the Sun's spectrum also allows us to determine the Sun's surface temperature, which averages about 5800 K, except in sunspots, where the gas is a cooler 4000 K. The temperature must increase with depth within the Sun, and theoretical models tell us that the temperature reaches an astounding 15 million K at the Sun's center.

The entire Sun is rotating, though unlike a spinning ball, different parts of the Sun rotate at different rates. The solar equator completes one rotation in about 25 days, and the rotation period increases with latitude to about 30 days near the solar poles. Table 8.1 summarizes the basic properties of the Sun.

Nuclear Fusion in the Sun The Sun's enormous energy output comes from **nuclear fusion**, a process in which two or more atomic nuclei slam together hard enough that they stick and become one larger nucleus. Fusion reactions are quite different from the nuclear reactions used in nuclear reactors on Earth; those reactions release energy by splitting large nuclei (such as those of uranium or plutonium) into smaller ones in a process called **nuclear fission**. Figure 8.2 summarizes the difference between fission and fusion.

Fusion occurs deep within the Sun's central **core**, where the 15-million-K plasma is like a "soup" of hot gas with positively charged atomic nuclei (and negatively charged electrons) whizzing about at extremely high speeds. At any time, some of these nuclei are on high-speed collision courses with each other.

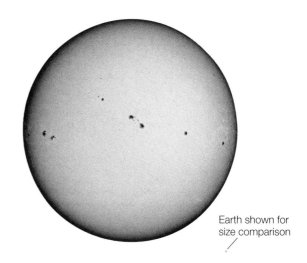

▲ **FIGURE 8.1**
This photo of the visible surface of the Sun shows several dark sunspots, each large enough to swallow our entire planet.

TABLE 8.1 Basic Properties of the Sun

Radius (R_{Sun})	696,000 km (about 109 times the radius of Earth)
Mass (M_{Sun})	2×10^{30} kg (about 300,000 times the mass of Earth)
Luminosity (L_{Sun})	3.8×10^{26} watts
Composition (by percentage of mass)	70% hydrogen, 28% helium, 2% heavier elements
Rotation rate	25 days (equator) to 30 days (poles)
Surface temperature	5800 K (average); 4000 K (sunspots)
Core temperature	15 million K

▲ **FIGURE 8.2**
Nuclear fission splits a nucleus into smaller nuclei, while nuclear fusion combines smaller nuclei into a larger nucleus.

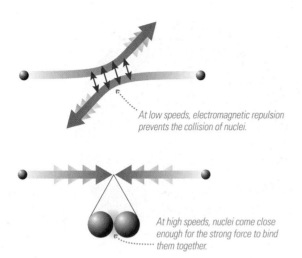

▲ FIGURE 8.3
Positively charged nuclei can fuse only if a high-speed collision brings them close enough for the strong force to come into play.

▲ FIGURE 8.4
In the Sun, four hydrogen nuclei (protons) fuse to make one helium nucleus (two protons and two neutrons).

common misconceptions

The Sun Is Not on Fire

We often say that the Sun is "burning," a term that conjures up images of a giant bonfire in the sky. However, the Sun does not burn in the same sense that a fire burns on Earth. Fires generate light through chemical changes that consume oxygen and produce a flame. The glow of the Sun has more in common with the glow of embers left over after the flames have burned out. Much like hot embers, the Sun's surface shines because it is hot enough to emit thermal radiation that includes visible light. Of course, while hot embers quickly stop glowing as they cool, the Sun keeps shining because its surface is kept hot by the energy rising from its core, where nuclear fusion (and not any form of chemical burning) converts mass into energy.

In most cases, electromagnetic forces deflect the nuclei, preventing collisions, because positive charges repel one another. However, if nuclei collide with sufficient energy, they can stick together (fuse) to form a heavier nucleus (Figure 8.3).

Sticking positively charged nuclei together is not easy. The **strong force**, which binds protons and neutrons together in atomic nuclei, is the only force in nature that can overcome the electromagnetic repulsion between two positively charged nuclei. In contrast to gravitational and electromagnetic forces, which drop off gradually as the distances between particles increase (by an inverse square law), the strong force is more like glue or Velcro: It overpowers the electromagnetic force over very small distances but is insignificant when the distances between particles exceed the typical sizes of atomic nuclei. The key to nuclear fusion is pushing the positively charged nuclei close enough together for the strong force to outmuscle electromagnetic repulsion. This can occur in the Sun's core, because the high temperature means the hydrogen nuclei collide at very high speeds.

A variety of different fusion reactions are always taking place simultaneously within the Sun. Overall, these fusion reactions end up transforming four individual hydrogen nuclei, which are protons, into a single helium nucleus containing two protons and two neutrons (Figure 8.4).

Fusing hydrogen nuclei into helium nuclei generates energy because a helium nucleus has a mass slightly less (by about 0.7%) than the combined mass of four hydrogen nuclei. That is, when four hydrogen nuclei fuse into a helium nucleus, a little bit of mass disappears. This disappearing mass becomes energy in accord with Einstein's famous formula $E = mc^2$, which tells us that an amount of mass m can be transformed into an amount of energy E equal to the mass times the speed of light squared. Fusion in the Sun converts about 600 million tons of hydrogen into 596 million tons of helium every second, which means that 4 million tons of matter is turned into energy each second. Although this sounds like a lot, it is such a small fraction of the Sun's total mass that it does not affect the overall mass of the Sun in a measurable way.

The fact that the Sun is gradually transforming its core hydrogen into helium means that the Sun will eventually run out of hydrogen. We can estimate the Sun's overall lifetime by dividing the total amount of hydrogen that the Sun originally had in its core by the rate at which it is currently fusing this hydrogen. We find that the Sun's lifetime is about 10 billion years. Since the Sun is now about $4\frac{1}{2}$ billion years old, it will continue to shine for about another 5 billion years. In the next chapter, we'll discuss what will happen to the Sun after it uses up its core supply of hydrogen.

◆ How does energy escape from the Sun?

Energy released by nuclear fusion in the Sun's core gradually makes its way to the surface, where it is radiated into space in the form of photons. Although we cannot look directly inside the Sun, we can investigate how energy flows outward by using mathematical models of the Sun's interior. These models agree well with observations of the Sun, successfully accounting for its size, surface temperature, and luminosity. This agreement gives us confidence in our understanding of the solar interior. We'll use this understanding to follow the journey of solar energy from the solar core to its release into space.

The Sun's Interior Nearly all the energy released by fusion in the core takes the form of gamma-ray photons. Although these photons travel at the speed of light, the plasma deep in the solar interior is so dense that a photon can travel only a fraction of a millimeter in any one direction before it interacts with an electron. Each time a photon collides with an electron, the photon gets deflected into a new and random direction. The photon thereby bounces around in a haphazard way (sometimes called a *random walk*) and only very gradually works its way outward (Figure 8.5). The region of the Sun in which energy is transported primarily by these randomly bouncing photons includes the core and the layer directly above the core, called the **radiation zone** (because photons represent electromagnetic radiation).

At the top of the radiation zone, where the temperature has dropped to about 2 million K, the solar plasma absorbs photons more readily (rather than just bouncing them around). This absorption creates the conditions needed for convection [Section 5.1] and hence marks the bottom of the Sun's **convection zone**. Recall that convection occurs because hot gas is less dense than cool gas. Like hot-air balloons, hot bubbles of plasma rise upward through the convection zone, while cooler plasma near the top slides around the rising bubbles and sinks. The rising of hot plasma and sinking of cool plasma form a cycle that transports energy outward from the base of the convection zone to the solar surface, or **photosphere**. This convecting gas explains the mottled appearance of the photosphere that we see in zoomed-in images of the Sun's surface (Figure 8.6).

Because of the slow outward progress of energy carried by radiation and convection, most of the energy released by fusion in the core takes several hundred thousand years to reach the photosphere and escape into space. However, a very small fraction of the fusion energy is released in the form of extremely lightweight subatomic particles called **neutrinos**. A defining characteristic of neutrinos is that they rarely interact with other matter in any way, even when the matter is as dense as that in the Sun's interior. As a result, neutrinos released by fusion travel directly outward from the core at nearly the speed of light. They can therefore reach Earth just minutes after they are produced, which means that scientists can use sensitive neutrino detectors to monitor what is happening in the Sun's core. Observations with neutrino detectors agree with predictions made by our models of fusion and the solar interior, giving us further confidence that we really do know what makes the Sun shine.

The Solar Atmosphere Once energy from the solar interior reaches the photosphere, it can escape directly into space as thermal radiation, because the gas layers above the photosphere are transparent to visible light. The photosphere

▲ **FIGURE 8.5**
A photon in the solar interior bounces randomly among electrons, slowly working its way outward. Actual photon paths are far more complex than those shown here; a typical photon bounces around inside the Sun for hundreds of thousands of years before finally reaching the surface.

a This diagram shows convection beneath the Sun's surface: Hot gas (yellow arrows) rises while cooler gas (black arrows) descends around it.

b This photograph shows the mottled appearance of the Sun's photosphere. The bright spots, each about 1000 kilometers across, correspond to the rising plumes of hot gas in the diagram in part a.

▲ **FIGURE 8.6**
The Sun's photosphere churns with rising hot gas and falling cool gas as a result of underlying convection.

is therefore the visible surface of the Sun. The transparent layer just above the photosphere, in which the temperature rises to over 9000 K, is called the **chromosphere**. Above that is the **corona**, in which temperatures reach more than 1 million K. Gas at the top of the corona is so hot that it emits x-rays, and some of it escapes the Sun in an outward flow of charged particles known as the **solar wind**. Figure 8.7 schematically shows the layers of the Sun.

tools of science — Spectroscopy

Three Types of Spectra

Spectroscopy is the study of light that has been spread out into a spectrum by a device such as a prism. The figure shows the three basic kinds of spectra that we observe: (1) a **continuous spectrum** contains light that is smoothly spread across a broad range of wavelengths; (2) an **emission line spectrum** has bright lines on a dark background; and (3) an **absorption line spectrum** has dark lines on a continuous background.

Spectra provide a wealth of information about the objects we study. We previously discussed how a *thermal radiation* spectrum—which is one type of continuous spectrum—tells us an object's surface temperature (see Tools of Science, page 79) and how *Doppler shifts* in the wavelengths of spectral lines tell us the speed of an object toward or away from us (see Tools of Science, page 116). We learn even more from the spectral lines themselves.

The laws of quantum mechanics (see Tools of Science, page 152) dictate that each different kind of atom, ion, or molecule produces spectral lines at a unique set of wavelengths—in essence, its "chemical fingerprint." We can therefore determine the composition of a distant object by identifying and comparing all the different "fingerprints" in its spectrum. It does not matter whether the lines are absorption lines or emission lines; the wavelengths of the lines are what tell us an object's composition.

We can put these ideas together to illustrate how the spectrum of a star like the Sun reveals its composition. The hot Sun emits thermal radiation, much like the hot light bulb in the figure, producing a continuous spectrum. The low-density, uppermost layers of the Sun act like a cloud of gas that the continuous spectrum must pass through on its way to Earth. Atoms in these layers absorb light at specific wavelengths, producing an absorption line spectrum. The wavelengths of those lines tell us the kinds of atoms that are present in the Sun, and a more detailed analysis tells us the relative proportions of those elements.

Examples of conditions under which we see the three basic types of spectra.

On and above the Sun's surface we observe a wide variety of phenomena including *sunspots* in the photosphere, gigantic loops of glowing gas that connect pairs of sunspots, and huge eruptions such as *solar flares* and *coronal mass ejections* (see the photo at the beginning of the chapter). All these features are created by magnetic fields, which form and change easily in the convecting plasma of the convection zone. Sunspots occur where magnetic fields suppress convection, allowing surface plasma to cool, and prevent the surrounding hot plasma from mixing into the cooler regions; that is why sunspots are significantly cooler (and appear darker) than the rest of the photosphere. The loops connecting sunspot pairs are lines of a magnetic field that has trapped glowing gas high in the corona (Figure 8.8). Flares and coronal mass ejections occur when pent-up magnetic field energy is suddenly released into the solar atmosphere. These eruptions can affect Earth because they feed energetic bursts of charged particles into the solar wind, which can end up producing the beautiful lights of the *aurora* [Section 5.1] as these charged particles interact with Earth's upper atmosphere.

The phenomena we observe in the solar atmosphere change over relatively short periods of time and therefore constitute what we call **solar activity** (or *solar weather*). The most notable pattern of solar activity is the **sunspot cycle**—a cycle in which the average number of sunspots gradually rises and falls (Figure 8.9). At the time of *solar maximum,* when sunspots are most numerous, we may see dozens of sunspots on the Sun at one time. In contrast, we may see few or no sunspots at the time of *solar minimum.* The length of time between maximums is 11 years on average, but we have observed it to be as short as 7 years and as long as 15 years. There have also been some periods during which no sunspots have been seen for many decades.

▲ FIGURE 8.7
The basic structure of the Sun.

Pairs of sunspots are connected by tightly wound magnetic field lines.

b This x-ray photo (from NASA's *TRACE* mission) shows hot gas trapped within looped magnetic field lines.

FIGURE 8.8
Strong magnetic fields keep sunspots cooler than the surrounding photosphere, while magnetic loops can arch from the sunspots to great heights above the Sun's surface.

▲ FIGURE 8.9
Sunspot cycle since about 1900.

Surveying the Stars (Learning Outcome 1)

8.2 Properties of Other Stars

As we turn our attention from the Sun to the other stars in the sky, it is natural to wonder what life would be like if Earth were orbiting another star. How bright would the star be in daytime? What color starlight would bathe Earth's surface? Would the star live long enough for life to develop? In this section, we'll touch on some of the answers to these questions as we study the properties of stars.

◆ How do we measure the properties of stars?

Much of what we know about stars other than our Sun comes from measurements of just three basic properties: *luminosity*, *surface temperature*, and *mass*. We have already seen how the Sun measures up in these categories. Let's examine how we determine these properties for other stars.

Measuring Luminosity If you go outside on any clear night, you'll immediately see that stars differ in brightness. Some stars are so bright that we can use them to identify constellations. Others are so dim that our naked eyes cannot see them at all. However, these differences in brightness do not by themselves tell us how much light the stars are generating, because brightness also depends on distance. For example, the stars Procyon and Betelgeuse appear about equally bright in our sky, but Betelgeuse actually emits about 15,000 times as much light as Procyon. It has about the same brightness in our sky because it is much farther away.

Because two similar-looking stars can be generating very different amounts of light, we need to distinguish clearly between a star's brightness in our sky and the actual amount of light that it emits into space (Figure 8.10):

- When we talk about how bright stars look in our sky, we are talking about **apparent brightness**—the brightness of a star as it appears to our eyes.
- When we talk about how bright stars are in an absolute sense, regardless of their distance, we are talking about **luminosity**—the total amount of power that a star emits into space.

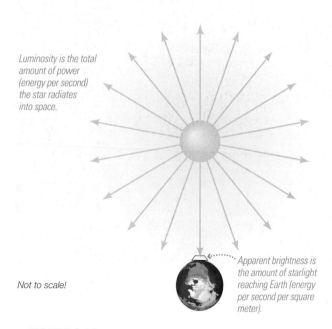

▲ FIGURE 8.10
Luminosity is a measure of power, and apparent brightness is a measure of power per unit area.

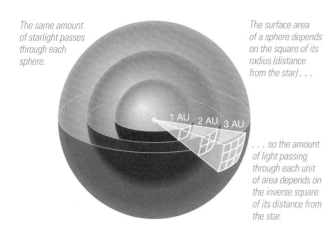

The same amount of starlight passes through each sphere.

The surface area of a sphere depends on the square of its radius (distance from the star)...

...so the amount of light passing through each unit of area depends on the inverse square of its distance from the star.

◀ **FIGURE 8.11**
The inverse square law for light: The apparent brightness of a star declines with the square of its distance.

The apparent brightness of a star (or any other light source) depends on both its luminosity and its distance. More specifically, apparent brightness obeys an *inverse square law* with distance (Figure 8.11), much like the inverse square law that describes the force of gravity [Section 3.3]. For example, if we viewed the Sun from twice Earth's distance, it would appear dimmer by a factor of $2^2 = 4$. If we viewed it from 10 times Earth's distance, it would appear $10^2 = 100$ times dimmer. We can express this relationship with a simple formula called the **inverse square law for light**:

$$\text{apparent brightness} = \frac{\text{luminosity}}{4\pi \times (\text{distance})^2}$$

Because the standard units of luminosity are watts, the units of apparent brightness are *watts per square meter*. [The 4π in the formula comes from the fact that the surface area of a sphere is given by $4\pi \times (\text{radius})^2$.]

think about it Suppose Star A is four times as luminous as Star B. How will their apparent brightnesses compare if both are the same distance from Earth? How will their apparent brightnesses compare if Star A is twice as far from Earth as Star B? Explain.

We can use the inverse square law for light to calculate a star's luminosity if we first measure its apparent brightness and distance. Apparent brightness is relatively easy to measure with modern instruments, because it only requires measuring the amount of light we receive from the star per square meter. The more difficult challenge is measuring a star's distance.

The most direct way to measure a star's distance is with *stellar parallax*, the small annual shifts in a star's apparent position caused by Earth's motion around the Sun (see Figure 2.20). As we'll discuss in more detail later, measuring the precise amount of a star's annual shift due to parallax allows us to calculate the star's distance [Section 12.1].

Once we have determined a star's luminosity from its apparent brightness and distance, we usually state the result in comparison to the Sun's luminosity, which we write as L_{Sun} for short. For example, Proxima Centauri, the nearest of the three stars in the Alpha Centauri system and the nearest star besides our Sun, is only about 0.0006 times as luminous as the Sun, or $0.0006 L_{Sun}$. Betelgeuse, the bright left-shoulder star of Orion, has a luminosity of $120,000 L_{Sun}$, meaning that it is 120,000 times as luminous as the Sun. Studies of the luminosities of many stars have taught us that stars have a wide range of luminosities, with our Sun somewhere in the middle. The dimmest stars have luminosities about 1/10,000 that of the Sun ($10^{-4} L_{Sun}$) and are extremely common, while the brightest stars are about 1 million times as luminous as the Sun ($10^6 L_{Sun}$) and are very rare.

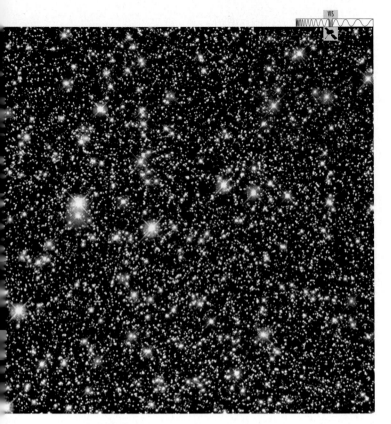

▲ **FIGURE 8.12**
This Hubble Space Telescope photo shows a wide variety of stars that differ in color and brightness. Most of the stars in this photo are at roughly the same distance, about 2000 light-years from the center of our galaxy.

common misconceptions

Photos of Stars

Astronomical photographs convey a great deal of information, but they also contain artifacts that are not real. For example, the spikes visible around bright stars are an artifact created by the way starlight interacts with the supports holding the secondary mirror in a telescope [Section 3.1], and stars appear to have different sizes in photographs only because bright stars are overexposed, causing their light to spill over a larger region of an image than the light of dimmer stars. These artifacts can sometimes be useful: The size issue makes it easy to identify the brighter stars, and the spikes generally occur only with point sources of light such as stars, which means you can use them to distinguish stars (which have the spikes) from distant galaxies (which don't) in photos like the Hubble eXtreme Deep Field on page 1 of this book.

Measuring Surface Temperature Measuring a star's surface temperature is somewhat easier than measuring its luminosity, because the star's distance doesn't affect the measurement. Note that we say *surface* temperature because it is the only stellar temperature that is directly measurable; interior temperatures must be inferred from mathematical models of stellar interiors.

We determine surface temperature from either the star's color or its spectrum. Figure 8.12 shows that stars come in a variety of colors. They come in different colors because they emit thermal radiation, and a thermal radiation spectrum depends only on the surface temperature of the object that emits it (see Tools of Science, page 79). For example, the Sun's 5800 K surface temperature causes it to emit most strongly in the middle of the visible portion of the spectrum, which is why the Sun looks yellow or white in color. A cooler star, such as Betelgeuse (surface temperature 3650 K), looks red because it emits much more red light than blue light. A hotter star, such as Sirius (surface temperature 9400 K), emits more blue light than red light and therefore has a bluish color.

We can measure surface temperature more precisely by carefully studying a star's spectral lines. Stars with spectral lines of highly ionized elements must be fairly hot, because it takes a high temperature to ionize atoms. Stars with spectral lines of molecules must be relatively cool, because molecules break apart into individual atoms unless they are at relatively cool temperatures. The types of spectral lines present in a star's spectrum therefore provide a direct measure of the star's surface temperature.

Astronomers classify stars according to surface temperature by assigning a **spectral type** determined from the spectral lines present in a star's spectrum. The hottest (bluest) stars are called spectral type O, followed in order of declining surface temperature by spectral types B, A, F, G, K, and M (Table 8.2). Astronomers sometimes subdivide the spectral types by following the letter with a number from 0 through 9. For example, the Sun's spectral type is G2, which means it is slightly hotter than a G3 star but slightly cooler than a G1 star.

think about it The traditional mnemonic for memorizing the OBAFGKM sequence was "Oh, Be A Fine Girl, Kiss Me!" However, since this may sound inappropriate to an enlightened 21st-century ear, try to invent a mnemonic of your own.

The range of surface temperatures for stars is much narrower than the range of luminosities. The coolest stars (spectral type M) have surface temperatures that are in some cases below 3000 K. The hottest stars (spectral type O) have surface temperatures that can exceed 40,000 K. Cool red stars are much more common than hot blue stars.

Measuring Mass Stellar mass is generally more difficult to measure than surface temperature or luminosity. The most dependable method for determining a star's mass relies on Newton's version of Kepler's third law (see Tools of Science, page 96). Recall that this law can be applied only when we can observe one object orbiting another, and it requires that we measure both the orbital period and the average orbital distance of the orbiting object. For stars, these requirements generally mean that we can apply the law to measure masses only in **binary star systems**, in which two stars orbit each other.

Binary star systems in which one star occasionally eclipses the other are particularly useful for mass measurements, because we know that the stars are orbiting in the same plane as our line of sight (Figure 8.13). Doppler shift measurements therefore tell us the true orbital velocities of the stars, because they are moving directly toward or directly away from us during part of the orbit. Combining these velocity measurements with the orbital period, which is simply the time between eclipses of one of the stars, gives the average orbital distance. Using the orbital period and an orbital distance, we can calculate the masses of the stars with Newton's version of Kepler's third law. (Note that this method is essentially the same as that used to find the masses of extrasolar planets that undergo transits [Section 7.1].)

TABLE 8.2 The Spectral Sequence

Spectral Type	Example(s)	Temperature Range	Key Absorption Line Features	Brightest Wavelength (color)	Typical Spectrum (selected lines labeled)
O	Stars of Orion's Belt	>33,000 K	Lines of ionized helium, weak hydrogen lines	<89 nm (ultraviolet)*	
B	Rigel	33,000 K–10,000 K	Lines of neutral helium and hydrogen	89–290 nm (ultraviolet)*	
A	Sirius	10,000 K–7500 K	Very strong hydrogen lines	290–390 nm (violet)*	
F	Polaris	7500 K–6000 K	Hydrogen lines, ionized calcium lines	390–480 nm (blue)*	
G	Sun, Alpha Centauri A	6000 K–5200 K	Weak hydrogen lines, strong lines of ionized calcium	480–560 nm (yellow)	
K	Arcturus	5200 K–3700 K	Lines of neutral and singly ionized atoms, some molecules	560–780 nm (red)	
M[†]	Betelgeuse, Proxima Centauri	<3700 K	Strong molecular lines	>780 nm (infrared)	

*All stars with temperatures above 6000 K look more or less white to the human eye because they emit plenty of radiation at all visible wavelengths.
[†]Astronomers also define spectral types L, T, and Y for starlike objects—usually *brown dwarfs* [Section 9.1]—cooler than stars of type M; these objects are not true stars because they are too low in mass to sustain nuclear fusion in their cores.

Careful observations of many different binary star systems containing stars of many different masses have helped to establish the overall mass range for stars. This range extends from as little as 0.08 times the mass of the Sun ($0.08M_{Sun}$) to at least 150 times the mass of the Sun ($150M_{Sun}$). We'll discuss the reasons for this mass range in Chapter 9.

◆ What patterns do we find in the properties of stars?

We have seen that stars come in a wide range of luminosities, surface temperatures, and masses. But are these characteristics randomly distributed among stars, or can we find patterns that provide clues to what determines the properties of stars?

Before reading any further, take another look at Figure 8.12 and think about how you would classify these stars. Almost all of them are at nearly the same distance from Earth, so we can compare their true luminosities by looking at their apparent brightnesses in the photograph. If you look closely, you'll notice a couple of important patterns:

- Most of the brightest stars are reddish in color.
- Aside from those relatively few bright red stars, there's a general trend to the luminosities and colors among all the rest of the stars: The brighter ones are blue or white, the more modest ones are yellowish like our Sun, and the dimmest ones are barely visible specks of red.

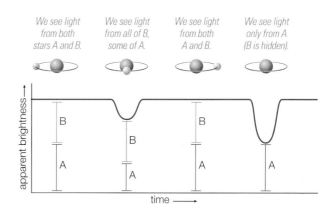

▲ FIGURE 8.13
If the stars in a binary star system orbit in the same plane as our line of sight, they will occasionally eclipse each other, causing a temporary drop in the apparent brightness we measure for the system. These types of binary systems are especially useful for measuring stellar masses.

Keeping in mind that colors tell us about surface temperature—blue is hotter and red is cooler—you can see that these patterns must be telling us about relationships between surface temperature and luminosity. Careful study of these relationships reveals that stars fall into three major groups.

The Main Sequence Although the very bright red stars stand out in Figure 8.12, the majority of the stars in the photograph follow the trend in which redder color, which indicates lower surface temperature, goes with lower luminosity, and bluer color, which indicates higher surface temperature, goes with higher luminosity. We'll see in the next chapter that the stars that follow this trend all have an important property in common: Like our Sun, these stars are generating energy by fusing hydrogen in their cores. This group is called the **main sequence**, for reasons that should become clear in Section 8.3. Because these stars follow a clear trend in which luminosity depends on surface temperature, and surface temperature is specified by spectral type, we can infer the luminosity of a main-sequence star simply by determining its spectral type.

Observations of main-sequence stars for which we can measure mass (those in binary systems) have shown that their luminosities and surface temperatures are closely related to their masses. Stars of spectral type G, like our Sun, have masses near $1M_{Sun}$, luminosities near $1L_{Sun}$, and surface temperatures close to the Sun's 5800 K, placing them approximately in the middle of the range for each property. The coolest main-sequence stars—those of spectral type M—have masses less than $0.3M_{Sun}$, luminosities less than $0.01L_{Sun}$, and surface temperatures less than 3500 K. At the other extreme, hot stars of spectral type O have masses greater than about $20M_{Sun}$, luminosities exceeding $30,000L_{Sun}$, and surface temperatures hotter than 30,000 K.

We can estimate the lifetimes of these stars from their masses and luminosities, just as we did for the Sun. Recall that the Sun was born with enough hydrogen fuel in its core for a lifetime of about 10 billion years. Now consider a star of spectral type B, with a mass of $10M_{Sun}$ and a luminosity of $10,000L_{Sun}$. Because it starts with 10 times the fuel of the Sun but consumes this fuel 10,000 times faster than the Sun does, its lifetime must be only about 1/1000 as long as the Sun's lifetime, or about 10 million years. Cosmically speaking, this is a remarkably short time and one reason massive stars are so rare: Most of the massive stars that were ever born have long since died.

On the other end of the scale, a 0.3-solar-mass star has a luminosity just 0.01 times that of the Sun and consequently lives roughly $0.3/0.01 = 30$ times as long as the Sun, or 300 billion years. In a universe that is now about 14 billion years old, even the most ancient of these small, dim, red stars of spectral type M still survive and will continue to shine faintly for hundreds of billions of years to come. Figure 8.14 compares these properties for a sample of four stars from the main sequence.

Giants and Supergiants The bright red stars in Figure 8.12 do not follow the trends we find for the main-sequence stars. In fact, because these stars are redder than the Sun, their surfaces must be cooler than that of the Sun, even though they are far more luminous. How can relatively cool stars have such high luminosities? Remember that a star's surface temperature determines the amount of light it emits per unit surface area: Hotter stars emit much more light per unit surface area than cooler stars. For example, a blue star emits far more total light than a red star of the same size. Therefore, a star that is red and cool can be very luminous only if it has a very large surface area.

We conclude that these very bright red stars must be far larger in size than our Sun. These large stars are known as either **giants** or **supergiants**, depending on their size. Giants have radii 10 to 100 times that of the Sun, and supergiants can be even larger. For example, Aldebaran is a giant star more than 40 times as large in radius as our Sun. Betelgeuse, the left shoulder in the constellation

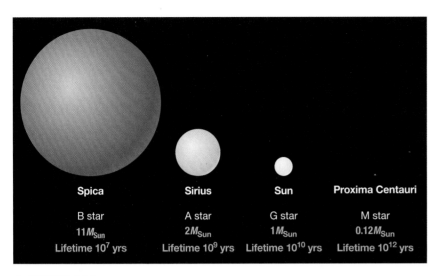

▲ FIGURE 8.14
Four typical hydrogen-fusing stars from the main sequence, shown to scale. Note that the more massive stars are hotter and more luminous than the less massive ones, but have shorter lifetimes.

Orion, is an enormous supergiant with a radius roughly 1000 times that of the Sun, equivalent to more than twice the Earth–Sun distance (Figure 8.15). As we'll discuss in Chapter 9, we now know that giant and supergiant stars have run out of hydrogen fuel in their cores and are nearing the ends of their lives.

White Dwarfs The third major group consists of what we call **white dwarfs**. These objects are so dim that none can be seen in Figure 8.12. A typical white dwarf, such as Procyon B, has a luminosity of only about $0.001 L_{Sun}$ but a surface temperature hotter than that of the Sun. To be so hot yet so dim, Procyon B must be much smaller in size than the Sun (see Figure 8.15). In fact, most white dwarfs are similar in radius to Earth, even though they are similar in mass to the Sun. This tells us that the matter in white dwarfs must be compressed to an extremely high density, unlike anything found on Earth.

▲ FIGURE 8.15
The relative sizes of stars. A supergiant like Betelgeuse would fill the inner solar system and extend more than 80% of the way to Jupiter's orbit. A giant like Aldebaran would fill the inner half of Mercury's orbit. The Sun is a hundred times larger in radius than a white dwarf, which is roughly the same size as Earth.

We'll see in Chapter 9 that white dwarfs are small and hot because they are the remaining embers of giants that have run out of usable fuel and blown off their outer layers. They are hot because they are essentially exposed stellar cores, but they are dim because they lack an ongoing energy source and radiate only their leftover heat into space. We'll discuss the nature of white dwarfs and other stellar corpses in Chapter 10.

Complete Stellar Classification We have discussed two key ways of classifying stars. First, we can classify a star by spectral type, which essentially tells us its surface temperature. The hottest stars have spectral type O, and temperatures decline in the order of the spectral sequence OBAFGKM. Second, we have found that stars fall into three major groupings: a *main sequence* consisting of stars that fuse hydrogen in their cores, for which higher surface temperature goes with higher luminosity; giants and supergiants, which are extraordinarily luminous and large in size; and white dwarfs, which are extremely dim and small in size. The complete classification of a star therefore consists of two parts: (1) its spectral type, stated as one of the letters OBAFGKM, and (2) its designation as a main-sequence star, a giant or supergiant, or a white dwarf. For example, we can say that our Sun is a spectral type G main-sequence star, while Betelgeuse is a spectral type M supergiant.

THE PROCESS OF SCIENCE IN ACTION

8.3 Visualizing Patterns Among Stars

Major discoveries in science sometimes come about when someone happens to look at the data in just the right way. The patterns among stars that we have just discussed were discovered about a century ago, and the story of their discovery is this chapter's case study in the process of science. As you will see, it demonstrates the importance of visualizing data, often through the use of graphs, in scientific discovery.

◆ How did we discover the patterns in stellar properties?

Our story of discovery begins at the end of the 19th century, when Edward Pickering was director of the Harvard College Observatory. He was interested in studying and classifying stellar spectra, but this was tedious and time-consuming work. He therefore hired numerous assistants, whom he called "computers." Many of his computers were women who had studied physics or astronomy at women's colleges such as Wellesley and Radcliffe (Figure 8.16). At the time, institutions like Harvard did not permit women to hold faculty positions, so Pickering's project offered a rare opportunity for these women to continue their work in astronomy.

Improving the Spectral Types One of the first computers was Williamina Fleming. Following Pickering's suggestions, Fleming classified stellar spectra according to the strength of the spectral lines of hydrogen: type A for the strongest hydrogen lines, type B for slightly weaker hydrogen lines, and so on, to type O for stars with the weakest hydrogen lines. As more stellar spectra were obtained and the spectra were studied in greater detail, it became clear that this classification scheme based solely on hydrogen lines was inadequate. Ultimately, the task of finding a better classification scheme fell to Annie Jump Cannon, who joined Pickering's team in 1896. Building on the work of Fleming and another of Pickering's computers, Antonia Maury, Cannon soon realized that the spectral classes fell into a natural order—but not the alphabetical order determined from hydrogen lines alone. Moreover, she found that some of the original classes overlapped others and could be eliminated. Cannon discovered that the natural sequence consisted of just a few

▲ **FIGURE 8.16**
Women astronomers pose with Edward Pickering at Harvard College Observatory in 1913. Annie Jump Cannon is fifth from the left in the back row.

of Pickering's original classes in the order OBAFGKM, which we now know to represent a sequence of surface temperatures. This breakthrough laid the foundation for our modern understanding of stellar properties.

The H-R Diagram

Hertzsprung-Russell Diagrams Around the same time that Cannon was reorganizing the spectral types, Danish astronomer Ejnar Hertzsprung and American astronomer Henry Norris Russell pioneered a new approach to analyzing the relationship between spectral type and luminosity. Building on the work of Cannon and others, they independently decided to make graphs of stellar properties by plotting stellar luminosity on one axis and spectral type on the other. These graphs ultimately revealed fundamental patterns among the properties of stars and are now called **Hertzsprung-Russell (H-R) diagrams**. These diagrams remain among the most important tools in astronomical research and are central to our study of stars. Figure 8.17 (pages 142–143) shows how we construct an H-R diagram, with a complete diagram on the right-hand page. Note that a star's position on an H-R diagram depends only on (1) its surface temperature and (2) its luminosity, which are plotted as follows:

- Surface temperature, which also corresponds to spectral type and color, is plotted on the horizontal axis. Note that temperature *decreases* from left to right, following the spectral sequence OBAFGKM.
- Luminosity is plotted on the vertical axis, in units of the Sun's luminosity (L_{Sun}). Because stellar luminosities span a wide range, each tick mark represents a luminosity 10 times as large as that of the tick mark just below it.

Each location on the H-R diagram represents a unique combination of spectral type and luminosity. For example, the dot representing the Sun in Figure 8.17 corresponds to the Sun's spectral type of G2 (or surface temperature 5800 K) and luminosity of $1L_{Sun}$. Because luminosity increases upward on the diagram and surface temperature increases leftward, stars near the upper left are hot and luminous. Similarly, stars near the upper right are cool and luminous, stars near the lower right are cool and dim, and stars near the lower left are hot and dim.

think about it Explain how the colors of the stars in Figure 8.17 help indicate stellar surface temperature. Do these colors tell us anything about interior temperatures? Why or why not?

Soon after Hertzsprung and Russell started plotting stars according to luminosity and spectral type, it became clear that stars cluster into three different regions of an H-R diagram:

- Most stars fall along the prominent streak known as the *main sequence*, which runs from the upper left to the lower right on the H-R diagram. All these stars are generating energy like the Sun, by fusing hydrogen into helium in their cores.
- *Giants and supergiants* are found to the upper right of the main sequence, because they are very luminous and tend to have relatively cool surfaces.
- *White dwarfs* are found to the lower left of the main sequence, because they are less luminous than main-sequence stars but have relatively high surface temperatures.

The H-R diagram also helps us visualize relationships among stellar radii. If two stars have the same surface temperature, one can be more luminous than the other only if it is larger in size. Stellar radii therefore increase as we go from the high-temperature, low-luminosity corner on the lower left of the H-R diagram to the low-temperature, high-luminosity corner on the upper right. Diagonal lines in Figure 8.17 indicate the radii of the stars in the graph.

cosmic context — FIGURE 8.17 Reading an H-R Diagram

Hertzsprung-Russell (H-R) diagrams are very important tools in astronomy because they reveal key relationships among the properties of stars. An H-R diagram is made by plotting stars according to their surface temperature and luminosities. This figure shows a step-by-step approach to building an H-R diagram.

1 An H-R Diagram Is a Graph: A star's position along the horizontal axis indicates its surface temperature, which is closely related to its color and spectral type. Its position along the vertical axis indicates its luminosity.

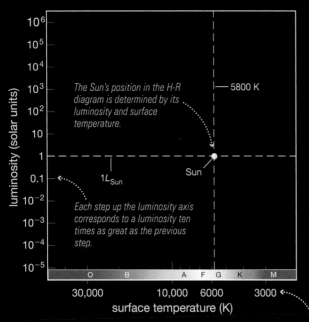

The Sun's position in the H-R diagram is determined by its luminosity and surface temperature.

Each step up the luminosity axis corresponds to a luminosity ten times as great as the previous step.

Temperature runs backward on the horizontal axis, with hot blue stars on the left and cool red stars on the right.

2 Main Sequence: Our Sun falls along the main sequence, a line of stars extending from the upper left of the diagram to the lower right. Most stars are main-sequence stars, which shine by fusing hydrogen into helium in their cores.

Star sizes on these diagrams indicate the general trend, but actual size differences are far greater than shown.

3 Giants and Supergiants: Stars in the upper right of an H-R diagram are more luminous than main-sequence stars of the same surface temperature. They must therefore be very large in radius, which is why they are known as *giants* and *supergiants*.

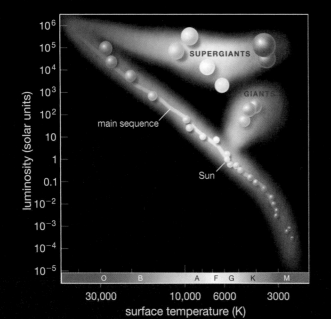

4 White Dwarfs: Stars in the lower left have high surface temperatures, dim luminosities, and small radii. These stars are known as *white dwarfs*.

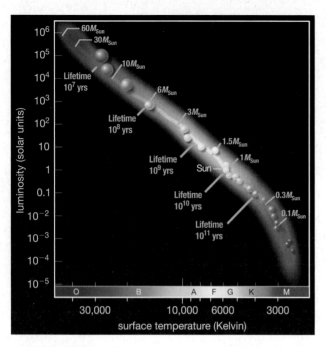

▲ **FIGURE 8.18**
The main sequence from Figure 8.17 is isolated here so that you can more easily see how masses and lifetimes vary along it. Notice that more massive hydrogen-fusing stars are brighter and hotter than less massive ones, but have shorter lifetimes. (Stellar masses are given in units of solar masses: $1M_{sun} = 2 \times 10^{30}$ kg.)

Mass and Lifetime Along the Main Sequence We have already noted that luminosity and surface temperature are closely related to mass for main-sequence stars. Purple labels in Figure 8.17 indicate the masses of stars along the main sequence and show that *stellar masses decrease downward along the main sequence*. To make the labels easier to see, Figure 8.18 repeats the same data but shows only the main sequence rather than the entire H-R diagram. At the upper end of the main sequence are the hot, luminous O stars, which have high masses. On the lower end are the cool, dim, and low-mass M stars.

From this close relationship between mass and luminosity, we can also determine a hydrogen-fusing star's lifetime. Green labels in Figures 8.17 and 8.18 indicate the lifetimes of main-sequence stars. Like masses, stellar lifetimes vary in an orderly way as we move along the main sequence: Massive stars near the upper end of the main sequence have *shorter* lives than less massive stars near the lower end.

The Bottom Line An H-R diagram visually represents an enormous amount of information about stars. From a single point representing a star on the diagram, we can infer the star's luminosity, surface temperature, spectral type, color, and radius. If that point is on the main sequence, it also tells us the mass and lifetime of the star. The first major discovery made with H-R diagrams was of the distinct patterns among these properties, but that was just the beginning. In the next chapter, we will investigate *why* the luminosity and surface temperature of a main-sequence star depend on its mass and why main-sequence stars become giants and supergiants after they run out of core hydrogen. As our investigation progresses, we will continue to make use of these remarkable diagrams.

summary of key concepts

8.1 Properties of the Sun

◆ **Why does the Sun shine?**

The Sun is a giant ball of plasma that shines with energy released by **nuclear fusion** of hydrogen into helium in its **core**. There, the temperature is 15 million K, and high density allows nuclei to slam into each other hard enough for the **strong force** to overcome their electromagnetic repulsion and hold them together.

◆ **How does energy escape from the Sun?**

Energy released by fusion in the Sun's core travels outward through the **radiation zone** and then through the **convection zone** to the **photosphere**, from which it escapes into space in the form of photons. Convection beneath the photosphere helps to generate magnetic fields that produce **sunspots** in the photosphere and other phenomena higher up in the **chromosphere** and **corona**.

8.2 Properties of Other Stars

◆ **How do we measure the properties of stars?**

We can determine a star's luminosity by measuring its **apparent brightness** and distance and then using the **inverse square law for light**. We can determine a star's surface temperature from either its color or its **spectral type**. We can measure the masses of stars in **binary star systems** by determining the orbital period and average orbital distance of the stars and then applying Newton's version of Kepler's third law.

◆ **What patterns do we find in the properties of stars?**

Most stars follow the same general trend, in which high-mass stars are very luminous and blue, with high surface temperatures, while lower-mass stars are less luminous and redder, with cooler surface temperatures. This group is known as the **main sequence**. The exceptions are **giants** and **supergiants**, which are much larger in radius and more luminous than normal stars, and **white dwarfs**, which are much smaller in radius and tend to have hotter than normal surface temperatures.

THE PROCESS OF SCIENCE IN ACTION

8.3 Visualizing Patterns Among Stars

- **How did we discover the patterns in stellar properties?**

Plotting the properties of stars on a **Hertzsprung-Russell (H-R) diagram**, which is a graph of surface temperature (or spectral type) versus luminosity, revealed the relationships between

stellar properties. Most stars fall along the *main sequence* of the diagram, which runs diagonally from the upper left to the lower right. Giants and supergiants appear to the upper right of the main sequence, while white dwarfs appear to the lower left.

Investigations

For instructor-assigned homework and other learning materials, go to www.MasteringAstronomy.com.

Quick Quiz

Choose the best answer to each of the following. Explain your reasoning.

1. At the center of the Sun, fusion converts hydrogen into (a) plasma. (b) radiation and elements like carbon and nitrogen. (c) helium and energy.
2. Which of these groups of particles has the greatest mass? (a) a helium nucleus with two protons and two neutrons (b) four electrons (c) four individual protons
3. Scientists estimate the central temperature of the Sun using (a) probes that measure changes in Earth's atmosphere. (b) mathematical models of the Sun. (c) telescopic observations of the Sun's interior.
4. Solar energy leaves the core of the Sun in the form of (a) photons. (b) rising hot gas. (c) sound waves.
5. Which of these layers of the Sun is coolest? (a) photosphere (b) chromosphere (c) corona
6. Why do sunspots appear darker than their surroundings? (a) They are cooler than their surroundings. (b) They block some of the sunlight from the photosphere. (c) They do not emit any light.
7. What do we need to measure in order to determine a star's luminosity? (a) apparent brightness and mass (b) apparent brightness and temperature (c) apparent brightness and distance
8. What two pieces of information would you need in order to measure the masses of stars in an eclipsing binary system? (a) the time between eclipses and the average distance between the stars (b) the period of the binary system and its distance from the Sun (c) the velocities of the stars and the Doppler shifts of their absorption lines
9. Which of these stars has the coolest surface temperature? (a) an A star (b) an F star (c) a K star
10. Which of these stars is the most massive? (a) a main-sequence A star (b) a main-sequence G star (c) a main-sequence M star
11. Which of these stars has the longest lifetime? (a) a main-sequence A star (b) a main-sequence G star (c) a main-sequence M star
12. Which of these stars has the largest radius? (a) a supergiant A star (b) a giant K star (c) a supergiant M star

Short-Answer/Essay Questions

Explain all answers clearly, with complete sentences and proper essay structure if needed. An asterisk (*) designates a quantitative problem, for which you should show all your work.

13. *A Really Strong Force.* How would the interior temperature of the Sun be different if the strong force that binds nuclei together were 10 times stronger?
14. *Covered with Sunspots.* Describe what the Sun would look like from Earth if the entire photosphere were the same temperature as a sunspot.
15. *Inside the Sun.* Describe how scientists determine what the interior of the Sun is like. Why can't we send a probe into the Sun to obtain direct information about the interior?
16. *An Angry Sun.* A *Time* magazine cover once suggested that an "angry Sun" was becoming more active as human activity changed Earth's climate. It's certainly possible for the Sun to become more active at the same time that humans are affecting Earth, but is it possible that the Sun could be responding to human activity? Why or why not?
17. *Interpreting the H-R Diagram.* Using the information in Figure 8.17, describe how Proxima Centauri differs from Sirius.
18. *Classification.* As discussed in the text, Annie Jump Cannon and her colleagues developed our modern system of stellar classification. Why do you think rapid advances in our understanding of stars followed so quickly on the heels of their efforts? What other areas in science have had huge advances in understanding follow directly from improved systems of classification?
19. *An Expanding Star.* Describe what would happen to the surface temperature of a star if its radius doubled in size with no change in luminosity.
20. *Colors of Eclipsing Binaries.* Figure 8.13 shows an eclipsing binary system consisting of a small blue star and a larger red star. Explain why the decrease in apparent brightness of the combined system is greater when the blue star is eclipsed than when the red star is eclipsed.
*21. *The Lifetime of the Sun.* The Sun's mass is about 2×10^{30} kg; this mass was about 70% hydrogen when the Sun formed, with about 13% of this hydrogen available for eventual fusion in the core.
   a. Use the given data to calculate the total mass of hydrogen available for fusion over the lifetime of the Sun.
   b. The Sun fuses about 600 billion kilograms of hydrogen each second. Based on your result from part a, calculate how long the Sun's initial supply of hydrogen can last. Give your answer in both seconds and years.
   c. Given that our solar system is now about 4.6 billion years old, when will the Sun run out of hydrogen for fusion?
*22. *The Inverse Square Law for Light.* Use Earth's distance of about 150 million kilometers from the Sun, the Sun's apparent brightness in our sky of about 1300 watts/m^2, and the inverse square law for light to calculate the apparent brightness that we would measure for the Sun *if* we were located at the following positions.
   a. Half Earth's distance from the Sun
   b. Twice Earth's distance from the Sun
   c. Five times Earth's distance from the Sun

9 Stellar Lives

LEARNING GOALS

9.1 Lives in Balance
- Why do stars shine so steadily?
- Why do a star's properties depend on its mass?

9.2 Star Death
- What will happen when our Sun runs out of fuel?
- How do high-mass stars end their lives?

 THE PROCESS OF SCIENCE IN ACTION

9.3 Testing Stellar Models with Star Clusters
- What do star clusters reveal about the lives of stars?

This photo shows the Orion Nebula, a huge interstellar cloud that acts as a stellar nursery, giving birth to thousands of stars over a period of a few million years. These stars then live out their lives, with fates dictated by their masses. Low-mass stars shine steadily for billions of years, sharing a fate similar to that of our Sun. High-mass stars shine so brightly that they burn out in just a few million years, dying in spectacular explosions that scatter their ashes into space. In this chapter, we'll study the how and why of star births, lives, and deaths. Along the way, we'll learn about the future of our own Sun and see how high-mass stars have produced the "star stuff" from which we and our planet are made.

▶ Star Stuff

9.1 Lives in Balance

The story of a star's life is in many ways the story of a battle between two opposing forces: pressure and gravity. A star is born when gravity becomes strong enough to overcome pressure within a gas cloud and compress it until nuclear fusion begins. A star lives while the energy generated by fusion can keep pressure and gravity in steady balance. A star dies as it exhausts its fuel for fusion and goes out of balance again. In this section, we examine how a star achieves the steady state of balance that fusion maintains.

◆ Why do stars shine so steadily?

Most stars shine steadily for most of their lives. Annual changes in our Sun's light output are less than 0.1%, and models indicate that the Sun's luminosity has changed by less than 30% over the last $4\frac{1}{2}$ billion years. This type of steady energy output has been critical for life on Earth, and perhaps for life on planets around other stars. But what keeps stars shining so steadily? To answer this question, let's first look at how stars are born.

Star Birth Stars form out of interstellar gas, within clouds in which the inward pull of gravity becomes stronger than the outward push of gas pressure. Two things can help gravity win out over pressure and start the contraction of a gas cloud: (1) high density, because packing the gas particles closer together makes the gravitational forces between them stronger, and (2) low temperature, because lowering a cloud's temperature reduces the gas pressure. We therefore expect star-forming clouds to be colder and denser than most other interstellar gas clouds.

Observations confirm this idea: Stars are indeed born within the coldest and densest clouds, called **molecular clouds** (Figure 9.1) because their low temperatures allow hydrogen atoms to pair up to form hydrogen molecules. These clouds typically have temperatures of only 10–30 K. And while their densities are still low enough that they would qualify as superb vacuums by earthly standards, these molecular clouds are hundreds to thousands of times as dense as other regions of

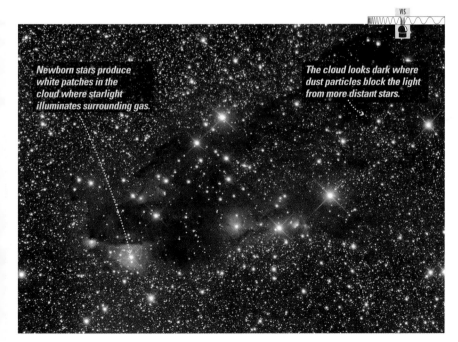

◀ **FIGURE 9.1**
A star-forming molecular cloud in the constellation Scorpius. The region pictured here is about 50 light-years across.

interstellar space. The molecular clouds that give birth to stars also tend to be quite large—typically containing enough material to form thousands of stars—because more total mass also helps gravity overcome gas pressure. The gas in these large clouds tends to form smaller clumps as they collapse, with each clump forming an individual star system. That is why stars are generally born in clusters. Our own Sun was presumably born in a cluster, but after $4\frac{1}{2}$ billion years the Sun's siblings have long since dispersed and mingled with the other stars in our galaxy.

We have already discussed how a clump of gas can give birth to an individual star system, when we discussed the birth of our Sun and solar system from the *solar nebula* [Section 4.2]. Gravity causes the cloud to shrink in size, a process called **gravitational contraction**. This contraction causes the gas temperature to rise, especially in the center of the cloud, because it converts gravitational potential energy into thermal energy. Gravitational contraction proceeds rapidly at first, because photons from the cloud quickly carry away the thermal energy, preventing a buildup of gas pressure that could counteract gravity. Eventually, however, the central region of the cloud becomes so dense that photons cannot easily escape, so the pressure builds and contraction slows. The cloud center is now a **protostar** that shines with energy generated by ongoing gravitational contraction. It is not yet a true star because its core is not yet hot enough to sustain nuclear fusion.

Despite the energy losses from its surface, a protostar continues to heat up, because much of the energy released by gravitational contraction remains trapped inside it. This process of ongoing shrinkage and heating continues until the central temperature and density grow high enough to sustain nuclear fusion. Gravitational contraction finally stops when the energy production from fusion in the core matches the energy losses from the star's surface. The thermal energy content of the star then remains constant, enabling pressure and gravity to come into steady balance. This is why we say that the star is "born" when its core becomes hot enough for sustained nuclear fusion.

Two Kinds of Balance Looking back at the story of star birth, we see that a star becomes a steady source of light when it achieves two kinds of balance. First, the outward push of internal gas pressure must balance the inward pull of gravity to prevent the star from expanding or contracting. A stack of acrobats provides a simple example of this kind of balance, which we call **gravitational equilibrium** (Figure 9.2). The bottom person supports the weight (due to gravity) of everyone above, so he must push upward with enough pressure to support all this weight. At each higher level, the overlying weight is less, so it's a little easier for each additional person to hold up the rest of the stack.

Gravitational equilibrium works much the same way within a star like the Sun, except the outward push against gravity comes from internal gas pressure. The internal pressure precisely balances gravity at every point within the star, keeping it stable in size (Figure 9.3). Because the weight of overlying layers is greater as we look deeper below the surface, the pressure increases with depth.

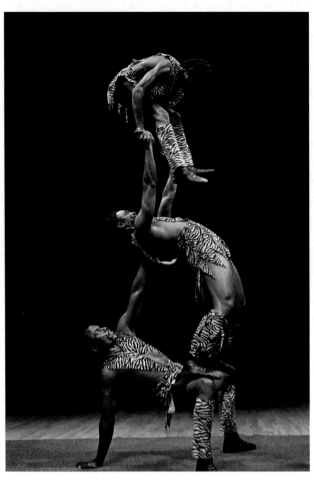

▲ FIGURE 9.2
An acrobat stack demonstrates gravitational equilibrium: The lowest person supports the most weight and feels the greatest pressure; the overlying weight and underlying pressure decrease for those higher up.

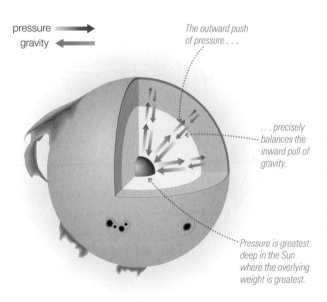

▲ FIGURE 9.3
Gravitational equilibrium in the Sun. At each point inside, the pressure pushing outward balances the weight of the overlying layers. This balance works the same way in other stars.

think about it Earth's atmosphere is also in gravitational equilibrium, with the pressure in lower layers supporting the weight of upper layers. Use this idea to explain why the air gets thinner at higher altitudes.

The second kind of balance is the **energy balance** between the rate at which fusion releases energy in the star's core and the rate at which the star's surface radiates this energy into space (Figure 9.4). Energy balance is important because without it, the balance between pressure and gravity cannot remain steady. If fusion in the core does not replace the energy radiated from the surface, thereby keeping the total thermal energy content constant, then gravitational contraction will cause the core to shrink and force its temperature to rise.

The Stellar Thermostat The interplay between these two kinds of balance enables stars to shine steadily for millions or billions of years. The key lies with a self-regulating **stellar thermostat** that quickly offsets any small change in temperature so that balance is restored (Figure 9.5). To see how it works, let's use our Sun as an example.

Suppose that for some reason the Sun's core temperature rose very slightly. The rate of nuclear fusion is very sensitive to temperature, so this slight temperature increase would cause the fusion rate to soar as protons in the core collided more frequently and with more energy. Because energy moves slowly through the Sun's interior, this extra energy would be bottled up in the core, causing an increase in the core pressure. The push of this pressure would temporarily exceed the pull of gravity, causing the core to expand and cool. This cooling, in turn, would cause the fusion rate to drop back down until the core returned to its original size and temperature, restoring both gravitational equilibrium and energy balance.

A slight drop in the Sun's core temperature would trigger an opposite chain of events. The reduced core temperature would lead to a decrease in the rate of nuclear fusion, causing a drop in pressure and contraction of the core. As the core shrank, its temperature would rise until the fusion rate returned to normal and restored the core to its original size and temperature.

To summarize, a star's luminosity will remain steady as long as hydrogen fusion continues in its core and keeps the star in both gravitational equilibrium and energy balance. A star like the Sun can shine steadily for about 10 billion years in this balanced state, because that is how long its core supply of hydrogen can last.

▲ **FIGURE 9.4**
Energy balance in the Sun: Fusion must supply energy in the core at the same rate as the Sun radiates energy from its surface.

◆ **Why do a star's properties depend on its mass?**

Now that we have seen how stars can shine steadily, we are ready to look at why the luminosity, surface temperature, mass, and lifetime of a main-sequence star are all so closely related. The stars on the main sequence of an H-R diagram are

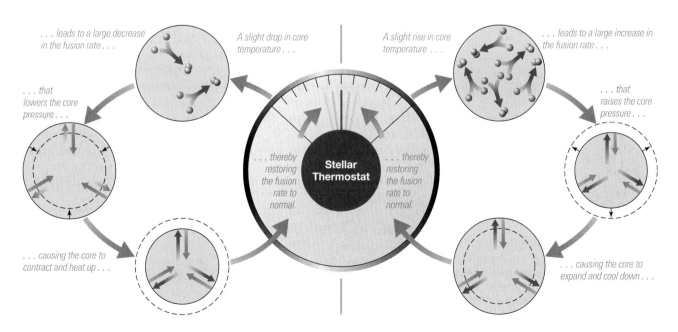

▲ **FIGURE 9.5**
The stellar thermostat. Gravitational equilibrium regulates a star's core temperature. Everything is in balance if the amount of energy leaving the core equals the amount of energy produced by fusion. A rise in core temperature causes the core to expand, lowering its temperature to its original value. A decrease in core temperature causes the core to contract, also restoring the original core temperature.

steadily fusing hydrogen into helium in their cores, just like the Sun, meaning that each one has achieved both gravitational equilibrium and energy balance. However, the core temperature at which a star achieves these two kinds of balance depends on its mass. Because mass determines a star's balancing point, it determines all the rest of a star's major properties and also determines which objects can become stars and which cannot.

Mass and the Main Sequence To see why a star's state of balance depends on its mass, consider what happens inside a protostar more massive than the Sun. Because this protostar has more mass than the Sun, gravitational contraction converts more gravitational potential energy into thermal energy during its protostar stage. Fusion therefore begins earlier in the process of contraction and brings this higher-mass protostar into balance with a higher core temperature and a greater luminosity. The luminosity of this newly formed star can be much greater than the Sun's because even a slightly higher core temperature leads to a much higher rate of fusion, enabling the star to achieve energy balance with both a greater radius and a hotter surface temperature.

All the trends we see with mass along the main sequence reflect the properties stars have when they first come into balance. Stars with less mass than the Sun need to contract more before fusion begins and come into balance with a lower core temperature, lower luminosity, cooler surface, and smaller radius. They will also live much longer than the Sun because their lower fusion rates consume the fuel for fusion more slowly. Stars with more mass than the Sun contract less before fusion begins and come into balance with a higher core temperature, greater luminosity, hotter surface, and larger radius. They have much shorter lifetimes than the Sun because they consume their fuel so rapidly.

Limits on Stellar Masses The interplay between mass and the two kinds of balance in stars also explains why stars have a relatively narrow range of masses. On the high-mass end, observations indicate that the maximum mass is at least $150 M_{Sun}$, and possibly as great as $300 M_{Sun}$. However, theoretical models suggest that stars with masses above about $100 M_{Sun}$ have such enormous energy output from their cores that the flow of photons through such a star rapidly drives its outer layers into space and therefore continually lowers its mass.

On the low-mass end, calculations show that a protostar cannot become a true star unless its mass is at least $0.08 M_{Sun}$, equivalent to about 80 times the mass of Jupiter. Below this mass, the protostar's gravity will be too weak to contract the core enough to reach the 10 million K threshold for efficient hydrogen fusion. In that case, the protostar will stabilize in size as a "failed star" known as a **brown dwarf**. Brown dwarfs radiate primarily in the infrared and actually look deep red or magenta in color rather than brown (Figure 9.6). Because they do not sustain steady fusion in their cores, brown dwarfs never achieve the energy balance that would allow them to shine steadily. Instead, they cool with time as they radiate away their internal thermal energy. In essence, brown dwarfs occupy a fuzzy gap between what we call a planet and what we call a star.

Pressure in Brown Dwarfs The source of the pressure that stops gravity from squeezing a brown dwarf's core to the point at which it could sustain fusion is quite different from the type of pressure we encounter in most other situations. Ordinary gas pressure is often called **thermal pressure** because it is closely linked to temperature: Raising the temperature increases the particle speeds and thereby raises the thermal pressure. However, calculations show that thermal pressure is insufficient to balance gravity in a brown dwarf. Instead, the

▲ **FIGURE 9.6**
Brown dwarfs are "failed stars" that have masses lower than the $0.08 M_{Sun}$ required to sustain nuclear fusion. This artist's conception shows a brown dwarf orbited by a planet (to its left) in a system with multiple stars. The reddish color approximates how a brown dwarf would appear to human eyes. The bands are shown because we expect brown dwarfs to look more like jovian planets than like stars.

a When there are many more available places (chairs) than particles (people), a particle is unlikely to try to occupy the same place as another particle. The only pressure comes from the temperature-related motion of the particles.

b When the number of particles (people) approaches the number of available places (chairs), finding an available place requires that the particles move faster than they would otherwise. The extra motion creates degeneracy pressure.

▲ **FIGURE 9.7**
An auditorium analogy to explain degeneracy pressure. Chairs represent available places (quantum states) for electrons, and people who must keep moving from chair to chair represent electrons.

crush of gravity in a brown dwarf is halted by **degeneracy pressure,** a type of pressure that does not depend on temperature at all. It depends instead on the laws of *quantum mechanics* that also give rise to distinct energy levels in atoms (see Tools of Science, page 152).

In much the same way that electrons in atoms can occupy only particular energy levels, the quantum laws restrict how closely electrons can be packed in a gas. Under most circumstances, these restrictions have little effect on the motions or locations of electrons, and hence little effect on the pressure. However, they become important inside brown dwarfs (and also in white dwarfs [Section 10.1]), because the electrons become packed together almost as closely as the quantum laws allow. A simple analogy illustrates the basic idea. Imagine an auditorium in which the laws of quantum mechanics dictate the spacing between chairs and in which people represent electrons (Figure 9.7). As though playing a game of musical chairs, the people are always moving from seat to seat, just as electrons must remain constantly in motion. Most objects are like auditoriums with many more available chairs than people, so the people (electrons) can easily find chairs as they move about. However, the cores of brown dwarfs are like much smaller auditoriums with so few chairs that the people (electrons) fill nearly all of them. Because there are virtually no open seats, the people (electrons) can't all squeeze into a smaller section of the auditorium.

This resistance to squeezing is the origin of degeneracy pressure. If the people were really like electrons, quantum laws would also require them to move faster and faster to find open seats as they were squeezed into a smaller section. However, their speeds would have nothing to do with temperature. Degeneracy pressure and the particle motion that goes with it arise *only* because of the restrictions on where the particles can go, which is why temperature does not affect them.

9.2 Star Death

A star will shine steadily as long as it can maintain balance between gravity and pressure and between the rates at which energy is generated in its core and released into space at its surface. Hydrogen fusion in the core can maintain this state of balance as long as the core contains enough hydrogen. But fusion cannot continue forever. In this section, we will investigate what happens when core hydrogen runs out and a star can no longer remain in balance.

Lives of Low-Mass Stars, Parts 1, 2

◆ What will happen when our Sun runs out of fuel?

Like everything else in a star's life, its fate when it runs out of core hydrogen depends on its mass. Scientists learn about the end stages of stellar lives through a combination of theoretical modeling and observations of stars that appear to be in these late stages. These studies show that stars divide into two major groups by mass: Low-mass stars, like our Sun, end their lives relatively quietly, while high-mass stars (more than $8M_{Sun}$) die in titanic explosions. Our Sun is a fairly typical low-mass star, so we will use it as a prototype to investigate the final life stages of low-mass stars.

Red Giant Stage When the Sun's core hydrogen is finally depleted, nuclear fusion in the core will temporarily cease. With no fusion to balance the thermal energy losses from the star's surface, the core will once again be out of balance, just as it was when the Sun was a protostar, and gravitational contraction of the core will resume. Somewhat surprisingly, the Sun's outer layers will expand outward during this time. Over a period of about a billion years—or about 10% of its main-sequence lifetime—the Sun will slowly grow in size and luminosity to become a **red giant**. At the peak of its red giant phase, the Sun will be more than 100 times as large in radius and more than 1000 times as luminous as it is today.

tools of science — Quantum Laws and Astronomy

Although astronomy is primarily concerned with very large things, the behavior of large objects sometimes depends on the laws that govern their constituent particles. The branch of physics that deals with the behavior of atoms and subatomic particles is called *quantum mechanics*, so we say that these particles obey *quantum laws*.

Perhaps the most important application of quantum laws in astronomy arises in spectroscopy (see Tools of Science, page 132). Consider a hydrogen atom, which has one proton and one electron. It's tempting to think of the electron as a tiny particle that orbits the proton, but that would imply that the electron could have any orbit with any amount of energy. This is not the case; quantum laws dictate that the electron can have only particular amounts of energy, and not other energies in between. As an analogy, suppose you're washing windows on a building. If you stand on an adjustable platform, you can stop at any height above the ground. But if you use a ladder, you can stand only at the particular heights of the rungs on the ladder, and not at other heights in between. The possible energies of electrons in atoms are like the specific heights on a ladder: Only a few particular energy levels are possible (see figure).

This fact explains why atoms produce emission and absorption lines with particular wavelengths. An electron can change its energy level only if it gains or loses the particular amount of energy that separates a pair of levels. Therefore, an atom can absorb only photons with the particular energies needed for electrons to rise from one energy level to another; because the energy of a photon depends on its wavelength (shorter wavelength means higher energy), the atom absorbs only particular wavelengths of light. An atom emits photons when an electron switches to lower energy levels, and therefore can emit only photons with the same set of particular wavelengths that it can absorb.

Another important quantum effect in astronomy is the degeneracy pressure discussed in this chapter, which arises from the combination of two quantum laws: (1) the uncertainty principle, which places a limit on the certainty with which we can know the combination of a subatomic particle's location and velocity, and (2) the exclusion principle, which places a limit on how closely electrons and other subatomic particles can be packed together. In a brown dwarf or white dwarf, electrons are packed in such a way that the exclusion principle limits how much more they can be packed, and this resistance to further packing creates degeneracy pressure. The uncertainty principle comes into play because the electrons are very close together, particularly in white dwarfs. This implies very little uncertainty in their locations, so their velocities must be highly uncertain, which means some of them must move with very high speeds. Because no particle can exceed the speed of light, there is also a limit to the ability of electron degeneracy pressure to resist the crush of gravity. Once that limit is surpassed, as is the case for a stellar core during a supernova, the electrons must combine with protons to make neutrons. The neutrons can then provide degeneracy pressure of their own—but only to a point, as we will discuss in the next chapter.

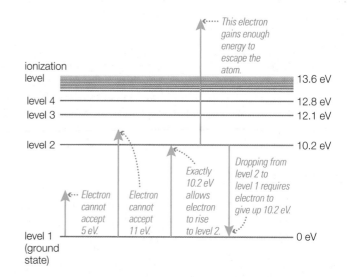

Energy levels for the electron in a hydrogen atom, with energies given in units of electron-volts, or eV. Electrons can move between energy levels only if they gain or lose the particular energies that separate the levels.

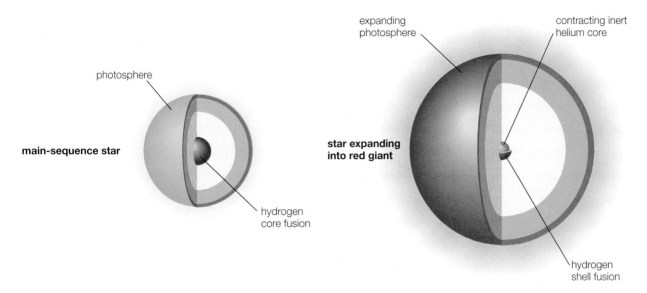

▲ **FIGURE 9.8**
After a star like the Sun ends its main-sequence life, its inert helium core contracts while hydrogen shell fusion begins. The high rate of fusion in the hydrogen shell forces the star's upper layers to expand outward. (Not to scale.)

To understand why the Sun will expand even while its core is shrinking, we need to think about the composition of the core at the end of the Sun's main-sequence life. After the core exhausts its hydrogen, it will be made almost entirely of helium, because helium is the product of hydrogen fusion. However, the gas surrounding the core will still contain hydrogen that has never previously undergone fusion. Because gravity shrinks both the *inert* (nonfusing) helium core and the surrounding *shell* of hydrogen, the hydrogen shell will soon become hot enough for **hydrogen shell fusion**—hydrogen fusion in a shell around the core (Figure 9.8). In fact, the shell will become so hot that hydrogen shell fusion will proceed at a much higher rate than core hydrogen fusion does today. The higher fusion rate will generate enough energy to dramatically increase the Sun's luminosity and enough pressure to push the surrounding layers of gas outward, expanding the Sun's size.

▲ **FIGURE 9.9**
Helium fusion converts three helium nuclei into one carbon nucleus, releasing energy in the process.

Helium Fusion Gravitational contraction of the inert helium core will continue—with the outer layers of the red giant Sun continuing to grow larger and more luminous—until the core temperature reaches 100 million K. At that point, **helium fusion** will ignite through a reaction that converts three helium nuclei into one carbon nucleus (Figure 9.9). Energy is released because the carbon-12 nucleus has a slightly lower mass than the three helium-4 nuclei, and the lost mass becomes energy in accord with $E = mc^2$. The temperature required for helium fusion is much greater than that for hydrogen fusion because helium nuclei, which contain two protons, have a greater positive charge than hydrogen nuclei, which contain just one proton. The greater charge means that helium nuclei repel one another more strongly than hydrogen nuclei, and therefore must slam into one another at much higher speeds in order to fuse [Section 8.1].

Once helium fusion begins, thermal pressure in the core will temporarily become stronger than gravity, forcing the core to expand. This core expansion will push the hydrogen-fusing shell outward, lowering its temperature and its fusion rate. As a result, even though core helium fusion *and* hydrogen shell fusion will be taking place simultaneously (Figure 9.10), the Sun's total energy production will fall from the peak it reached during the red giant stage. This reduction in energy production means that the Sun, now a *helium-fusing star*,

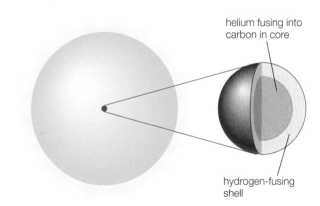

▲ **FIGURE 9.10**
Core structure of the Sun after helium fusion begins. Helium fusion will cause the core and hydrogen-fusing shell to expand and slightly cool, thereby reducing the overall energy generation rate below what it was during the red giant stage. The outer layers will then shrink back, making the helium-fusing Sun smaller than the Sun was during its red giant stage.

will be somewhat smaller in size, lower in luminosity, and yellower in color than it was as a red giant. With fusion once again operating in the core, the helium-fusing Sun will regain the same sort of balance it had as a main-sequence star.

Last Gasps It is only a matter of time until a helium-fusing star fuses all its core helium into carbon. In the Sun, the core helium will run out after about 100 million years—only about 1% as long as the Sun's 10-billion-year hydrogen-fusing lifetime. When the core helium is exhausted, fusion will again cease. The core, now made of the carbon produced by helium fusion, will begin to shrink once again under the crush of gravity.

The exhaustion of core helium will cause the Sun to expand once again, just as it did in becoming a red giant. This time, the trigger for the expansion will be helium fusion in a shell around the inert carbon core. Meanwhile, hydrogen fusion will continue in a shell atop the helium layer. The Sun will have become a *double shell–fusion star*. Both shells will contract along with the inert core, driving their temperatures and fusion rates so high that the Sun will expand to an even greater size and luminosity than in its first red giant stage.

The furious fusion rates in the helium and hydrogen shells cannot last for more than a few million years. The Sun's only hope of extending its life will then lie with its carbon core, but this is a false hope for a low-mass star like the Sun. Carbon fusion is possible only at temperatures above about 600 million K, and degeneracy pressure will halt the contraction of the Sun's inert carbon core before it ever gets that hot (in the same way that it halts the contraction of a brown dwarf). With the carbon core unable to undergo fusion and provide a new source of energy, the Sun will finally have reached the end of its life.

think about it Suppose the universe contained only low-mass stars. Would elements heavier than carbon exist? Why or why not?

The Sun's last act will be to eject its outer layers into space, creating a huge shell of gas expanding away from the inert carbon core. The exposed core will still be very hot and will therefore emit intense ultraviolet radiation. This radiation will ionize the gas in the expanding shell, making it glow brightly as a **planetary nebula**. We see many examples of planetary nebulae around other low-mass stars that have recently died in this same way (Figure 9.11). The glow of a planetary nebula fades as the exposed core cools and the ejected gas disperses into space, disappearing within a million years or less.

The fate of the leftover carbon core is even more interesting. Degeneracy pressure will halt its collapse when it is about the size of Earth, but it will still have much of the Sun's original mass. It will be quite hot, but its small size will mean it is very dim. In other words, it will be the type of object that we call a *white dwarf* [Section 8.2]. White dwarfs are small in radius but high in mass because they are the compressed, exposed cores of dead stars, supported against the crush of gravity by degeneracy pressure.

The Sun's Life on an H-R Diagram Mathematical models enable astronomers to determine how the Sun's surface temperature and luminosity should change as it evolves. Plotting the results on an H-R diagram produces a **life track** (also called an *evolutionary track*) that shows the Sun's luminosity and surface temperature at each point in its life (Figure 9.12).

Notice that the Sun's life track goes generally upward as it becomes a red giant, because the Sun expands in size and luminosity while its inert helium core is contracting. The track also goes slightly to the right during this stage, because its surface temperature falls a bit. At the tip of the red giant stage, helium fusion begins and the Sun contracts in size, as shown by the track going down and left a Helix Nebula. The central white dot is the hot white dwarf.

b Butterfly Nebula. The hot white dwarf is hidden in the dark band of dusty gas crossing the center.

▲ FIGURE 9.11
Hubble Space Telescope photos of planetary nebulae, which form when low-mass stars in their final death throes cast off their outer layers of gas, leaving behind the hot core that ejected the gas. The hot core ionizes and energizes the shells of gas that surround it. As the nebula gas disperses into space, the hot core remains as a white dwarf.

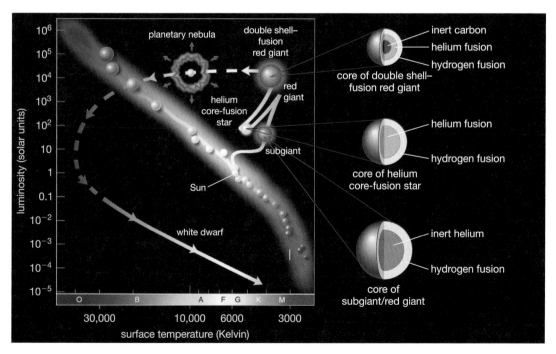

▲ FIGURE 9.12
The life track of our Sun from when it first becomes a hydrogen-fusing, main-sequence star to when it dies as a white dwarf. Core structure is shown at key stages.

as the Sun becomes a helium-fusing star. The life track again turns upward as the Sun enters its second red giant phase, this time with energy generated by fusion in shells of both helium and hydrogen. The dashed curve indicates that, as the Sun ejects its planetary nebula, we shift from plotting the surface temperature of a red giant to plotting the surface temperature of the exposed core left behind. The curve becomes solid again near the lower left, indicating that this core is a hot white dwarf. From that point, the curve continues downward and to the right as the white dwarf cools and fades.

Star Stuff — Origin of Elements

◆ How do high-mass stars end their lives?

We now turn our attention to the end stages in the lives of high-mass stars. At first, these end stages are very similar to those of a low-mass star like the Sun, except they proceed much more rapidly. For example, a 25-solar-mass star lasts only a few million years as a hydrogen-fusing main-sequence star before it goes out of balance. As its core hydrogen runs out, a hydrogen-fusing shell forms around a shrinking helium core, generating so much energy that the star's outer layers expand outward until the star becomes a **supergiant**. Gravitational contraction of the helium core continues until it becomes hot enough to fuse helium into carbon. The high-mass star fuses helium into carbon so rapidly that it is left with an inert carbon core after no more than a few hundred thousand years.

After that, the life stages of this 25-solar-mass star become quite different from those of the Sun. Gravitational contraction simply continues after helium fusion ends, the inert carbon core shrinks, and the core pressure, temperature, and density all rise. The shrinking core gets hotter and hotter, and it soon becomes hot enough for carbon fusion. As we'll discuss shortly, the core and shells go through several more phases of fusion of increasingly heavier elements—producing the "star stuff" that makes our lives possible—while the star's outer layers continue to swell in size.

Despite the dramatic events taking place in its interior, the high-mass star's outer appearance changes slowly. As each stage of core fusion ceases, the surrounding shell fusion intensifies and further inflates the star's outer layers. Each time the core flares up, the outer layers contract somewhat, but the star's overall luminosity remains about the same. The result is that the star's life track zigzags across the top of the H-R diagram (Figure 9.13). In the most massive stars, the core changes happen so quickly that the outer layers don't have time to respond, and the star progresses steadily toward becoming a red supergiant.

One of these massive red supergiant stars happens to be relatively nearby: Betelgeuse, the upper-left shoulder of Orion. Its radius is about 900 solar radii, or some four times the distance from the Sun to Earth. We have no way of knowing exactly what stage of nuclear fusion is now taking place in its core. Betelgeuse may have a few thousand years of fusion still ahead, or it may right now be in its final stages of life. If the latter is the case, then sometime soon we will witness one of the most dramatic events that ever occurs in the universe: a supernova.

Advanced Nuclear Fusion A high-mass star like Betelgeuse can make elements heavier than carbon because degeneracy pressure is never able to halt the contraction of its core. The crush of gravity within the star is simply too strong. Each time the core exhausts one source of fuel for fusion, it contracts and heats up until it can fuse even larger nuclei, while fusion of lighter elements continues in multiple shells around the core. The complete set of nuclear reactions in a high-mass star's final stages of life is quite complex, and many different reactions

▲ **FIGURE 9.13**
Life tracks on the H-R diagram from main-sequence star to red supergiant for selected high-mass stars. Labels on the tracks give the star's mass at the beginning of its main-sequence life. (Based on models from A. Maeder and G. Meynet.)

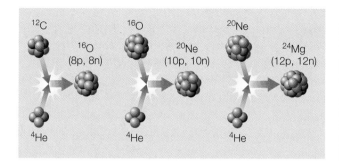
a Many reactions proceed through *helium capture*, in which fusion joins a helium nucleus to some other nucleus.

b At extremely high temperatures, fusion of even heavier nuclei can occur.

▲ FIGURE 9.14
A few of the many nuclear reactions that occur in the final stages of a high-mass star's life.

may take place simultaneously, making elements like oxygen, silicon, and sulfur (Figure 9.14). Near the end, the star's central region resembles the inside of an onion, with layer upon layer of shells fusing different elements (Figure 9.15).

The core continues shrinking, heating, and fusing new elements until iron starts to pile up. If iron were like the other elements in prior stages of nuclear fusion, core contraction would stop when iron fusion ignited. However, iron is unique among the elements: It is the one element from which it is *not* possible to generate any kind of nuclear energy.

To understand why iron is unique, remember that only two basic processes can release nuclear energy: *fusion* of light elements into heavier ones and *fission* of heavy elements into not-so-heavy ones (see Figure 8.2). Recall that hydrogen fusion converts four protons (hydrogen nuclei) into a helium nucleus that consists of two protons and two neutrons, which means that the total number of *nuclear particles* (protons and neutrons combined) does not change. However, this fusion reaction generates energy (in accord with $E = mc^2$) because the *mass* of the helium nucleus is less than the combined mass of the four hydrogen nuclei that fused to create it—despite the fact that the *number* of nuclear particles is unchanged.

In other words, fusing hydrogen into helium generates energy because helium has a lower *mass per nuclear particle* than hydrogen. Similarly, fusing three helium-4 nuclei into one carbon-12 nucleus generates energy because carbon has a lower mass per nuclear particle than helium, which means that some mass disappears and becomes energy in this fusion reaction. The decrease in mass per nuclear particle from hydrogen to helium to carbon is part of a general trend shown in Figure 9.16.

The mass per nuclear particle tends to decrease as we go from light elements to iron, which means that fusion of light nuclei into heavier nuclei generates energy. This trend reverses beyond iron: The mass per nuclear particle tends to *increase* as we look to heavier elements. As a result, elements heavier than iron can generate nuclear energy only through fission into lighter elements.

Iron has the lowest mass per nuclear particle of all nuclei and therefore cannot release energy by either fusion or fission. Once the matter in a stellar core turns to iron, it can generate no more energy. For the iron core, the only hope of resisting the crush of gravity lies with degeneracy pressure, but iron keeps piling up until even degeneracy pressure cannot support the core. What ensues is the ultimate nuclear waste catastrophe: The star explodes, scattering all of its newly made elements into interstellar space.

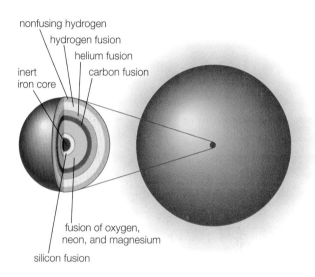

▲ FIGURE 9.15
The multiple layers of nuclear fusion in the core of a high-mass star during the final days of its life.

think about it How would the universe be different if hydrogen, rather than iron, had the lowest mass per nuclear particle? Why?

▶ **FIGURE 9.16**
Overall, the average mass per nuclear particle declines from hydrogen to iron and then increases. Selected nuclei are labeled to provide reference points. (This graph shows only the most general trends. A more detailed graph would show numerous up-and-down bumps superimposed on the general trends. The vertical scale is arbitrary, but shows the general idea.)

▲ **FIGURE 9.17**
During the final, catastrophic collapse of a high-mass stellar core, electrons and protons combine to form neutrons, in the process releasing neutrinos.

The Supernova Explosion Degeneracy pressure cannot support the inert iron core because the immense gravity of a high-mass star pushes electrons past their quantum mechanical limit. Once they get too close together, they can no longer exist freely. In an instant, the electrons disappear by combining with protons to form neutrons (Figure 9.17), releasing tiny subatomic particles known as *neutrinos* [Section 8.1] in the process. With the degeneracy pressure gone, gravity has free rein. In a fraction of a second, an iron core with a mass comparable to that of our Sun and a size larger than that of Earth collapses into a ball of neutrons just a few kilometers across, making the type of stellar corpse that we call a *neutron star*. The collapse halts only because the neutrons have a degeneracy pressure of their own. In some cases, the remaining core may be massive enough that gravity also overcomes the degeneracy pressure of the neutrons, in which case the core continues to collapse until it becomes a *black hole*. We'll discuss both neutron stars and black holes in the next chapter.

The gravitational collapse of the core releases an enormous amount of energy—more than 100 times what the Sun will radiate over its entire 10-billion-year lifetime. This energy drives the outer layers of the star off into space in a titanic explosion called a **supernova**. The heat of the explosion makes the gas shine with dazzling brilliance. For about a week, the supernova blazes as powerfully as 10 billion Suns, rivaling the luminosity of a moderate-size galaxy. The ejected gases slowly cool and fade in brightness over the next several months, but they continue to expand outward until they eventually mix with other gases in interstellar space (Figure 9.18).

Star Stuff Our high-mass star has died, but the variety of elements produced in the star's nuclear furnace are now scattered throughout the gas clouds of interstellar space. Millions or billions of years later, a new round of star formation may incorporate this supernova debris into a new generation of stars. Some of the heavy elements that came from this supernova will become the building blocks of new planets—and perhaps even new life forms—around those newborn stars. In fact, the building blocks of Earth and of our very own bodies came from supernovae in the distant past.

We are confident that the majority of the heavy elements within the solar nebula that made our planet came from earlier supernovae, because our observations of those elements match the predictions of supernova models. Figure 9.19 shows the measured abundances of elements in our solar system. Hydrogen and helium are by far the most abundant, because all other elements (except lithium) have been produced only by stars. Even-numbered elements are more common than adjacent odd-numbered elements, because advanced fusion reactions

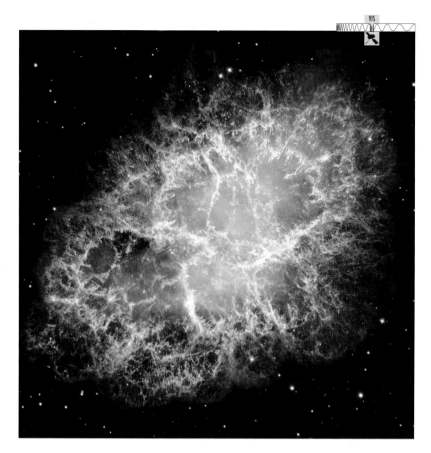

▲ **FIGURE 9.18**
This Hubble Space Telescope photograph shows the Crab Nebula, the remnant of the supernova observed in A.D. 1054.

▲ **FIGURE 9.19**
This graph shows the observed abundances of elements in our solar system relative to the abundance of hydrogen (set to 1 in this comparison). For example, the abundance of nitrogen is about 10^{-4}, or 1/10,000, which means that there are about 10,000 times as many hydrogen atoms as nitrogen atoms.

typically fuse helium nuclei (atomic number = 2) with other even-numbered elements. Abundances drop off even further beyond iron, because the heavier elements are produced only in small quantities during or after supernova explosions. The telltale patterns we see among these abundances therefore show that we ourselves are "star stuff," built from the debris of stars that exploded long ago.

9.3 Testing Stellar Models with Star Clusters

You might be wondering how scientists can be so confident about their models of the Sun and other stars. After all, the lives of stars are so much longer than human lifetimes that all our observations of stars amount to only very brief snapshots of their lives. The answer, as always in science, is that those models make very specific predictions about our observations of stars that have been confirmed through repeated testing.

Observations of star clusters have been particularly useful in testing stellar models for two key reasons:

1. All the stars in a cluster lie at about the same distance from Earth, meaning that the apparent brightness of each star directly tells how its luminosity compares with that of other stars in the cluster.

▲ FIGURE 9.20
The Pleiades is a nearby open cluster of stars, visible in the constellation Taurus. The Pleiades is often called the *Seven Sisters*, although only six of the cluster's several thousand stars are easily visible to the naked eye. (This cluster is known as Subaru in Japanese, which is why it is used as the emblem for the auto maker.) The region shown is about 11 light-years across.

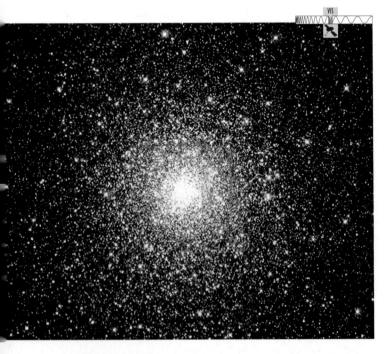

▲ FIGURE 9.21
The globular cluster M80 is more than 12 billion years old. The prominent reddish stars in this Hubble Space Telescope photo are red giant stars nearing the ends of their lives. The central region pictured here is about 15 light-years across.

2. All the stars in a cluster formed at about the same time from the same large molecular cloud, meaning that they are all now about the same age.

We will therefore devote this chapter's exploration of the process of science in action to investigating how star clusters enable us to test our models of stellar lives.

◆ What do star clusters reveal about the lives of stars?

Star clusters come in two basic types: modest-size **open clusters** and densely packed **globular clusters**. The two types differ not only in how densely they are packed with stars but also in their locations and ages. Open clusters, such as the Pleiades (Figure 9.20), are found in the galaxy's relatively flat disk; they typically contain a few hundred to a few thousand stars in a region about 30 light-years across, and all the stars are relatively young.

Globular clusters are generally found high above or below the disk of the galaxy (in the region called the *halo* [Section 11.1]); they are much more densely packed than open clusters and contain some of the oldest stars in the universe. A globular cluster can contain more than a million stars in a ball-shaped region no more than about 150 light-years across. Its center may have 10,000 stars packed into a space just a few light-years across (Figure 9.21). The view from a planet in a globular cluster would be marvelous, with thousands of stars lying closer than Alpha Centauri is to the Sun.

The value of star clusters for testing our models becomes apparent when we plot their stars on an H-R diagram. Figure 9.22 shows an H-R diagram for the Pleiades. Most of the stars fall along the main sequence, with one important exception: The Pleiades' stars trail away to the right of the main sequence at the upper end. That is, the hot, short-lived stars of spectral type O are missing from the main sequence. This fact is consistent with what we expect from the assumption that hot, massive stars do indeed have shorter lifetimes than cooler, less massive stars. In this case, we conclude that the Pleiades cluster is old enough for its main-sequence O stars to have already ended their hydrogen-fusing lives, but not quite old enough for all of its stars of spectral type B to have exhausted their core hydrogen.

The precise point on the H-R diagram at which the Pleiades' stars diverge from the main sequence is called its **main-sequence turnoff point**. Theoretical models indicate that the main-sequence lifetime of stars at this point is roughly 100 million years, so we conclude that this must be the age of the Pleiades. Stars with lifetimes longer than 100 million years are still fusing hydrogen in their cores and hence remain main-sequence stars.

The models also predict what should happen over the next few billion years: The B stars in the Pleiades will die out first, followed by the A stars and the F stars. If we could make an H-R diagram for the Pleiades every few million years, we would expect to see the main sequence gradually grow shorter. This is exactly what we see in Figure 9.23, which shows the Pleiades along with stars from three other open clusters plotted on the same H-R diagram. Two of the clusters are older than the Pleiades; one has no more B stars and the other has no more A stars. The remaining cluster is much younger than the Pleiades, with a main-sequence turnoff point indicating an age of only 14 million years. Only its most massive O stars have died out, and the next most massive have shifted to the upper-right part of the H-R diagram, indicating that they have become supergiants, just as predicted by our models of high-mass stars.

think about it Suppose a star cluster is precisely 10 billion years old. Where would you expect to find its main-sequence turnoff point? Would you expect this cluster to have any main-sequence stars of spectral type A? Would you expect it to have main-sequence stars of spectral type K? Explain. (*Hint:* What is the lifetime of our Sun?)

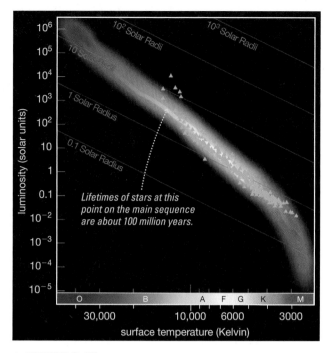

▲ FIGURE 9.22
An H-R diagram for the stars of the Pleiades. Triangles represent individual stars. The Pleiades cluster is missing its upper main-sequence stars, indicating that core hydrogen fusion in these stars has already ended. The main-sequence turnoff point at about spectral type B6 tells us that the cluster is about 100 million years old.

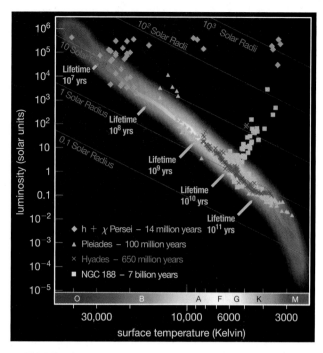

▲ FIGURE 9.23
This H-R diagram shows stars from four clusters. Their differing main-sequence turnoff points indicate different ages.

Eventually, we would expect the H-R diagram of the Pleiades to look something like the one in Figure 9.24, which shows the stars of a 13-billion-year-old globular cluster. Within this diagram, you can see stars in many of the different life stages that the Sun will go through. Stars along the lower right remain on the main sequence, meaning that they are still fusing hydrogen in their cores. Just above and to the right of the main-sequence turnoff point, we see stars that have just begun their expansion into red giants as their cores have shut down and hydrogen shell fusion has begun. The upward line of stars to the right of the turnoff point shows that red giants gradually grow in radius and luminosity. At the tip of this line are the red giants on the verge of helium fusion. Stars that have already begun helium fusion appear below and to the left of the red giants, because they are somewhat smaller, hotter, and less luminous. Finally, to the lower left of the main sequence we see stars that ended their lives and became white dwarfs.

Comparisons between diagrams like Figure 9.24 and computer models of stars give us great confidence that these models realistically reflect the lives of stars. You can do a quick check for yourself by comparing Figure 9.24 with Figure 9.12, which shows model predictions for the Sun's life track. The match is not exact, because these stars are not exactly the same as the Sun (they contain a smaller proportion of elements besides hydrogen and helium), but all the general features are indeed similar. Astronomers can make more detailed comparisons by computing models for entire star clusters and plotting the results on an H-R diagram. These efforts both confirm the stories of stellar lives that we have told in this chapter and continue to help us improve our stellar models.

Figure 9.25 summarizes the life stories of both low- and high-mass stars, as determined from our star models.

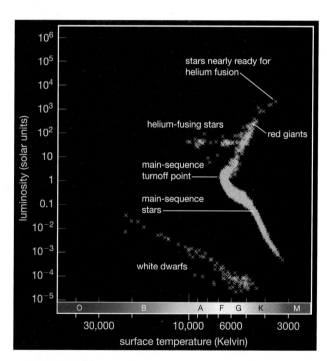

▲ FIGURE 9.24
This H-R diagram shows stars from the globular cluster M4, with an age of around 13 billion years. These stars are currently going through many of the same stages that models predict our Sun will go through near the end of its life.

cosmic context   FIGURE 9.25 Stellar Lives

All stars spend most of their time as main-sequence stars and then change dramatically near the ends of their lives. This figure shows the life stages of a high-mass star and a low-mass star, using the cosmic calendar from Chapter 1 to illustrate the relative lengths of these life stages. On this calendar, the 14-billion-year lifetime of the universe corresponds to a single year.

LIFE OF A HIGH-MASS STAR ($25 M_{Sun}$)

This high-mass star goes from protostar to supernova in about 6 million years, corresponding to less than 4 hours on the cosmic calendar.

(1) **Protostar:** A star system forms when a cloud of interstellar gas collapses under gravity.

(2) **Blue main-sequence star:** In the core of a high-mass star, four hydrogen nuclei fuse into a single helium nucleus by the series of reactions known as the CNO cycle.

(3) **Red supergiant:** After core hydrogen is exhausted, the core shrinks and heats. Hydrogen fusion begins around the inert helium core, causing the star to expand into a red supergiant.

Actual Length of Stage	40,000 years	5 million years	100,000 years
Time on Cosmic Calendar	12:00:00 AM → 12:01:30 AM	12:01:30 AM → 3:10:00 AM	3:10:00 AM → 3:14:00 AM

These times correspond to the life stages of a $25 M_{Sun}$ star born around midnight on a typical day of the cosmic calendar.

LIFE OF A LOW-MASS STAR ($1 M_{Sun}$)

This low-mass star goes from protostar to planetary nebula in about 11.5 billion years, corresponding to 10 months on the cosmic calendar.

(1) **Protostar:** A star system forms when a cloud of interstellar gas collapses under gravity.

(2) **Yellow main-sequence star:** In the core of a low-mass star, four hydrogen nuclei fuse into a single helium nucleus by the series of reactions known as the proton–proton chain.

(3) **Red giant star:** After core hydrogen is exhausted, the core shrinks and heats. Hydrogen fusion begins around the inert helium core, causing the star to expand into a red giant.

Actual Length of Stage	30 million years	10 billion years	1 billion years
Time on Cosmic Calendar	March 1 → March 2	March 2 → November 30	November 30 → December 27

These dates correspond to the life stages of a $1 M_{Sun}$ star born in early March on the cosmic calendar.

④ **Helium core-fusion supergiant:** Helium fusion begins when the core temperature becomes hot enough to fuse helium into carbon. The core then expands, slowing the rate of hydrogen fusion and allowing the star's outer layers to shrink.

⑤ **Multiple shell–fusion supergiant:** After the core runs out of helium, it shrinks and heats until fusion of heavier elements begins. Late in life, the star fuses many different elements in a series of shells while iron collects in the core.

⑥ **Supernova:** Iron cannot provide fusion energy, so it accumulates in the core until degeneracy pressure can no longer support it. Then the core collapses, leading to the catastrophic explosion of the star.

⑦ **Neutron star or black hole:** The core collapse forms a ball of neutrons, which may remain as a neutron star or collapse further to make a black hole.

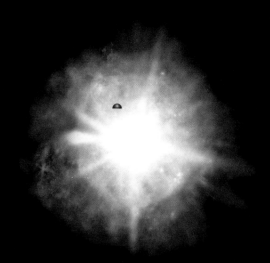

1 million years	10,000 years	a few months	indefinite
3:14:00 AM → 3:52:00 AM	3:52:00 AM → 3:52:23 AM	3:52:23 AM	—

④ **Helium core-fusion star:** Helium fusion begins when the core becomes hot enough to fuse helium into carbon. The core then expands, slowing the rate of hydrogen fusion and allowing the star's outer layers to shrink.

⑤ **Double shell–fusion red giant:** Helium fusion begins around the inert carbon core after the core helium is exhausted. The star then enters its second red giant phase, with fusion in both a hydrogen shell and a helium shell.

⑥ **Planetary nebula:** The dying star expels its outer layers in a planetary nebula, leaving behind the exposed inert core.

⑦ **White dwarf:** The remaining white dwarf is made primarily of carbon and oxygen because the core of the low-mass star never grows hot enough to produce heavier elements.

100 million years	30 million years	10,000 years	indefinite
December 27 → December 30	December 30 → December 31	December 31	—

The lifetime of this $1M_{Sun}$ star is almost 2000 times as long as that of a $25M_{Sun}$ star.

summary of key concepts

9.1 Lives in Balance

◆ **Why do stars shine so steadily?**

A star is born when **gravitational contraction** causes a cloud of interstellar gas to contract until hydrogen fusion ignites in its core. It can then shine steadily as long as it remains in two kinds of balance: (1) **gravitational equilibrium,** in which the outward push of pressure balances the inward pull of gravity, and (2) **energy balance,** in which the energy produced by fusion in its core matches the energy lost in the form of radiation from its surface.

◆ **Why do a star's properties depend on its mass?**

A star's mass determines the core temperature at which fusion brings it into balance. Stars less massive than the Sun come into balance with a cooler core temperature, lower luminosity, and smaller radius. Stars more massive than the Sun come into balance with a hotter core temperature, greater luminosity, and larger radius. Energy balance is possible for stars between about $150 M_{Sun}$ and $0.08 M_{Sun}$. Below this mass, **degeneracy pressure** prevents gravity from making the star hot enough for efficient hydrogen fusion, and the object becomes a "failed star" known as a **brown dwarf**.

9.2 Star Death

◆ **What will happen when our Sun runs out of fuel?**

When hydrogen fusion stops in the core, the Sun will go out of balance. The core will then start contracting, and **hydrogen shell fusion** will begin around it, causing the Sun to become a **red giant**. When the core becomes hot enough for **helium fusion,** making carbon in the core, the Sun will temporarily return to a state of balance. After the core runs out of helium, fusion of hydrogen and helium will continue in shells around a contracting carbon core, but degeneracy pressure will stop the contraction before carbon fusion begins. During this time, the star will increase in size and luminosity until it ejects its outer layers in a **planetary nebula,** leaving behind a **white dwarf** supported by degeneracy pressure.

◆ **How do high-mass stars end their lives?**

After a high-mass star (more than $8 M_{Sun}$) runs out of core hydrogen, it goes through hydrogen shell fusion and helium fusion stages similar to those of the Sun, and expands into a **supergiant**. However, fusion does not stop with carbon, because degeneracy pressure cannot hold off the crushing power of gravity within a high-mass star. Its core continues to contract, heat up, and produce even heavier elements until it starts to make iron. Because iron fusion cannot release energy, iron continues to pile up in the core until gravity causes the core to collapse and the star explodes as a **supernova,** dispersing the heavy elements necessary for life into the galaxy.

THE PROCESS OF SCIENCE IN ACTION

9.3 Testing Stellar Models with Star Clusters

◆ **What do star clusters reveal about the lives of stars?**

Star clusters come in two types. **Open clusters** are loose collections of up to several thousand stars and are generally younger than our Sun. **Globular clusters** are more densely packed, contain hundreds of thousands of stars, and are generally older than our Sun. Because all the stars in a cluster were born about the same time, we can determine a star cluster's age from its **main-sequence turnoff point**. Comparing clusters of different ages shows how the properties of stars change as they age, enabling us to test our models of stellar lives.

Investigations

For instructor-assigned homework and other learning materials, go to www.MasteringAstronomy.com.

Quick Quiz

Choose the best answer to each of the following. Explain your reasoning.

1. What slows down the contraction of a star-forming cloud when it makes a protostar? (a) production of fusion energy (b) magnetic fields (c) trapping of thermal energy inside the protostar
2. Which of these stars has the hottest core? (a) a white main-sequence star (b) an orange main-sequence star (c) a red main-sequence star
3. Which of these stars has the hottest core? (a) a blue main-sequence star (b) a red supergiant (c) a red main-sequence star
4. Which of these stars does not have fusion occurring in its core? (a) a red giant (b) a red main-sequence star (c) a blue main-sequence star
5. What happens to a low-mass star after helium fusion begins? (a) Its luminosity goes up. (b) Its luminosity goes down. (c) Its luminosity stays the same.
6. What would stars be like if hydrogen had the smallest mass per nuclear particle? (a) Stars would be brighter. (b) All stars would be red giants. (c) Nuclear fusion would not occur in stars of any mass.
7. What would stars be like if carbon had the smallest mass per nuclear particle? (a) Supernovae would be more common. (b) Supernovae would never occur. (c) High-mass stars would be hotter.
8. What would you be most likely to find if you returned to the solar system in 10 billion years? (a) a neutron star (b) a white dwarf (c) a black hole
9. In which of the following objects does degeneracy pressure fail to stop gravitational contraction? (a) a brown dwarf (b) a white dwarf (c) the core of a high-mass star
10. What happens to the core of a high-mass star after it runs out of hydrogen? (a) It shrinks and heats up. (b) It shrinks and cools down. (c) Helium fusion begins right away.
11. Which of these star clusters is youngest? (a) a cluster whose brightest main-sequence stars are white (b) a cluster whose brightest stars are red (c) a cluster containing stars of all colors
12. Which of these star clusters is oldest? (a) a cluster whose brightest main-sequence stars are white (b) a cluster whose brightest main-sequence stars are yellow (c) a cluster containing stars of all colors

Short-Answer/Essay Questions

Explain all answers clearly, with complete sentences and proper essay structure if needed. An asterisk (*) designates a quantitative problem, for which you should show all your work.

13. *Homes to Civilization?* We do not yet know how many stars have Earth-like planets, nor do we know the likelihood that such planets harbor advanced civilizations like our own. However, some stars can probably be ruled out as candidates for advanced civilizations. For example, given that it took a few billion years for humans to evolve on Earth, it seems unlikely that advanced life would have had time to evolve around a star that is only a few million years old. For each of the following stars, decide whether you think it is possible that it could harbor an advanced civilization. Explain your reasoning in one or two paragraphs.
    a. A $10 M_{Sun}$ main-sequence star
    b. A $1.5 M_{Sun}$ main-sequence star
    c. A $1.5 M_{Sun}$ red giant
    d. A $1 M_{Sun}$ helium-fusing star
    e. A red supergiant
14. *Brown Dwarfs.* How are brown dwarfs like jovian planets? In what ways are brown dwarfs like stars?
15. *Rare Elements.* Lithium, beryllium, and boron are elements with atomic numbers 3, 4, and 5, respectively. Despite their being three of the five simplest elements, Figure 9.19 shows that they are rare compared to many heavier elements. Suggest a reason for their rarity. (*Hint:* Consider the process by which helium fuses into carbon.)
16. *Future Skies.* As a red giant, the Sun will have an angular size in Earth's sky of about 30°. What will sunset and sunrise be like? About how long will they take? Do you think the color of the sky will be different from what it is today? Explain.
17. *Research: Historical Supernovae.* Historical accounts exist of supernovae in the years 1006, 1054, 1572, and 1604. Choose one of these supernovae and learn more about historical records of the event. Did the supernova influence human history in any way? Write a two- to three-page summary of your research findings.
*18. *Supernova Betelgeuse.* The best current estimate of the distance from Earth to the red supergiant Betelgeuse is about 600 light-years. If it were to explode as a supernova, it would be one of the brightest stars in the sky. Right now, the brightest star other than the Sun is Sirius, with a luminosity of $26 L_{Sun}$ and a distance of 8.6 light-years. How much brighter in our sky than Sirius would the Betelgeuse supernova be if it reached a maximum luminosity of $10^{10} L_{Sun}$?

10 The Bizarre Stellar Graveyard

LEARNING GOALS

10.1 White Dwarfs and Neutron Stars
- What are white dwarfs?
- What are neutron stars?

10.2 Black Holes
- What are black holes?
- What is the evidence for black holes?

THE PROCESS OF SCIENCE IN ACTION

10.3 Gravitational Waves
- What are gravitational waves and how have astronomers detected them?

Welcome to the afterworld of stars, the fascinating domain of white dwarfs, neutron stars, and black holes. The x-ray photograph above shows a neutron star surrounded by hot gas from the supernova explosion that created it. Dead stars like this one behave in unusual ways, because fusion energy can no longer keep them in balance. Only the quantum mechanical effect of degeneracy pressure can stave off the crushing power of their gravity. But even degeneracy pressure cannot save the most massive stellar cores, which collapse to become *black holes* that drastically curve space around them and make time appear to grind to a halt. Prepare to be amazed by the eerie inhabitants of the stellar graveyard!

▶ The Bizarre Stellar Graveyard

10.1 White Dwarfs and Neutron Stars

In Chapter 9, we saw that stars of different masses leave different types of stellar corpses. Low-mass stars like the Sun leave behind white dwarfs when they die. Higher-mass stars die in the titanic explosions known as supernovae, leaving behind neutron stars or black holes. We'll begin our study of stellar corpses by considering white dwarfs and neutron stars.

◆ What are white dwarfs?

A **white dwarf** is essentially the exposed core of a low-mass star that has died and shed its outer layers in a planetary nebula [Section 9.2]. It is quite hot when it first forms, because it was recently the inside of a star, but slowly cools with time. White dwarfs are stellar in mass but small in size (radius) [Section 8.2], which is why they are generally quite dim compared to stars like the Sun. However, the hottest white dwarfs can shine brightly in high-energy ultraviolet and x-ray light (Figure 10.1).

A white dwarf's starlike mass and small size make gravity very strong near its surface. If gravity were unopposed, it would crush the white dwarf to an even smaller size, so some source of pressure must be pushing back equally hard to keep the white dwarf stable. For a white dwarf, which has no fusion to maintain thermal pressure, the source is *degeneracy pressure*—the same type of pressure that supports brown dwarfs and that arises when subatomic particles are packed as closely as the laws of quantum mechanics allow [Section 9.1]. More specifically, the degeneracy pressure in white dwarfs arises from closely packed *electrons*, so we call it **electron degeneracy pressure.** A white dwarf exists in a state of balance in which the outward push of electron degeneracy pressure matches the inward crush of gravity.

White Dwarf Composition, Density, and Size Because a white dwarf is the core left over after a star has ceased nuclear fusion, its composition reflects the products of the star's final fusion stage. The white dwarf left behind by a $1M_{Sun}$ star consists mostly of carbon, since stars like the Sun fuse helium nuclei to create carbon in their final stage of life.

Despite its ordinary-sounding composition, a scoop of matter from a white dwarf would be unlike anything seen on Earth. A typical white dwarf has a mass similar to that of the Sun compressed into a volume the size of Earth. If you recall that Earth is smaller than a typical sunspot, you'll realize that packing the entire mass of the Sun into the volume of Earth is no small feat. The density of a white dwarf is so high that a teaspoon of its material would weigh as much as a small truck if you could bring that material to Earth.

More massive white dwarfs are actually smaller in size than less massive ones. For example, a $1.3M_{Sun}$ white dwarf is half the diameter of a $1.0M_{Sun}$ white dwarf (Figure 10.2). The more massive white dwarf is smaller because its stronger gravity compresses its matter to a greater density. According to the laws of quantum mechanics, the electrons in a white dwarf respond to this compression by moving faster, which makes the degeneracy pressure strong enough to resist the greater force of gravity. The most massive white dwarfs are therefore the smallest.

The White Dwarf Limit The fact that electron speeds are higher in more massive white dwarfs leads to a fundamental limit on the maximum mass of a white dwarf. Theoretical calculations show that electron speeds would reach the speed of light in a white dwarf with a mass of about 1.4 times the mass of the Sun

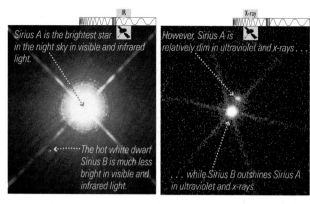

a Sirius as seen in infrared light by the Hubble Space Telescope.   **b** Sirius as seen by the Chandra X-Ray Telescope.

▲ **FIGURE 10.1**
Sirius, the brightest star in the night sky, is actually a binary system consisting of a main-sequence star (Sirius A) and a white dwarf (Sirius B). The main-sequence star is much brighter in infrared and visible light, but the hot white dwarf shines more brightly in high-energy light. (The spikes in the images are artifacts of the telescope optics.)

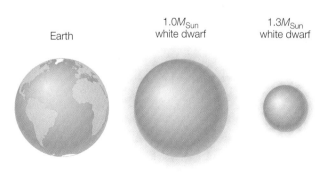

▲ **FIGURE 10.2**
More massive white dwarfs are actually *smaller* (and therefore denser) than less massive white dwarfs. Earth is shown for scale.

167

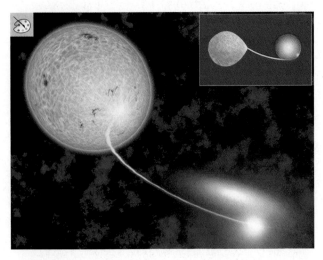

▲ FIGURE 10.3
This artist's conception shows how mass spilling from a companion star (left) toward a white dwarf (right) forms an accretion disk. The white dwarf itself is in the center of the accretion disk—too small to be seen on this scale. Matter streaming onto the disk creates a hot spot where the stream joins the disk. The inset shows how the system looks from above rather than from the side.

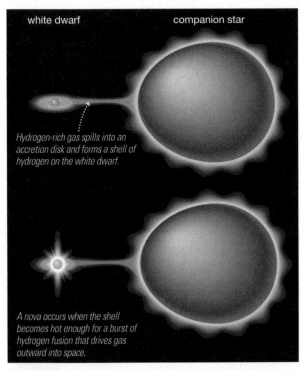

▲ FIGURE 10.4
A nova occurs when hydrogen fusion ignites on the surface of a white dwarf in a binary star system.

($1.4M_{Sun}$). Because neither electrons nor anything else can travel faster than the speed of light, no white dwarf can have a mass greater than this $1.4M_{Sun}$ **white dwarf limit** (also called the *Chandrasekhar limit*, after its discoverer).

Strong observational evidence supports this theoretical limit on the mass of a white dwarf. Many known white dwarfs are members of binary systems, and hence we can measure their masses [Section 8.2]. In every observed case, the white dwarfs have masses below $1.4M_{Sun}$, just as theoretical models predict.

White Dwarfs and Accretion Left to itself, a white dwarf will never again shine as brightly as the star it once was. With no source of fuel for fusion, it will cool with time. Its size will never change, because electron degeneracy pressure will forever keep it stable against the crush of gravity. However, the situation can be quite different for a white dwarf in a binary system in which the two stars orbit close together.

A white dwarf in a close binary system can gradually gain mass if its companion is a main-sequence or giant star (Figure 10.3). When a clump of mass first spills over from the companion star to the white dwarf, it has some small orbital velocity. The law of conservation of angular momentum (see Tools of Science, page 62) dictates that the clump must orbit faster and faster as it falls toward the white dwarf's surface. The infalling matter therefore forms a whirlpool-like disk around the white dwarf. Because the process in which matter collects into a larger body is called *accretion*, this rapidly rotating disk is called an **accretion disk**.

Novae Accretion can provide a "dead" white dwarf with a new energy source. The inward-spiraling gas in the accretion disk becomes very hot as its gravitational potential energy is converted into thermal energy, causing it to shine with intense ultraviolet or x-ray radiation. Moreover, the hydrogen gas spilling toward the white dwarf eventually falls onto the surface of the white dwarf, where the white dwarf's strong gravity compresses it into a thin surface layer. Both the pressure and the temperature rise as the layer builds up with more accreting gas. When the temperature at the bottom of the layer reaches about 10 million K, hydrogen fusion suddenly ignites.

The white dwarf blazes back to life with the energy from this hydrogen-fusing surface layer, causing the binary system to shine for a few glorious weeks as a **nova** (Figure 10.4). A nova is far less luminous than a supernova, but still can shine as brightly as 100,000 Suns. It generates heat and pressure, ejecting most of the material that has accreted onto the white dwarf. This material expands outward, creating a remnant that sometimes remains visible years after the nova explosion. Accretion resumes after a nova explosion subsides, so the entire process can repeat itself. In some cases, novae have been observed to repeat after just a few decades. More commonly, thousands of years pass between nova explosions.

White Dwarf Supernovae In rare cases, a white dwarf in a close binary will undergo an event far more dramatic than a nova. Either by gaining mass through accretion or by merging with its binary companion, the white dwarf one day reaches the $1.4M_{Sun}$ white dwarf limit. This day is the white dwarf's last. Remember that most white dwarfs are made largely of carbon. As a white dwarf's mass closes in on the $1.4M_{Sun}$ limit, its temperature rises and ultimately becomes great enough for carbon fusion. The carbon fusion ignites almost instantly throughout the white dwarf, and the white dwarf explodes completely in what we call a **white dwarf supernova**.

think about it According to our understanding of novae and white dwarf supernovae, can either of these events ever occur with a white dwarf that is *not* a member of a binary star system? Explain.

The carbon detonation that creates a white dwarf supernova is quite different from the iron catastrophe that leads to a supernova ending the life of a high-mass star [Section 9.2], which we call a **massive star supernova**.* Astronomers can distinguish between the two types of supernova by studying their light. Both types shine brilliantly, with peak luminosities about 10 billion times that of the Sun ($10^{10}L_{Sun}$), but the luminosities of white dwarf supernovae fade steadily, while the decline in brightness of a massive star supernova is often more complicated (Figure 10.5). In addition, spectra of white dwarf supernovae always lack hydrogen lines (because white dwarfs contain very little hydrogen), while these lines are prominent in the spectra of most massive star supernovae.

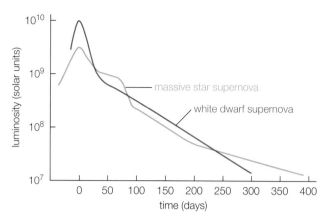

▲ **FIGURE 10.5**
The curves on this graph show how the luminosities of two different supernovae fade with time. The white dwarf supernova fades quickly at first and then more gradually a few weeks after its peak, while the massive star supernova fades in a more complicated pattern.

◆ What are neutron stars?

White dwarfs with densities of tons per teaspoon may seem incredible, but neutron stars are stranger still. The possibility that neutron stars might exist was first proposed in the 1930s, but many astronomers thought it preposterous that nature would make anything so bizarre. Nevertheless, a vast amount of evidence now makes it clear that neutron stars really exist.

Neutron Star Properties A **neutron star** is the ball of neutrons created by the collapse of the iron core in a massive star supernova (Figure 10.6). Typically just 10 kilometers in radius yet more massive than the Sun, neutron stars are like giant atomic nuclei made almost entirely of neutrons and held together by gravity. Like white dwarfs, neutron stars resist the crush of gravity with degeneracy pressure that arises when particles are packed as closely as nature allows. In the case of neutron stars, however, it is *neutrons* rather than electrons that are closely packed, so **neutron degeneracy pressure** supports them against the crush of gravity.

The force of gravity at the surface of a neutron star is awe-inspiring. A rocket ship trying to escape from a neutron star's surface would need to move at about half the speed of light. If you foolishly chose to visit a neutron star's surface, gravity would immediately squash you into a microscopically thin pancake of subatomic particles.

It's also interesting to contemplate what would happen if a neutron star came to Earth (there is no danger of this actually happening). Because it would be only about 10 kilometers in radius, the neutron star would probably fit in your hometown. However, since it would be 300,000 times as massive as Earth, the neutron star's immense surface gravity would quickly destroy your hometown and the rest of civilization. By the time the dust settled, the former Earth would be squashed into a shell no thicker than your thumb on the surface of the neutron star.

▲ **FIGURE 10.6**
This x-ray image from the Chandra X-Ray Observatory shows the supernova remnant G11.2-03, the remains of a supernova observed by Chinese astronomers in A.D. 386. The white dot at the center represents x-rays from the neutron star left behind by the supernova. The different colors correspond to different x-ray wavelength bands. The region pictured is about 23 light-years across.

---

*Observationally, astronomers classify supernovae as *Type II* if their spectra show hydrogen lines, and *Type I* otherwise. All Type II supernovae are assumed to be massive star supernovae. However, a Type I supernova can be either a white dwarf supernova or a massive star supernova in which the star lost all its hydrogen before exploding. Type I supernovae fall into three classes, called *Type Ia, Type Ib,* and *Type Ic*. Only Type Ia supernovae are thought to be white dwarf supernovae.

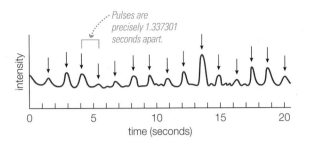

▲ **FIGURE 10.7**
About 20 seconds of data from the first pulsar discovered by Jocelyn Bell in 1967.

▲ **FIGURE 10.8**
This time-lapse image of the pulsar at the center of the Crab Nebula, a supernova remnant, shows its main pulse recurring every 0.033 second. The fainter pulses are thought to come from the pulsar's other lighthouse-like beam. (Photo from the Very Large Telescope of the European Southern Observatory.)

Pulsars The first observational evidence for neutron stars came in 1967, when a graduate student named Jocelyn Bell discovered a strange source of radio waves. The radio waves pulsed on and off at intervals of precisely 1.337301 seconds (Figure 10.7). No known astronomical source produced such regular pulses, and for a while astronomers half-jokingly called the new radio source "LGM" for Little Green Men. Today we refer to such rapidly pulsing radio sources as **pulsars.**

The mystery of pulsars was soon solved. By the end of 1968, astronomers had found pulsars at the center of two supernova remnants, the Vela Nebula and the Crab Nebula (Figure 10.8). They concluded that pulsars are neutron stars left behind by supernova explosions.

Pulsars "pulse" because the neutron star is spinning rapidly as a result of the conservation of angular momentum: As an iron core collapses into a neutron star, its rotation rate must increase as it shrinks in size. The collapse also bunches the magnetic field lines running through the core far more tightly, greatly amplifying the strength of the magnetic field. As a result, a neutron star's magnetic field can be a trillion times as strong as Earth's. This intense magnetic field directs beams of radiation out along the neutron star's magnetic poles. If a neutron star's magnetic poles are not aligned with its rotation axis, each beam of radiation sweeps round and round like a lighthouse beam, so we see a pulse of light each time the beam sweeps past Earth (Figure 10.9).

Pulsars are not quite perfect clocks. The continual twirling of a pulsar's magnetic field generates electromagnetic radiation that carries away energy and angular momentum, causing a gradual slowing in the neutron star's rotation rate. The pulsar in the Crab Nebula, for example, currently spins about 30 times per second. Two thousand years from now, it will spin less than half as fast. Eventually, a pulsar's spin slows so much and its magnetic field becomes so weak that we can no longer detect it. In addition, some spinning neutron stars may be oriented so that their beams do not sweep past our location. Therefore, we have the following rule: *All pulsars are neutron stars, but not all neutron stars are pulsars.*

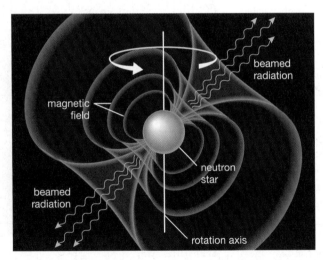

a A pulsar is a rotating neutron star that beams radiation along its magnetic axis.

b If the magnetic axis is not aligned with the rotation axis, the pulsar's beams sweep through space like lighthouse beams. Each time one of the pulsar's beams sweeps across Earth, we see a pulse of radiation.

▲ **FIGURE 10.9**
Radiation from a rotating neutron star can appear to pulse like the beams from a lighthouse.

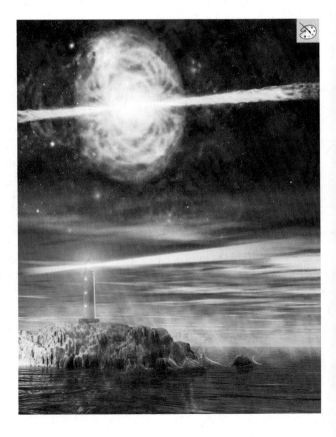

think about it Suppose we observe no pulses of radiation from a neutron star. Is it possible that a civilization in some other star system would see this neutron star as a pulsar? Explain.

We know that pulsars must be neutron stars because no other massive object could spin so fast. A white dwarf, for example, can spin no faster than about once per second. A faster spin would tear it apart because its gravity would no longer be strong enough to hold it together. Pulsars have been discovered that rotate as fast as 625 times per second. Only an object as small and dense as a neutron star could spin so fast without breaking apart.

Neutron Stars in Binary Systems Much as with a white dwarf in a close binary system, gas overflowing from a companion main-sequence or giant star can create a hot, swirling accretion disk around a neutron star in a close binary system. However, the much stronger gravity of a neutron star makes its accretion disk much hotter and denser than an accretion disk around a white dwarf.

The high temperatures in the inner regions of the accretion disk cause it to radiate powerfully in x-rays. Some close binaries with neutron stars emit 100,000 times as much energy in x-rays as our Sun emits in all wavelengths of light combined, so we call them **x-ray binaries.** Occasionally, the gas accumulating on the neutron star's surface undergoes a burst of nuclear fusion similar to a nova, but with most of the light emerging in the form of x-rays.

10.2 Black Holes

The story of stellar corpses would be strange enough if it ended with white dwarfs and neutron stars, but it doesn't. Sometimes, the gravity of a stellar corpse becomes so strong that nothing can prevent it from collapsing under its own weight. The stellar corpse collapses without end, crushing itself out of existence and forming perhaps the most bizarre type of object in the universe: *black hole.*

◆ What are black holes?

The basic idea behind a black hole originated near the end of the 18th century. Newton's laws of motion and gravity were well known by that time, and scientists were starting to think about the force of gravity at the surface of an object with a large amount of mass but a small radius. In particular, they wondered whether the gravity of such an object could be so strong that nothing, not even light, could escape its gravitational pull.

Escape Velocity To illustrate the conditions required for gravity to keep something from escaping, we'll use the example of a spacecraft orbiting Earth. Firing its rockets to give it more kinetic energy will increase the size of its orbit. If it gains enough kinetic energy, it can end up in an unbound orbit that allows it to escape Earth completely (Figure 10.10). Although it might make more sense to say that the spacecraft achieves "escape energy," we instead say that it achieves **escape velocity,** because the velocity it needs in order to escape does not depend on the spacecraft's mass. The escape velocity from Earth's surface is about 40,000 km/hr, or 11 km/s. This is the minimum speed required to escape Earth's gravity for any object that starts near Earth's surface.

Now think about what would happen if we could make Earth smaller while keeping its mass the same. Making Earth more compact would raise the escape velocity, because the strength of gravity at Earth's surface depends on the inverse square of its radius [Section 3.3]. If we could squeeze Earth down to about the size of a golf ball, its escape velocity would reach the speed of light. Scientists of

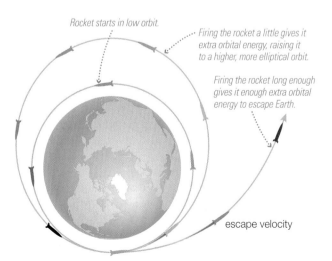

▲ **FIGURE 10.10**
If an object orbiting Earth gains energy, it moves to a higher or more elliptical orbit. With enough extra orbital energy, it may achieve escape velocity. Escape velocity from Earth's surface is about 11 km/s.

the 18th century assumed that the light emitted by such an object would behave like a rock thrown upward, eventually slowing to a stop and falling back down.

Einstein's general theory of relativity (see Tools of Science, below) shows that extremely compact, massive objects are considerably stranger than people imagined in the 18th century, but the basic idea was correct: It is indeed possible for an object's gravity to be so strong that not even light can escape. We call such objects "black holes." The "black" part of the name comes from the fact that no light can escape. The "hole" part of the name captures the idea that a black hole is like a hole in the observable universe in the following sense: If you enter a black hole, you leave the region of the universe that we can observe with a telescope and can never return.

Black Hole Formation A black hole forms when a massive object is somehow squeezed down to a much smaller size. We have already seen how extreme squeezing can happen inside a massive star when its iron core reaches the white dwarf limit of $1.4 M_{Sun}$. Electron degeneracy pressure cannot resist gravity above that mass, and the core collapses to form a neutron star, causing a supernova explosion [Section 9.2].

tools of science — Einstein's Theories of Relativity

Einstein developed his theory of relativity in two parts. The special theory of relativity was published in 1905 and the general theory in 1915. Both are now indispensable tools of modern science.

Special Relativity. Contrary to a common myth, the special theory of relativity does not say that *everything* is relative; rather, it states that *motion* is relative but two things in the universe are absolute:

1. The laws of nature are the same for everyone.
2. The speed of light is the same for everyone.

Statement 1 implies that no one can claim to be standing absolutely still, because the only kinds of motion that matter to the laws of nature are *relative* motions of one object with respect to another. Otherwise, observers moving through space at different speeds would disagree about the laws of physics. Statement 2, that the speed of light is the same for everyone, is much more surprising. Ordinarily, we expect speeds to add and subtract. If you watched someone throw a ball forward from a moving car, you'd see the ball traveling at the speed at which it was thrown *plus* the speed of the car. But if a person shined a light beam from a moving car, you'd see it moving at precisely the speed of light (about 300,000 kilometers per second), no matter how fast the car was going. This strange fact has been experimentally verified countless times.

The fact that the speed of light is always the same leads to several famous consequences. First, it means that no physical object can move faster than light. You can see why by imagining a test ride in the most incredible rocket possible and thinking about your headlight beams. Because everyone measures the speed of light to be the same, both you and someone back on Earth will agree that your headlight beams are traveling at the speed of light: 300,000 kilometers per second. And since you'll also both agree that the headlight beams are shining *ahead* of you, the person on Earth will always conclude that you are going slower than the speed of light. Other consequences of the constant speed of light include Einstein's famous equation $E = mc^2$ and the fact that an object moving by you seems to have slower time, shorter length, and greater mass than it would at rest.

General Relativity. Einstein's general theory extended special relativity by including the effects of gravity, and it led to a radical revision in how we think about space and time. Special relativity showed that, instead of thinking about the three dimensions of space and the one dimension of time as separate, we should think of them as a seamless, four-dimensional entity known as *spacetime*. The general theory of relativity showed that what we perceive as gravity arises from curvature of this four-dimensional spacetime. Mass curves spacetime, and greater spacetime curvature means stronger gravity. We cannot visualize curvature in four dimensions, but we can visualize two-dimensional analogies with rubber sheet diagrams like the one shown in the figure. According to general relativity, planets orbit the Sun because of the way spacetime is curved by the Sun: Each planet is going as straight as it can, but spacetime is curved in a way that keeps the planets going round and round like marbles in a salad bowl.

Given that we cannot actually perceive all four dimensions of spacetime at once, you may wonder why scientists think spacetime curvature is real. The answer is that Einstein's theory predicts measurable effects from such curvature, and all observations made to date are in accordance with those predictions. Strange as it may seem, we live in a four-dimensional universe in which space and time are intertwined and can never be disentangled.

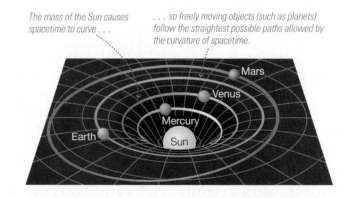

According to the general theory of relativity, planets orbit the Sun for much the same reason that you can make a marble go around in a salad bowl: The planet is going as straight as it can, but the curvature of spacetime causes its path through space to curve.

Calculations show that the mass of a neutron star has a limit that lies somewhere between about 2 and 3 solar masses. Above this mass, neutron degeneracy pressure cannot hold off the crush of gravity in a collapsing stellar core. Most supernovae are thought to leave neutron stars behind because the explosion ejects all the matter surrounding the core, keeping its mass below this limit. However, theoretical models show that very massive stars might not succeed in blowing away all their upper layers. If enough matter falls back onto the core, its mass may rise above the neutron star limit. As soon as the core exceeds the neutron star limit, gravity overcomes neutron degeneracy pressure and the core collapses once again. This time, no known force can keep the core from collapsing into oblivion.

According to Einstein's theory of relativity, it is highly unlikely that any as-yet-unknown force could intervene and prevent the collapse. Einstein's theory tells us that energy is equivalent to mass ($E = mc^2$), implying that energy, like mass, must exert some gravitational attraction. The gravity of pure energy is usually very small, but that is not the case for a stellar core collapsing beyond the neutron star limit. Here, the energy associated with the rapidly rising temperature and pressure acts like additional mass, making the crushing power of gravity even stronger. The more the core collapses, the stronger gravity gets. To the best of our understanding, *nothing* can halt the crush of gravity at this point. The core collapses without end, forming a black hole.

The Event Horizon The boundary between the inside of the black hole and the universe outside is called the **event horizon.** The event horizon essentially marks the point of no return for objects entering a black hole: It is the boundary around a black hole at which the escape velocity equals the speed of light. Nothing that passes within this boundary can ever escape. The event horizon gets its name from the fact that we have no hope of learning about any events that occur within it.

We usually think of the "size" of a black hole as the radius of its event horizon. This radius is known as the **Schwarzschild radius,** named for the man who derived its formula in 1916. The Schwarzschild radius of a black hole depends only on its mass. A black hole with the mass of the Sun has a Schwarzschild radius of about 3 kilometers—only a little smaller than the radius of a neutron star of the same mass. More massive black holes have larger Schwarzschild radii. For example, a black hole with 10 times the mass of the Sun has a Schwarzschild radius of about 30 kilometers.

You may wonder what happens inside a black hole's event horizon. Einstein's theory of relativity says that nothing can stop the crush of gravity in a black hole, implying that all the matter that forms a black hole must ultimately be crushed to an infinitely tiny and dense point in the black hole's center. We call this point a **singularity.** Unfortunately, this idea of a singularity pushes up against the limits of scientific knowledge today. The problem is that two very successful theories make very different predictions about the nature of a singularity. Einstein's general theory of relativity, which seems to explain successfully how gravity works throughout the universe, predicts that spacetime should grow infinitely curved as it enters the singularity. Quantum physics, which successfully explains the nature of atoms and the spectra of light, predicts that spacetime should fluctuate chaotically near the singularity. These are very different claims, and no theory that can reconcile them has yet been developed.

▶ **Voyage to a Black Hole**

Space and Time Near a Black Hole Einstein discovered that space and time are actually bound up together as four-dimensional **spacetime,** and that gravity arises from *curvature of spacetime* (see Tools of Science, page 172). Figure 10.11 represents all four dimensions of spacetime with a two-dimensional rubber sheet. In this analogy, the sheet is flat in a region far from any mass (Figure 10.11a). Near a massive object with strong gravity, the sheet becomes curved (Figure 10.11b); the stronger the gravity, the more curved it gets. In this analogy, a black hole

a A two-dimensional representation of "flat" spacetime. Each pair of circles is separated by the same radial distance.

b A mass affects the rubber sheet similarly to the way gravity curves spacetime. The circles become more widely separated—indicating greater curvature—as we look closer to the mass.

event horizon

c The curvature of spacetime becomes greater and greater as we approach a black hole, and a black hole itself is a bottomless pit in spacetime.

▲ **FIGURE 10.11**
We can use two-dimensional rubber sheets to show an analogy to curvature in four-dimensional spacetime.

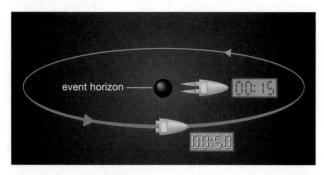

▲ **FIGURE 10.12**
Time runs more slowly on the clock nearer to the black hole, and a gravitational redshift makes its glowing blue numerals appear red from your orbiting spaceship.

common misconceptions

Black Holes Don't Suck

What would happen if our Sun suddenly became a black hole? For some reason, the idea that Earth and the other planets would inevitably be "sucked in" by the black hole has become part of our popular culture, but this is not true. Although the sudden disappearance of light and heat from the Sun would be bad news for life, Earth's orbit would not change. Newton's law of gravity tells us that the allowed orbits in a gravitational field are ellipses, hyperbolas, and parabolas. Note that "sucking" is not on the list! Earth would get into trouble only if it came so close to the black hole—within about three times its Schwarzschild radius—that the force of gravity deviated significantly from what Newton's law predicts. Otherwise, it would simply continue on an ordinary orbit.

Space travelers also need not fear running into black holes, because most are so small. Their typical Schwarzschild radii are far smaller than the radius of any star or planet, so a black hole is actually one of the most difficult things in the universe to fall into by accident.

is like a bottomless pit in spacetime (Figure 10.11c). Gravity gets stronger and stronger as we get closer and closer to the black hole, so the sheet curves more and more. Keep in mind that the illustration is only an analogy, and black holes are actually spherical, not funnel-shaped.

see it for yourself Curved space has different rules of geometry than "flat" space, an idea you can see with a plastic ball and a marker. Draw a straight line one-quarter of the way around the ball, then make a right-angle turn and go one-quarter of the way around the ball in your new direction, and finish a triangle with a straight line back to where you started. Now measure all three angles and add them up. How does the sum of the three angles of a triangle on a sphere compare with the 180° sum you would find for triangles in a flat plane?

Einstein's general theory of relativity also predicts that time should run more slowly as the force of gravity grows stronger. This prediction has been tested by experiments with clocks in Earth's gravitational field and through observations of *gravitational redshift,* a redshift that occurs in spectra of the Sun and other stars because the slowdown in time near the strong gravity of their surfaces means an increase in the wavelength of their emitted light. The effect on time should become quite extreme near a black hole. In principle, you could test this prediction with the aid of two identical clocks whose numerals glow with blue light. Suppose you were orbiting a black hole with a mass of $10M_{Sun}$ and a Schwarzschild radius of 30 kilometers on a circular orbit a few thousand kilometers from the event horizon. This orbit would be stable—there would be no need to worry about getting "sucked in"—so you could perform your experiment by keeping one of the identical clocks aboard your spaceship and pushing the other one toward the black hole with a small rocket that would fire its engines just enough so that the clock would fall only gradually toward the event horizon (Figure 10.12). According to relativity, the clock on the rocket would tick more slowly as it headed toward the black hole, so its light would become increasingly redshifted. When the clock reached a distance of about 10 kilometers above the event horizon, you would see it ticking only half as fast as the clock on your spaceship, and its numerals would appear red instead of blue.

When the rocket ran out of fuel, the clock would start to plunge toward the black hole. From your safe vantage point inside the spaceship, you would see the clock ticking more and more slowly as it fell. However, you would soon need a radio telescope to "see" it, as the light from the clock face would shift from the red part of the visible spectrum to the infrared and on into the radio. The redshift would soon become so great that no telescope could detect the clock's light. As the clock vanished from view, you would see the time on its face freezing to a stop.

For one further illustration of how strange black holes are, imagine that one of your friends on the spaceship climbs into a space suit, grabs the other clock, resets it, and jumps out of the air lock on a trajectory aimed straight for the black hole. Down he falls, clock in hand. He watches the clock, but because he and the clock are traveling together, its time seems to run normally and its numerals stay blue. From his point of view, time seems to neither speed up nor slow down. When his clock reads, say, 00:30, he and the clock pass through the event horizon. There is no barrier, no wall, no hard surface. The event horizon is a mathematical boundary, not a physical one. From his point of view, the clock keeps ticking. He is inside the event horizon, the first human being ever to vanish into a black hole.

Back on the spaceship, you watch in horror as your friend plunges to his death. Yet, from your point of view, he will *never* cross the event horizon. You'll see time coming to a stop for him and his clock just as he vanishes from view because of the huge gravitational redshift of light. When you return home, you can play a video for the judges at your trial, proving that your friend is still outside the black hole. Strange as it may seem, all of these events are in accordance with Einstein's theory. From your point of view, your friend takes *forever* to cross the event horizon (even though he vanishes from view because of his ever-increasing redshift). From his point of view, it is but a moment's plunge before he passes into oblivion.

The truly sad part of this story is that your friend did not live to experience the crossing of the event horizon. The force of gravity grew so quickly as he approached the black hole that it pulled much harder on his feet than on his head, simultaneously stretching him lengthwise and squeezing him from side to side (Figure 10.13). In essence, your friend was stretched in the same way the oceans are stretched by the tides, except that the *tidal force* near the black hole is trillions of times stronger than the tidal force of the Moon on Earth. No human could survive it.

◆ What is the evidence for black holes?

The fact that black holes emit no light might make it seem as if they should be impossible to detect. However, a black hole's gravity can influence its surroundings in ways that reveal its presence.

The earliest observational evidence for the existence of black holes came from studies of x-ray binaries. Some of the x-ray binaries that astronomers discovered contained objects too massive to be neutron stars. One example was an x-ray binary called Cygnus X-1 (Figure 10.14), which consists of an extremely luminous star with an estimated mass of $19M_{Sun}$ and a compact, unseen companion which must have a mass of about $15M_{Sun}$, based on the orbital velocity inferred from Doppler shifts. Although there is some uncertainty in these estimates, the mass of the invisible accreting object clearly exceeds the $3M_{Sun}$ neutron star limit. It is therefore too massive to be a neutron star, so according to current knowledge it must be a black hole. A few dozen other x-ray binaries offer similar evidence for black holes formed from the collapse of massive stellar cores.

Observations also indicate that *supermassive black holes*—some with masses millions or billions of times that of our Sun—reside at the centers of many galaxies. As we will discuss in the next chapter, these black holes must have formed differently from the stellar-mass black holes in x-ray binaries, and they are thought to power some of the most luminous objects in the universe. However, the strongest evidence for black holes comes from recent observations of gravitational waves from distant mergers of stellar-mass black holes, which we will discuss in the next section.

The bottom line is that while white dwarfs, neutron stars, and black holes may seem bizarre by earthly standards, we have strong evidence that all three types of objects really exist and play important roles in our universe. This fact means we can now bring the story of the battle between pressure and gravity, which began in the previous chapter, to its ultimate conclusion. Figure 10.15 shows how a star's fate depends on its birth mass. Objects with less than $0.08M_{Sun}$ never become stars because degeneracy pressure halts their contraction before fusion can bring them into balance. Stars beginning with less than $8M_{Sun}$ end up as white dwarfs because degeneracy pressure stops their cores from contracting before fusion starts to make iron. Stars with greater masses become either neutron stars or black holes, depending on whether the mass of the core remaining after the supernova explosion exceeds the neutron star limit.

▲ **FIGURE 10.13**
Tidal forces would be lethal near a black hole formed by the collapse of a star. The black hole would pull more strongly on the astronaut's feet than on his head, stretching him lengthwise and squeezing him from side to side.

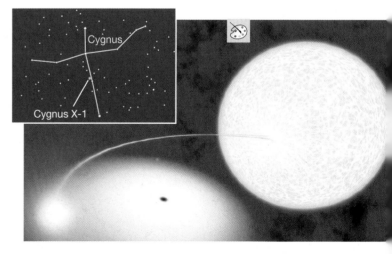

▲ **FIGURE 10.14**
Artist's conception of the Cygnus X-1 system, so named because it is the brightest x-ray source in the constellation Cygnus. The x-rays come from the high-temperature gas in the accretion disk surrounding the black hole. The inset shows the location of Cygnus X-1 in the sky.

cosmic context FIGURE 10.15 Balancing Pressure and Gravity

We can understand the entire life cycle of a star in terms of the changing balance between pressure and gravity. This illustration shows how that balance changes over time and why those changes depend on a star's birth mass. *(Stars not to scale.)*

Key
pressure ⟶
gravity ⟵

(2) Thermal pressure comes into steady balance with gravity when the core becomes hot enough for hydrogen fusion to replace the thermal energy the star radiates from its surface.

(3) The balance tips in favor of gravity after the core runs out of hydrogen. Fusion of hydrogen into helium temporarily stops supplying thermal energy in the core. The core again contracts and heats up. Hydrogen fusion begins in a shell around the core. The outer layers expand and cool, and the star becomes redder.

Main-sequence star

Pressure balances gravity at every point within a main-sequence star.

> $0.08 M_{Sun}$

Hydrogen shell-fusion star

(1) Gravity overcomes pressure inside a protostar, causing the core to contract and heat up. A protostar cannot achieve steady balance between pressure and gravity because nuclear fusion is not replacing the thermal energy it radiates into space.

The balance between pressure and gravity acts as a thermostat to regulate the core temperature:

A drop in core temperature decreases fusion rate, which lowers core pressure, causing the core to contract and heat up.

Stellar Thermostat

A rise in core temperature increases fusion rate, which raises core pressure, causing the core to expand and cool down.

Luminosity continually rises because core contraction causes the temperature and fusion rate in the hydrogen shell to rise.

Protostar

Contraction converts gravitational potential energy into thermal energy.

The balancing point between pressure and gravity depends on a star's mass:

Balance between pressure and gravity in high-mass stars results in a higher core temperature, a higher fusion rate, greater luminosity, and a shorter lifetime.

more luminous ↑

← hotter

Balance between pressure and gravity in low-mass stars results in a cooler core temperature, a slower fusion rate, less luminosity, and a longer lifetime.

Degeneracy pressure balances gravity in objects of less than $0.08 M_{Sun}$ before their cores become hot enough for steady fusion. These objects never become stars and end up as brown dwarfs.

< $0.08 M_{Sun}$

④ Balance between thermal pressure and gravity is restored when the core temperature rises enough for helium fusion into carbon, which can once more replace the thermal energy radiated from the core.

⑤ Gravity again gains the upper hand over pressure after the core helium is gone. Just as before, fusion stops replacing the thermal energy leaving the core. The core therefore resumes contracting and heating up, and helium fusion begins in a shell around the carbon core.

⑥ In high-mass stars, core contraction continues, leading to multiple shell fusion that terminates with iron and a supernova explosion.

⑦ At the end of a star's life, either degeneracy pressure has come into permanent balance with gravity or the star has become a black hole. The nature of the end state depends on the mass of the remaining core.

Helium core-fusion star

Double shell–fusion star

Multiple shell–fusion star

HIGH-MASS

Black hole

Degeneracy pressure cannot balance gravity in a black hole.

Neutron star ($M < 3M_{Sun}$)

Neutron degeneracy pressure can balance gravity in a stellar corpse with less than about 2–3M_{Sun}.

Luminosity remains steady because helium core fusion restores balance.

White dwarf ($M < 1.4M_{Sun}$)

LOW-MASS

Electron degeneracy pressure balances gravity in the core of a low-mass star before it gets hot enough to fuse heavier elements. The star ejects its outer layers and ends up as a white dwarf.

Electron degeneracy pressure can balance gravity in a stellar corpse of mass $< 1.4M_{Sun}$.

Brown dwarf ($M < 0.08M_{Sun}$)

Degeneracy pressure keeps a brown dwarf stable in size even as it cools steadily with time.

The ripples represent gravitational waves carrying energy away from the system.

The loss of orbital energy means the orbit must decay . . .

. . . until the two objects ultimately merge, producing a particularly strong pulse of gravitational waves.

▲ **FIGURE 10.16**
This sequence uses a rubber sheet analogy to illustrate the gravitational waves produced by a binary system with two neutron stars. The gravitational waves carry away orbital energy, causing the orbit to decay until the two neutron stars collide and merge together. The time required for this sequence to play out (from left to right) would typically be a few tens of millions of years.

THE PROCESS OF SCIENCE IN ACTION

10.3 Gravitational Waves

Einstein's general theory of relativity makes many predictions that have been tested and verified, including the effects of gravity on time discussed in the previous section. Another important prediction is the existence of what we call *gravitational waves*. Einstein himself doubted that we would ever detect such waves, but we have, and how we detect them is this chapter's case study in the process of science in action. As you will see, gravitational-wave detectors have given astronomers an entirely new way to observe the universe.

▶ **Gravitational Waves**

◆ What are gravitational waves and how have astronomers detected them?

Look back at Figure 10.11, which uses rubber sheet analogies to illustrate the idea that masses cause curvature of spacetime. On a rubber sheet, any major change—such as suddenly changing one of the masses or having two large masses orbit each other closely—will change the curvature and create ripples moving outward from the point of the change. Einstein predicted that similar ripples must be created by changes in spacetime, and he called these ripples **gravitational waves.** He predicted that gravitational waves would be like light waves in carrying energy (but no mass) through space at the speed of light, but expressed doubt that we would ever be able to detect them. Fortunately for science, nature provided gravitational wave sources stronger than Einstein anticipated and our technology has advanced beyond what he had imagined.

To understand the evidence for gravitational waves, consider a close binary system containing two neutron stars. Such a system should generate strong gravitational waves that carry orbital energy away from the system, causing the orbits to decay (Figure 10.16). The process is gradual at first, but the strength and frequency of the gravitational waves should increase as the orbits decay further, so that the process accelerates and the two neutron stars ultimately merge. The merger should produce a particularly strong pulse of gravitational waves. A similar process should occur with other closely orbiting massive objects, such as two orbiting black holes.

The first binary neutron star was discovered in 1974, and observations soon showed the orbits in this system to be decaying at precisely the rate predicted by general relativity (Figure 10.17). This match between observation and theory

▲ **FIGURE 10.17**
This graph shows data for the first discovered neutron star binary (known as the Hulse-Taylor system after its discoverers). The dots are observed data points, and the red curve shows what general relativity predicts under the assumption that the system emits gravitational waves. The close match provided the first strong evidence for the existence of gravitational waves. (Data courtesy of Joel Weisberg and Yuping Huang.)

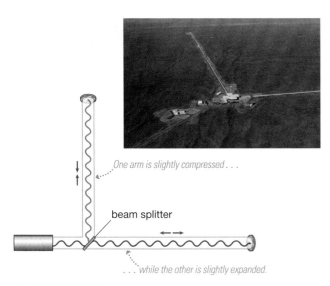

▲ FIGURE 10.18
This figure shows the arrangement of the LIGO detector arms. Laser beams are used to measure extremely tiny changes in the arm lengths as gravitational waves pass by. The inset shows the LIGO detector in Hanford, Washington.

▲ FIGURE 10.19
Gravitational-wave signal observed on September 14, 2015 by both detectors of the LIGO experiment, which are located in Hanford, Washington and Livingston, Louisiana. In order to measure the signal, the LIGO detectors needed to be sensitive enough to record vibrations (strain) of less than one part in 10^{21}. Analysis of the signal shows that it came from the merger of two black holes, each about 30 solar masses, at a distance of 1.3 billion light-years.

gave scientists great confidence that gravitational waves really exist, but the evidence was still indirect.

Direct detection of gravitational waves proved far more challenging, but a multi-decade effort to develop the necessary technology recently came to fruition with the *Laser Interferometer Gravitational-Wave Observatory,* or LIGO (Figure 10.18). Gravitational waves should alternately compress and expand objects as they pass by, but those changes in shape are extremely tiny. LIGO therefore searches for such compression and expansion with laser beams that travel back and forth along a pair of 4-kilometer-long arms arranged in an L-shape. The device is called an *interferometer* because it uses interference patterns created by the two beams to measure tiny changes in the light-travel times caused by changes in the arm lengths. LIGO is so sensitive that it can measure changes in the arm lengths that are far smaller than the size of a proton! To be confident that such small changes are not just a local artifact, scientists must record the same signal at more than one location, so LIGO began operations with two detectors (one in Washington and one in Louisiana).

LIGO first detected gravitational waves just weeks after it became fully operational in 2015 (Figure 10.19). Based on the frequency and strength of the gravitational waves, scientists concluded that this first pulse came from the merger of two black holes. Measurements of the difference in the time at which the signal reached the two LIGO detectors confirmed that the gravitational waves were traveling at the speed of light. Several more detections soon followed, and in 2017 LIGO was joined by a third detector (called Virgo), located in Italy. Shortly thereafter, scientists announced the first detection of a neutron star merger in a close binary system like the one shown in Figure 10.16.

Because gravitational waves carry information that differs from what light waves carry, our brand-new ability to detect them effectively opens a new window on the universe. Gravitational-wave astronomy is only in its infancy, but astronomers have high hopes for it in the future and are already in the process of developing detectors that will be even more sensitive than LIGO.

think about it How many detections of gravitational waves has LIGO recorded to date? Briefly describe at least three important results from gravitational-wave astronomy.

summary of key concepts

10.1 White Dwarfs and Neutron Stars

◆ What are white dwarfs?

A white dwarf is supported against the crush of gravity by **electron degeneracy pressure** and cannot have a mass greater than $1.4 M_{Sun}$. If a white dwarf is close enough to another star in a binary system, it can acquire hydrogen from its companion star and trigger a burst of nuclear fusion that produces a **nova**. In extreme cases, a white dwarf may accrete enough matter to exceed the **white dwarf limit** of $1.4 M_{Sun}$, at which point it will explode as a **white dwarf supernova**.

◆ What are neutron stars?

A neutron star is supported against the crush of gravity by **neutron degeneracy pressure**. Some neutron stars emit

beams of radiation that sweep through space as the neutron star spins. We call such neutron stars **pulsars**, and they provided the first direct evidence for the existence of neutron stars. Additional evidence comes from x-ray-emitting accretion disks around neutron stars in **x-ray binaries**. In some of these systems, nuclear fusion can ignite on the neutron star's surface, producing a burst of x-rays.

10.2 Black Holes

◆ What are black holes?

A black hole is an object so massive and compact that nothing can ever escape it, not even light; one way a black hole can form is in a supernova that leaves behind a core with a mass above the neutron star limit of $2 M_{Sun}$ to $3 M_{Sun}$. The black hole's **event horizon** marks the boundary between our observable universe and the inside of the black hole; the size of a black hole is characterized by its **Schwarzschild radius**. Einstein's theory of relativity predicts that space and time should become highly curved near a black hole, with time running slower nearer to the event horizon.

◆ What is the evidence for black holes?

Black holes emit no light, but they reveal their presence through their gravitational effects. Some x-ray binaries include compact objects far too massive to be neutron stars; it is likely that they are black holes. We have also found supermassive black holes in the centers of galaxies, and recent detections of gravitational waves from black-hole mergers have provided even stronger evidence for their existence.

THE PROCESS OF SCIENCE IN ACTION

10.3 Gravitational Waves

◆ What are gravitational waves and how have astronomers detected them?

General relativity predicts the existence of **gravitational waves** that travel at the speed of light; the strongest gravitational waves should come from closely orbiting massive objects or massive explosions. Indirect evidence for gravitational waves came from observations of the orbital decay of binary neutron stars. The first direct detections came from the instrument called LIGO, which is now detecting pulses of gravitational waves produced when orbiting neutron stars or black holes merge.

Investigations

For instructor-assigned homework and other learning materials, go to www.MasteringAstronomy.com.

Quick Quiz

Choose the best answer to each of the following. Explain your reasoning.

1. A *white dwarf* is (a) what most stars become when they die. (b) a slowly rotating neutron star. (c) a type of supernova.
2. If you had something the size of a sugar cube that was made of white dwarf matter, it would weigh about as much as (a) an average person. (b) a truck. (c) the entire Earth.
3. Why does matter falling toward a white dwarf, neutron star, or black hole in a binary system form an accretion disk? (a) because the central object spins very rapidly (b) because the infalling matter has some angular momentum (c) because the gas emits x-rays.
4. If we see a *nova*, we know that we are observing (a) a rapidly rotating neutron star. (b) a gamma-ray-emitting supernova. (c) a white dwarf in a binary system.
5. What happens if a white dwarf reaches the 1.4 M_{Sun} limit? (a) It explodes as a white dwarf supernova. (b) It collapses to become a neutron star. (c) It collapses to become a black hole.
6. A *neutron star* is (a) the remains of a star that died in a supernova. (b) the remains of a star that died in a planetary nebula. (c) an object made of elements with high atomic mass numbers.
7. A *pulsar* is (a) an unstable high-mass star. (b) an accreting white dwarf. (c) a rapidly rotating neutron star.
8. What is the basic definition of a black hole? (a) a dead star that has faded from view (b) an object with gravity so strong that not even light can escape (c) an object made from dark matter
9. How would a flashing red light appear as it fell into a black hole? (a) Its flashes would look the same. (b) Its flashes would appear bluer. (c) Its flashes would shift to the infrared part of the spectrum.
10. What makes astronomers think that Cygnus X-1 contains a black hole? (a) We can directly observe that one member of the system emits no light. (b) The unseen object orbited by a luminous star is too massive to be a neutron star. (c) The strong x-ray emission from the system means it must contain a black hole.
11. What predicted the existence of gravitational waves? (a) Einstein's general theory of relativity (b) Newton's universal law of gravitation (c) the nebular theory
12. The first gravitational waves that were detected directly came from (a) the merger of two black holes. (b) the birth of a galaxy. (c) the supernova of a very massive star.

Short-Answer/Essay Questions

Explain all answers clearly, with complete sentences and proper essay structure if needed. An asterisk (*) designates a quantitative problem, for which you should show all your work.

13. *Life Stories of Stars.* Write a one- to two-page life story for each of the following scenarios. Each story should be detailed and scientifically correct but also creative. That is, it should be entertaining while at the same time proving that you understand stellar evolution. Be sure to state whether you are a member of a binary system.
    a. You are a white dwarf of $0.8M_{Sun}$.
    b. You are a neutron star of $1.5M_{Sun}$.
    c. You are a black hole of $10M_{Sun}$.
    d. You are a white dwarf in a binary system that is accreting matter from its companion star.
14. *Census of Stellar Corpses.* Which kind of object do you think is most common in our galaxy: white dwarfs, neutron stars, or black holes? Explain your reasoning.
15. *Gravitational Wave Detectors.* The gravitational-wave detections made with the Laser Interferometer Gravitational-Wave Observatory (LIGO) have stimulated enormous international interest in building more detectors. Look up the nations other than the United States that have now built or are currently building similar gravitational-wave detectors. Find out why having more detectors will help us better determine the directions from which gravitational-wave bursts are coming. Write a short report on the number of detectors that will be coming on line in the next decade, where they will be located, and how they will help us identify the sources of these fantastic events.
16. *Why Black Holes Are Safe.* Explain why the principle of conservation of angular momentum makes it very difficult to fall into a black hole.
17. *Gravitational-Wave Detection.* Go to the website for the Laser Interferometer Gravitational-Wave Observatory. Look up the detections that it has made of gravitational waves. How many detections have there been so far? What kinds of objects have made the detected waves? What are the approximate distances of those objects? Write a short report on what you learn.
*18. *A Black Hole I?* You've just discovered a new x-ray binary, which we will call Hyp-X1 ("Hyp" for hypothetical). The system Hyp-X1 contains a bright, B2 main-sequence star orbiting an unseen companion. The separation of the stars is estimated to be 20 million kilometers, and the orbital period of the visible star is 4 days.
    a. Use Newton's version of Kepler's third law to calculate the sum of the masses of the two stars in the system. (*Hint:* See Tools of Science, page 96.) Give your answer in both kilograms and solar masses ($M_{Sun} = 2.0 \times 10^{30}$ kg).
    b. Determine the mass of the unseen companion. Is it a neutron star or a black hole? Explain. (*Hint:* A B2 main-sequence star has a mass of about $10M_{Sun}$.)
*19. *A Black Hole II?* You've just discovered another new x-ray binary, which we will call Hyp-X2 ("Hyp" for hypothetical). The system Hyp-X2 contains a bright, G2 main-sequence star orbiting an unseen companion. The separation of the stars is estimated to be 12 million kilometers, and the orbital period of the visible star is 5 days.
    a. Use Newton's version of Kepler's third law to calculate the sum of the masses of the two stars in the system. (*Hint:* See Tools of Science, page 96.) Give your answer in both kilograms and solar masses ($M_{Sun} = 2.0 \times 10^{30}$ kg).
    b. Determine the mass of the unseen companion. Is it a neutron star or a black hole? Explain. (*Hint:* A G2 main-sequence star has a mass of $1M_{Sun}$.)
*20. *White Dwarf Density.* A typical white dwarf has a mass of about $1.0M_{Sun}$ and the radius of Earth (about 6400 kilometers). Calculate the average density of a white dwarf, in *kilograms per cubic centimeter*. How does this compare to the mass of familiar objects?
*21. *Neutron Star Density.* A typical neutron star has a mass of about $1.5M_{Sun}$ and a radius of 10 kilometers.
    a. Calculate the average density of a neutron star, in *kilograms per cubic centimeter*.
    b. Compare the mass of 1 cm³ of neutron star material to the mass of Mount Everest ($\approx 5 \times 10^{10}$ kg).
*22. *Energy of a Black Hole Merger.* Models indicate that the gravitational-wave signal shown in Figure 10.19 came from the merger of two black holes with masses of $29M_{Sun}$ and $36M_{Sun}$ and resulted in a single black hole with a mass of $62M_{Sun}$. The difference in total mass between the start and finish of the merger corresponds to the amount of energy carried away in the form of gravitational waves. Use Einstein's formula $E = mc^2$ to calculate this amount of energy. Be sure to put the mass in units of kilograms and the speed of light in meters per second, so that the resulting energy has units of joules. Compare your answer to the energy released by the Sun each year, which is about 10^{34} joules.

11 Galaxies

LEARNING GOALS

11.1 Our Galaxy: The Milky Way
- What does our galaxy look like?
- How did the Milky Way form?

11.2 Galaxies Beyond the Milky Way
- What are the major types of galaxies?
- Why do galaxies differ?

THE PROCESS OF SCIENCE IN ACTION

11.3 Seeking Supermassive Black Holes
- What is the evidence for supermassive black holes at the centers of galaxies?

This photo shows a galaxy—an island of stars bound together by gravity. Our own Milky Way Galaxy would look similar if we could see it from a few million light-years away. More than a hundred billion other large galaxies populate the observable universe. Within these galaxies, stars form in dusty clouds of hydrogen gas, fuse new elements out of that hydrogen, and add those new elements to the galaxy's gas clouds when they die. Galaxies therefore function like huge ecosystems in which the cycling of gas through stars gradually builds up the elements necessary to make planets and support life. In this chapter, we'll study how galaxies work, paying special attention to how the history of star formation shapes the appearance of the galaxies we see today.

Our Galaxy

11.1 Our Galaxy: The Milky Way

We'll begin our study of galaxies with our own Milky Way Galaxy, the only one we can observe in great detail. We'll look at the structure and motion of the galaxy, investigate the galactic processes that maintain the ongoing cycle of stellar life and death, examine our galaxy's center, and examine the clues we have collected about how our galaxy formed. Through it all, we'll see that we are not only "star stuff" but also "galaxy stuff"—the product of eons of complex recycling and reprocessing of matter and energy in the Milky Way Galaxy.

◆ What does our galaxy look like?

From Earth, the Milky Way Galaxy looks like a band of light crossing the sky (Figure 11.1). The name *Milky Way* comes from the appearance of this band, which looked to ancient people like a ribbon of flowing milk. In fact, even the word *galaxy* comes from the Greek word for milk: *galactos*.

Because we live inside the Milky Way Galaxy, trying to determine its true size and shape is somewhat like trying to draw a picture of your house without ever leaving your bedroom. The task is difficult because much of our galaxy's visible light is hidden from view. Only in the last few decades have we had the technology to observe the galaxy in other wavelengths of light (see Tools of Science, page 184). Nevertheless, by carefully observing our galaxy and comparing it to others, we have developed a good understanding of its structure and of the processes that shape it.

Structure of the Milky Way Figure 11.2 shows the Milky Way Galaxy as it would look from the outside. We say that it is a *spiral galaxy* because of the spectacular **spiral arms** that would be visible in a face-on view. From the side, the spiral arms blend into a thin **disk** that contains more than 100 billion stars. At the center of the disk, we find a bright central **bulge**. Surrounding the disk is a vast, roughly spherical **halo**. The halo is nearly invisible, except for its approximately 200 *globular clusters* of stars [Section 9.3].

The disk is about 100,000 light-years in diameter, but only about 1000 light-years thick. Our Sun is located in the disk about 27,000 light-years from the galactic center—a little more than halfway out from the center to the edge of the disk. Remember that this distance is incredibly vast [Section 1.1]: The few thousand stars visible to the naked eye would fit inside a tiny dot in a picture like that in Figure 11.2.

The galactic disk is filled with interstellar gas and dust—known collectively as the **interstellar medium**—that obscures our view when we try to peer directly through it. The dusty, smoglike nature of the interstellar medium hides most of our galaxy from view when we try to observe it in visible light, and as a result it long fooled astronomers into believing that we lived near our galaxy's center. As we saw in Chapter 9, the dusty gas clouds of the interstellar medium provide the raw material from which all the stars in our galaxy have formed.

Orbits of Disk Stars The Milky Way's structure reflects how stars move within it. Stars in the disk orbit in roughly circular paths that all go in the same direction in nearly the same plane. If you could stand outside the Milky Way and watch it for a few billion years, the disk would resemble a huge merry-go-round. Like horses on a merry-go-round, individual stars bob up and down through the disk as they orbit. The general orbit of a star around the galaxy arises from its gravitational attraction toward the galactic center, while the bobbing arises from

▲ FIGURE 11.1
This photo shows the Milky Way over Haleakala crater on the island of Maui, Hawaii. The bright spot just below (and slightly left of) the center of the band is the planet Jupiter.

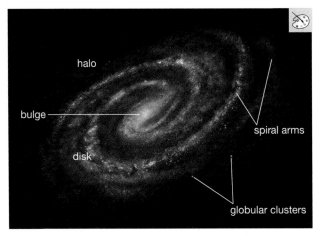

a Artist's conception of the Milky Way viewed from the outside.

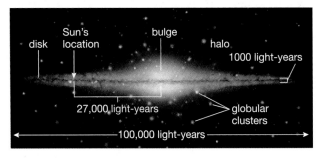

b Edge-on schematic view of the Milky Way.

▲ FIGURE 11.2
The Milky Way Galaxy.

tools of science | Observing Different Kinds of Light

Much of modern astronomy depends on our ability to detect not just visible light from space but light from across the electromagnetic spectrum (see Tools of Science, page 79). Two technological advances have given us this ability. The first is the development of specialized instruments capable of detecting forms of light that our eyes cannot see. The second is the development of spacecraft that can carry these instruments high above Earth's atmosphere, which prevents most kinds of light from reaching Earth's surface.

Figure 1 gives an example of how much we can learn by observing wavelengths of light that are invisible to human eyes. It shows seven views of the disk of our Milky Way Galaxy, each made by photographing the Milky Way's disk in every direction from Earth in a particular wavelength band. Four of the images (c, e, f, and g) come from observatories in space. The other three (a, b, and d) are from ground-based telescopes.

Figure 1a is a visible-light view that shows a haze of starlight interrupted by dark blotches running primarily down the center of the disk. These dark blotches occur where dusty clouds of interstellar gas block our view of stars behind them. Figure 1b shows the same view as it appears in infrared light. Infrared light can pass through the dusty gas clouds, which is why the center of the disk is bright and we can see the Milky Way's bulge in the center of the image. Figure 1c shows long-wavelength infrared light, which comes primarily from dust grains in the interstellar medium. Moving to even longer wavelengths, Figure 1d shows radio emission from atomic hydrogen gas, and Figure 1e shows radio emission from carbon monoxide molecules in molecular clouds. Notice how the regions of bright carbon monoxide emission correspond to the dark patches in Figure 1a.

On the opposite end of the electromagnetic spectrum, at very short wavelengths, we observe the most energetic photons. Figure 1f is an x-ray image showing the distribution of hot gas and x-ray binaries in our galaxy's disk. Figure 1g is a gamma-ray image showing where extremely energetic particles known as *cosmic rays* are colliding with atomic nuclei in the interstellar medium.

You can see from this sequence of images that the ability to observe many different kinds of light allows us to gather much more information about astronomical objects than we would be able to gather from visible light alone. That is why astronomers today observe the universe in many wavelengths of light.

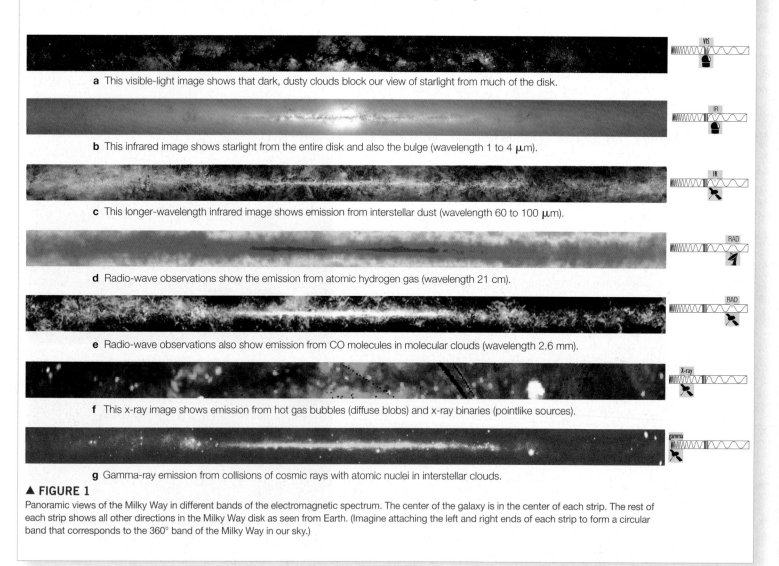

a This visible-light image shows that dark, dusty clouds block our view of starlight from much of the disk.

b This infrared image shows starlight from the entire disk and also the bulge (wavelength 1 to 4 μm).

c This longer-wavelength infrared image shows emission from interstellar dust (wavelength 60 to 100 μm).

d Radio-wave observations show the emission from atomic hydrogen gas (wavelength 21 cm).

e Radio-wave observations also show emission from CO molecules in molecular clouds (wavelength 2.6 mm).

f This x-ray image shows emission from hot gas bubbles (diffuse blobs) and x-ray binaries (pointlike sources).

g Gamma-ray emission from collisions of cosmic rays with atomic nuclei in interstellar clouds.

▲ FIGURE 1
Panoramic views of the Milky Way in different bands of the electromagnetic spectrum. The center of the galaxy is in the center of each strip. The rest of each strip shows all other directions in the Milky Way disk as seen from Earth. (Imagine attaching the left and right ends of each strip to form a circular band that corresponds to the 360° band of the Milky Way in our sky.)

the localized pull of gravity within the disk itself (Figure 11.3). A star that is "too far" above the disk is pulled back into the disk by gravity. Because the density of interstellar gas is too low to slow the star, it flies through the disk until it is "too far" *below* the disk on the other side. Gravity then pulls it back in the other direction. This ongoing process produces the bobbing of the stars.

The up-and-down motions of the disk stars give the disk its thickness of about 1000 light-years—a great distance by human standards, but only about 1% of the disk's 100,000-light-year diameter. In the vicinity of our Sun, each star's orbit takes more than 200 million years, and each up-and-down bob takes a few tens of millions of years.

The galaxy's rotation is unlike that of a merry-go-round in one important respect: On a merry-go-round, horses near the edge move much faster than those near the center. But in our galaxy's disk, the orbital velocities of stars near the edge and those near the center are about the same. As we will discuss in Chapter 14, the surprisingly high velocities of stars near the edge provide important evidence for the existence of *dark matter*.

Orbits of Halo and Bulge Stars The orbits of stars in the halo are much less organized than those of disk stars. Individual halo stars travel around the galactic center on more or less elliptical paths, but the orientations of these paths are relatively random (see Figure 11.3). Neighboring halo stars can circle the galactic center in opposite directions. They swoop from high above the disk to far below it and back again, plunging through the disk at velocities so great that the disk's gravity hardly alters their trajectories. These swooping orbits explain why the halo is much puffier than the disk. Bulge-star orbits are somewhat similar to halo-star orbits in that bulge stars swoop above and below the disk, but observations indicate that many bulge stars orbit in the same general direction as disk stars.

think about it Is there much danger that a halo star swooping through the disk of the galaxy will someday hit the Sun or Earth? Why or why not? (*Hint:* Consider the typical distances between stars on the 1-to-10-billion scale introduced in Chapter 1.)

Galactic Recycling Closer inspection of the Milky Way's disk reveals that stars are continually forming and dying within it through a process we will call the **star–gas–star cycle** (Figure 11.4). Several of the steps in this cycle should already be familiar from Chapter 9: Stars are born when gravity causes the collapse of molecular clouds within the interstellar medium. They shine for millions or billions of years with energy produced by nuclear fusion, dying only when they've exhausted their fuel for fusion. As they die, they return much of their material to the interstellar medium, through a planetary nebula in the case of a low-mass star or through a supernova in the case of a high-mass star.

The remaining stages of the star–gas–star cycle take place in the interstellar medium. Gas ejected by stars, particularly the gas from supernova explosions, often enters the interstellar medium in the form of a hot bubble. Figure 11.5 shows the bubble produced by a 400-year-old supernova explosion, which contains ionized gas hot enough to emit x-rays. This million-degree gas takes many thousands of years to cool, but eventually reaches a temperature of around 10^4 K, which is still quite warm but cool enough for hydrogen atoms to remain neutral rather than being ionized. We refer to this warm gas as **atomic hydrogen gas**, although it contains a significant amount of helium and small amounts of other elements as well. Mapping the distribution of atomic hydrogen gas in the Milky Way using radio observations of its spectral line at a wavelength of 21 centimeters tells us that the gas is distributed throughout the galactic disk. Matter remains in the warm atomic hydrogen stage for millions of years, and during

▲ **FIGURE 11.3**
Characteristic orbits of disk stars (yellow), bulge stars (red), and halo stars (green) around the galactic center. (The yellow path exaggerates the up-and-down motion of the disk star orbits.)

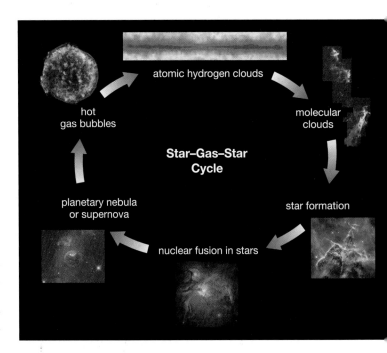

▲ **FIGURE 11.4**
The star–gas–star cycle.

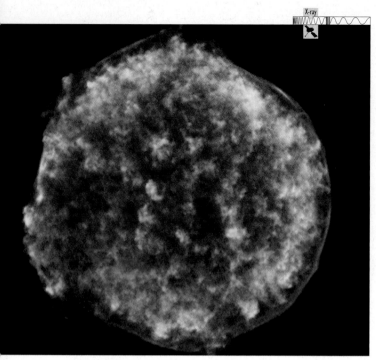

▲ FIGURE 11.5
This image from the Chandra X-Ray Observatory shows x-ray emission from hot gas in a young supernova remnant. The most energetic x-rays (blue) come from 20-million-degree gas just behind the expanding shock wave. Less energetic x-rays (green and red) come from the 10-million-degree debris ejected by the exploded star. The remnant is about 20 light-years across.

▲ FIGURE 11.6
Hubble Space Telescope photo of the Orion Nebula, an ionization nebula energized by ultraviolet photons from hot stars.

this time gravity slowly draws blobs of atomic gas together into tighter clumps, which radiate energy and cool more efficiently as they grow denser. Even so, the interstellar medium remains a nearly perfect vacuum by earthly standards: On average, each cubic centimeter contains only one hydrogen atom.

The temperature of these gas clumps eventually drops well below 100 K, allowing the hydrogen atoms to pair up into hydrogen molecules. The cool, dense clumps then become *molecular clouds,* which go on to form stars, thereby completing the star–gas–star cycle. Molecular hydrogen (H_2) is by far the most abundant molecule in molecular clouds, but it is difficult to detect because temperatures are usually too cold for the gas to produce H_2 emission lines. Most of what we know about molecular clouds therefore comes from observing spectral lines of molecules that make up only a tiny fraction of a cloud's mass. Carbon monoxide (CO) is the most abundant of these molecules and produces strong emission lines in the radio portion of the spectrum at the 10–30 K temperatures of molecular clouds.

In addition to making new generations of stars possible, this galactic recycling process gradually changes the chemical composition of the interstellar medium. Recall that the early chemical composition of the universe was almost 100% hydrogen and helium; heavier elements have been produced by stars. The newly created elements mix with other interstellar gas and become incorporated into new generations of stars. That is how our solar system came to have the elements from which our planet was made. Today, thanks to more than 12 billion years of galactic recycling, elements heavier than helium constitute about 2% of the galaxy's gaseous content by mass. The remaining 98% consists of hydrogen (about 70%) and helium (about 28%).

think about it Recall from Chapter 9 that stars in our galaxy's globular clusters are extremely old, while stars in open clusters are relatively young. Based on this fact, which stars do you think contain a higher proportion of heavy elements: stars in globular clusters or stars in open clusters? Explain.

Star-Forming Regions The star–gas–star cycle has operated continuously since the Milky Way's birth, yet new stars are not spread evenly across the galaxy. Some regions seem much more fertile than others. Galactic environments rich in molecular clouds tend to spawn new stars easily, while gas-poor environments do not. However, molecular clouds are dark and hard to see, so we often have to look for other signs of star formation.

Wherever we see hot, massive stars, we know that we have spotted a region of active star formation. Because these stars live fast and die young, they never get a chance to move very far from their birth mates. They therefore signal the presence of star clusters in which many of their lower-mass companions are still forming.

Near these hot stars, we often find colorful, wispy blobs of glowing gas known as **ionization nebulae** (sometimes called *emission nebulae* or *H II regions*). These nebulae glow because electrons in their atoms are raised to high energy levels or ionized when they absorb ultraviolet photons from the hot stars, so they emit light as the electrons fall back to lower energy levels (see Tools of Science, page 152). The Orion Nebula, about 1500 light-years away in the "sword" of the constellation Orion, is among the most famous (Figure 11.6).

Spiral Arms Looking more broadly at our galaxy, we can see that its spiral arms must be full of newly forming stars because they bear all the hallmarks of star formation. They are home both to molecular clouds and to numerous clusters of young, bright blue stars surrounded by ionization nebulae. Photos of other spiral galaxies show these characteristics clearly (Figure 11.7). Hot blue stars and ionization nebulae trace out the arms, while the stars between the arms

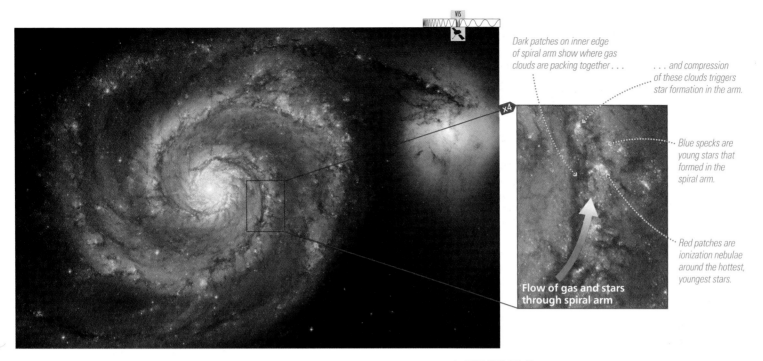

▲ **FIGURE 11.7**
This photo from the Hubble Space Telescope shows Galaxy M51's two magnificent spiral arms along with a smaller galaxy that is currently interacting with one of those arms. Notice that the spiral arms are much bluer than the central bulge. Because massive blue stars live for only a few million years, the blueness of the spiral arms tells us that stars must be forming more actively within them than elsewhere in the galaxy. (The large image shows a region roughly 90,000 light-years across.)

are generally redder and older. We also see enhanced amounts of molecular and atomic gas in the spiral arms, indicating that the spiral arms contain the material necessary to make new stars.

At first glance, spiral arms look as if they ought to move with the stars, like the fins of a giant pinwheel in space. However, observations indicate that stars throughout most of the disk orbit at approximately the same speed, which means that stars near the center (which travel in circles with smaller radii) complete each orbit in less time than stars far from the center. As a result, if the spiral arms moved along with the stars, the central parts of the arms would wind around the galaxy many more times than the outer parts. After a few galactic rotations, the arms would become wound into a tight coil. We do not see such tightly wound coils of arms in galaxies, so we conclude that spiral arms must be more like swirling ripples in a whirlpool than like the fins of a giant pinwheel.

In fact, spiral arms appear to be enormous waves of star formation that propagate through the disks of spiral galaxies. Theoretical models suggest that disturbances called *spiral density waves* are responsible for the spiral arms. According to these models, spiral arms are places in a galaxy's disk where stars and gas clouds get more densely packed. Pushing the stars closer together has little effect on the stars themselves—they are still much too widely separated to collide with each other. However, large gas clouds do collide, and packing the clouds closer together enhances the force of gravity within them, triggering the formation of new star clusters. Supernova explosions from massive stars in these clusters can compress the surrounding clouds further, triggering even more star formation.

common misconceptions

The Sound of Space

In many science fiction movies, a thunderous sound accompanies the demolition of a spaceship. If the moviemakers wanted to be more realistic, they would silence the explosion. On Earth, we perceive sound when sound waves—which are waves of alternately rising and falling pressure—cause trillions of gas atoms to push our eardrums back and forth. Although sound waves can and do travel through interstellar gas, the extremely low density of this gas means that only a handful of atoms per second would collide with something the size of a human eardrum. It would therefore be impossible for a human ear (or a similar-size microphone) to register any sound, which is why the real sound of space is silence.

▶ **The Galactic Center**

The Galactic Center The center of the Milky Way Galaxy lies in the direction of the constellation Sagittarius and does not look particularly special to our unaided eyes. However, if we could remove the interstellar dust that obscures our view, the galaxy's central bulge would be one of the night sky's most spectacular sights. And deep within the bulge, at the very center of the Milky Way, evidence indicates the presence of a gigantic black hole.

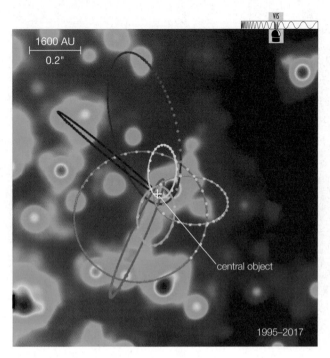

▲ **FIGURE 11.8**
Evidence for a black hole at the center of our galaxy. Each colored ellipse represents the calculated orbit for a particular star, based on the star's positions (dots) as observed with the Keck Telescope at 1-year intervals. By applying Newton's version of Kepler's third law, we infer that the central object has a mass about 4 million times that of our Sun, packed into a space no larger than our solar system. (The 1600 AU shown on the scale bar is equivalent to about 9 light-days.)

The central object of our galaxy is called Sagittarius A* (pronounced "Sagittarius A-star"), or Sgr A* for short. Observations of radio waves and x-rays from Sgr A* indicate that it is no larger across than our solar system, yet it is massive enough to hold several hundred stars in orbit within just 1 light-year of it. By applying Newton's version of Kepler's third law to the orbits of the stars passing closest to it (Figure 11.8), astronomers have determined that Sgr A* has a mass of about 4 million solar masses, and so much mass packed into such a small space means that the light we see from Sgr A* is almost certainly coming from accretion of gas onto a gigantic black hole. In Section 11.3, we'll see that many other galaxies also have massive central black holes, and that some are far more massive than the 4 million solar masses of Sgr A*.

◆ How did the Milky Way form?

The major features of our Milky Way Galaxy that we have discussed, including its structure and the operation of the star–gas–star cycle, provide clues to how it formed. However, some of the most important clues come from a detailed comparison of disk stars with halo stars.

Clues from Disk Stars and Halo Stars We have already seen how the disorderly orbits of halo stars differ from the orbits of disk stars. Two other differences between halo stars and disk stars also stand out. First, all halo stars are very old, while disk stars can be of any age. Second, spectra show that halo stars contain smaller proportions of heavy elements than disk stars. Based on these differences, astronomers divide the Milky Way's stars into two distinct populations.

1. The **disk population** (sometimes called *Population I*) consists of stars that follow the orderly orbital patterns of the disk. This population includes both young stars and old stars, all of which have heavy-element proportions of about 2%, like our Sun.
2. The **halo population** (sometimes called *Population II*) consists of stars with orbits through the halo. Stars in this population are all old and therefore low in mass, and can have heavy-element proportions as low as 0.02%—meaning that heavy elements are about 100 times rarer in these stars than in the Sun.

We can understand why halo stars differ from disk stars by looking at how the Milky Way's gas is distributed. The halo does not contain the cold, dense molecular clouds required for star formation. Because star-forming molecular clouds are found only in the disk, new stars can be born only in the disk and not in the halo.

The relative lack of heavy elements in halo stars indicates that they must have formed early in the galaxy's history—before many supernovae had exploded and added heavy elements to star-forming clouds. We conclude that the halo has lacked the gas needed for star formation for a very long time. Apparently, all the Milky Way's cool gas settled into the disk long ago. The only stars that still survive in the halo are old, low-mass stars.

think about it How does the halo of our galaxy resemble the distant future fate of the galactic disk? Explain.

A Model of Galaxy Formation Any model of our galaxy's formation must account for the differences between disk stars and halo stars. The simplest such model, shown in Figure 11.9, starts with a giant **protogalactic cloud** of hydrogen and helium gas.

Early on, the gravity associated with this protogalactic cloud would have drawn in matter from all directions, creating a cloud that was blobby in shape with little or no measurable rotation. The orbits of stars forming within such a cloud could have had any orientation, accounting for the randomly oriented orbits of stars in the halo. This simple model also explains the old ages of halo stars, since it tells us that they were born as the galaxy first began to form.

As time passed, the remaining gas contracted under the force of gravity and settled into a flattened, spinning disk because of conservation of angular momentum (Figure 11.9). This process was much like the process that led to the formation of the solar nebula [Section 4.2], but on a much larger scale. Collisions among gas particles tended to average out their random motions, causing them to orbit in the same direction and in the same plane. Stars born after the formation of the spinning disk therefore share the disk's organized motion. Their ages vary because star formation has continued ever since the disk formed, thanks to the ongoing star–gas–star cycle.

Although the model in Figure 11.9 successfully explains the basic differences between halo and disk stars, detailed studies of the proportions of heavy elements in stars suggest that this model is a bit too simplistic. If the Milky Way had formed from a single protogalactic cloud, it would have steadily accumulated heavy elements during its inward collapse as stars formed and exploded within it. In that case, the outermost stars in the halo would be the oldest and would have the smallest proportions of heavy elements, and the heavy-element proportions would steadily rise as we looked at stars orbiting closer to the disk and bulge. But that is not the pattern we observe.

Instead, we find variations in heavy-element proportions, suggesting that the Milky Way's oldest stars formed in relatively small protogalactic clouds, each with a few globular clusters. These clouds later collided and combined to create a single galaxy, the Milky Way (Figure 11.10). Similar processes may still be under way. Two small galaxies (the Sagittarius and Canis Major dwarf galaxies) are currently being torn apart as they crash through the Milky Way's disk. A billion years from now, their stars will be indistinguishable from halo stars, because they will all be circling the Milky Way on orbits that carry them high above the disk. Evidence indicates that this process has also occurred in the past: Some halo stars move in organized streams that are probably the remnants of small galaxies torn apart long ago by the Milky Way's gravity.

The distribution of elements also suggests that when the disk finally formed, it began with only about 10% of the amount of heavy elements that it has today, though with a greater concentration of heavy elements toward the bulge. As time passed, the star–gas–star cycle gradually increased the disk's heavy-element content in an orderly fashion, which is why young stars tend to have higher proportions of heavy elements than older stars.

11.2 Galaxies Beyond the Milky Way

Take another look at the picture on page 1 of this book. This Hubble Space Telescope photo shows a typical patch of the universe, including galaxies of many sizes, colors, and shapes. Some look large, some small. Some are reddish, some whitish. Some appear round, and some appear flat. We would like to understand why these galaxies differ, but this is no easy task. Just as with stars, our observations capture only the briefest snapshot of any galaxy's life, leaving us to piece together the life story of a typical galaxy from the clues we find in pictures of many different galaxies. Let's look more closely at the different types of galaxies and see what their differences tell us about their life histories.

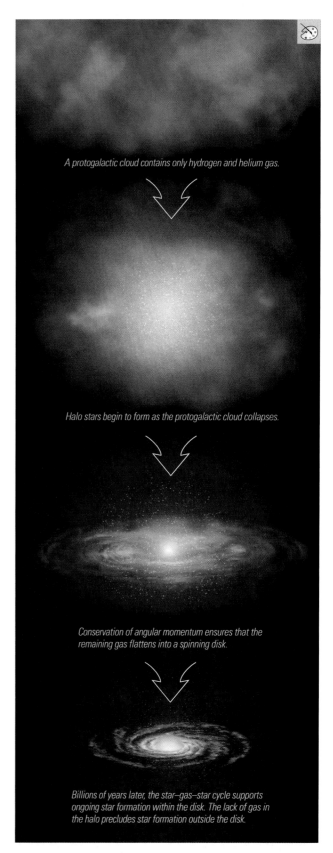

▲ FIGURE 11.9
This four-picture sequence illustrates a simple schematic model of galaxy formation, showing how a spiral galaxy might develop from a protogalactic cloud of hydrogen and helium gas.

▲ **FIGURE 11.10**
This painting shows a model of how the Milky Way's halo may have formed. The characteristics of stars in the Milky Way's halo suggest that several smaller gas clouds, already containing some stars and globular clusters, may have merged to form the Milky Way's protogalactic cloud. These stars and star clusters remained in the halo while the gas settled into the Milky Way's disk.

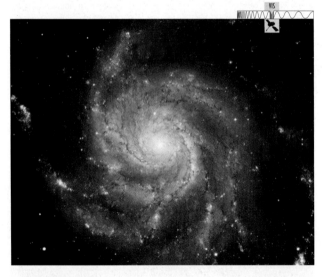

▲ **FIGURE 11.11**
The giant spiral galaxy M101. It is about 170,000 light-years in diameter.

▲ **FIGURE 11.12**
NGC 1300, a barred spiral galaxy about 110,000 light-years in diameter.

▶ **Galaxies and the Foundation of Modern Cosmology (Learning Outcome 1)**

◆ What are the major types of galaxies?

Astronomers classify galaxies into three major categories:

- **Spiral galaxies**, such as our own Milky Way, look like flat white disks with yellowish bulges at their centers. The disks are filled with cool gas and dust, interspersed with hotter ionized gas, and usually display beautiful spiral arms.
- **Elliptical galaxies** are redder, rounder, and often elongated like a football. Compared with spiral galaxies, elliptical galaxies contain very little cool gas and dust, though they often contain very hot, ionized gas.
- **Irregular galaxies** appear neither disklike nor round.

Galaxies have different colors because of the different kinds of stars that populate them: Spiral and irregular galaxies look white because they contain stars of all different colors and ages, while elliptical galaxies look redder because old, reddish stars produce most of their light. Galaxies also come in a wide range of sizes, from *dwarf galaxies* containing less than a billion (10^9) stars to *giant galaxies* with more than 1 trillion (10^{12}) stars.

think about it   Take a moment to classify the larger galaxies in the photo on page 1. How many appear spiral? Elliptical? Irregular? Do the colors of galaxies seem related to their shapes?

Spiral Galaxies   Like the Milky Way, other spiral galaxies also have a thin *disk* and a central *bulge* (Figure 11.11). The bulge merges smoothly into a nearly invisible *halo* that can extend to a radius of more than 100,000 light-years. A spiral galaxy's halo is more difficult to see than its bulge and disk because halo stars are generally dim and spread over a large volume of space.

Some spiral galaxies, known as *barred spirals*, appear to have a straight bar of stars cutting across the center, with spiral arms curling away from the ends of the bar (Figure 11.12). Astronomers suspect that the Milky Way is a barred spiral galaxy, because observations of our galaxy's bulge indicate that it is cigar-shaped rather than spherical. Other galaxies have disk and halo populations like spiral galaxies but appear to lack spiral arms. These *lenticular galaxies* (*lenticular* means "lens-shaped") are sometimes considered an intermediate class between spirals and ellipticals, because they tend to have less cool gas than normal spirals but more than ellipticals.

Among large galaxies in the universe, most (75–85%) are spiral or lenticular; these types are much rarer among small galaxies. Both spiral and lenticular galaxies tend to be found in loose collections of up to a few dozen galaxies, called *groups*. Our Local Group is one example (see Figure 1.1).

Elliptical Galaxies   Elliptical galaxies differ from spiral galaxies primarily in that they lack a significant disk population. They therefore look much like the bulge and halo of a spiral galaxy that is missing its disk.

Large elliptical galaxies can contain more than a trillion stars (Figure 11.13). They are most common in clusters of galaxies, which can contain hundreds and sometimes thousands of galaxies extending over more than 10 million light-years. Elliptical galaxies make up about half the large galaxies in the central regions of clusters, but represent only a small minority (about 15%) of the large galaxies found outside clusters. On the other end of the size scale, very small elliptical galaxies (including classes known as *dwarf ellipticals* and *dwarf spheroidals*) appear to be the most common type of galaxy in the universe.

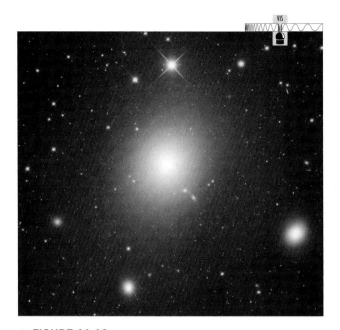

▲ FIGURE 11.13
M87, a giant elliptical galaxy in the Virgo Cluster, is one of the most massive galaxies in the universe. The region shown is more than 300,000 light-years across.

▲ FIGURE 11.14
The Large Magellanic Cloud, an irregular galaxy that is a small companion to the Milky Way. It is about 30,000 light-years across.

Elliptical galaxies usually contain very little dust or cool gas, although some have relatively small and cold gaseous disks rotating at their centers. However, some large elliptical galaxies contain substantial amounts of very hot gas that can emit x-rays. The lack of cool gas in elliptical galaxies means that, like the halo of the Milky Way, they generally have little or no ongoing star formation. Elliptical galaxies therefore tend to look red or yellow because they do not have any of the hot, young, blue stars found in the disks of spiral galaxies.

Irregular Galaxies Some galaxies do not fit either of the two major categories. The class of *irregular galaxies* is a miscellaneous class that includes small galaxies such as the Magellanic Clouds (Figure 11.14) and larger galaxies that appear to be in disarray. Irregular galaxies are usually white and dusty, like the disks of spirals, and contain many young, massive stars.

Irregular galaxies represent only a small fraction of the nearby large galaxies, but appear more common among distant galaxies. Because the light of more distant galaxies has taken longer to reach us, these observations tell us that irregular galaxies were more common when the universe was younger [Section 12.3].

Hubble's Galaxy Classes Edwin Hubble invented a system for classifying galaxies that organizes the galaxy types into a diagram shaped like a tuning fork (Figure 11.15). Elliptical galaxies appear on the "handle" at left, designated by the letter E and a number. The larger the number, the flatter the elliptical galaxy: An E0 galaxy is a sphere, and the numbers increase to the highly flattened type E7. The two prongs of the fork show spiral galaxies, designated by the letter S for ordinary spirals and SB for barred spirals, followed by a lowercase a, b, or c: The bulge size decreases from a to c, while the amount of dusty gas increases. Lenticular galaxies are designated S0, and irregular galaxies are designated Irr. Unfortunately for astronomers, Hubble's classification scheme has not led to easy answers about why galaxies differ.

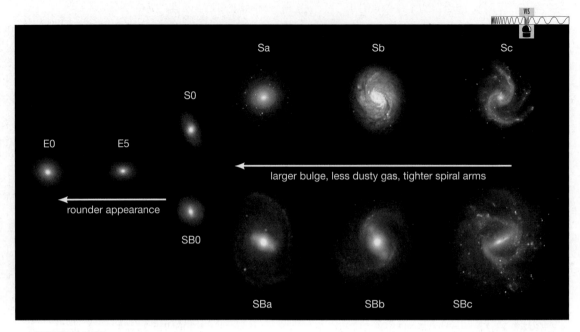

▲ FIGURE 11.15
This "tuning fork" diagram illustrates Hubble's galaxy classes.

Galaxy Evolution (Learning Outcome 2)

◆ Why do galaxies differ?

The distinct differences among spiral, elliptical, and irregular galaxies tell us that their life stories must be quite different. We do not yet understand the life stories of galaxies as well as we understand the life stories of stars, but we are rapidly learning more. That is why the formation and development of galaxies, or **galaxy evolution**, is one of the most active areas of research in astronomy.

The early stages in the formation of all galaxies are thought to have been similar to those discussed for our Milky Way, with gravity pulling matter into a protogalactic cloud system that contracted and began to form stars. In that case, we might expect all galaxies to be similar to the Milky Way: spirals with gas-rich disks. However, we know that many galaxies are ellipticals that lack such disks. Therefore, a key issue in studying galaxy evolution is understanding why some galaxies ended up elliptical and others spiral. Detailed models suggest two general possibilities: (1) Galaxies may have ended up looking different because they began with slightly different birth conditions in their protogalactic clouds, or (2) galaxies may have begun their lives similarly but later changed because of interactions with other galaxies.

Birth Conditions The first general category of explanation traces a galaxy's type back to the protogalactic clouds from which it formed. For example, if the original protogalactic cloud system had a significant amount of angular momentum, it would have rotated quickly as it collapsed, thereby forming a disk and making the resulting galaxy a spiral galaxy. In contrast, if the protogalactic cloud had little or no angular momentum, its gas might not have formed a disk at all, and the resulting galaxy would be elliptical.

Alternatively, birth conditions might affect a galaxy's type through the density of the protogalactic clouds from which it formed. Protogalactic clouds with relatively high gas density would have radiated energy more effectively and cooled more quickly, thereby allowing more rapid star formation. If star formation proceeded fast enough, all the gas could have been turned into stars before any of it had time to settle into a disk. The resulting galaxy would therefore lack a disk, making it an elliptical galaxy. In contrast, lower-density clouds would have formed stars more slowly, leaving plenty of gas to form the disk of a spiral galaxy.

Later Interactions Differences in birth conditions probably play an important role in the overall story of why some galaxies have gas-rich disks and others do not. However, differing birth conditions probably are not the whole story, because of one key fact: Galaxies rarely evolve in perfect isolation.

Think back to our scale-model solar system in Chapter 1. On a scale on which the Sun is the size of a grapefruit, the nearest star is like another grapefruit a few thousand kilometers away. Because the average distances between stars are so huge compared to the sizes of stars, collisions between stars are extremely rare. However, if we rescale the universe so that our *galaxy* is the size of a grapefruit, the Andromeda Galaxy is like another grapefruit only about 3 meters away, and a few smaller galaxies lie considerably closer. The average distances between galaxies are not much larger than the sizes of galaxies, so collisions between galaxies are inevitable.

Galaxy collisions are spectacular events that unfold over hundreds of millions of years (Figure 11.16). During our short lifetimes, we can view only a snapshot of a collision in progress, distorting the shapes of the colliding galaxies. We can learn much more about galactic collisions with the aid of computer simulations that allow us to "watch" collisions unfold. These computer models show that a collision between two spiral galaxies can create an elliptical galaxy (Figure 11.17). Tremendous tidal forces between the colliding galaxies tear apart the two disks, randomizing the orbits of their stars. Meanwhile, a large fraction of their gas sinks to the center of the collision and rapidly forms new stars. Supernovae and stellar winds eventually blow away the rest of the gas. When the cataclysm finally settles down, the merger of the two spirals has produced a single elliptical galaxy. Little gas is left for a disk, and the orbits of the stars have random orientations.

Observations support the idea that at least some elliptical galaxies result from collisions that lead to mergers. Elliptical galaxies dominate the galaxy populations at the cores of dense clusters of galaxies, where collisions should be most frequent. This fact may indicate that any spirals once present became ellipticals through collisions.

▲ **FIGURE 11.16**
A pair of colliding spiral galaxies known as the Antennae (NGC 4038/4039). The photo taken from the ground (left) reveals their vast tidal tails, while the zoom-in from the Hubble Space Telescope shows the burst of star formation at the center of the collision.

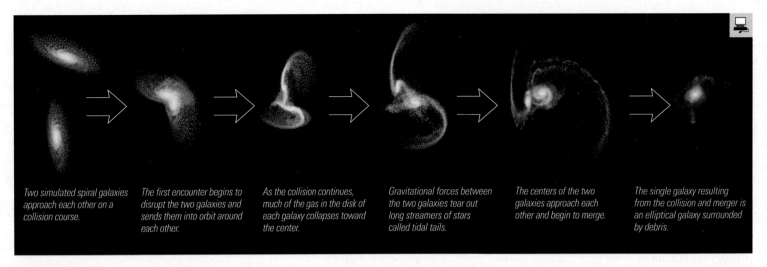

▲ **FIGURE 11.17**
Several stages in a supercomputer simulation of a collision between two spiral galaxies that results in an elliptical galaxy. At least some of the elliptical galaxies in the present-day universe formed in this way. The whole sequence spans about 1.5 billion years.

An Incomplete Answer Birth conditions and subsequent interactions probably both play important roles in galaxy evolution. The birth-condition scenarios may explain why the vast majority of galaxies are either spiral or elliptical in shape. The interaction scenarios may explain why ellipticals are more common in clusters while spirals are more common outside of clusters. These ideas can also account for the relatively small fraction of galaxies that are irregular: At least some irregulars are probably galaxies undergoing a disruptive interaction.

These ideas appear to explain the observed patterns of galaxy color and galaxy luminosity illustrated in Figure 11.18. All galaxies are thought to begin as star-forming systems that are blue-white in color, making them members of the *blue cloud*

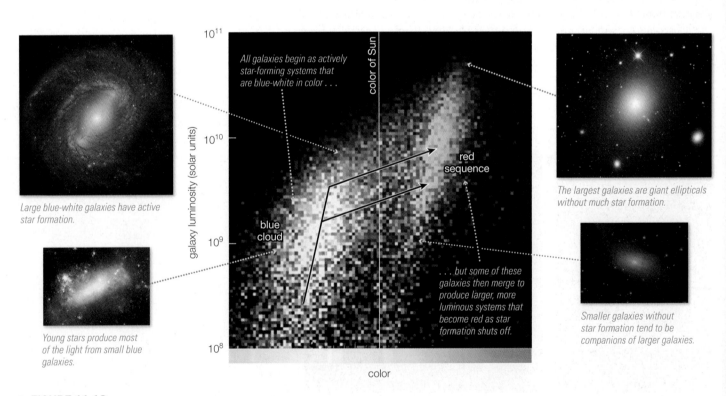

▲ **FIGURE 11.18**
This figure plots galaxies according to luminosity on the vertical axis and color on the horizontal axis. All galaxies begin as star-forming systems, putting them in the blue cloud at the left. The black arrows show how later mergers can produce larger galaxies, some of which cease forming new stars and so become part of the red sequence on the right. (The brightness of each square on the graph reflects the number of galaxies with the corresponding color and luminosity, based on data from the Sloan Digital Sky Survey.)

region of the graph. However, some of them later transition to the *red sequence*. The upper portion of the red sequence forms through mergers that produce large, red elliptical galaxies. The lower portion of the red sequence consists of smaller galaxies that have used up their cool gas, so that star formation has ceased, leaving them with only cool, red stars. Astronomers are conducting and planning further observations to test these ideas and to help complete our understanding of galaxy evolution.

THE PROCESS OF SCIENCE IN ACTION

11.3 Seeking Supermassive Black Holes

Galaxy collisions are spectacular, but even more awe-inspiring events occur deep in the centers of galaxies. In a small percentage of galaxies, we observe extreme amounts of radiation emanating from a tiny central region, occasionally accompanied by powerful jets of material shooting outward at nearly the speed of light (Figure 11.19). What could possibly release so much energy from within such a small region of space? In this section, we'll investigate the evidence that has led scientists to conclude that the culprits are **supermassive black holes**, much like the one at the Milky Way's center but in many cases far more massive.

Galaxy Evolution (Learning Outcomes 3, 4)

◆ **What is the evidence for supermassive black holes at the centers of galaxies?**

Galaxies with unusually bright centers are called *active galaxies,* and the bright central regions themselves are known as **active galactic nuclei**. The most luminous examples are known as **quasars** and can produce luminosities 1000 times that of the Milky Way (Figure 11.20). Activity in the center of a galaxy seems to be linked to galaxy evolution because quasars were much more common early in time than they are now. For some reason, the engines responsible for supplying the awesome luminosities of quasars have become dormant as galaxies have aged.

Sizes of Active Galactic Nuclei Long before astronomers suspected the existence of supermassive black holes, they had noticed that about 1% of galaxies (sometimes called *Seyfert galaxies*) have extremely bright centers. Our best visible-light images show these active galactic nuclei to be smaller than 100 light-years across, and higher-resolution radio observations limit their size to no more than about 3 light-years across. Rapid changes in luminosity show them to be even smaller than that.

To understand how variations in luminosity give us clues about an object's size, imagine that you were a master of the universe and you wanted to signal one of your fellow masters a billion light-years away. An active galactic nucleus would make an excellent signal beacon because it is so bright. However, suppose the smallest active galactic nucleus you could find was 1 light-year across. Each time you flashed it on, the photons from the front end would reach your fellow master a full year before the photons from the back end. If you flashed it on and off more than once a year, your signal would be smeared out. Similarly, if you used a source that was 1 light-day across, you could transmit signals that could flash on and off no more than once a day. Using this logic and the fact that the luminosities of real active galactic nuclei sometimes double in a matter of hours, we conclude that their light sources must be less than a few light-hours across, not much bigger than our solar system.

Explaining how such small objects can produce so much light became even more difficult after quasars were first discovered in the early 1960s. Standard

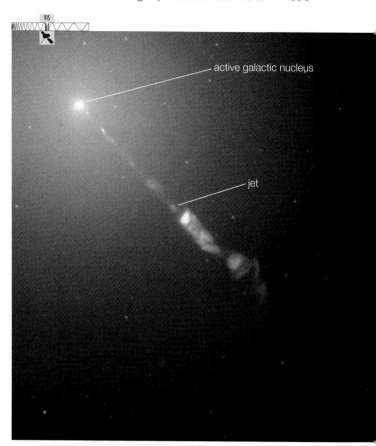

▲ **FIGURE 11.19**
The active galactic nucleus in the elliptical galaxy M87. The bright yellow spot is the active nucleus, and the blue streak is a jet of particles shooting outward from the nucleus at nearly the speed of light.

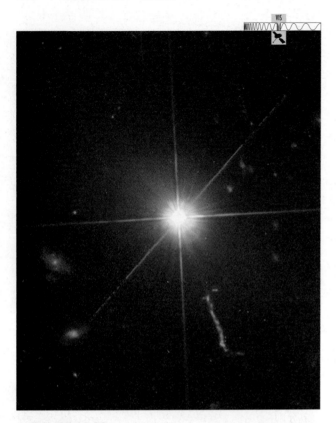

▲ FIGURE 11.20
This Hubble Space Telescope photo shows a quasar named 3C 273, the first quasar to be discovered. Its luminosity is more than 1 trillion times that of the Sun. (The photo shows a region of space 275,000 light-years across at the distance of this quasar.)

techniques of distance measurement [Section 12.1] indicated that quasars are generally billions of light-years from Earth, and with these distances the inverse square law for light [Section 8.2] indicates that the luminosities of many quasars are trillions of times the Sun's luminosity. These results were so astonishing that some scientists feared the distances of quasars had somehow been grossly overestimated. For a while there was great debate over the basic nature of quasars, but observations with more powerful telescopes eventually verified their distances and proved that quasars are the bright centers of very distant galaxies.

The Black Hole Hypothesis The only known way to account for the extreme luminosities of active galactic nuclei is with a model in which the central object is a supermassive black hole surrounded by a swirling accretion disk of very hot gas. This idea is much like the idea we used earlier to explain the emission from x-ray binary star systems [Section 10.1]. The gravitational potential energy of matter falling toward a black hole is converted into kinetic energy, and collisions between infalling particles convert kinetic energy into thermal energy. The resulting heat causes this matter to emit the intense radiation we observe. As in x-ray binaries, we expect that the infalling matter swirls through an accretion disk before it disappears within the event horizon of the black hole.

Matter falling into a black hole can generate awesome amounts of energy. As a chunk of matter falls to the event horizon of a black hole, as much as 10–40% of its mass-energy ($E = mc^2$) can be converted into thermal energy and ultimately into radiation. Accretion by black holes can therefore produce light far more efficiently than nuclear fusion, which converts less than 1% of mass-energy into photons. This mechanism is so efficient that it can account for the enormous luminosity of a quasar if we assume that the black hole swallows the equivalent of about one star per year.

The supermassive black hole model also accounts for the small sizes implied by quasar variability. A black hole accreting a solar mass of matter per year could eventually attain a mass of a billion solar masses if accretion continued at that rate for a billion years. But the event horizon of such a massive black hole, with a radius of about 3 light-hours, would still fit within the orbit of Neptune.

Mounting Evidence Demonstrating that supermassive black holes are indeed the energy sources for active galactic nuclei has been difficult, largely because black holes themselves do not emit any light. We therefore need to infer their existence from the ways in which they alter their surroundings. In the vicinity of a black hole, matter should be orbiting at high speed around something invisible. We have already seen such evidence for a supermassive black hole at the center of our own Milky Way [Section 10.3], but what about the centers of other galaxies?

Detailed observations of matter orbiting at the centers of nearby galaxies, performed during the last two decades, are showing that supermassive black holes are actually quite common. In fact, it is possible that *all* galaxies contain supermassive black holes at their centers. One prominent example is the giant elliptical galaxy M87 (see Figure 11.19). Images and spectra of M87's central regions reveal that gas some 60 light-years from the center is orbiting an unseen object at a speed of hundreds of kilometers per second (Figure 11.21). This high-speed orbital motion indicates that the central object has a mass at least 3 billion times that of our Sun. A supermassive black hole is the only thing we know of that could be so massive yet so small.

Even galaxies whose nuclei are not currently active show evidence of supermassive black holes, and their mass measurements follow a very interesting pattern: The mass of the black hole at the center of a galaxy appears to be closely related to the properties of the surrounding galaxy. More specifically, the mass of a galaxy's central black hole is typically about $\frac{1}{500}$ of the mass of the galaxy's bulge (Figure 11.22). This relationship appears to hold for galaxies with a wide range of sizes, from small

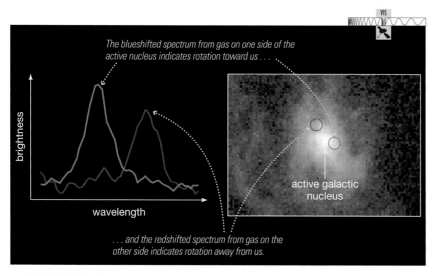

▲ **FIGURE 11.21**
This Hubble Space Telescope photo shows gas near the center of the galaxy M87, and the graph shows Doppler shifts of spectra from gas 60 light-years from the center on opposite sides (the circled regions in the photo). These Doppler shifts tell us that gas is orbiting around the galactic center, and precise measurements tell us that its orbital speed is about 800 km/s. Using the orbital speed and Newton's laws, we calculate that the central object must have a mass at least 3 billion times that of the Sun.

spiral galaxies to giant elliptical galaxies. We conclude that the growth of a central black hole must be closely linked with the process of galaxy formation.

The link between supermassive black holes and galaxy formation is clear but not yet fully understood. Some scientists have suggested that the black holes form very early in a galaxy's history and that their energy output then regulates the growth of the galaxy around them through sporadic outbursts of energy that disrupt the supply of star-forming gas. Recent observations are showing that this feedback process may be necessary to prevent the largest galaxies from forming even more stars than they already have. If this interpretation of the observations is correct, large central black holes are *required* in large galaxies, in order to explain why current rates of star formation within them are relatively low.

The Bottom Line Taken as a whole, the evidence to date strongly favors the supermassive black hole hypothesis for powering quasars and other active galactic nuclei. Moreover, we now suspect that many or even all large galaxies may harbor supermassive black holes at their centers, and that these black holes may play a major role in galaxy evolution. Scientists look forward to additional observations that will help us understand both the mechanisms through which supermassive black holes lead to the enormous power output of quasars and the linkages between supermassive black holes and galaxy evolution.

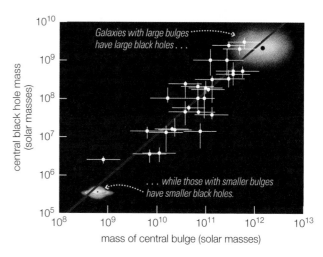

▲ **FIGURE 11.22**
The relationship between the mass of a galaxy's bulge and the mass of its supermassive black hole.

summary of key concepts

11.1 Our Galaxy: The Milky Way

◆ **What does our galaxy look like?**

The Milky Way Galaxy consists of a thin **disk** about 100,000 light-years in diameter, with a central **bulge** and a spherical region called the **halo** that surrounds the entire disk. The disk contains most of the gas and dust of the **interstellar medium**, in which the **star–gas–star cycle** continually generates new stars and increases the heavy-element content of the galaxy. Active star-forming regions, marked by the presence of hot, massive stars and **ionization nebulae**, are found primarily in **spiral arms**.

◆ **How did the Milky Way form?**

Our galaxy probably began as a huge blob of gas called a **protogalactic cloud**. Gravity caused the cloud to shrink

in size, and conservation of angular momentum caused the gas to form the spinning disk of the galaxy. Stars in the halo formed before the gas finished collapsing into the disk.

11.2 Galaxies Beyond the Milky Way

◆ **What are the major types of galaxies?**

Spiral galaxies have prominent disks and spiral arms, and they tend to collect in groups of up to several dozen galaxies. **Elliptical galaxies** are rounder and redder than spiral galaxies and contain less cool gas and dust. They are more common in large clusters of galaxies than elsewhere in the universe. **Irregular galaxies** are neither disklike nor rounded in appearance.

◆ **Why do galaxies differ?**

Differences between present-day galaxies probably arise both from conditions in their protogalactic clouds and from collisions with other galaxies. Slowly rotating or high-density protogalactic clouds may form elliptical rather than spiral galaxies. Ellipticals may also form through the collision and merger of two spiral galaxies.

THE PROCESS OF SCIENCE IN ACTION

11.3 Seeking Supermassive Black Holes

◆ **What is the evidence for supermassive black holes at the centers of galaxies?**

Some galaxies have bright centers known as **active galactic nuclei**. **Quasars** are the most luminous active galactic nuclei, with light outputs than can reach 1000 times the luminosity of the Milky Way, all coming from a region the size of our solar system. The most likely source for this energy is accretion of gas onto a **supermassive black hole**. Observations of orbiting gas and stars at the centers of nearby galaxies indicate that every large galaxy contains a supermassive black hole with a mass closely related to the total mass of the stars in the galaxy's bulge.

Investigations

For instructor-assigned homework and other learning materials, go to www.MasteringAstronomy.com.

Quick Quiz

Choose the best answer to each of the following. Explain your reasoning.

1. What do we mean by the *halo* of a galaxy? (a) It is the central region that is crowded with stars. (b) It is a large, roughly spherical region surrounding the disk. (c) It is the pancake-shaped region in which spiral arms are found.
2. Where are most of the Milky Way's globular clusters found? (a) in the disk (b) in the bulge (c) in the halo
3. What is the typical hydrogen content (by mass) of stars that are forming right now in the vicinity of the Sun? (a) 100% hydrogen (b) 75% hydrogen (c) 50% hydrogen
4. Compared to gas in the rest of the galaxy, the gas in star-forming regions is typically (a) colder and denser. (b) hotter and denser. (c) hotter and of lower density.
5. Where would you *most* expect to find an ionization nebula? (a) in the halo (b) in the bulge (c) in a spiral arm
6. What do we think resides in the center of our galaxy (that is, in Sgr A*)? (a) more than 1 million closely packed stars (b) a 4-million-solar-mass black hole (c) a 100-billion-solar-mass black hole
7. Which of these galaxies would you be most likely to find at the center of a large cluster of galaxies? (a) a large spiral galaxy (b) a large elliptical galaxy (c) a small irregular galaxy
8. About how many galaxies are there in a typical cluster of galaxies? (a) about 10 (b) a few dozen (c) a few hundred
9. A collision between two large spiral galaxies is likely to produce (a) a large elliptical galaxy. (b) a large spiral galaxy. (c) one large spiral galaxy and one large elliptical galaxy.
10. The luminosity of a quasar is generated in a region the size of (a) the Milky Way. (b) a star cluster. (c) the solar system.
11. The primary source of a quasar's energy is (a) chemical energy. (b) nuclear energy. (c) gravitational potential energy.
12. Observations indicate that galaxies with more massive central black holes tend also to have (a) a greater mass of stars in their central bulges. (b) a greater overall luminosity. (c) a more elliptical shape.

Short-Answer/Essay Questions

Explain all answers clearly, with complete sentences and proper essay structure if needed. An asterisk (*) designates a quantitative problem, for which you should show all your work.

13. *Research: Discovering the Milky Way.* Humans have been looking at the Milky Way since long before recorded history, but only in the past century did we verify the true shape of the galaxy and our location within it. Learn more about how conceptions of the Milky Way developed through history. What names did different cultures give to the band of light they saw? What stories did they tell about it? How have ideas about the galaxy changed in the past few centuries? Write a two- to three-page summary of your findings.
14. *Future of the Milky Way.* Describe in one or two paragraphs how the Milky Way would look from the outside if you could watch it for the next 100 billion years. How would its appearance change during this time?
15. *High-Velocity Star.* The average speed of stars in the solar neighborhood relative to the Sun is about 20 km/s (i.e., the speed at which we see stars moving toward or away from the Sun—*not* their orbital speed around the galaxy). Suppose you discover a star in the solar neighborhood that is moving relative to the Sun at a much higher speed—say, 200 km/s. What kind of orbit does this star probably have around the Milky Way? In what part of the galaxy does it spend most of its time? Explain.
16. *A Nonspinning Galaxy.* How would the development of the Milky Way Galaxy have been different if its original protogalactic cloud had had no angular momentum? Describe how you think our galaxy would look today and explain your reasoning.
17. *Unenriched Stars.* Suppose you discovered a star made purely of hydrogen and helium. How old do you think it would be? Explain.
18. *Enrichment of Star Clusters.* The gravitational pull of an isolated globular cluster is rather weak—a single supernova can blow all the interstellar gas out of a globular cluster. How might this fact be related to observations indicating that stars ceased to form in globular clusters long ago? How might it be related to the fact that globular clusters are deficient in elements heavier than hydrogen and helium? Summarize your answers in one or two paragraphs.
19. *Supernovae in Other Galaxies.* In which type of galaxy would you be most likely to observe a massive star supernova: a giant elliptical galaxy or a large spiral galaxy? Explain.
20. *Hubble's Galaxy Types.* How would you classify the following galaxies using the system illustrated in Figure 11.15? Justify your answers.
    a. Galaxy M101 (Figure 11.11)
    b. Galaxy NGC 1300 (Figure 11.12)
    c. Galaxy M87 (Figure 11.13)
21. *Black Holes in Other Galaxies.* In which type of galaxy would you expect to find the most massive black hole: a dwarf irregular galaxy, a large spiral galaxy with a small bulge, or a large elliptical galaxy? Explain.
22. *Weighing Supermassive Black Holes.* Which of these devices would be a better tool for measuring the masses of black holes at the centers of other galaxies: a camera that takes extremely sharp pictures or a spectrograph that produces precise measurements of emission-line wavelengths? Explain.
*23. *Counting Galaxies.* Estimate how many galaxies are shown in the photo on page 1. Explain the method you used to arrive at your estimate. This picture shows about 1/50,000,000 of the sky, so multiply your estimate by 50,000,000 to obtain an estimate of how many galaxies fill the entire sky.

12 Galaxy Distances and Hubble's Law

LEARNING GOALS

12.1 Measuring Cosmic Distances
- How do we measure the distances to galaxies?
- What is Hubble's law?

12.2 The Implications of Hubble's Law
- In what sense is the universe expanding?
- How do distance measurements tell us the age of the universe?

THE PROCESS OF SCIENCE IN ACTION

12.3 Observing Galaxy Evolution
- What do we see when we look back through time?

The image to the right shows a galaxy containing a white dwarf supernova (indicated by the arrow) that was observed in 2011. Measurements indicate that this galaxy is 21 million light-years away, meaning that this supernova exploded 21 million years ago. As we'll discuss in this chapter, this type of supernova provides an important link in the chain of measurements that astronomers use to establish the distances of galaxies throughout the observable universe. We'll also discuss how such measurements led to the revelation that we live in an expanding universe, which in turn has helped us learn answers to some of the most fundamental questions we can ask: How old is the universe? How big is it? How is it changing with time?

12.1 Measuring Cosmic Distances

Much of what we know about the universe as a whole depends on our ability to measure the distances to galaxies. These measurements are quite challenging, and astronomers are continually working to refine and improve distance measurement techniques. In this section, we'll discuss the key techniques used in modern astronomy and then look at how Edwin Hubble used similar techniques to discover the expansion of the universe.

▶ **Galaxies and the Foundation of Modern Cosmology (Learning Outcome 2)**

◆ **How do we measure the distances to galaxies?**

Our measurements of astronomical distances depend on a chain of methods in which each step allows us to measure greater distances in the universe. Let's follow this chain, which builds link by link from our solar system to the outermost reaches of the observable universe.

Radar Ranging   The first link in our astronomical distance chain is the Sun–Earth distance, or *astronomical unit* (AU) [Section 1.1]. The precise value of the AU has been determined through **radar ranging**, in which radio waves are transmitted from Earth and bounced off Venus. Because radio waves travel at the speed of light, the round-trip travel time for the radar signals tells us Venus's distance from Earth. We can then use Kepler's laws and a little geometry to calculate the length of an AU.

Parallax   The next link in the distance chain depends on *stellar parallax* [Section 2.3]. Recall that we observe stellar parallax when looking at a relatively nearby star against the more distant background from different points in Earth's orbit (Figure 12.1); it occurs for the same reason your finger seems to shift back and forth when you hold it at arm's length and alternately close one eye and then the other. The maximum shift occurs when we observe the star from opposite sides of Earth's orbit, through observations 6 months apart.

We can calculate a star's distance if we know the precise amount of the star's annual shift due to parallax. This means measuring the angle p in Figure 12.1, which we call the star's *parallax angle*. Notice that this angle would be smaller if the star were farther away. That is, more distant stars have *smaller* parallax angles. Real stellar parallax angles are very small: Even the nearest stars have parallax angles smaller than 1 arcsecond, well below the approximately 1 arcminute angular resolution of the naked eye. By definition, the distance to an object with a parallax angle of 1 arcsecond is 1 parsec (pc), which is equivalent to 3.26 light-years. You'll often hear astronomers state distances in parsecs, kiloparsecs (1000 parsecs), or megaparsecs (1 million parsecs), but in this book we'll stick to giving distances in light-years.

Parallax is our most accurate technique for measuring stellar distances, but parallax angles are measurable only within our own Milky Way Galaxy. As of 2018, astronomers using the European *GAIA* spacecraft had made parallax measurements for more than 1 billion stars, out to distances of tens of thousands of light-years.

Standard Candles   The next links in the distance-measurement chain rely on light sources that can serve as **standard candles**, meaning objects whose luminosity we already know. Such objects are useful because we can use their known luminosity and apparent brightness to calculate their distance with the *inverse square law for light* [Section 8.2].

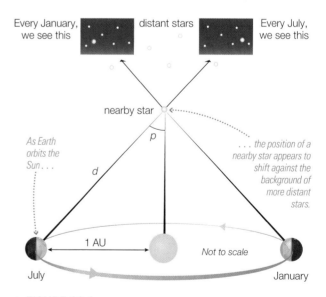

▲ **FIGURE 12.1**
Parallax makes the apparent position of a nearby star shift back and forth with respect to distant stars over the course of each year. The angle p, called the *parallax angle*, represents half the total parallax shift each year. If we measure p in arcseconds, the distance d to the star is $1/p$ in units of parsecs and $3.26/p$ in units of light-years. The angle in this figure is greatly exaggerated: All stars have parallax angles of less than 1 arcsecond.

▲ FIGURE 12.2
These streetlights are standard candles because they all put out the same amount of light, allowing us to determine their relative distances from their relative brightnesses.

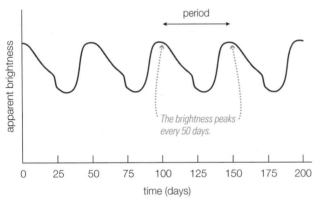

▲ FIGURE 12.3
This graph shows how the brightness of a Cepheid varies with time. The period is the time from one peak of brightness to the next.

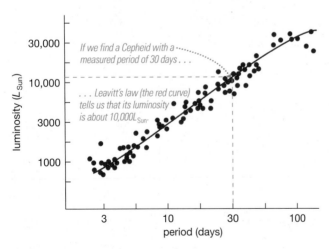

▲ FIGURE 12.4
The data (red dots) on this graph show that the period of a Cepheid variable star is very closely related to its luminosity, leading to the relationship (red curve) known as *Leavitt's law*. Measuring a Cepheid's period therefore allows us to determine its luminosity and calculate its distance. (Cepheids actually come in two types; the relation shown here is for Cepheids with heavy-element content similar to that of our Sun, or "Type I Cepheids.")

Figure 12.2 shows how standard candles work. All of the lights in the illustration are streetlamps of identical luminosity, but the lamps at greater distance appear dimmer (by the square of their distance). We can therefore determine each lamp's distance from its known luminosity. For example, suppose we know that the lamps all have a luminosity of 1000 watts of light. Then if one lamp appears only $\frac{1}{4}$ as bright as another, we know that it must be twice as far away as the other.

Unlike light bulbs, astronomical objects do not come marked with wattage. However, an astronomical object can still serve as a standard candle if we have some other way of knowing its true luminosity. Many astronomical objects meet this requirement. For example, any star that is a twin of our Sun—that is, a main-sequence star with spectral type G2—should have about the same luminosity as the Sun. Therefore, if we measure the apparent brightness of a Sun-like star, we can assume it has about the same luminosity as the Sun and can apply the inverse square law for light to estimate its distance (see Tools of Science, page 203).

Beyond the distances to which we can measure parallax, we use standard candles for most cosmic distance measurements. These measurements always have some uncertainty, because no astronomical object is a perfect standard candle. The challenge of measuring astronomical distances comes down to the challenge of finding the objects that make the best standard candles. The more confident we are about an object's true luminosity, the more certain we are of its distance.

Cepheid Variables The key to measuring distances beyond the Milky Way is to identify objects that can serve as standard candles and that are also bright enough to be detected at great distances. The most useful such objects for relatively nearby galaxies are a special class of extremely luminous stars known as **Cepheid variable stars**, or **Cepheids** for short. These stars vary in luminosity (and hence also in apparent brightness), alternately becoming dimmer and brighter with periods ranging from a few days to a few months. Figure 12.3 shows the brightness variations of a Cepheid with a period of about 50 days.

Cepheids are valuable as standard candles because a Cepheid's period is closely related to its luminosity: The longer the period, the more luminous the star (Figure 12.4). This relation is often called the *Cepheid period-luminosity relation*, but we will call it **Leavitt's law**[*] because it was first discovered by Henrietta Leavitt in 1912 through careful observations of Cepheids in the Large Magellanic Cloud. The distance to this small galaxy was not yet known at the time, so Leavitt was unable to determine the true luminosities of the Cepheids. However, because all the stars in the Large Magellanic Cloud are at approximately the same distance from us, she was able to discover her law by comparing the apparent brightness of these Cepheids. Today, we know the relationship between Cepheid periods and distances embodied in Leavitt's law quite accurately, because we have measured the distances to nearby Cepheids with parallax and used these distances and the Cepheids' apparent brightnesses to calculate true luminosities.

As an example of how we use Cepheids to measure distance, look at the point marked with dashed lines in Figure 12.4. Notice that this star's period of 30 days corresponds to a luminosity of about 10,000 Suns. In other words,

---

[*]The American Astronomical Society has recommended this change in terminology to honor the critical role that Henrietta Leavitt's work played in Hubble's subsequent discoveries.

the fact that this Cepheid's brightness peaks every 30 days means it is effectively screaming out, "Hey, everybody, my luminosity is 10,000 times that of the Sun!" More generally, once we measure a Cepheid's period, we can use Leavitt's law to learn its luminosity (within about 10%) and can then use the inverse square law for light to determine its distance. Today, observations of Cepheids can be used to measure distances to galaxies up to about 100 million light-years away.

Distant Standard Candles Astronomers have identified several types of objects that can be used as standard candles for measuring distances beyond those for which we can observe Cepheids. The most valuable of these are white dwarf supernovae.

Recall that white dwarf supernovae are thought to be exploding white dwarf stars that have reached the 1.4-solar-mass limit [Section 10.1]. These supernovae all have nearly the same luminosity, probably because they all come from stars of similar mass that explode in nearly the same way. Although white dwarf supernovae are rare in any individual galaxy, numerous examples have been observed within about 50 million light-years of the Milky Way, including the one in the photo that opens this chapter. We can determine the true luminosities of these supernovae by using Cepheids to measure the distances to the galaxies in which they occurred. These measurements not only tell us the true luminosities of white dwarf supernovae but also confirm that their luminosities are all nearly the same and therefore that we can use them as standard candles.

Because white dwarf supernovae are so bright—about 10 billion solar luminosities at their peak—we can detect them even when they occur in galaxies billions of light-years away (Figure 12.5). Although the number of galaxies whose distances we can measure with this technique is relatively small (because white dwarf supernovae occur only once every few hundred years in a typical galaxy), these galaxies have allowed us to calibrate yet another technique—one that relies on the expansion of the universe.

▲ **FIGURE 12.5**
White dwarf supernovae observed by the Hubble Space Telescope. White arrows in the lower images indicate the supernovae, and the upper images show what these galaxies looked like without supernovae. The first two (from left) are supernovae that exploded when the universe was approximately half its current age; the one to the right exploded about 9 billion years ago.

tools of science — Measuring Distances with Standard Candles

The standard-candle technique for distance measurement is astronomy's most important tool for measuring large distances in the universe. It is based on the inverse square law for light, which relates an object's apparent brightness to its distance and luminosity [Section 8.2]. Rearranging this law with a little algebra gives us a formula for calculating an object's distance from its apparent brightness and luminosity:

$$\text{distance} = \sqrt{\frac{\text{luminosity}}{4\pi \times (\text{apparent brightness})}}$$

If we can measure the apparent brightness of an object of known luminosity, we can use this formula to calculate the object's distance. However, this technique works only for objects that are good *standard candles*, meaning that we know their luminosities without knowing their distances.

Many different kinds of astronomical objects can be used as standard candles, but if we want to measure large distances, we need to use very luminous objects that can be detected from far away. Cepheid variable stars are particularly useful standard candles because they can have luminosities as great as $30,000 L_{\text{Sun}}$, making them clearly visible from distances as large as 100 million light-years.

EXAMPLE: You observe a Cepheid variable star that varies in brightness with a period of 30 days and has an average apparent brightness of 1.0×10^{-18} watt/m². According to Leavitt's law (see Figure 12.4), a Cepheid with a 30-day period has a luminosity of about $10,000 L_{\text{Sun}}$. How far away is this star? (*Note:* The Sun's luminosity, $1 L_{\text{Sun}}$, is 3.8×10^{26} watts.)

Solution: We are given both the star's apparent brightness and its luminosity, so we have all the information we need to calculate its distance using the modified inverse square law for light. However, before we can use this law, we must convert the star's luminosity to units of watts. Multiplying the Sun's luminosity in watts by 10,000 (or 10^4) gives a luminosity of 3.8×10^{30} watts for the Cepheid. Using this luminosity and the apparent brightness in the formula then gives

$$\text{distance} = \sqrt{\frac{3.8 \times 10^{30} \text{ watts}}{4\pi \times \left(1.0 \times 10^{-18} \frac{\text{watt}}{\text{m}^2}\right)}} \approx 5.5 \times 10^{23} \text{ m}$$

The distance to the star is about 5.5×10^{23} meters, or 58 million light-years (dividing by 9.5×10^{15} meters/light-year).

▲ FIGURE 12.6
Edwin Hubble at the Mount Wilson Observatory.

▲ FIGURE 12.7
Redshifted galaxy spectrum.

◆ What is Hubble's law?

The ability to measure distances to galaxies is the key to much of our modern understanding of the size and age of the universe. We can trace the beginning of this understanding back to discoveries made by Edwin Hubble, including the law that bears his name.

Hubble and the Andromeda Galaxy

Before the 1920s, no one knew for certain whether the spiral-shaped objects they saw in the sky were merely clouds of gas within the Milky Way—and therefore that the Milky Way represented the entire universe—or distant and distinct galaxies. The opinions of astronomers were split on this issue, which became a subject of great debate.

Edwin Hubble put the debate to rest in 1924. Using the new 100-inch telescope atop California's Mount Wilson (Figure 12.6)—the largest telescope in the world at that time—he identified Cepheid variables in the Andromeda Galaxy by comparing photographs of the galaxy taken on different nights. These comparisons allowed him to measure the Cepheids' periods, from which he could estimate their luminosities with Leavitt's law and then calculate the galaxy's distance with the inverse square law for light. He thereby proved that the Andromeda Galaxy resides far beyond the outer reaches of the Milky Way.

This single scientific discovery dramatically changed our perspective. Rather than wondering if the Milky Way might be the entire universe, we suddenly knew that it is just one among many galaxies in an enormous universe. The stage was set for Hubble to make an even greater discovery.

Distance and Redshift

Astronomers were already aware that the spectra of most spiral galaxies tend to be *redshifted* (Figure 12.7), indicating that they are moving away from us (see Tools of Science, page 116). However, no one understood the true significance of these motions, because Hubble had not yet proved that the spiral galaxies were separate from the Milky Way.

In the years following his discovery of Cepheids in Andromeda, Hubble sought to estimate more galaxy distances and measure the redshifts of those galaxies. Because even Cepheids were too dim to be seen in most of these galaxies, Hubble needed brighter standard candles for his distance estimates. One of his favorite techniques was to use the brightest object he could see in each galaxy as a standard candle, because he assumed these objects to be very bright stars that would always have about the same luminosity. In fact, we later learned that Hubble's bright objects were star *clusters*, which are not very good standard candles, but they were good enough that he still got the correct general result.

In 1929, Hubble announced his conclusion: The more distant a galaxy, the greater its redshift and hence the faster it is moving away from us. As we will discuss in the next section, this relationship implies that galaxies all across the universe are moving away from one another. Hubble had discovered that the universe is expanding.

Hubble's Law

We express the idea that more distant galaxies move away from us faster with a simple formula now known as **Hubble's law**:

$$v = H_0 \times d$$

where v stands for a galaxy's velocity away from us (sometimes called a *recession velocity*), d stands for its distance, and H_0 (pronounced "H-naught") is a number called **Hubble's constant**. We usually write Hubble's law in this form to express the idea that speeds of galaxies depend on their distances. However, astronomers more often use the law in reverse—measuring a galaxy's speed from its redshift and then using Hubble's law to estimate its distance.

Today, Hubble's law is our most useful technique for determining distances to galaxies that are very far away. Nevertheless, it's important to keep two practical limitations of the law in mind:

1. Galaxies do not obey Hubble's law perfectly. Hubble's law would give an exact distance only for a galaxy with a speed determined solely by the expansion of the universe. In reality, nearly all galaxies experience gravitational tugs from other galaxies, and these tugs alter their speeds from the values predicted by Hubble's law.
2. Even when galaxies obey Hubble's law well, the distances we find with it are only as accurate as our best measurement of Hubble's constant.

The first problem is most serious for nearby galaxies. Within the Local Group, for example, Hubble's law does not work at all: The galaxies in the Local Group are gravitationally bound together with the Milky Way and therefore are *not* moving away from us in accord with Hubble's law. However, Hubble's law works well for more distant galaxies, which have recession speeds so great that any motions caused by the gravitational tugs of neighboring galaxies are small in comparison.

The second problem means that, even for distant galaxies, we can know only *relative* distances until we pin down the true value of H_0. For example, Hubble's law tells us that a galaxy moving away from us at 20,000 km/s is twice as far away as one moving at 10,000 km/s, but we can determine the actual distances of the two galaxies only if we know H_0.

One of the main missions of the Hubble Space Telescope was to measure an accurate value of H_0. Astronomers used the telescope to measure distances to Cepheids in galaxies out to about 100 million light-years, and this in turn allowed them to determine the luminosities of distant standard candles such as white dwarf supernovae. Plotting galactic distances measured with those distant standard candles against the velocities indicated by their redshifts has pinned down the value of H_0 to somewhere between 21 and 23 *kilometers per second per million light-years* (Figure 12.8). That is, a galaxy's speed away from us is between 21 and 23 km/s for every million light-years it is from us. For example, with this range of values for Hubble's constant, Hubble's law predicts that a galaxy located 100 million light-years away would be moving away from us at a speed of between 2100 and 2300 km/s.

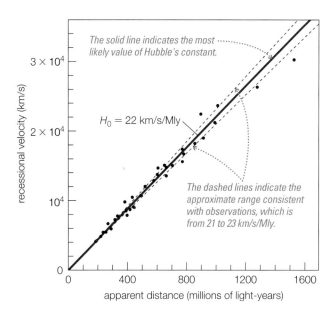

▲ **FIGURE 12.8**
White dwarf supernovae can be used as standard candles to establish Hubble's law out to very large distances. The points on this figure show the apparent distances of white dwarf supernovae and the recession velocities of the galaxies in which they exploded. The fact that these points all fall close to a straight line demonstrates that these supernovae are good standard candles.

Distance Chain Summary

Figure 12.9 summarizes the techniques we have discussed, showing how they make up the chain of measurements that allows us to determine distances across the universe:

- *Radar ranging.* We measure the Earth–Sun distance by bouncing radio waves off Venus and using some geometry.
- *Parallax.* We measure the distances to stars by observing how their positions appear to change as Earth orbits the Sun. These distances rely on our knowledge of the Earth–Sun distance, determined with radar ranging.
- *Cepheid variables.* Leavitt's law allows us to determine the luminosities of Cepheids from their measured periods, which in turn allows us to calculate their distances using the inverse square law for light. We learn the precise relationship between Cepheids' periods and luminosities from relatively nearby Cepheids whose distances have been measured using parallax.
- *Distant standards.* By measuring distances to relatively nearby galaxies with Cepheids, we learn the true luminosities of white dwarf supernovae. We can then use those supernovae as standard candles to measure great distances throughout the universe.
- *Hubble's law.* Distances measured to galaxies with white dwarf supernovae allow us to measure Hubble's constant, H_0. Once we know H_0, we can use Hubble's law to determine a galaxy's distance from its redshift.

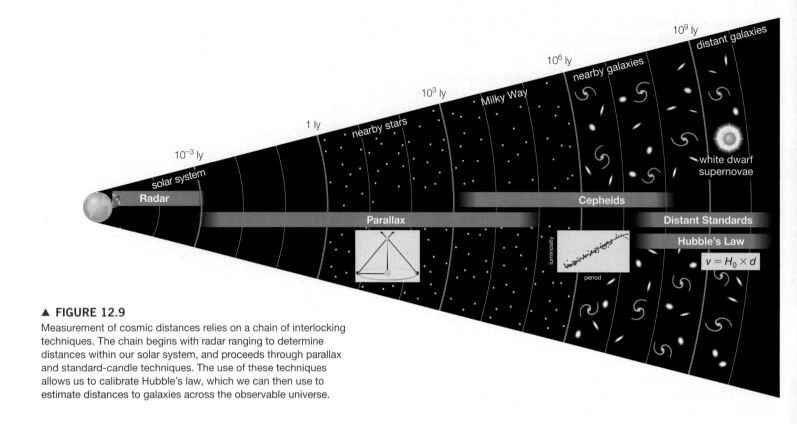

▲ **FIGURE 12.9**
Measurement of cosmic distances relies on a chain of interlocking techniques. The chain begins with radar ranging to determine distances within our solar system, and proceeds through parallax and standard-candle techniques. The use of these techniques allows us to calibrate Hubble's law, which we can then use to estimate distances to galaxies across the observable universe.

Note that the uncertainty increases with each link in the distance chain. As a result, although we know the Earth–Sun distance at the beginning of the chain extremely accurately, distances to the farthest reaches of the observable universe remain uncertain by about 5%.

12.2 The Implications of Hubble's Law

Hubble's law is a powerful tool for understanding the universe. Not only does it tell us that the universe is expanding and give us a way to measure galactic distances, but it also has enormous implications: Tracing the motions of galaxies back through time indicates that all the matter in the observable universe started very close together and that the entire universe we observe today came into being at a single moment, approximately 14 billion years ago. In this section, we'll explore how these implications arise from Hubble's discovery.

◆ In what sense is the universe expanding?

One way to visualize the expansion of the universe is to imagine making a raisin cake in which the distance between adjacent raisins is 1 centimeter. You place the cake in the oven, where it expands as it bakes. After 1 hour, you remove the cake, which has expanded so that the distance between adjacent raisins has increased to 3 centimeters (Figure 12.10). The expansion of the cake seems fairly obvious. But what would you see if you lived *in* the cake, as we live in the universe?

Pick any raisin, call it the Local Raisin, and identify it in the pictures of the cake both before and after baking. Figure 12.10 shows one possible choice for the Local Raisin, with three nearby raisins labeled. The accompanying table summarizes what you would see if you lived within the Local Raisin. Notice, for example, that Raisin 1 starts out at a distance of 1 centimeter before baking and ends up at a distance of 3 centimeters after baking, which means it moves a distance of 2 centimeters away from the Local Raisin during the hour of baking. Its speed as seen from the Local Raisin is therefore 2 centimeters per hour. Raisin 2 moves from a distance of 2 centimeters before baking to a distance of 6 centimeters after baking, which means it moves a distance of 4 centimeters away from the Local Raisin during the hour. Its speed is therefore 4 centimeters per hour, twice the speed of Raisin 1. The fact that the cake is expanding means that all raisins are moving away from the Local Raisin, with more distant raisins moving away faster.

Hubble's discovery that galaxies are moving in much the same way as the raisins in the cake, with most moving away from us and more distant ones moving away faster, implies that the universe in which we live is expanding much like the raisin cake. If you now imagine the Local Raisin as representing our Local Group of galaxies and the other raisins as representing more distant galaxies or clusters of galaxies, you have a basic picture of the expansion of the universe. Like the expanding batter between the raisins in the cake, *space itself* is growing between galaxies. More distant galaxies move away from us faster because they are carried along with this expansion like the raisins in the expanding cake.

However, there's an important difference between the raisin cake and our universe: A cake has a center and edges, but we do not think the same is true of the entire universe. To the best of our knowledge, the universe is not expanding *into* anything. As far as we can tell, there is no edge to the distribution of galaxies in the universe. On very large scales, the distribution of galaxies appears to be relatively smooth, meaning that the overall appearance of the universe around you would be more or less the same no matter where you were located. The idea that the matter in the universe is evenly distributed, without a center or an edge, is often called the **Cosmological Principle**. Although we cannot prove it to be true, it is consistent with all our observations of the universe.

So how can the universe be expanding if it's not expanding *into* anything? A better analogy than the raisin cake would be something that can expand but has no center and no edges. We cannot use a three-dimensional object for this purpose, since all such objects have a center and edges. However, the two-dimensional *surface* of a balloon fits the bill, as does an infinite surface such as a sheet of rubber that extends to infinity in all directions.

Because it's hard to visualize infinity, let's use the surface of a balloon as our analogy to the expanding universe (Figure 12.11). Note that this analogy uses the balloon's two-dimensional *surface* to represent all three dimensions of space. The surface of the balloon therefore represents the entire universe, and the spaces inside and outside the balloon have no meaning in this analogy. Aside from the reduced number of dimensions, the analogy works well because the balloon's spherical surface has no center and no edges, just as no city is the center of Earth's surface and no edges exist where you could walk or sail off Earth. Now, remember that although the universe as a whole is expanding, individual galaxies and galaxy clusters do not expand, because gravity holds them together. We can therefore represent galaxies (or clusters of galaxies) with plastic dots attached to the balloon, and we can make our model universe expand by inflating the balloon. The dots move apart as the surface of the balloon expands, but they themselves do not grow in size.

see it for yourself Make a *one*-dimensional model of the expanding universe with a rubber band and some paper clips. Cut the rubber band so that you can stretch it out along a line, then attach four paper clips along it. Pin down the ends of the rubber band and measure the distance from one paper clip to each of the others. Then unpin one of the ends, stretch the rubber band more, pin it down again, and remeasure the distances. How much have they changed? How do your measurements illustrate Hubble's law?

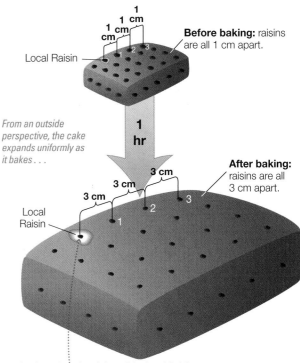

Distances and Speeds as Seen from the Local Raisin

Raisin Number	Distance Before Baking	Distance After Baking (1 hour later)	Speed
1	1 cm	3 cm	2 cm/hr
2	2 cm	6 cm	4 cm/hr
3	3 cm	9 cm	6 cm/hr

▲ **FIGURE 12.10**
An expanding raisin cake offers an analogy to the expanding universe. Someone living in one of the raisins inside the cake could figure out that the cake was expanding by noticing that all the other raisins were moving away, with more distant raisins moving away faster. In the same way, we know that we live in an expanding universe because all galaxies outside our Local Group are moving away from us, with more distant ones moving faster.

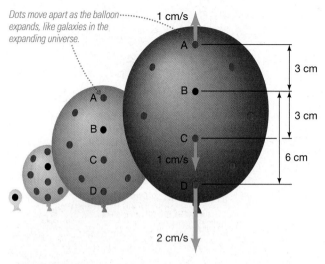

▲ **FIGURE 12.11**
The expansion of a balloon's surface shows how a finite universe can expand without having a center or edges. As the balloon expands, dots on it move apart in the same way that galaxies move apart in our expanding universe.

common misconceptions

What Is the Universe Expanding Into?

When you first learn about the expansion of the universe, it's natural to assume that the universe must be expanding "into" something. Moreover, the name "Big Bang" evokes a mental image of a huge explosion that sent matter flying outward into a vast space that was previously empty. However, the scientific view of the Big Bang is very different. According to modern science, the Big Bang filled *all of* space with matter and energy.

How can the Big Bang have started "all of space" expanding without its expanding into some empty void? A complete answer requires Einstein's general theory of relativity, which tells us that space can be "curved" in ways that we can measure but cannot visualize (see Tools of Science, page 172). But you can understand the key ideas with the balloon analogy. Remember that the balloon's *surface* represents all of space; just as the surface has no center and no edges, space has no center and no edges. In the analogy, the expanding universe is like the expanding surface of the balloon, and the Big Bang was the moment when an extremely tiny balloon first started growing in size. As the balloon expands, any two points on the surface get farther apart because the surface itself is stretching, not because they are moving into a previously empty region above the surface. In the same way, galaxies move apart in the expanding universe because the space between them is stretching, not because they move into a space that was previously empty.

Galaxies and the Foundation of Modern Cosmology (Learning Outcome 3)

◆ How do distance measurements tell us the age of the universe?

We are now ready to discuss how Hubble's law leads us to an age estimate for the universe. Imagine that some miniature scientists are living on dot B on the balloon in Figure 12.11. Suppose that, 3 seconds after the balloon begins to expand, they measure the following:

- Dot A is 3 centimeters away and moving at 1 cm/s.
- Dot C is 3 centimeters away and moving at 1 cm/s.
- Dot D is 6 centimeters away and moving at 2 cm/s.

They could summarize these observations as follows: *Every dot is moving away from us with a speed that is 1 cm/s for each 3 centimeters of distance.* Because the expansion of the balloon is uniform, scientists living on any other dot would come to the same conclusion. Each scientist living on the balloon would determine that the following formula relates the distances and velocities of other dots on the balloon:

$$v = \left(\frac{1}{3\,\text{s}}\right) \times d$$

where v and d are the velocity and distance of any dot, respectively.

think about it Confirm that this formula gives the correct values for the speeds of dots C and D, as seen from dot B, 3 seconds after the balloon begins expanding. How fast would a dot located 9 centimeters from dot B move, according to the scientists on dot B?

If the miniature scientists think of their balloon as a bubble, they might call the number relating distance to velocity—the term $\frac{1}{3\,\text{s}}$ in the preceding formula—the "bubble constant." An especially insightful miniature scientist might flip over the "bubble constant" and find that it is exactly equal to the time elapsed since the balloon started expanding. That is, the "bubble constant" $\frac{1}{3\,\text{s}}$ tells them that the balloon has been expanding for 3 seconds. Perhaps you see where we are heading.

Just as the inverse of the "bubble constant" tells the miniature scientists that their balloon has been expanding for 3 seconds, the inverse of the Hubble constant, or $1/H_0$, tells us something about how long our universe has been expanding. The "bubble constant" for the balloon depends on when it is measured, but it is always equal to 1 divided by the time since the balloon started expanding. Similarly, the Hubble constant changes with time, but it stays roughly equal to 1 divided by the age of the universe. We call it a *constant* because it is the same at all locations in the universe and because its value does not change noticeably on the time scale of human civilization.

A simple estimate based only on the value of Hubble's constant puts the age of the universe at a little less than 14 billion years. To derive a more precise value for the universe's age, we need to know whether the expansion has been speeding up or slowing down over time, a question we will examine more closely in Chapter 14. If the rate of expansion has slowed with time (because of gravity, for example), then the universe's age is somewhat less than $1/H_0$.

▲ **FIGURE 12.12**
Distances to faraway objects keep changing in an expanding universe. This figure shows how the distance to a galaxy changes during the time it takes for the light from a supernova explosion in that galaxy to reach us. Because the distance changes, the most useful way to describe a galaxy's "distance" is to state its *lookback time*—the time it took for the light to make its journey to Earth.

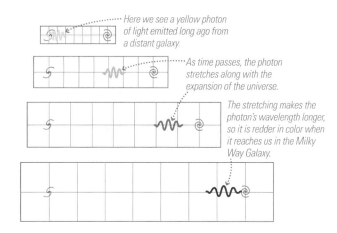

▲ **FIGURE 12.13**
As the universe expands, photon wavelengths stretch. This *cosmological redshift* shifts them toward the red side of the spectrum.

If the expansion rate has accelerated with time—and current evidence indicates a small acceleration—then the universe's age is somewhat more than $1/H_0$. The best available evidence suggests that the universe is very close to 14 billion years old.

Lookback Time and Cosmological Redshift The expansion of the universe leads to a complication when we try to specify the distance of a faraway galaxy, because its distance is always changing. Here's the problem: Suppose you are trying to determine the distance to a galaxy using observations of a supernova that actually occurred 1 billion years ago. The photons you observe from the supernova and the galaxy it resides in must have traveled a total distance of 1 billion light-years, because they travel at the speed of light and their journey took 1 billion years (Figure 12.12). However, this 1 billion light-year distance is *smaller* than the current distance to the supernova's galaxy and *larger* than its distance at the time the supernova occurred, because the galaxy's distance has been continually increasing during the time it took for the supernova's light to travel to us. So what should we say if someone asks us the "distance" to the galaxy?

Astronomers often find that the clearest way to describe the galaxy's distance is in terms of the time that its light has taken to reach us. We call this time the galaxy's **lookback time**, because it means we are seeing the galaxy as it was that many years ago. For example, when astronomers say that a galaxy is "1 billion light-years away," they usually mean that the galaxy has a lookback time of 1 billion light-years.

Redshifts of distant galaxies call for a similar shift in thinking. One way to interpret a galaxy's redshift is in terms of how fast the galaxy is moving away from us (the *v* in Hubble's law). However, our balloon analogy suggests an alternative interpretation: We can think of galaxies as remaining essentially stationary—like the dots glued to the balloon—while the space between them grows. From this perspective, the expansion of the universe causes photon wavelengths to become longer (redder) with time in much the same way that wavy lines on a rubber sheet stretch out as the sheet expands (Figure 12.13).

In essence, we have a choice when we interpret the redshift of a distant galaxy: We can think of the redshift either as being caused by the Doppler effect as the galaxy moves away from us or as being a **cosmological redshift** that arises

common misconceptions

Beyond the Horizon

Perhaps you're thinking there must be something beyond the cosmological horizon. After all, we can see farther and farther with each passing second. The new matter we see had to come from somewhere, didn't it? The problem with this reasoning is that the cosmological horizon, unlike a horizon on Earth, is a boundary in time, not in space.

At any one moment, in whatever direction we look, the cosmological horizon lies at the beginning of time and encompasses a certain volume of the universe. At the next moment, the cosmological horizon still lies at the beginning of time, but it encompasses a slightly larger volume. When we peer into the distant universe, we are looking back in both space and time. We cannot look "past the horizon" because we cannot look back to a time before the universe began.

from photon stretching in an expanding universe. For distant galaxies, astronomers find it more useful to adopt the latter perspective, in which space itself is considered to be expanding with time, carrying galaxies along for the ride and stretching the wavelengths of photons. A galaxy's redshift therefore tells us how much space has expanded during the time that the galaxy's light traveled to us, and its lookback time tells us the distance its light has traveled.

The Horizon of the Universe In our discussion of the expanding universe, we stressed that the universe as a whole does not seem to have an edge. However, as we discussed in Chapter 1, the age of the universe limits the size of our *observable* universe (see Figure 1.3): We cannot see any object beyond our **cosmological horizon**, which represents the distance at which lookback times equal the 14-billion-year age of the universe.

Note that the cosmological horizon is a boundary in time, not in space. The reason we cannot see past it is *not* because it's any kind of limit on the distances at which galaxies can exist, but rather because anything beyond it has a lookback time that is greater than the age of the universe. Light hasn't had time to reach us from those locations in time and space, so the observable universe ends at the cosmological horizon.

 THE PROCESS OF SCIENCE IN ACTION

12.3 Observing Galaxy Evolution

The relationship among distance, cosmological redshift, and lookback time is helpful to modern astronomers because it means that powerful telescopes can be used like time machines to observe the life histories of galaxies. We'll conclude this chapter with a brief overview of how the ability to look back through time enables astronomers to test models of galaxy evolution. Through this example, we'll see how technological progress is helping to extend the frontiers of our knowledge of the universe.

◆ What do we see when we look back through time?

As in most areas of astronomy, we study galaxy evolution through a combination of observations and theoretical modeling. Observations allow us to see key aspects of how galaxies evolve through time, while modeling is particularly important for learning about the earliest stages of galaxy formation, which we cannot observe directly.

▶ **Galaxy Evolution (Learning Outcome 1)**

Observing Galaxies at Different Ages Powerful telescopes can be used like time machines to observe the life histories of galaxies, because the farther we look into the universe, the further back we look in time [Section 1.1]. The most distant galaxies already had some stars in place by about 13 billion years ago, which means the ages of these stars match those of the oldest stars in our own galaxy. It therefore seems safe to assume that most galaxies began to form at about this time, in which case most galaxies would be roughly the same age today. A galaxy's lookback time is therefore linked directly to its age. A galaxy with a lookback time of 13 billion years must be newly born in a 14-billion-year-old universe, while most nearby galaxies share the 13-billion-year age of the Milky Way.

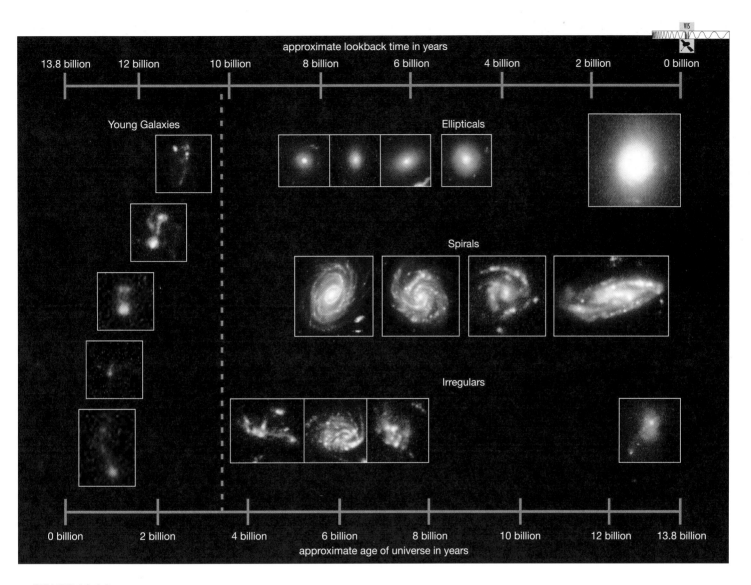

▲ **FIGURE 12.14**
Family albums for elliptical, spiral, and irregular galaxies of different ages, plus some very young galaxies shown on the far left. These photos are all zoomed-in images of galaxies from the Hubble eXtreme Deep Field, a portion of which is shown in the photo on page 1 of this book. We see more distant galaxies as they were when they were younger; the approximate age of the universe is indicated along the bottom horizontal axis, with lookback time shown at the top. (The figure assumes an age of 13.8 billion years for the universe, which is the best current estimate.)

This linkage between lookback time and age gives us a remarkable opportunity: Simply by photographing galaxies at different distances, we can assemble "family albums" showing galaxies in different stages of development. Figure 12.14 shows partial family albums for elliptical, spiral, and irregular galaxies. Each individual photograph shows a single galaxy at a single stage in its life. Grouping these photographs by galaxy type allows us to see how galaxies of a particular type have changed through time.

think about it We've used the term *today* in a very broad sense. For example, a relatively nearby galaxy may be located, say, 20 million light-years away, so we see it as it was 20 million years ago. In what sense is this "today"?

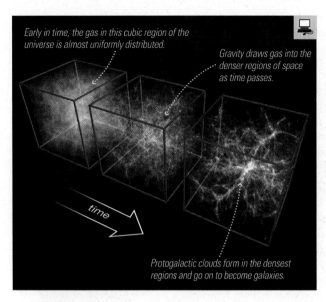

▲ FIGURE 12.15
A computer simulation of the formation of protogalactic clouds. The simulated region of space is about 500 million light-years wide and goes on to form numerous galaxies.

One of the most interesting things we learn when we arrange galaxy photos along a time line is that the youngest galaxies are distinctly different from mature galaxies like the Milky Way. For example, look at the "Young Galaxies" in Figure 12.14. They all look disrupted, suggesting that they are in the midst of colliding and merging like the galaxies in Figure 11.16. Disrupted galaxies like these apparently were extremely common before the universe was about 3 billion years old. In contrast, mature galaxies generally look more organized, like the spirals and ellipticals in Figure 12.14.

Models of Galaxy Evolution These findings support models of galaxy evolution in which galaxies build up gradually over time through collisions and mergers with other galaxies and protogalactic clouds. The most successful models of this kind assume the following:

- Hydrogen and helium gas filled all of space more or less uniformly when the universe was very young, in the first million years after its birth.
- The distribution of matter in the universe was not perfectly uniform—certain regions of the universe started out ever so slightly denser than others.

Beginning from these assumptions, for which we have strong observational evidence [Section 13.2], we can model galaxy formation by using well-established laws of physics to trace how the denser regions in the early universe grew into galaxies (Figure 12.15). The models show that the regions of enhanced density originally expanded along with the rest of the universe. However, the slightly greater pull of gravity in these regions gradually slowed their expansion. Within about a billion years, the expansion of these denser regions halted and reversed, and the material within them began to contract into *protogalactic clouds*, much like those that eventually formed our Milky Way [Section 11.1].

From this point onward, the models predict that galaxy evolution depended critically on how often protogalactic clouds and newly formed galaxies collided with their neighbors. Early in the universe's history, when all the galaxies were much closer together, these collisions should have been frequent. Later in time, after the average distances between galaxies became much greater, collisions became less common, allowing galaxies to settle into the spiral and elliptical shapes typical of the present-day universe.

Next Steps The predictions of galaxy evolution models agree well with what we see in Figure 12.14. Galaxies do indeed look more disrupted early in time and more settled later in the universe's history. However, even the most successful models of galaxy evolution do not yet match our observations in every detail. Further progress in this area will depend on both computer technology and a new generation of telescopes. Current efforts to model galaxy evolution require some of the world's most powerful supercomputers, and as computing power improves, our models should become increasingly accurate.

Our observations of galaxy evolution should also improve in the coming decade. Current technology allows us to see some galaxies at a time when the universe was only about 500 million years old, meaning that we cannot yet observe the earliest stages of galaxy formation. Scientists hope to see farther with the James Webb Space Telescope, scheduled for launch in 2021. This telescope will be particularly sensitive to infrared light, which means it will be ideally suited to observing the hugely redshifted light of the most distant galaxies. Astronomers therefore hope to see galaxies in the earliest stages of formation, allowing the process of science to lead us one step closer to the beginning of time.

summary of key concepts

12.1 Measuring Cosmic Distances

◆ **How do we measure the distances to galaxies?**

Our measurements of galaxy distances depend on a chain of methods. The chain begins with radar ranging in our own solar system and parallax measurements of distances to stars in our own galaxy; then it relies on **standard candles** to measure distances to other galaxies.

◆ **What is Hubble's law?**

Hubble's law tells us that more distant galaxies are moving away faster: $v = H_0 \times d$, where H_0 is **Hubble's constant**. It allows us to determine a galaxy's distance from the speed at which it is moving away from us, which we can measure from its Doppler shift.

12.2 The Implications of Hubble's Law

◆ **In what sense is the universe expanding?**

The average distances between galaxies are continually increasing, implying that space itself is expanding with time. Our expanding universe does not appear to have either a center or an edge. In that sense, its expansion is like the stretching of the surface of an inflating balloon: Objects attached to the surface gradually move farther apart, but the surface itself has no center or edges.

◆ **How do distance measurements tell us the age of the universe?**

Dividing a galaxy's current distance by its current velocity gives us an estimate of how long it took the galaxy to reach that distance. Such measurements allow us to determine **Hubble's constant**, which tells us that it took approximately 14 billion years for the universe's galaxies to reach their current distances, giving us an estimate for the age of the universe. Because distances between galaxies are always changing, it is best to express the distance to a faraway galaxy in terms of its **lookback time**—the time it has taken for the galaxy's light to reach us. The expansion of the universe during that time stretches the light coming from the galaxy, leading to a **cosmological redshift** directly related to the galaxy's lookback time.

THE PROCESS OF SCIENCE IN ACTION

12.3 Observing Galaxy Evolution

◆ **What do we see when we look back through time?**

Today's telescopes are powerful enough to detect light from galaxies with lookback times almost as large as the age of the universe. Assembling "family albums" of galaxies with different lookback times shows that galaxies were more likely to look disrupted when the universe was very young than at later times. This finding supports models in which galaxies build up gradually through mergers of smaller galaxies and protogalactic clouds. During the next decade, improved telescopes and supercomputers will help us observe and model even younger galaxies.

Investigations

For instructor-assigned homework and other learning materials, go to www.MasteringAstronomy.com.

Quick Quiz

Choose the best answer to each of the following. Explain your reasoning.

1. Which of these galaxies is likely to be oldest? (a) a galaxy in the Local Group (b) a galaxy observed at a distance of 5 billion light-years (c) a galaxy observed at a distance of 10 billion light-years
2. Which set of star-position measurements is likely to show the largest parallax shift for nearby stars? (a) measurements made 1 month apart (b) measurements made 6 months apart (c) measurements made 1 year apart
3. If Earth's orbital radius doubled in size, (a) the parallax shifts of nearby stars would double in size. (b) the parallax shifts of nearby stars would be only half their current size. (c) the parallax shifts of nearby stars would stay the same.
4. According to Leavitt's law, we can determine the luminosity of a Cepheid in a distant galaxy by (a) measuring its parallax. (b) measuring its period of brightening and dimming. (c) measuring its recession speed.
5. If you observe two Cepheid variable stars to have the same period and one is brighter than the other, (a) the brighter one is closer. (b) the brighter one is farther away. (c) there is not enough information to determine which star is closer.
6. Which kind of object is the best standard candle for measuring distances to extremely distant galaxies? (a) a white dwarf (b) a Cepheid variable star (c) a white dwarf supernova
7. When the ultraviolet light from hot stars in very distant galaxies finally reaches us, it arrives at Earth in the form of (a) x-rays. (b) slightly more energetic ultraviolet light. (c) visible or infrared light.
8. Why do virtually all the galaxies in the universe appear to be moving away from our own? (a) because we are located near where the Big Bang happened (b) because we are located near the center of the universe (c) because observers in all galaxies see a similar phenomenon due to the universe's expansion
9. If you observed the redshifts of galaxies at a given distance to be twice as large as they are now, the value you would determine for Hubble's constant would be (a) twice its current value. (b) equal to its current value. (c) half its current value.
10. The distance to the cosmological horizon depends on (a) the age of the universe. (b) the number of galaxies in the universe. (c) the location of the outer edge of the universe.
11. When we observe a distant galaxy whose photons have traveled for 10 billion years before reaching Earth, we are seeing that galaxy as it was when the universe was about (a) 10 billion years old. (b) 7 billion years old. (c) 4 billion years old.
12. Which of these statements is a key assumption in our most successful models for galaxy formation? (a) The distribution of matter was perfectly uniform early in time. (b) Some regions of the universe were slightly denser than others. (c) Galaxies formed around supermassive black holes.

Short-Answer/Essay Questions

Explain all answers clearly, with complete sentences and proper essay structure if needed. An asterisk (*) designates a quantitative problem, for which you should show all your work.

13. *Distance Measurements.* The techniques astronomers use to measure distances are not so different from the ones you use every day. Describe how using standard-candle measurements is similar to the way you estimate the distance to an oncoming car at night.
14. *Cepheids as Standard Candles.* Suppose you are observing Cepheids in a nearby galaxy. You observe one Cepheid with a period of 8 days between peaks in brightness, and another with a period of 35 days. Estimate the luminosity of each star. Explain how you arrived at your estimate. (*Hint:* See Figure 12.4.)
15. *Galaxies at Great Distances.* The most distant galaxies that astronomers have observed are much easier to study with infrared light than with visible light. Explain why this is the case.
16. *Universe on a Balloon.* In what ways does the surface of a balloon offer a good analogy to the universe? In what ways is this analogy limited? Explain why a miniature scientist living in a polka dot on the balloon would observe all other dots to be moving away, with more distant dots moving away faster.
17. *Lookback Time and the Ages of Galaxies.* Suppose you observe a distant galaxy with a lookback time of 10 billion years. What was the maximum possible age of that galaxy when the light we are now observing from it began its journey to Earth? Explain your reasoning. (*Hint:* Assume the galaxy was born less than a billion years after the Big Bang.)
*18. *Cepheids in M100.* Consider these actual data for three Cepheids in the galaxy M100 (measured with observations from the Hubble Space Telescope). The luminosities were determined from the observed periods using Leavitt's law.

- Cepheid 1: luminosity = 3.9×10^{30} watts

  apparent brightness = 9.3×10^{-19} watt/m^2
- Cepheid 2: luminosity = 1.2×10^{30} watts

  apparent brightness = 3.8×10^{-19} watt/m^2
- Cepheid 3: luminosity = 2.5×10^{30} watts

  apparent brightness = 8.7×10^{-19} watt/m^2

Compute the distance to M100 with data from each of the three Cepheids. Do the three distance computations agree? Based on your results, estimate the uncertainty in the distance you have found.

*19. *Estimating the Universe's Age.* Suppose that Hubble's constant were $H_0 = 33$ km/s/Mly (which is *not* its actual value). What would the approximate age of the universe be in that case?

*20. *Distances Between Galaxies.* If you were to divide the present-day universe up into cubes with sides 10 million light-years long, each cube would contain, on average, about one galaxy similar in size to the Milky Way. Now suppose you traveled back in time, to an era when the average distance between galaxies was one-quarter of its current value. How many galaxies similar in size to the Milky Way would you expect to find, on average, in cubes of the same size? In order to simplify the problem, assume that the total number of galaxies of each type has not changed between then and now. Based on your answer, would you expect collisions to be much more frequent at that time or only moderately more frequent?

The Birth of the Universe   13

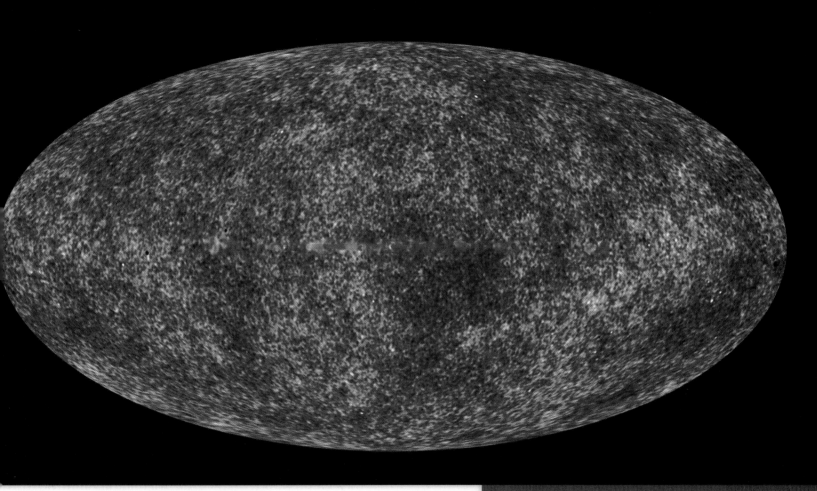

The image on this page is a record of light from far beyond the most distant galaxies. It shows microwave radiation released at a time when the universe was so young that no stars or galaxies had yet been born. We see the entire sky as it appears in all directions from Earth; the color differences represent slight differences in the temperature of the universe at the time the radiation was released. This image is important because it provides strong evidence for the Big Bang theory, which says that the universe began in a very hot and dense state and has been expanding and cooling ever since. In this chapter, we'll explore the Big Bang theory and the evidence that supports it.

LEARNING GOALS

13.1 The Big Bang Theory
- What were conditions like in the early universe?
- How did the early universe change with time?

13.2 Evidence for the Big Bang
- How do observations of the cosmic microwave background support the Big Bang theory?
- How do the abundances of elements support the Big Bang theory?

 THE PROCESS OF SCIENCE IN ACTION

13.3 Inflation
- Did the universe undergo an early episode of inflation?

215

▲ **FIGURE 13.1**
The universe cools as it expands. By using the laws of physics and the current temperature of the universe (about 3 K), we can calculate how hot the universe must have been in the past. This graph shows the results. Notice that both axis scales use powers of 10, so although most of the graph shows temperatures during the first second after the Big Bang, the far right extends to the present (14 billion years = 4×10^{17} s). (The small kinks in the graph at temperatures of 10^{12} K and 10^9 K correspond to the moments when the universe became too cold to produce new protons and new electrons, respectively.)

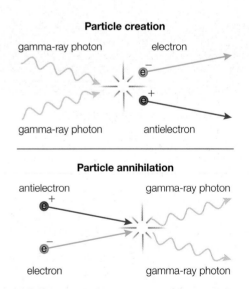

▲ **FIGURE 13.2**
Electron-antielectron creation and annihilation. Reactions like these continually converted photons to particles, and vice versa, in the early universe.

The Birth of the Universe

13.1 The Big Bang Theory

Throughout this book, we have discussed how the matter produced in the early universe gradually assembled into galaxies, stars, and planets. However, we have not yet answered one big question: Where did *matter itself* come from? To answer this question, we must go beyond even the most distant galaxies. We must go back not only to the origins of matter and energy but also to the beginning of time itself.

The scientific theory that describes what the universe was like early in time is called the **Big Bang theory**. It is based on applying known and tested laws of physics to the idea that all we see today, from Earth to the cosmological horizon, began as an incredibly hot and dense collection of matter and radiation. The Big Bang theory successfully describes how the expansion and cooling of this intense mixture of particles and photons could have led to the present universe of stars and galaxies, and it explains key aspects of today's universe with impressive accuracy. Our main goal in this chapter is to understand the evidence supporting the Big Bang theory, but first we must explore what the theory tells us about the early universe.

◆ What were conditions like in the early universe?

Observations demonstrate that the universe is expanding and cooling with time, implying that it must have been hotter and denser in the past. Calculating exactly how hot and dense the universe must have been when it was more compressed is much like calculating how the temperature and density of gas in a balloon change when you squeeze it, except that the conditions become much more extreme. Figure 13.1 shows how the temperature of the universe has changed with time, according to such calculations.

Particle Creation and Annihilation The universe was so hot during the first few seconds after the Big Bang that photons could transform themselves into matter, and vice versa, in accordance with Einstein's formula $E = mc^2$. Reactions that create and destroy matter are now relatively rare in the universe, but physicists can reproduce many such reactions with particle accelerators such as the Large Hadron Collider (see Tools of Science, page 217).

One such reaction is the creation or destruction of an *electron-antielectron pair* (Figure 13.2). When two photons collide with a total energy greater than twice the mass-energy of an electron (the electron's mass times c^2), they can create two brand-new particles: a negatively charged electron and its positively charged twin, the *antielectron* (also known as a *positron*). The electron is a particle of **matter**, and the antielectron is a particle of **antimatter**. The reaction that creates an electron-antielectron pair also runs in reverse. When an electron and an antielectron meet, they *annihilate* each other totally, transforming their mass-energy back into photon energy.

Similar reactions can produce or destroy any particle-antiparticle pair, such as a proton and antiproton or a neutron and antineutron. The early universe therefore was filled with an extremely hot and dense mix of photons, matter, and antimatter, converting furiously back and forth. Despite all these vigorous reactions, describing conditions in the early universe is straightforward, at least in principle. We simply need to use the laws of physics to calculate the proportions of the various forms of radiation and matter at each moment in the universe's early history. The only difficulty is our incomplete understanding of the laws of physics.

To date, physicists have investigated the behavior of matter and energy at temperatures as high as those that existed in the universe just *one ten-billionth* (10^{-10}) of a second after the Big Bang, giving us confidence that we actually understand what was happening at that early time. Our understanding of physics under the more extreme conditions that prevailed even earlier is less certain, but we have some ideas about what the universe was like when it was a mere 10^{-38} second old, and perhaps a glimmer of what it was like at the age of just 10^{-43} second. These tiny fractions of a second are so small that, for all practical purposes, we are studying the very moment of creation—the Big Bang itself.

▶ Fundamental Forces

Fundamental Forces To understand the changes that occurred in the early universe, it helps to think in terms of *forces*. Everything that happens in the universe today is governed by four distinct forces: *gravity*, the *electromagnetic force*, the *strong force*, and the *weak force*. We have already encountered examples of some of these forces in action.

Gravity is the most familiar of the four forces, providing the "glue" that holds planets, stars, and galaxies together. The electromagnetic force, which depends on the electrical charge of a particle instead of its mass, is far stronger than gravity. It is therefore the dominant force between particles in atoms and

tools of science | Particle Accelerators

Much of what we know about the universe's earliest moments comes from devices known as *particle accelerators*. Physicists can generate many unusual particles with the aid of these machines, including the Large Hadron Collider, which is currently the world's most powerful. Inside a particle accelerator, charged particles are channeled along a narrow tube—typically about as thick as garden hose—by large, powerful magnets. These magnets can accelerate familiar particles such as electrons or protons to very high speeds—often extremely close to the speed of light. When these particles collide with one another or with a stationary target, they release a substantial amount of energy within a very small space. For an instant, the conditions within this small space are similar to those that prevailed a fraction of a second after the Big Bang. The energy released is so great that much of it spontaneously turns into mass, producing a shower of matter and antimatter particles. Recall that energy can turn into mass, just as mass can turn into energy, in accord with Einstein's formula $E = mc^2$.

Scientists identify different types of particles produced in particle accelerators by carefully observing the paths they follow in response to magnetic fields within the accelerator. These paths depend on properties such as the mass and electrical charge of the particles. Whenever scientists observe a particle that has previously unseen characteristics, they catalog it as a new type of particle and give it a name. Dozens of different particles have been discovered in accelerators, but only a few of them, such as *electrons* and *quarks*, appear to be truly fundamental. All the remaining particles can be constructed from the fundamental few according to a scheme known as the *standard model* for the structure of matter. The standard model has proved very successful and has even predicted the existence of new particles that were later discovered in particle accelerators. The standard model has been well tested in particle accelerators at energies equivalent to a temperature of 10^{15} K, giving scientists confidence that we understand how particles behaved a mere 10^{-10} second after the Big Bang. Perhaps most spectacularly, in 2012, scientists using the Large Hadron Collider identified a particle called the *Higgs boson*, which the standard model had predicted some four decades earlier. Such successes give scientists great confidence in our understanding of the early universe.

Aerial photograph of the Large Hadron Collider, which lies on the border between Switzerland and France. The large circle traces the path of the main particle acceleration ring, which lies underground and has a circumference of 27 kilometers.

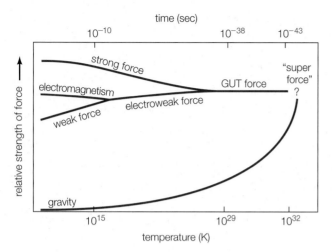

▲ FIGURE 13.3
The four forces are distinct at low temperatures but may merge at very high temperatures, such as those that prevailed during the first fraction of a second after the Big Bang.

molecules, responsible for all chemical and biological reactions. However, the balance between positive and negative electrical charges in large objects makes them electrically neutral and causes the electromagnetic force to lose out to gravity on large scales. Gravity therefore becomes the dominant force for such objects, because more mass always means more gravity.

The strong and weak forces operate only over extremely short distances, making them important within atomic nuclei but not on larger scales. The strong force binds atomic nuclei together [Section 8.1]. The weak force plays a crucial role in nuclear reactions such as fission and fusion, and it is the only force besides gravity that affects weakly interacting particles such as neutrinos.

Although the four forces behave quite differently from one another, current models of fundamental physics predict that they are actually different aspects of a smaller number of more fundamental forces. These models predict that at the high temperatures that prevailed in the very early universe, the four forces would not have been as distinct as they are today (Figure 13.3).

As an analogy, think about ice, liquid water, and water vapor. These three substances are quite different from one another in appearance and behavior, yet they are all different phases of the single substance H_2O. In a similar way, experiments have shown that the electromagnetic and weak forces lose their separate identities under conditions of very high temperature or energy and merge into a single **electroweak force**. At even higher temperatures and energies, the electroweak force may merge with the strong force and ultimately with gravity. Models that predict the merger of the electroweak and strong forces are often called **grand unified theories**, or **GUTs** for short. The merger of the strong, weak, and electromagnetic forces is therefore often called the *GUT force*. Many physicists suspect that at even higher energies, the GUT force and gravity merge into a single "super force" that governs the behavior of everything. (Current ideas for linking all four forces go by names that include *supersymmetry, superstrings,* and *supergravity*.)

If these ideas are correct, then the universe was governed solely by the super force in the first instant after the Big Bang. As the universe expanded and cooled, the super force split into gravity and the GUT force, which then split further into the strong and electroweak forces. Ultimately, all four forces became distinct. As we'll see shortly, these changes in the fundamental forces probably occurred before the universe was one ten-billionth of a second old.

think about it Based on Figure 13.3, about how long after the Big Bang did the strong force first become distinct, as it is in the universe today? At what time did all four forces first have their own individual identities?

◆ How did the early universe change with time?

The Big Bang theory uses scientific understanding of particles and forces to reconstruct the history of the universe. Here we will outline this history as a series of eras, or time periods. Each era is distinguished from the next by some major change in physical conditions as the universe cools. You'll find it useful to refer to the time line shown in Figure 13.4 as you read along; notice that the time scale in this figure runs by powers of 10, which means that early eras were very brief, even though they appear spread out on the figure. It will take you longer to read this chapter than it took the universe to progress through the first five eras we will discuss, by which point the chemical composition of the universe had already been determined.

The Planck Era The first era after the Big Bang is called the **Planck era**, named for physicist Max Planck; it represents times before the universe was 10^{-43} second old. We cannot say much about this instant in time, because current theories of physics cannot make firm predictions about the conditions that prevailed then. Nevertheless, we have at least some idea of how the Planck era ended. If you look back at Figure 13.3, you'll see that all four forces are thought to merge into the single, unified super force at temperatures above 10^{32} K—the temperatures that prevailed during the Planck era. In that case, the Planck era would have been a time of ultimate simplicity, when just a single force operated in nature, and it came to an end when the temperature dropped low enough for gravity to become distinct from the other three forces, which were still merged as the GUT force. By analogy to the way ice crystals form as a liquid cools, we say that gravity "froze out" at the end of the Planck era.

The GUT Era The next era is called the **GUT era**, named for the grand unified theories (GUTs) that predict the merger of the strong, weak, and electromagnetic forces into a single GUT force at temperatures above 10^{29} K (see Figure 13.3). The GUT era is the era during which two forces—gravity and the GUT force—operated in the universe. It came to an end when the GUT force split into the strong and electroweak forces, which happened when the universe was a mere 10^{-38} second old.

Our current understanding of physics allows us to say only slightly more about the GUT era than about the Planck era, and none of our ideas about the GUT era have been sufficiently tested to give us great confidence about what occurred during that time. However, if the grand unified theories are correct, the freezing out of the strong and electroweak forces may have released an enormous amount of energy, causing a sudden and dramatic expansion of the universe called **inflation**. In a mere 10^{-36} second—a trillion-trillion-trillionth of a second—pieces of the universe the size of an atomic nucleus may have grown to the size of our solar system. Inflation sounds bizarre, but as we will discuss later, it explains several important features of today's universe.

The Electroweak Era The splitting of the GUT force marked the beginning of an era during which three distinct forces operated: gravity, the strong force, and the electroweak force. We call this time the **electroweak era**, indicating that the electromagnetic and weak forces were still merged together. Intense radiation continued to fill all of space, as it had since the Planck era, spontaneously producing matter and antimatter particles that almost immediately annihilated each other and turned back into photons.

The universe continued to expand and cool throughout the electroweak era, dropping to a temperature of 10^{15} K when it reached an age of 10^{-10} second. This temperature is still 100 million times hotter than the temperature in the core of the Sun today, but it was low enough for the electromagnetic and weak forces to freeze out from the electroweak force. After this instant (10^{-10} second), all four forces were forever distinct in the universe.

The Particle Era As long as the universe was hot enough for the spontaneous creation and annihilation of particles, the total number of particles was roughly in balance with the total number of photons. Once it became too cool for this spontaneous exchange of matter and energy to continue, photons became the dominant form of energy in the universe. We refer to the time between the end of the electroweak era and the moment when spontaneous particle production ceased as the **particle era**, to emphasize the importance of subatomic particles during this period.

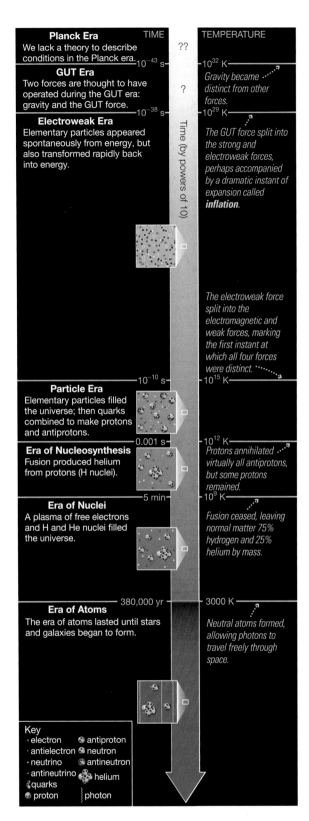

▲ FIGURE 13.4
A time line for the eras of the early universe. The only era not shown is the era of galaxies, which began with the birth of stars and galaxies when the universe was a few hundred million years old.

During the early parts of the particle era (and earlier eras), photons turned into all sorts of exotic particles that we no longer find freely existing in the universe today, including *quarks*—the building blocks of protons and neutrons. By the end of the particle era, all quarks had combined into protons and neutrons, which then shared the universe with other particles such as electrons and neutrinos.

The particle era came to an end when the universe reached an age of 1 millisecond (0.001 second), at which point it was no longer hot enough to produce protons and antiprotons spontaneously from energy. If the universe had contained equal numbers of protons and antiprotons at that time, they all would have annihilated each other, creating photons and leaving essentially no matter in the universe. From the obvious fact that the universe contains a significant amount of matter today, we conclude that protons must have slightly outnumbered antiprotons at the end of the particle era.

We can estimate the ratio of matter to antimatter by comparing the present numbers of protons and photons in the universe. The two numbers should have been similar in the very early universe, but today photons outnumber protons by about a billion to one. This ratio indicates that for every billion antiprotons in the early universe, there must have been about a billion and one protons. That is, for each 1 billion protons and antiprotons that annihilated each other at the end of the particle era, a single proton was left over. This slight excess of matter over antimatter makes up all the ordinary matter in the present-day universe.

The Era of Nucleosynthesis The eras we have discussed so far all occurred within the first 0.001 second of the universe's existence—less time than it takes you to blink an eye. At this point, the protons and neutrons left over after the annihilation of antimatter began to fuse and make heavier nuclei. However, the temperature of the universe remained so high that most nuclei were blasted apart by gamma rays as fast as they formed. This dance of fusion and demolition marked the **era of nucleosynthesis**, which ended when the universe was about 5 minutes old. By this time, the density in the expanding universe had dropped so much that fusion no longer occurred, even though the temperature was still about a billion Kelvin (10^9 K)—much hotter than the temperature of the Sun's core.

When fusion ceased at the end of the era of nucleosynthesis, the chemical content of the universe had become (by mass) about 75% hydrogen and 25% helium, along with trace amounts of deuterium (hydrogen with a neutron) and lithium (the next heaviest element after hydrogen and helium). Except for the small proportion of matter that stars later forged into heavier elements, the chemical composition of the universe remains the same today.

The Era of Nuclei After fusion ceased, the universe consisted of a very hot plasma of hydrogen nuclei, helium nuclei, and free electrons. The fully ionized nuclei moved independently of electrons during this period (rather than being bound with electrons in neutral atoms), so we call it the **era of nuclei**. Throughout this era, photons bounced rapidly from one electron to the next, just as they do deep inside the Sun today [Section 8.1], never traveling far between collisions. Any time a nucleus managed to capture an electron to form a complete atom, one of the photons quickly ionized it.

The era of nuclei came to an end when the expanding universe was about 380,000 years old. At this point the temperature had fallen to about 3000 K—roughly half the temperature of the Sun's surface today. Hydrogen and helium nuclei finally captured electrons for good, forming stable, neutral atoms for the first time. With electrons now bound into atoms, the universe became transparent, as if a thick fog had suddenly lifted. Photons, formerly trapped among the electrons, began to stream freely through the universe. We still see these photons today as the *cosmic microwave background*, which we will discuss shortly.

The Eras of Atoms and Galaxies We've already discussed the rest of the universe's history in earlier chapters. The end of the era of nuclei marked the beginning of the **era of atoms**, when the universe consisted of a mixture of neutral atoms and plasma (ions and electrons), along with a large number of photons. Because the density of matter in the universe differed slightly from place to place, gravity slowly drew atoms and plasma into the higher-density regions, which assembled into protogalactic clouds [Section 12.3]. Stars formed in these clouds, and the clouds subsequently merged to form galaxies.

The first full-fledged galaxies had formed by the time the universe was about 1 billion years old, beginning what we call the **era of galaxies**, which continues to this day. Generation after generation of star formation in galaxies steadily builds elements heavier than helium and incorporates them into new star systems. Some of these star systems develop planets, and on at least one of these planets life burst into being a few billion years ago. Now here we are, thinking about it all.

Early Universe Summary Figure 13.5 (pages 222–223) summarizes the major ideas from our brief overview of the history of the universe as it is described by the Big Bang theory. In the rest of this chapter, we will discuss the evidence that supports this theory.

13.2 Evidence for the Big Bang

Like any scientific theory, the Big Bang theory is a model of nature designed to explain a set of observations. The model was inspired by Edwin Hubble's observation that the universe is expanding, which implies that the universe must have been much denser and hotter in the past. However, like any scientific model, the Big Bang model had to make testable predictions about other observable features of the universe, and it gained wide scientific acceptance because of two major predictions that have been verified:

- The Big Bang theory predicts that the radiation that began to stream across the universe at the end of the era of nuclei should still be present today. Sure enough, we find that the universe is filled with what we call the **cosmic microwave background**. Its characteristics precisely match those predicted by the theory.
- The Big Bang theory predicts that some of the original hydrogen in the universe should have fused into helium during the era of nucleosynthesis. Observations of the actual helium content of the universe closely match the value predicted by the Big Bang theory.

◆ **How do observations of the cosmic microwave background support the Big Bang theory?**

The discovery of the cosmic microwave background was announced in 1965. Arno Penzias and Robert Wilson, two physicists working at Bell Laboratories in New Jersey, were calibrating a sensitive microwave antenna designed for satellite

▲ FIGURE 13.6
Arno Penzias and Robert Wilson, discoverers of the cosmic microwave background, with the Bell Labs microwave antenna.

communications (Figure 13.6). Much to their chagrin, they kept finding unexpected "noise" in every measurement they made. The noise was the same no matter where they pointed their antenna, indicating that it came from all directions in the sky and ruling out the possibility that it came from any particular astronomical object or from any place on Earth.

Meanwhile, physicists at nearby Princeton University were busy calculating the expected characteristics of the radiation left over from the heat of the Big Bang. They concluded that, if the Big Bang had really occurred, this radiation should be permeating the entire universe and should be detectable with a microwave antenna. The Princeton group soon met with Penzias and Wilson to compare notes, and both teams realized that the "noise" from the Bell Labs antenna was the predicted cosmic microwave background—the first strong evidence that the Big Bang really did happen.

Origin of the Cosmic Microwave Background

The cosmic microwave background consists of microwave photons that have traveled through space since the end of the era of nuclei, when most of the electrons in the universe joined with nuclei to make neutral atoms. With very few free electrons left to block them, most of the photons from that time have traveled unobstructed through the universe ever since (Figure 13.7). When we observe the cosmic microwave background, we essentially are seeing back to the end of the era of nuclei, when the universe was only 380,000 years old.

Note that the cosmic microwave background represents a limit on how far back we can see in time. The universe was filled with an impenetrable fog of electrons before this radiation was released, so we cannot see light from earlier times. However, the cosmic microwave background provides clues about those earlier eras, because conditions in the universe at the time the background radiation was released were determined by events that occurred even earlier in time.

Characteristics of the Cosmic Microwave Background

The Big Bang theory predicts that the cosmic microwave background should have an essentially perfect thermal radiation spectrum (see Tools of Science, page 79), because it came from the heat of the universe itself. Because the cosmic microwave background broke free when the universe had cooled to a temperature of about 3000 K, similar to the surface temperature of a red giant star, its spectrum should have originally peaked at a wavelength of about 1000 nanometers (just like the thermal radiation from a red star). Because the universe has since expanded by a factor of about 1000, the wavelengths of these photons should by now have stretched to about 1000 times their original values [Section 12.2]. We therefore expect the peak wavelength of the cosmic microwave background now to be about a millimeter, squarely in the microwave portion of the spectrum and corresponding to a temperature of a few degrees above absolute zero.

In the early 1990s, a NASA satellite called the *Cosmic Background Explorer (COBE)* was launched to test these ideas about the cosmic microwave background. The results were a stunning success for the Big Bang theory. As shown in Figure 13.8, the cosmic microwave background does indeed have a perfect thermal radiation spectrum, with a peak corresponding to a temperature of 2.73 K.

▲ FIGURE 13.7
Photons (yellow squiggles) frequently collided with free electrons during the era of nuclei and thus could travel freely only after electrons became bound into atoms. This transition was something like the transition from a dense fog to clear air. The photons released at the end of the era of nuclei, when the universe was about 380,000 years old, make up the cosmic microwave background. Precise measurements of these microwaves tell us what the universe was like at this moment in time.

think about it Suppose the cosmic microwave background did not really come from the heat of the universe itself, but instead came from many individual stars and galaxies. Explain why, in that case, we would not expect it to have a perfect thermal radiation spectrum. How does the spectrum of the cosmic microwave background lend support to the Big Bang theory?

COBE and its successor missions, the *Wilkinson Microwave Anisotropy Probe* (*WMAP*) and the European *Planck* satellite, have also mapped the temperature of the cosmic microwave background in all directions (Figure 13.9). The temperature turns out to be extraordinarily uniform throughout the universe—just as the Big Bang theory predicts it should be—with variations from one place to another of only a few parts in 100,000. Moreover, these slight variations also represent a predictive success of the Big Bang theory. Recall that our theory of galaxy formation depends on the assumption that the early universe was *not quite* perfectly uniform; some regions of the universe must have started out slightly denser than other regions, so that they could serve as seeds for galaxy formation [Section 12.3]. The small variations in the temperature of the cosmic microwave background indicate that the density of the early universe really did differ slightly from place to place.

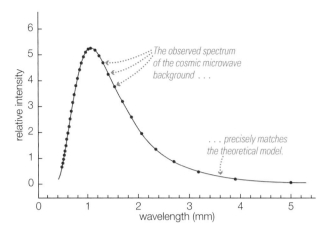

▲ **FIGURE 13.8**
This graph shows the spectrum of the cosmic microwave background recorded by NASA's *COBE* satellite. A theoretically calculated thermal radiation spectrum (smooth curve) for a temperature of 2.73 K perfectly fits the data (dots). This excellent fit is important evidence in favor of the Big Bang theory.

◆ How do the abundances of elements support the Big Bang theory?

The Big Bang theory also solves what had previously been another long-standing astronomical problem: the origin of cosmic helium. Everywhere in the universe, about three-quarters of the mass of ordinary matter (not including dark matter) is hydrogen and about one-quarter is helium. Helium fusion in stars cannot account for this large abundance of helium, because even small galaxies that have experienced very little star formation still have helium fractions of 25%. We therefore conclude that the majority of the helium in the universe must already have been present in the protogalactic clouds that preceded the formation of galaxies. In other words, the universe itself must once have been hot enough to fuse hydrogen into helium. The current microwave background temperature of 2.73 K tells us precisely how hot the universe was in the distant past and exactly how much helium it should have made. The result—25% helium—is another impressive success of the Big Bang theory.

Helium Formation in the Early Universe To see why 25% of ordinary matter became helium, we need to understand what protons and neutrons were doing during the 5-minute-long era of nucleosynthesis. Early in this era, the universe's temperature was so high (about 10^{11} K) that nuclear reactions could convert protons into neutrons, and vice versa. These reactions kept the numbers of protons and neutrons nearly equal. But as the universe cooled, neutron–proton conversion reactions began to favor protons.

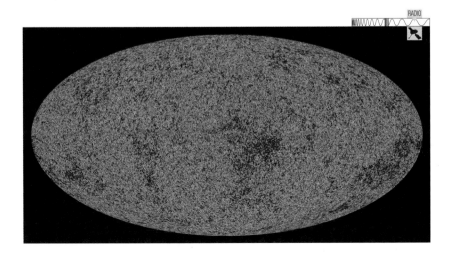

◀ **FIGURE 13.9**
This all-sky map shows temperature differences in the cosmic microwave background measured by *Planck*. The background temperature is about 2.73 K everywhere, but the brighter regions of this picture are about 0.0001 K hotter than the darker regions—indicating that the early universe was very slightly lumpy at the end of the era of nuclei. We are essentially seeing what the universe was like at the surface marked "380,000 years" in Figure 13.5. Gravity later drew matter toward the centers of these lumps, forming galaxies and other structures that we see in the universe today.

▲ FIGURE 13.10
During the 5-minute-long era of nucleosynthesis, virtually all the neutrons in the universe fused with protons to form helium. This figure illustrates one of the ways in which helium formed.

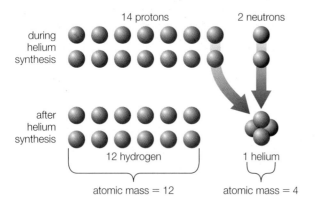

▲ FIGURE 13.11
Calculations show that protons outnumbered neutrons 7 to 1, which is the same as 14 to 2, during the era of nucleosynthesis. The result was 12 hydrogen nuclei (individual protons) for each helium nucleus. As shown, this means a hydrogen-to-helium mass ratio of 12 to 4, which is the same as 75% to 25%.

Neutrons are slightly more massive than protons, so reactions that convert protons to neutrons require energy to proceed (in accordance with $E = mc^2$). As the temperature fell below 10^{11} K, the energy required for neutron production was no longer readily available, so the rate of these reactions slowed. In contrast, reactions that convert neutrons into protons release energy and therefore are unhindered by cooler temperatures. By the time the temperature of the universe fell to 10^{10} K, protons had begun to outnumber neutrons because the conversion reactions ran only in one direction. Neutrons changed into protons, but the protons didn't change back.

For the next few minutes, the universe was still hot and dense enough for nuclear fusion. Protons and neutrons combined to form *deuterium*—a rare form of hydrogen whose nucleus contains a neutron in addition to a proton—and deuterium nuclei fused to form helium (Figure 13.10). At first, the helium nuclei were almost immediately blasted apart by the many gamma rays that filled the early universe. But fewer gamma rays remained as the universe cooled, and helium nuclei were able to survive by the time the universe was about 1 minute old.

Calculations show that the proton-to-neutron ratio at this time should have been about 7 to 1. Moreover, almost all the available neutrons should have become incorporated into nuclei of helium. Figure 13.11 shows that, based on the 7-to-1 ratio of protons to neutrons, the universe should have had a composition of 75% hydrogen and 25% helium by mass at the end of the era of nucleosynthesis. This match between the predicted and observed helium ratios provides strong support for the Big Bang theory.

think about it Briefly explain why it should not be surprising that some galaxies (including the Milky Way) contain a little more than 25% helium, but why it would be very surprising if some galaxies contained less.

Abundances of Other Light Elements Why didn't the Big Bang produce heavier elements? By the time stable helium nuclei formed, when the universe was about a minute old, the temperature and density of the rapidly expanding universe had already dropped too low to sustain a process like carbon production (three helium nuclei fusing into carbon [Section 9.2]). Reactions between protons, deuterium nuclei, and helium were still possible, but most of these reactions led nowhere, because the resulting nuclei were unstable and quickly split apart.

A few reactions involving hydrogen-3 (also known as *tritium*) or helium-3 can create long-lasting nuclei. However, the contributions of these reactions to the overall composition of the universe were minor because hydrogen-3 and helium-3 were so rare. Models of element production in the early universe show that, before the cooling of the universe halted fusion entirely, such reactions generated only trace amounts of lithium, the next lightest element after helium. Aside from hydrogen, helium, and this trace of lithium, all other elements were forged much later by stars.

THE PROCESS OF SCIENCE IN ACTION

13.3 Inflation

The Big Bang theory has gained wide acceptance because of the strong evidence from the cosmic microwave background and the abundance of helium in the universe. However, the theory in its simplest form leaves several major features of our universe unexplained. The three most pressing questions are the following:

- *Where do structures like galaxies come from?* Recall that successful models of galaxy formation start from the assumption that gravity could collect matter together in regions of the early universe that had slightly enhanced density [Section 12.3]. We know from our observations of variations in the cosmic microwave background that such regions of enhanced density were present in the universe at an age of 380,000 years, but we have not yet explained how these density variations came to exist.

- *Why is the large-scale universe nearly uniform?* Although the slight variations in the cosmic microwave background show that the universe is not *perfectly* uniform on large scales, the fact that it is smooth to within a few parts in 100,000 is remarkable enough that we would not expect it to have occurred by pure chance.

- *Why is the geometry of the universe flat?* Einstein's general theory of relativity tells us that the overall geometry of the universe can be curved, like the surface of a balloon or a saddle. However, observational efforts to measure the large-scale geometry of the universe have not yet detected any curvature. As far as we can tell, the large-scale geometry of the universe is flat, suggesting a precisely balanced geometry that would be unlikely to have arisen by chance.

The process of science demands that we seek natural explanations for these features of the universe. We will therefore conclude this chapter by examining the hypothesis of *inflation*, which may provide the necessary explanations.

◆ Did the universe undergo an early episode of inflation?

The idea of inflation first emerged in 1981, when physicist Alan Guth was considering how the separation of the strong force from the GUT force at the end of the GUT era might have affected the expansion of the universe. Some theories of high-energy physics predict that this separation of forces would have released enormous amounts of energy, and Guth realized that this energy could have caused the universe to expand dramatically, perhaps by a factor of 10^{30} in a time of less than 10^{-36} second. This period of rapid expansion is what scientists refer to as *inflation*, and it can explain the three key mysteries as follows.

Structure: Giant Quantum Fluctuations To understand how inflation can explain the origin of large structures such as galaxies, which require "seed regions" of enhanced density, we need to recognize a special feature of energy fields. Laboratory-tested principles of quantum mechanics tell us that on very small scales, the energy fields at any point in space are always fluctuating. The distribution of energy through space is therefore very slightly irregular, even in a complete vacuum. The tiny quantum "ripples" that make up the irregularities can be characterized by a wavelength that corresponds roughly to their size. In principle, quantum ripples in the very early universe could have been the seeds for density enhancements that later grew into galaxies. However, the wavelengths of the original ripples were far too small to explain density enhancements like those that left imprints on the cosmic microwave background.

Inflation would have dramatically increased the wavelengths of these quantum fluctuations. The rapid growth of the universe during the period of inflation would have stretched tiny ripples from a size smaller than that of an atomic nucleus to the size of our solar system (Figure 13.12), making them large enough to become the density enhancements from which galaxies and larger structures later formed. If that's the case, then the structure of today's universe all started as tiny quantum fluctuations just before the period of inflation.

▲ **FIGURE 13.12**
During inflation, quantum ripples in spacetime would have stretched by a factor of perhaps 10^{30}. The peaks of these ripples then would have become the density enhancements that produced all the structure we see in the universe today.

Uniformity: Equalizing Temperatures and Densities The remarkable uniformity of the cosmic microwave background might at first seem quite natural, but with further thought it becomes difficult to explain. Imagine observing the cosmic microwave background in a certain part of the sky. You are seeing microwaves that have traveled through the universe since the end of the era of nuclei, which means you are seeing a region of the universe as it was nearly 14 billion years ago, when the universe was only 380,000 years old. Now imagine turning around and looking at the cosmic microwave background coming from the opposite direction. You are also seeing this region as it was nearly 14 billion years ago, and it looks virtually identical in temperature and density. The surprising part is this: The two regions are billions of light-years apart on opposite sides of our observable universe, but we are seeing them as they were when they were only 380,000 years old. They couldn't possibly have exchanged light or any other information, because a signal traveling at the speed of light from one region to the other would barely have started its journey. So how did they come to have the same temperature and density?

The inflation hypothesis answers this question by saying that even though the two regions cannot have had any contact *since* the time of inflation, they were in contact prior to that time. Before the onset of inflation, when the universe was 10^{-38} second old, the two regions were less than 10^{-38} light-second away from each other. Radiation traveling at the speed of light would therefore have had time to bounce between the two regions, and this exchange of energy equalized their temperatures and densities. Inflation then pushed these equalized regions to much greater distances, far out of contact with each other. Like criminals getting their stories straight before being locked in separate jail cells, the two regions (and all other parts of the observable universe) came to the same temperature and density before inflation spread them far apart.

Because inflation caused different regions of the universe to separate so far in such a short period of time, many people wonder whether it violates Einstein's theories, which say that nothing can move faster than the speed of light. It does not, because nothing actually *moves* through space as a result of inflation or the ongoing expansion of the universe. Instead, the expansion of the universe is the expansion of *space itself*. Objects may be separating from one another at a speed faster than the speed of light, but no matter or radiation is able to travel between them during that time. In essence, inflation opens up a huge gap in space between objects that were once close together. The objects get very far apart, but nothing ever travels between them at a speed that exceeds the speed of light.

Geometry: Balancing the Universe The third question asks why the overall geometry of the universe is "flat." To understand this question, we must consider the overall geometry of the universe in a little more detail.

Recall that Einstein's general theory of relativity tells us that the presence of matter can curve spacetime [Section 10.2]. We cannot visualize this curvature in all four dimensions of spacetime, but we can detect its presence by its effects on light traveling through the universe. Although the curvature of the universe can vary from place to place, the universe as a whole must have some overall shape. Almost any shape is possible, but all the possibilities fall into just three general categories (Figure 13.13). Using analogies to objects that we can see in three dimensions, scientists refer to these three categories of shape as *flat*, *spherical*, and *saddle shaped*.

According to general relativity, the overall geometry depends on the average density of matter and energy in the universe, and the geometry can be flat only if the combined density of matter plus energy is precisely equal to a value known as the **critical density**. If the universe's average density is less than the critical density, then the overall geometry is saddle shaped. If its average density is greater than the critical density, then the overall geometry is spherical.

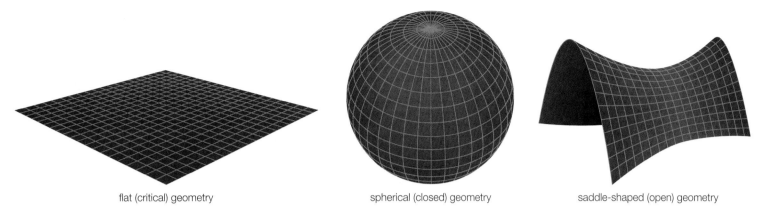

▲ **FIGURE 13.13**
The three possible categories of overall geometry for the universe. Keep in mind that the real universe has these shapes in more dimensions than we can see.

Inflation can explain why the overall geometry is so close to being flat. In terms of Einstein's theory, the effect of inflation on spacetime curvature is similar to the flattening of a balloon's surface when you blow the balloon up (Figure 13.14). The flattening of space caused by inflation would have been so enormous that any curvature the universe might have had previously would be noticeable only on size scales much larger than that of the observable universe. Inflation therefore makes the observable universe appear flat, which means that the average density of matter plus energy must be very close to the critical density.

Evidence for Inflation We've seen that inflation offers natural answers to our three key questions about the universe, but did it really happen? We cannot directly observe the universe at the very early time when inflation is thought to have occurred. Nevertheless, we can test the idea of inflation by exploring whether its predictions are consistent with our observations of the universe at later times. Scientists are only beginning to make observations that test inflation, but the findings to date are consistent with the idea that inflation made the universe uniform and flat while planting the seeds of structure formation.

The strongest tests of inflation to date come from detailed studies of the cosmic microwave background, and in particular the maps made by the *WMAP* and *Planck* satellites (see Figure 13.9). Remember that these maps show tiny temperature differences corresponding to density variations in the universe at the end of the era of nuclei, when the universe was about 380,000 years old. The inflation hypothesis predicts that these density variations were actually created much earlier, when tiny quantum ripples expanded to much longer wavelengths. Careful observations of the temperature variations in the microwave background can therefore tell us about the structure of the universe at that very early time.

The graph in Figure 13.15 shows how temperature differences between patches of sky depend on their angular size on the celestial sphere. The dots represent data from observations by *Planck* and other microwave telescopes, and the red curve shows the inflation-based model that best fits the observations. Notice the close agreement between the model predictions and the data. Moreover, the same model also makes specific predictions about other characteristics of our universe, such as its composition and age, and these also agree well with observations.

The bottom line is that, all things considered, inflation does a remarkable job of explaining features of our universe that are otherwise unaccounted for in the Big Bang theory. Many astronomers and physicists therefore suspect that some process akin to inflation really did occur in the early universe, though many of the details remain unclear. If these details can be worked out successfully, we face an amazing prospect—a breakthrough in our understanding of the very smallest particles, achieved by studying the universe on the largest observable scales.

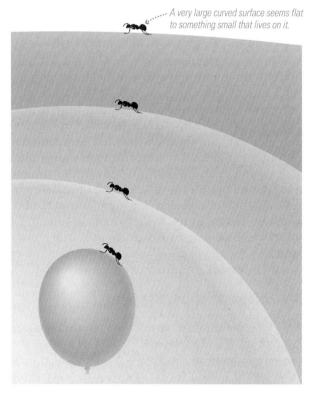

▲ **FIGURE 13.14**
As a balloon expands, its surface seems increasingly flat to an ant crawling along it. Inflation is thought to have made the universe seem flat in a similar way.

▶ **FIGURE 13.15**
All-sky maps of the cosmic microwave background like the one in Figure 13.9 allow scientists to measure temperature differences between different patches of the sky. This graph shows how the typical sizes of those temperature differences depend on the angular separation of the patches of sky. The data points represent actual measurements of the cosmic microwave background, while the red curve is the prediction of a model that relies on inflation to produce slight variations of temperature and density in the universe. Note the close agreement between the data and the model. (Bars indicate the uncertainty range in the data points.)

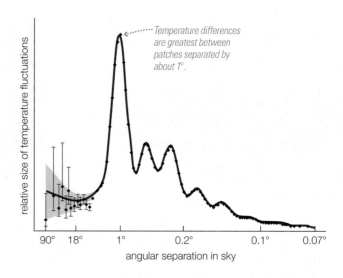

summary of key concepts

13.1 The Big Bang Theory

◆ **What were conditions like in the early universe?**

 The early universe was filled with radiation and elementary particles. It was so hot and dense that the energy of radiation could turn into particles of **matter** and **antimatter**, which then collided and turned back into radiation.

◆ **How did the early universe change with time?**

According to the Big Bang theory, the universe's history is marked by a number of eras, each with unique physical conditions. We know little about the **Planck era**, when the four forces may have all behaved as one. Gravity became distinct at the start of the **GUT era**, which may have ended with a period of rapid expansion called **inflation**. Electromagnetism and the weak force became distinct at the end of the **electroweak era**. Matter particles annihilated all the antimatter particles at the end of the **particle era**. Fusion of protons and neutrons into helium ceased at the end of the **era of nucleosynthesis**. Hydrogen nuclei captured all the free electrons, forming hydrogen atoms, at the end of the **era of nuclei**. Galaxies began to form at the end of the **era of atoms**. The **era of galaxies** continues to this day.

13.2 Evidence for the Big Bang

◆ **How do observations of the cosmic microwave background support the Big Bang theory?**

 Microwave telescopes allow us to observe the **cosmic microwave background**—radiation left over from the Big Bang. Its spectrum matches the characteristics expected of the radiation released at the end of the era of nuclei, confirming a key prediction of the Big Bang theory.

◆ **How do the abundances of elements support the Big Bang theory?**

 The Big Bang theory predicts the ratio of protons to neutrons during the era of nucleosynthesis and, from this, predicts that the chemical composition of the universe should be about 75% hydrogen and 25% helium by mass. This prediction matches observations of the element abundances in the universe, providing another piece of evidence to support the Big Bang theory.

THE PROCESS OF SCIENCE IN ACTION

13.3 Inflation

◆ **Did the universe undergo an early episode of inflation?**

 Scientists hypothesize that the universe underwent an early episode of inflation, thereby explaining (1) the density enhancements that led to galaxy formation, (2) the smoothness of that cosmic microwave background, and (3) the flat geometry of the observable universe. We can test the idea of inflation because it makes specific predictions about patterns we observe in the cosmic microwave background, and observations made by microwave telescopes so far match those predictions.

Investigations

For instructor-assigned homework and other learning materials, go to www.MasteringAstronomy.com.

Quick Quiz

Choose the best answer to each of the following. Explain your reasoning.

1. What is the current temperature of the universe? (a) absolute zero (b) a few degrees K (c) a few thousand degrees K
2. What is the charge of an antielectron? (a) positive (b) negative (c) neutral
3. What happens when a proton collides with an antiproton? (a) They repel each other. (b) They fuse together. (c) They convert into two photons.
4. Which of the following does *not* provide strong evidence for the Big Bang theory? (a) observations of the cosmic microwave background (b) observations of the amount of hydrogen in the universe (c) observations of the ratio of helium to hydrogen in the universe
5. What kinds of new particles are produced in particle accelerators? (a) only matter (b) only antimatter (c) both matter and antimatter
6. What kinds of particles were present in the universe during the particle era? (a) mostly matter (b) mostly antimatter (c) approximately equal amounts of matter and antimatter
7. What kinds of particles were present in the universe during the era of nucleosynthesis? (a) mostly matter (b) mostly antimatter (c) approximately equal amounts of matter and antimatter
8. What kinds of atomic nuclei were present in the universe during the era of nuclei? (a) all kinds of nuclei (b) mostly hydrogen and helium nuclei (c) only hydrogen nuclei
9. If you had been present in the universe at the beginning of the era of atoms, what color light would you have seen? (a) white light (b) red light (c) no visible light, because only microwave light was present
10. Which of the following does inflation help to explain? (a) the uniformity of the cosmic microwave background (b) the amount of helium in the universe (c) the temperature of the cosmic microwave background
11. Adding the idea of inflation to the Big Bang theory accounts for (a) the origin of hydrogen. (b) the origin of galaxies. (c) the origin of atomic nuclei.
12. Which of these pieces of evidence supports the idea that inflation really happened? (a) observations showing that the universe is expanding (b) measurements of the abundance of helium in the universe (c) observations of the cosmic microwave background that indicate a flat geometry for the universe

Short-Answer/Essay Questions

Explain all answers clearly, with complete sentences and proper essay structure if needed. An asterisk (*) designates a quantitative problem, for which you should show all your work.

13. *Life Story of a Proton.* Tell the life story of a proton from its formation shortly after the Big Bang to its presence in the nucleus of an oxygen atom you have just inhaled. Your story should be creative and imaginative, but it should also demonstrate your scientific understanding of as many stages in the proton's life as possible. Your story should be three to five pages long.

14. *Creative History of the Universe.* The story of creation as envisioned by the Big Bang theory is quite dramatic, but it is usually told in a fairly straightforward, scientific way. Write a more dramatic version of the story, in the form of a short story, play, or poem. Be as creative as you wish, but be sure to remain scientifically accurate.

15. *Re-creating the Big Bang.* Particle accelerators on Earth can push particles to extremely high speeds. When these particles collide, the amount of energy associated with the colliding particles is much greater than the mass-energy these particles have when at rest. As a result, these collisions can produce many other particles out of pure energy. Explain in your own words how the conditions that occur in these accelerators are similar to the conditions that prevailed shortly after the Big Bang. Also point out some of the differences between what happens in particle accelerators and what happened in the early universe.

16. *Wavelength of the Microwave Background.* Suppose you could have observed the spectrum of the cosmic microwave background radiation (as in Figure 13.8) 1 billion years after the Big Bang. How would the wavelength of radiation at the peak of that spectrum compare with the peak wavelength of today's background spectrum? Explain your reasoning.

17. *"Observing" the Early Universe.* The only way we have of studying the era of nucleosynthesis is through the abundances of the nuclei that it left behind. Explain why we will never be able to observe that era through direct detection of photons emitted at that time.

18. *Element Production in the Big Bang.* Nucleosynthesis in the early universe was unable to produce more than trace amounts of elements heavier than helium. Using the information in Figure 9.19, which shows the number of protons per nuclear particle for many different elements, explain why producing elements like lithium (3 protons), boron (4 protons), and beryllium (5 protons) was so difficult.

19. *Evidence for the Big Bang.* Make a list of at least six observed features of the universe that are satisfactorily explained by the Big Bang theory when it is combined with the idea of inflation.

*20. *Energy from Antimatter.* The total annual U.S. power consumption is about 2×10^{20} joules. Suppose you could supply that energy by combining pure matter with pure antimatter. Estimate the total mass of matter-antimatter fuel you would need in order to provide the United States with energy for 1 year. How does that mass compare with the amount of matter in your car's gas tank? (A gallon of gas has a mass of about 4 kilograms.) (*Hint:* Consider the meaning of the formula $E = mc^2$ and use the fact that 1 joule $= 1$ kg $\times$ m^2/s^2.)

*21. *Uniformity of the Cosmic Microwave Background.* The temperature of the cosmic microwave background differs by only a few parts in 100,000 across the sky. Compare this level of uniformity to that of the surface of a square table that is 1 meter on each side: How big would the largest bumps on that table be if its surface were smooth to 1 part in 100,000? Could you see bumps of that size on the table's surface with your naked eyes?

14 Dark Matter and Dark Energy

LEARNING GOALS

14.1 Evidence for Dark Matter
- What is the evidence for dark matter?
- What might dark matter be made of?

14.2 Gravity versus Expansion
- How did structures like galaxies form?
- Will the universe continue expanding forever?

THE PROCESS OF SCIENCE IN ACTION

14.3 Evidence for Dark Energy
- What is the evidence for dark energy?

This composite image shows two galaxy clusters undergoing an enormous collision. The many yellow and white objects are individual galaxies, but they represent only a small fraction of the cluster's total mass. More mass is contained in very hot gas, which emits x-rays that are mapped in red on this image. But most of the mass of the cluster seems to consist of *dark matter* that is completely invisible—blue regions show where it is located, based on observations of its gravitational effects. In this chapter, we'll discuss the evidence for dark matter and for the even more mysterious *dark energy* that is overpowering the gravity of dark matter on large scales and causing the expansion of the universe to accelerate. As we'll see, dark matter and dark energy are critical to understanding both the past evolution and future fate of the universe.

Dark Matter, Dark Energy, and the Fate of the Universe

14.1 Evidence for Dark Matter

What is the universe made of? You might think this would be a simple question for an astronomer to answer, but based on all the evidence available today, the answer is "We do not yet know."

It might seem incredible that we still do not know the composition of most of the universe, and you might wonder why we should be so clueless. After all, astronomers can measure the chemical composition of distant stars and galaxies from their spectra, so we know that stars and gas clouds are made almost entirely of hydrogen and helium, with small amounts of heavier elements mixed in. But notice the key words "chemical composition." When we say these words, we are talking about the composition of material built from atoms of elements such as hydrogen, helium, carbon, and iron.

While it is true that all familiar objects—including people, planets, and stars—are built from atoms, the same may not be true of the universe as a whole. In fact, we now have good reason to think that the universe is *not* composed primarily of atoms. Instead, observations indicate that the universe consists largely of a mysterious form of mass known as *dark matter* and a mysterious form of energy known as *dark energy*. In this section, we'll focus on dark matter.

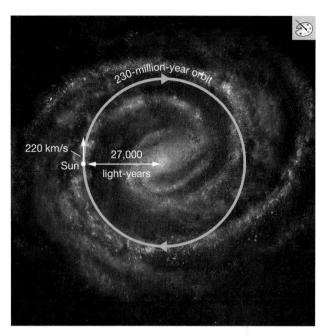

▲ **FIGURE 14.1**
Our Sun orbits the Milky Way at a distance 27,000 light-years from the center and at a speed of 220 km/s. From these orbital measurements, we determine that the mass of the portion of the Milky Way lying within the Sun's orbit is about 2×10^{41} kg, equivalent to 100 billion times the mass of the Sun.

◆ **What is the evidence for dark matter?**

It might seem natural to assume that galaxies and galaxy clusters are held together by the combined gravity of their stars and gas. However, careful observations reveal that stars and gas together have too little mass to provide the needed gravity. In other words, there must be a great deal of mass besides that in stars and gas. Because we do not observe light of any wavelength from this mass, we call it **dark matter**. Evidence for the existence of dark matter comes both from our own galaxy and from others.

Dark Matter in the Milky Way Recall that Newton's law of gravity determines the speeds (or velocities) at which objects orbit one another, allowing us to measure the mass of an orbiting system. In the case of the Milky Way Galaxy, we can use the Sun's orbital velocity and its distance from the galactic center to determine the mass lying *within* the Sun's orbit (Figure 14.1). Similarly, we can use the orbital motion of any other star or gas cloud to measure the mass of the galaxy within the orbit of that star or gas cloud (see Tools of Science, page 235). Because interstellar dust obscures our view of stars throughout most of our galaxy's disk, most measurements of motion in the Milky Way are based on observations of atomic hydrogen gas clouds, which emit radio waves that penetrate interstellar dust [Section 11.1].

We can summarize the results of these orbital velocity measurements with a diagram that plots the orbital speed of objects in the galaxy against their orbital distances. As a simple example of such a diagram, sometimes called a *rotation curve*, consider a plot of the rotation speed of a merry-go-round. Every object on a merry-go-round goes around the center in the same amount of time, but objects farther from the center move in larger circles, so they must move at faster speeds. The rotation curve for a merry-go-round is therefore a straight line that rises steadily with distance (Figure 14.2a).

In contrast, orbital speeds in our solar system *decrease* with distance from the Sun (Figure 14.2b). This drop-off in speed with distance occurs because virtually all the mass of the solar system is concentrated in the Sun. The gravitational force holding a planet in its orbit therefore decreases with distance from the Sun, and a smaller force means a lower orbital speed. Orbital speeds must drop similarly with distance in any other astronomical system that has its mass concentrated at its center.

233

a The rotation curve for a merry-go-round is a rising straight line.

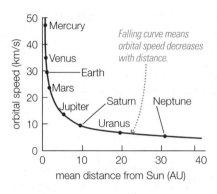
b The rotation curve for the planets in our solar system.

c The rotation curve for the Milky Way Galaxy. Dots represent actual data points for stars or gas clouds.

▲ **FIGURE 14.2**
These graphs show how the orbital speed of a system depends on distance from its center.

Figure 14.2c shows how orbital speed depends on distance in the Milky Way Galaxy. Each individual dot represents the orbital speed and distance from the galactic center of a particular star or gas cloud, and the curve running through the dots represents a best fit to the data. Notice that the orbital velocities remain approximately constant beyond the inner few thousand light-years, so most of the curve is relatively flat. Because this curve is so different from the declining rotation curve of the solar system, we conclude that most of the Milky Way's mass is *not* concentrated at its center. Instead, the orbits of progressively more distant gas clouds must encircle more and more mass. The Sun's orbit encompasses about 100 billion solar masses, but a circle twice as large surrounds twice as much mass, and a larger circle surrounds even more mass.

To summarize, orbital speeds in the Milky Way imply that most of our galaxy's mass lies well beyond the orbit of our Sun. A more detailed analysis suggests that most of this mass is distributed throughout the spherical halo that surrounds the disk of our galaxy, and that the total amount of this mass is more than *10 times* the total mass of all the stars in the disk. In other words, evidence suggests that the luminous part of the Milky Way's disk is like the tip of an iceberg, marking only the center of a much larger clump of unseen dark matter (Figure 14.3).

think about it Suppose we made a graph of orbital speeds and distances for the moons orbiting Jupiter. Which graph in Figure 14.2 would it most resemble? Why?

Dark Matter in Other Galaxies Other galaxies also seem to contain vast quantities of dark matter. We can determine the amount of dark matter in a galaxy by comparing the galaxy's mass to its luminosity. The procedure is simple in principle. First, we use the galaxy's luminosity to estimate the amount of mass that the galaxy contains in the form of stars. Next, we determine the galaxy's total mass by applying the law of gravity to observations of the orbital velocities of stars and gas clouds. If this total mass is larger than the mass that we can attribute to stars, then we infer that the excess mass must be dark matter.

Measuring the galaxy's total mass requires measuring orbital speeds as far from the galaxy's center as possible. For spiral galaxies, we can detect atomic hydrogen gas clouds much farther from the galactic center than stars, so most of our data come from using Doppler shifts to determine how fast these clouds are moving toward us or away from us. We then use the orbital speeds of these clouds and their distances from the center of the galaxy to determine the mass within those orbits. Figure 14.4 shows a few examples of graphs of orbital speed versus distance, illustrating that, like the Milky Way, most other spiral galaxies have orbital speeds that remain high even at great distances from their centers. Detailed analysis tells us that most spiral galaxies also have

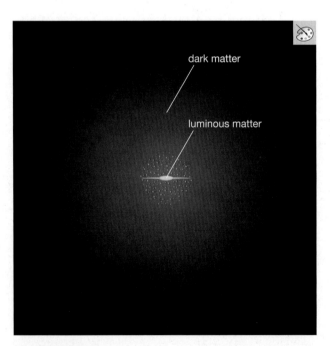

▲ **FIGURE 14.3**
The dark matter associated with the Milky Way occupies a much larger volume than the galaxy's luminous matter. The radius of this dark-matter halo may be 10 times as large as that of the galaxy's halo of stars.

at least 10 times as much mass in dark matter as they do in stars, with most of this mass far out in their halos.

Elliptical galaxies contain very little atomic hydrogen gas, so we generally measure the masses of elliptical galaxies by observing the motions of their stars. When we compare the Doppler shifts of spectral lines from different regions of an elliptical galaxy, we find that the orbital speeds of stars remain fairly constant as we look farther from the galaxy's center. Just as in spirals, we conclude that most of the matter in elliptical galaxies must lie beyond the distance where the light trails off and hence must be dark matter. The evidence for dark matter is even more convincing for cases in which we can measure the speeds of globular star clusters orbiting at large distances from the center of an elliptical galaxy. These measurements suggest that elliptical galaxies, like spirals, contain 10 or more times as much mass in dark matter as they do in the form of stars.

Dark Matter in Clusters of Galaxies Observations of galaxy clusters suggest that the total proportion of dark matter is even greater than that in individual galaxies. The evidence for dark matter in clusters comes from three different methods for measuring cluster masses: using the speeds of galaxies orbiting the center of the cluster, measuring the temperature of hot gas between the cluster's galaxies, and observing how the cluster bends light as a *gravitational lens*.

The first method is similar to the methods we've discussed for individual galaxies. We measure the speeds of galaxies orbiting within the cluster from the Doppler shifts of their spectral lines. We then combine this information with the distances of those galaxies from the cluster center to calculate how much mass the cluster contains.

think about it What would happen to a cluster of galaxies if you could remove all the dark matter without changing the velocities of the galaxies?

The second method relies on the fact that clusters of galaxies generally contain vast amounts of very hot, x-ray-emitting gas (Figure 14.5). We can measure the temperature of this gas from its x-ray spectrum, and this temperature tells us

▲ **FIGURE 14.4**
Graphs of orbital speed versus distance for four spiral galaxies. In each galaxy, the orbital speeds remain nearly constant over a wide range of distances from the center, indicating that dark matter is common in spiral galaxies.

tools of science — The Orbital Velocity Formula

Newton's laws of motion and gravity are astronomers' primary tools for measuring mass. We have already seen how Newton's version of Kepler's third law can tell us the mass of an orbiting system if we know its period and average orbital distance (see Tools of Science, page 96). With a little algebra, we can transform this law into a related formula that allows us to calculate the amount of mass contained within any object's orbit if we know the radius and velocity of that orbit. For an object on a circular orbit, this *orbital velocity formula* is

$$M_r = \frac{r \times v^2}{G},$$

where v is the object's orbital velocity, r is its orbital radius, M_r is the amount of mass contained within its orbit, and $G = 6.67 \times 10^{-11} \frac{m^3}{kg \times s^2}$ is the gravitational constant. Notice that for a given radius, a larger orbital velocity indicates a larger amount of mass, reflecting the fact that a stronger gravitational pull is necessary to hold a faster-moving object in orbit.

This formula is one of our best tools for measuring the amount of dark matter in the universe. When we apply it to the orbits of stars and gas clouds on the outskirts of spiral galaxies, we find that these galaxies contain at least 10 times as much matter as we can see in the form of stars. When we apply it to the orbits of galaxies within clusters of galaxies, we find that the total amount of matter in these clusters is as much as 50 times the amount of matter in the form of stars.

EXAMPLE: Calculate the mass of the Milky Way Galaxy within the Sun's orbit.

Solution: The Sun orbits the center of the Milky Way Galaxy with velocity $v = 220$ km/s $= 2.2 \times 10^5$ m/s. The Sun's orbital radius (distance) is 27,000 light-years, equivalent to (27,000 ly) $\times$ (9.46 $\times 10^{15}$ m/ly) $= 2.6 \times 10^{20}$ m. Using these values in the orbital velocity formula, we find

$$M_r = \frac{r \times v^2}{G} = \frac{(2.6 \times 10^{20}\ \mathrm{m}) \times \left(2.2 \times 10^5\ \frac{\mathrm{m}}{\mathrm{s}}\right)^2}{6.67 \times 10^{-11}\ \frac{\mathrm{m}^3}{\mathrm{kg} \times \mathrm{s}^2}}$$

$$= 1.9 \times 10^{41}\ \mathrm{kg}$$

The mass of the Milky Way Galaxy within the Sun's orbit is about 2×10^{41} kg, which we can convert to solar masses by dividing by the Sun's mass of about 2×10^{30} kg; it is equivalent to about 10^{11} (or 100 billion) solar masses. Note that most of this mass is not dark matter; evidence for dark matter comes from using the same equation for stars or gas clouds orbiting much farther from the center of the galaxy than our Sun.

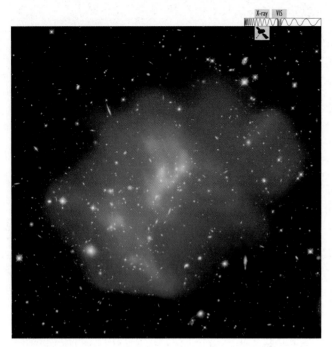

▲ **FIGURE 14.5**
A distant cluster of galaxies in both visible light and x-ray light. The visible-light photo (from the Hubble Space Telescope) shows the individual galaxies. The blue-violet overlay (from the Chandra X-Ray Observatory) represents different levels of x-ray emission from extremely hot gas in the cluster. Evidence for dark matter comes both from the observed motions of the visible galaxies and from the temperature of the hot gas. (The region shown is about 8 million light-years across.)

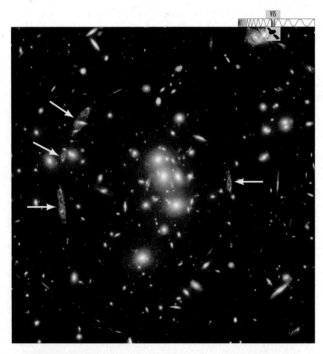

▲ **FIGURE 14.6**
This Hubble Space Telescope photo shows a galaxy cluster acting as a gravitational lens. The yellow elliptical galaxies are cluster members. The four small blue ovals (indicated by the arrows) are multiple images of a single galaxy that lies almost directly behind the cluster's center. (The picture shows a region about 1.4 million light-years across.)

about dark matter because it depends on the total mass of the cluster. The gas in most clusters is nearly in a state of *gravitational equilibrium*—that is, the outward gas pressure balances gravity's inward pull [Section 9.1]. In this state of balance, the average kinetic energies of the gas particles are determined primarily by the strength of gravity and hence by the amount of mass within the cluster. Because the temperature of a gas reflects the average kinetic energies of its particles, it tells us the average speeds of the gas particles, which we can then use to determine the cluster's total mass. The mass measurements obtained with this technique agree with those found by studying the orbital motions of the cluster's galaxies.

The third method relies on Einstein's general theory of relativity (see Tools of Science, page 172), which tells us that masses distort spacetime—the "fabric" of the universe. Massive objects can therefore act as **gravitational lenses** that bend light beams passing by them. Because the light-bending angle of a gravitational lens depends on the mass of the object doing the bending, we can measure the masses of objects by observing how much they distort light paths.

Figure 14.6 shows a striking example of how a cluster of galaxies can act as a gravitational lens. Notice the multiple blue ovals found at several positions around the central clump of yellow galaxies. These ovals all represent gravitationally lensed images of a single blue galaxy that lies almost directly behind the center of the cluster, at a much greater distance. We see multiple images because photons from the more distant galaxy do not follow straight paths to Earth. Instead, the cluster's gravity bends the photon paths, allowing light from the galaxy to arrive at Earth from a few slightly different directions (Figure 14.7). Each alternative path produces a separate, distorted image of the blue galaxy.

Multiple images of a gravitationally lensed galaxy are rare. They occur only when a distant galaxy lies directly behind the lensing cluster. However, single distorted images of gravitationally lensed galaxies are quite common. Figure 14.8 shows a typical example. This picture shows numerous normal-looking galaxies and several arc-shaped galaxies. The oddly curved galaxies are not members of the cluster, nor are they really curved. Rather, each is a normal galaxy lying far beyond the cluster, whose image has been distorted by the cluster's gravity.

Careful analyses of these images enable us to measure how massive these clusters must be to generate the observed distortions. It is reassuring that cluster masses derived in this way generally agree with those derived from galaxy velocities and x-ray temperatures. All three methods indicate that clusters of galaxies hold substantial quantities of dark matter, amounting to more than 40 times the total mass of the stars in the cluster's galaxies.

Alternatives to Dark Matter Astronomers have made a strong case for the existence of dark matter, but is it possible that there's a completely different explanation for the observations we've discussed? All the evidence for dark matter relies on our understanding of gravity. For individual galaxies, the case for dark matter rests primarily on applying Newton's laws of motion and gravity to observations of the orbital speeds of stars and gas clouds. We've used the same laws to make the case for dark matter in clusters, along with additional evidence based on gravitational lensing predicted by Einstein's general theory of relativity. It therefore seems that one of the following must be true:

1. Dark matter really exists, and we are observing the effects of its gravitational attraction.
2. There is something wrong with our understanding of gravity that is causing us to mistakenly infer the existence of dark matter.

We cannot completely rule out the second possibility, but most astronomers consider it very unlikely. Newton's laws of motion and gravity are among the most trustworthy tools in science. They have proven to make extremely accurate predictions of orbital motions in the solar system and in binary star systems.

We have therefore used them time and again to measure masses of celestial objects from their orbital properties. They have revealed the general relationships between the masses of stars and their outward appearances and have told us about many things we can't see directly, such as the masses of orbiting neutron stars in x-ray binaries and of black holes in active galactic nuclei. Einstein's general theory of relativity likewise stands on solid ground, having been repeatedly tested and verified to high precision in many observations and experiments. We therefore have good reason to trust our current understanding of gravity. While we should always keep an open mind about the possibility of future changes in our understanding, we will proceed with the rest of the chapter under the assumption that dark matter really does exist.

think about it Should the fact that we have three different ways of measuring cluster masses give us confidence that we really do understand gravity and that dark matter really does exist? Why or why not?

◆ What might dark matter be made of?

We have strong evidence that dark matter really exists, but what is it? There are two basic possibilities: (1) It could be made of *ordinary matter*, meaning the familiar type built from protons, neutrons, and electrons, but in forms too dark for us to detect with current technology; or (2) it could be made of *exotic matter*, meaning particle types that are different from those in ordinary atoms and that are dark because they do not interact with light at all.

A first step in distinguishing between the two possibilities is to measure the total amount of matter that is out there. Rather than focusing on total mass, astronomers usually focus on the average *density* of each of the forms of matter and energy in the universe, expressed as a percentage of the *critical density*—the density of mass-energy needed to make the geometry of the universe flat [Section 13.3]. The observations of galaxies and clusters that we have discussed indicate that the amount of matter in stars is only about 0.5% of the critical density, while the total amount of matter represents about 30% of the critical density. Clearly, there is a lot of dark matter that needs to be accounted for.

Ordinary Matter: Not Enough Could all this dark matter simply be ordinary matter in some hard-to-observe form? After all, matter doesn't necessarily need to be exotic to be dark. Astronomers consider matter to be "dark" as long as it is too dim for us to see at the great distances of the halo of our galaxy and beyond. Your body is dark matter, because our telescopes could not detect you if you were somehow flung into the halo of our galaxy. Planets, the "failed stars" known as brown dwarfs [Section 9.1], and even some faint red main-sequence stars also qualify as dark matter, because they are too dim for current telescopes to see in the halo.

However, calculations made with the Big Bang model allow scientists to place limits on the total amount of ordinary matter in the universe. Recall that, during the era of nucleosynthesis, protons and neutrons first fused to make deuterium nuclei (nuclei consisting of one proton and one neutron) and these then fused into helium [Section 13.2]. The fact that some deuterium nuclei still exist in the universe indicates that this process stopped before all the deuterium nuclei were used up. The amount of deuterium in the universe today therefore tells us about the density of protons and neutrons (ordinary matter) during the era

▲ **FIGURE 14.7**
A cluster's powerful gravity bends light paths from background galaxies to Earth. If light arrives from several different directions, we see multiple images of the same galaxy.

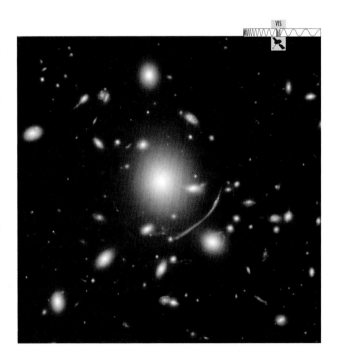

▶ **FIGURE 14.8**
Hubble Space Telescope photo of the cluster Abell 383. The thin, elongated galaxies are images of background galaxies distorted by the cluster's gravity. By measuring these distortions, astronomers can determine the total amount of mass in the cluster. (The region pictured is about 1 million light-years across.)

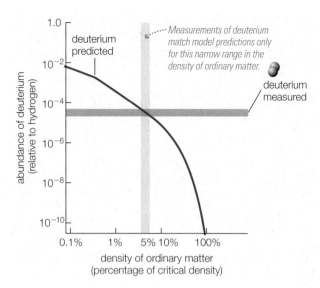

▲ **FIGURE 14.9**
This graph shows how the measured abundance of deuterium leads to the conclusion that the density of ordinary matter is about 5% of the critical density. The horizontal band shows the measured abundance of deuterium. The curve shows predictions for the deuterium abundance based on the Big Bang theory, and how the prediction depends on the density of ordinary matter in the universe. Notice that the prediction matches up with the measurements only in the gray vertical strip, which represents a density of about 5% of the critical density.

of nucleosynthesis: The higher the density, the more efficiently fusion would have proceeded. A higher density in the early universe would have therefore left less deuterium in the universe today, and a lower density would have left more deuterium.

Calculations based on the observed deuterium abundance indicate that the overall density of ordinary matter in the universe is around 5% of the critical density (Figure 14.9). Detailed studies of temperature patterns in the cosmic microwave background give the same result. Because this 5% is much too small to account for the one-quarter of the critical density that is dark, we conclude that most of the dark matter cannot be made from ordinary matter.

Exotic Matter: The Leading Hypothesis We are left with the idea that dark matter is made of exotic particles, and probably of a type of exotic particle that has not yet been discovered. Let's begin to explore this possibility by taking another look at a type of exotic particle that we first encountered in connection with nuclear fusion in the Sun: *neutrinos* [Section 8.1]. Neutrinos are dark by nature because they have no electrical charge and cannot emit electromagnetic radiation of any kind. Moreover, they are never bound together with charged particles in the way that neutrons are bound in atomic nuclei, so their presence cannot be revealed by associated light-emitting particles. In fact, neutrinos interact with other forms of matter through only two of the four forces: gravity and the *weak force* [Section 13.1]. For this reason, neutrinos are said to be *weakly interacting particles*.

The dark matter in galaxies cannot be made of neutrinos, because these very-low-mass particles travel through the universe at enormous speeds and can easily escape a galaxy's gravitational pull. But what if other weakly interacting particles exist that are similar to neutrinos but considerably heavier? They, too, would evade direct detection, but they would move more slowly, which means that their mutual gravity could hold together a large collection of them. Such hypothetical particles are called **weakly interacting massive particles**, or **WIMPs** for short. Note that they are subatomic particles, so the "massive" in their name is relative—they are massive only in comparison to lightweight particles like neutrinos. Such particles could make up most of the mass of a galaxy or cluster of galaxies, but they would be completely invisible in all wavelengths of light. Most astronomers now consider it likely that WIMPs make up the majority of dark matter, and hence the majority of all matter in the universe.

Searching for Dark Matter Particles The case for the existence of WIMPs seems fairly strong but is still circumstantial. Detecting the particles directly would be much more convincing, and physicists are currently searching for them in two different ways. The first and most direct way is with detectors that can potentially capture WIMPs from space. Because these particles are thought to interact only very weakly, the search requires building large, sensitive detectors deep underground, where they are shielded from other particles from space. As of 2018, these detectors have provided some tantalizing signals, but so far no clear evidence that dark matter particles really exist.

The second way scientists are currently searching for dark matter particles is with particle accelerators, and in particular with the Large Hadron Collider (see Tools of Science, page 217). None of the particles found as of 2018 has the characteristics of a WIMP, but the search is continuing.

think about it What do you think of the idea that much of the universe is made of as-yet-undiscovered particles? Can you think of other instances in the history of science in which the existence of something was predicted before it was discovered?

14.2 Gravity versus Expansion

The nature of dark matter remains enigmatic, but we are rapidly learning more about its role in the universe. We've already seen that dark matter appears to be the primary source of gravity in galaxies and galaxy clusters. We therefore suspect that the gravitational attraction of dark matter is what formed structures like galaxies and clusters of galaxies in the first place. Without dark matter, expansion would have dispersed all the matter in the universe before any galaxies could form. In this section, we'll explore the role of dark matter in producing galaxies and then consider an even bigger question: Could the gravity of dark matter be powerful enough that it will someday halt the expansion of the universe as a whole? In trying to answer that question, we will encounter something even more enigmatic than dark matter: indications that a mysterious form of *dark energy* is causing the expansion to accelerate.

◆ How did structures like galaxies form?

During the first few million years after the Big Bang, the universe expanded everywhere. However, the density of the universe was not completely uniform; recall that the temperature patterns we observe in the cosmic microwave background (see Figure 13.9) show that the density of matter varied slightly from place to place. Over time, gravity drew matter toward higher-density regions and away from lower-density regions, which made the density differences greater. Gradually, the stronger gravity in regions of enhanced density stopped the expansion of these regions, which then contracted to form protogalactic clouds, even as the universe as a whole continued (and still continues) to expand.

Formation of Galaxies and Galaxy Clusters   Dark matter completes the picture of galaxy formation we first discussed in Section 12.3. The protogalactic clouds that formed around density enhancements contained both ordinary matter, consisting of the hydrogen and helium gas produced by the Big Bang, and dark matter. Collisions among the ordinary gas particles could convert some of their orbital energy into radiative energy that escaped the cloud in the form of photons. This loss of orbital energy allowed the gas particles to sink toward the center of the cloud, where they formed a rotating disk. In contrast, dark matter made of weakly interacting particles could not produce photons and rarely interacted or exchanged energy with other particles. These dark matter particles therefore remained stuck in orbits far out in the galactic halo, in agreement with observations indicating that most dark matter is located at large distances from the centers of galaxies.

Galaxy clusters probably formed similarly. Early on, all the galaxies that would eventually constitute a cluster were flying apart with the expansion of the universe, but the overall gravity of the cluster—most of which is associated with dark matter—eventually reversed the trajectories of these galaxies. The galaxies ultimately fell back inward to form the cluster. On even larger scales, clusters themselves seem to be tugging on one another, hinting that even larger structures, called **superclusters**, may still be in the early stages of formation (Figure 14.10).

Large-Scale Structures   Over the past few decades, astronomers have measured the redshifts of millions of galaxies. These redshifts can be converted to distances with Hubble's law [Section 12.1], allowing astronomers to make three-dimensional maps of the distribution of galaxies in space. Such maps have revealed the existence of **large-scale structures** much vaster than clusters or superclusters of galaxies.

▶ **FIGURE 14.10**
This diagram shows how galaxies move within the Local Supercluster (or Laniakea), which is the region within the orange line. White dots represent locations of individual galaxies, with distances from the Milky Way inferred from Hubble's law. Galaxies flow toward denser regions of the supercluster along the white lines, which have been inferred from a combination of observations and modeling.

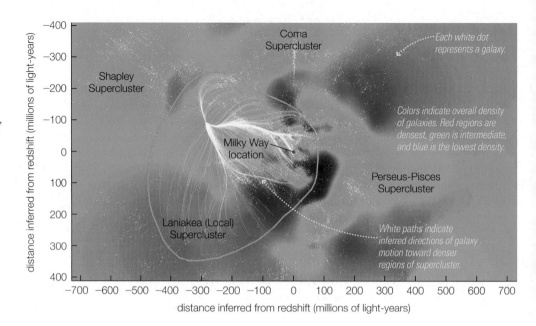

Figure 14.11 shows the distribution of galaxies in three slices of the universe, each extending farther out in distance than the one before. Our Milky Way Galaxy is located at the vertex at the far left, and each dot represents another galaxy. The slice at the left comes from one of the first surveys of large-scale structures, performed in the 1980s. This map showed that galaxies are not scattered randomly through space, but are instead arranged in huge chains and sheets that span many millions of light-years. Clusters of galaxies are located at the intersections of these chains. Between these chains and sheets of galaxies lie giant empty regions called **voids**. The other two slices show data from the more recent Sloan Digital Sky Survey, which has measured redshifts for more than a million galaxies spread across about one-fourth of the sky. Some of the structures in these pictures are amazingly large: The so-called *Sloan Great Wall*, clearly visible in the center slice, extends more than 1 billion light-years from end to end.

The universe may still be growing structures on these very large scales. However, there seems to be a limit to the size of the largest structures. If you look closely at the rightmost slice in Figure 14.11, you'll notice that the overall distribution of galaxies appears nearly uniform on scales larger than about a billion light-years. In other words, on very large scales the universe looks much the same everywhere, in agreement with what we expect from the *Cosmological Principle* [Section 12.2].

Why is gravity collecting matter on such enormous scales? Just as we suspect that galaxies formed from regions of slightly enhanced density in the early universe, we suspect that these larger structures formed from regions of enhanced density. Galaxies, clusters, superclusters, and the Sloan Great Wall probably all started as mildly high-density regions of different sizes. The voids in the distribution of galaxies probably started as mildly low-density regions.

If this picture of structure formation is correct, then the structures we see in today's universe mirror the original distribution of dark matter very early in time. Supercomputer models of structure formation in the universe can now simulate the growth of galaxies, clusters, and larger structures from tiny density enhancements as the universe evolves (Figure 14.12). The results of these models look remarkably similar to the slices of the universe in Figure 14.11, giving us confidence that the models are on the right track. Moreover, the patterns of mass distribution are consistent with the patterns of density enhancements revealed in maps of the cosmic microwave background. Overall, we now seem to have a basic picture, summarized in Figure 14.13, of how galaxies and larger structures developed from slight density enhancements in the early universe.

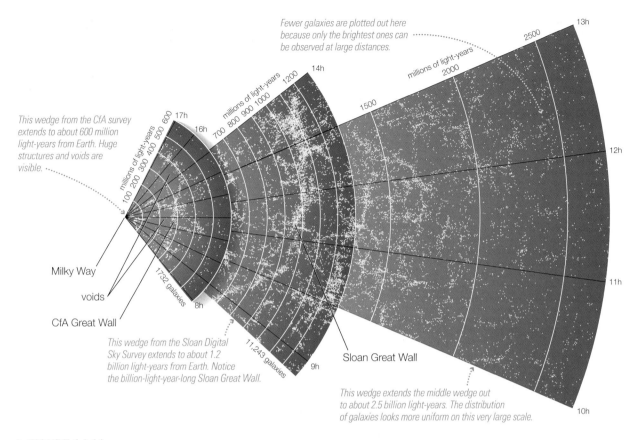

▲ **FIGURE 14.11**
Each of these three wedges shows a slice of the universe extending outward from our own Milky Way Galaxy. The dots represent galaxies, shown at their measured distances from Earth. We see that galaxies trace out long chains and sheets, surrounded by huge voids containing very few galaxies. (The wedges are shown flat but actually are a few angular degrees in thickness; the CfA wedge at left does not actually line up with the two Sloan wedges.)

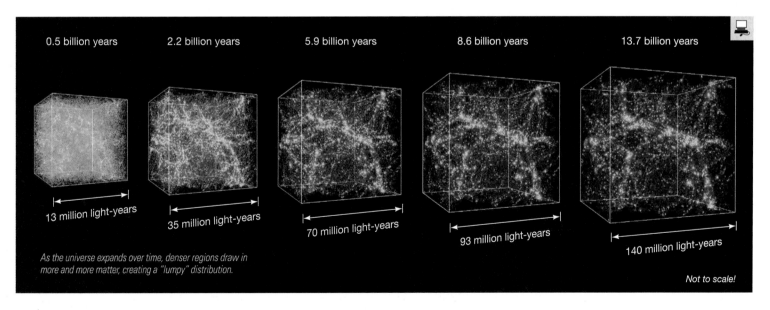

▲ **FIGURE 14.12**
Frames from a supercomputer simulation of structure formation. These five boxes depict the development of a cubical region that is now 140 million light-years across. The labels above the boxes give the age of the universe, and the labels below give the size of the box as it expands with time. Notice that the distribution of matter is only slightly lumpy when the universe is young (left frame). Structures grow more pronounced with time as the densest lumps draw in more and more matter.

cosmic context — FIGURE 14.13 Galaxy Evolution

All galaxies, including our Milky Way, developed as gravity pulled together matter in regions of the universe that started out slightly denser than surrounding regions. The central illustration depicts how galaxies formed over time, starting from the Big Bang in the upper left and proceeding to the present day in the lower right, as space gradually expanded according to Hubble's law.

Big Bang

380,000 years

1 billion years

TIME

(1) Dramatic inflation early in time is thought to have produced large-scale ripples in the density of the universe. All the structure we see today formed as gravity drew additional matter into the peaks of these ripples.

Inflation may have stretched tiny quantum fluctuations into large-scale ripples.

(2) Observations of the cosmic microwave background show us what the regions of enhanced density were like about 380,000 years after the Big Bang.

Photo by *Planck*

Variations in the cosmic microwave background show that regions of the universe differed in density by only a few parts in 100,000.

(3) Large-scale surveys of the universe show that gravity has gradually shaped early regions of enhanced density into a web-like structure, with galaxies arranged in huge chains and sheets.

Sloan Digital Sky Survey

The web-like patterns of structure observed in large-scale galaxy surveys agree with those seen in large-scale computer simulations of structure formation.

◆ Will the universe continue expanding forever?

We have seen that gravity has already halted and reversed the expansion of regions as large as clusters of galaxies, and also seems to be slowing the expansion of structures as large as superclusters. Could gravity be strong enough to someday halt the expansion of the universe and cause it to collapse in a cataclysmic "big crunch"? It is time to consider what modern astronomy can tell us about the ultimate fate of the universe.

If we think only about gravity (due to both ordinary and dark matter), the fact that the universe is expanding would seem to suggest only two possible fates for the universe: If gravity were strong enough, the expansion would someday halt and reverse; if the total strength of gravity were too weak, the expansion would continue forever, though gravity would be expected to slow it with time. However, much to astronomers' surprise, nearly two decades of observation now show that the expansion is not slowing at all, but instead is *accelerating* with time. It is this acceleration that has led astronomers to conclude that the universe must contain some repulsive force that counteracts gravity. The source of this repulsive force is not known, but it has been given a name: **dark energy**.

Four Expansion Models To understand the evidence that has led astronomers to conclude that the expansion is accelerating, let's consider how different amounts of gravity and repulsion cause the expansion rate to change with time, according to the following four general models (Figure 14.14):

- A **recollapsing universe**. In the case of extremely strong gravitational attraction and no repulsive force, the expansion would continually slow down with time and eventually would stop entirely and then reverse. Galaxies would come crashing back together, and the universe would end in a fiery "big crunch." We call this a *recollapsing* universe, because the final state, with all matter collapsed together, would look much like the state in which the universe began in the Big Bang.

- A **critical universe**. In the case of gravitational attraction that was not quite strong enough to reverse the expansion in the absence of a repulsive force, the expansion would decelerate forever, leading to a universe that would never collapse but would expand ever more slowly as time progressed. We call this a *critical* universe, because calculations show that it is what we would expect if the total density of the universe were the critical density and only matter (and not dark energy) contributed to this density.

- A **coasting universe**. In the case of weak gravitational attraction and no repulsive force, galaxies would always move apart at approximately the speeds they have today. We call this a *coasting* universe, because it is what we would find if no forces acted to change the expansion rate, much as a spaceship can coast through space at constant speed if no forces act to slow it down or speed it up.

- An **accelerating universe**. In the case of a repulsive force strong enough to overpower gravity, the expansion would *accelerate* with time, causing galaxies to recede from one another with ever-increasing speed.

The Fate of the Universe As we'll discuss in the next section, current evidence indicates that the accelerating model provides a much better fit to the data than any other model. Moreover, an inventory of all the matter known to exist in the universe—both ordinary and dark—shows that it adds up to enough to make only about 30% of the critical density. In other words, even if there were no repulsive force, there does not seem to be enough matter in the universe for gravity ever to stop the expansion. The fact that the expansion seems to be accelerating only strengthens the case arguing that the universe will continue to expand forever.

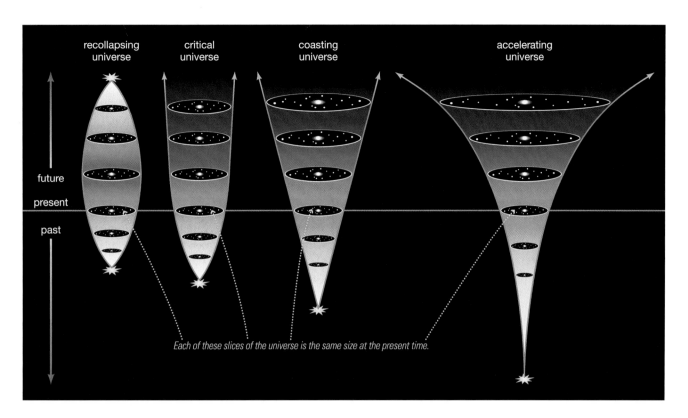

▲ **FIGURE 14.14**
Four general models for how the universal expansion rate could in principle change with time. Each diagram shows how the size of a circular slice of the universe changes with time in a particular model. The slices are the same size at the present time, marked by the red line, but the models make different predictions about the sizes of the slices in the past and future.

It may therefore seem that the ultimate fate of the universe is to expand forever, with stars ultimately all burning out so that all of existence will become ever colder and darker. But before you take this fate too seriously, keep in mind that "forever" is a very long time. It is only in the past century that we learned that we live in an expanding universe, and only in the past couple of decades that we learned that the expansion is accelerating. It doesn't seem too difficult to imagine that we'll make other surprising discoveries in the future, and that these might lead us to rethink what might happen between now and the end of time.

THE PROCESS OF SCIENCE IN ACTION

14.3 Evidence for Dark Energy

We built the case for the existence of dark matter on several strong lines of evidence, all based on our well-tested understanding of gravity through either Newton's laws or Einstein's general theory of relativity. The case for the existence of dark energy is somewhat more circumstantial, but has been growing stronger with each passing year. We will therefore use the evidence for dark energy as this chapter's example of the process of science in action, illustrating how careful observations sometimes require scientists to change their thinking about major features of our universe.

◆ **What is the evidence for dark energy?**

Two major lines of evidence support the case for the existence of dark energy: the accelerating expansion and the flatness of the universe. Let's investigate each in a little more detail.

The evidence for dark energy is based on observations of distant white dwarf supernovae that have enabled us to probe the fate of the universe in an entirely new way. Recall that white dwarf supernovae are excellent standard candles that allow us to measure the distances of very distant galaxies [Section 12.1]. We can therefore use observations of white dwarf supernovae to determine how the expansion rate of the universe has changed over long periods of time. In the 1990s, astronomers began seeking out and measuring white dwarf supernovae, hoping to learn the rate at which gravity has been slowing the universe's expansion over time. Much to their surprise, they discovered that the expansion is not slowing down at all, but speeding up.

▶ **Evidence for Acceleration**

The Accelerating Expansion The first key line of evidence for the existence of dark energy comes from the observations that indicate an accelerating expansion. After all, dark energy was proposed in order to explain the source of the repulsive force counteracting gravity, so there would be no need to invoke dark energy if the expansion were not accelerating. Figure 14.15 shows the evidence that has led astronomers to favor the accelerating model. The four solid curves show how the four general models (the same ones shown in Figure 14.14) predict that average distance between galaxies should have changed with time. The curves for the accelerating, coasting, and critical universes always continue upward as time increases, because in these cases the universe is always expanding. The steeper the slope, the faster the expansion. For the recollapsing universe, the curve begins on an upward slope but eventually turns around and declines as the universe begins to contract. All the curves pass through the same point and have the same slope at the moment labeled "now," because the current separation between galaxies and the current expansion rate in each case must agree with observations of the present-day universe.

▶ **FIGURE 14.15**
Data from white dwarf supernovae are shown along with four possible models for the expansion of the universe. Each curve shows how the average distance between galaxies changes with time for a particular model. A rising curve means that the universe is expanding, and a falling curve means that the universe is contracting. Notice that the supernova data fit the accelerating universe better than the other models.

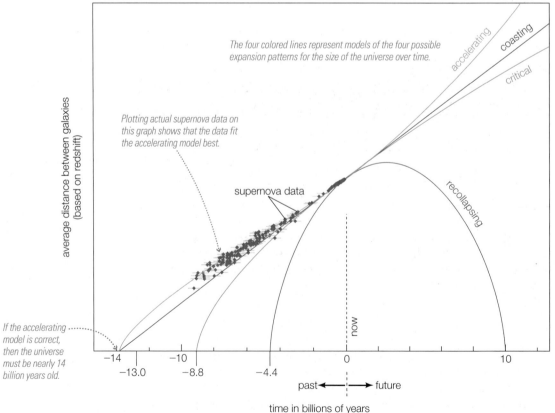

The black dots in Figure 14.15 show actual data, which are based on distance measurements using white dwarf supernovae as standard candles [Section 12.1]. (The horizontal line through each dot indicates the range of uncertainty in the measured lookback time.) Although there is some scatter in the data points, they clearly fit the curve for the accelerating model better than any of the other models. In other words, the observations agree best with a model of the universe in which the expansion is accelerating with time. We conclude that some force must counteract gravity, providing evidence for the existence of what we call *dark energy*.

see it for yourself Toss a ball in the air, and observe how it rises and falls. Then make a graph to illustrate your observations, with time on the horizontal axis and height on the vertical axis. Which universe model does your graph most resemble? What is the reason for that resemblance? How would your graph look different if Earth's gravity were not as strong? Would the time for the ball to rise and fall be longer or shorter?

Flatness and Dark Energy Additional support for the existence of dark energy comes from the overall geometry of the universe. Recall that Einstein's general theory of relativity tells us that the overall geometry can take one of three general forms—spherical, flat, or saddle shaped (see Figure 13.13)—and careful observations of the cosmic microwave background provide strong evidence that the actual geometry is flat (see Figure 13.15). As we discussed in Chapter 13, a flat geometry implies that the total density of *matter plus energy* in the universe is exactly equal to the critical density.

Now, recall that observations indicate that the total density of matter amounts to only about 30% of the critical density. In that case, the remaining 70% (approximately) of the universe's density must be in the form of energy. Tellingly, the amount of dark energy required to explain the observed acceleration of the expansion also is about 70% of the critical density. The startling conclusion: About 70% of the total mass-energy of the universe takes the form of dark energy.

A more detailed analysis of the temperature patterns in the cosmic microwave background has led astronomers to conclude that the inventory of matter and energy in the universe is as follows (Figure 14.16):

- Ordinary matter (made up of protons, neutrons, electrons) makes up about 5% of the total mass-energy of the universe. Note that this model prediction agrees with what we find from observations of deuterium in the universe. Some of this matter is in the form of stars (about 0.5% of the universe's mass-energy). The rest is presumed to be in the form of intergalactic gas, such as the hot gas found in galaxy clusters.
- Some form of exotic dark matter—most likely weakly interacting massive particles (WIMPs)—makes up about 27% of the mass-energy of the universe, in close agreement with what we infer from measurements of the masses of clusters of galaxies.
- Dark energy makes up the remaining 68% of the mass-energy of the universe, accounting both for the observed acceleration of the expansion and for the pattern of temperatures in the cosmic microwave background.

Moreover, the same models that give us the above inventory also lead to precise conclusions about the age of the universe. According to the model that best agrees with the data, the current age of the universe is about 13.8 billion years, with an uncertainty of about 0.1 billion years (100 million years). That is why, throughout this book, we have said that the universe is "about 14 billion years old." Note that this age is in good agreement with what we infer from Hubble's constant and observed changes in the expansion, and also agrees well with the fact that the oldest stars in the universe appear to be about 13 billion years old.

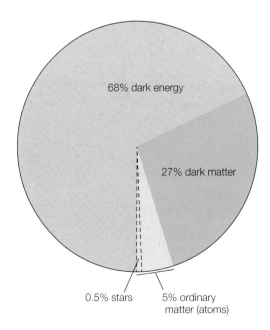

▲ FIGURE 14.16
This pie chart shows the proportion of each of the major components of matter and energy in the universe, based on current evidence.

cosmic context   FIGURE 14.17   Dark Matter and Dark Energy

Scientists suspect that most of the matter in the universe is *dark matter* we cannot see, and that the expansion of the universe is accelerating because of a *dark energy* we cannot directly detect. Both dark matter and dark energy have been proposed to exist because they bring our models of the universe into better agreement with observations, in accordance with the process of science. This figure presents some of the evidence supporting the existence of dark matter and dark energy.

① Dark Matter in Galaxies: Applying Newton's laws of gravity and motion to the orbital speeds of stars and gas clouds suggests that galaxies contain much more matter than we observe in the form of stars and glowing gas.

Orbital speeds of stars and gas clouds remain high even quite far from our galaxy's center...

...indicating that the visible portion of our galaxy lies at the center of a much larger volume of dark matter.

HALLMARK OF SCIENCE **A scientific model must seek explanations for observed phenomena that rely solely on natural causes.** Orbital motions within galaxies demand a natural explanation, which is why scientists proposed the existence of dark matter.

② Dark Matter in Clusters: Further evidence for dark matter comes from studying galaxy clusters. Observations of galaxy motions, hot gas, and gravitational lensing all suggest that galaxy clusters contain far more matter than we can directly observe in the form of stars and gas.

This cluster of galaxies acts as a gravitational lens to bend light from a single galaxy behind it into the multiple blue shapes in this photo. The amount of bending allows astronomers to calculate the total amount of matter in the cluster.

HALLMARK OF SCIENCE **Science progresses through creation and testing of models of nature that explain the observations as simply as possible.** Dark matter accounts for our observations of galaxy clusters more simply than alternative hypotheses.

③ Structure Formation: If dark matter really is the dominant source of gravity in the universe, then its gravitational force must have been what assembled galaxies and galaxy clusters in the first place. We can test this prediction using supercomputers to model the formation of large-scale structures both with and without dark matter. Models with dark matter provide a better match to what we observe in the real universe.

④ Universal Expansion and Dark Energy: The expansion of a universe consisting primarily of dark matter would slow down over time (because of gravity), but observations suggest that the expansion is actually speeding up. Scientists hypothesize that a mysterious *dark energy* is causing the expansion to accelerate. Models that include both dark matter and dark energy agree more closely with observations than models containing dark matter alone.

Supercomputer models in which dark matter is the dominant source of gravity show galaxies organized into strings and sheets similar in size and shape to those we observe in the real universe.

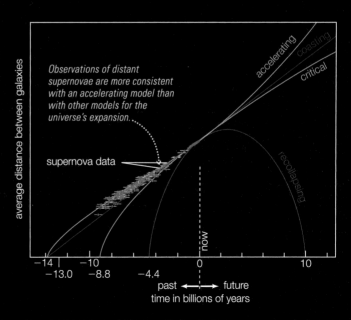

Observations of distant supernovae are more consistent with an accelerating model than with other models for the universe's expansion.

HALLMARK OF SCIENCE A scientific model makes testable predictions about natural phenomena. **If predictions do not agree with observations, the model must be revised or abandoned.** Observations of the universe's expansion have forced us to further modify our models of the universe to include dark energy along with dark matter.

The Nature of Dark Energy Together, the accelerating expansion and the apparent flatness of the universe make a strong case for the existence of what we have dubbed *dark energy*. However, scientists still have little idea of what dark energy might actually be. None of the four known forces in nature could provide a force to oppose gravity, and while some theories of fundamental physics suggest ways in which energy could fit the bill, no known type of energy produces the right amount of acceleration.

Continued observations of distant supernovae have the potential to tell us exactly how large an effect dark energy has had throughout cosmic history and whether the strength of this effect has changed with time. Already there are some intriguing hints. For example, it appears that the expansion was not accelerating immediately after the Big Bang, indicating that gravity was strong enough to slow the expansion for the first few billion years until dark energy became dominant. (The curve for the accelerating model in Figure 14.15 shows this scenario.) Interestingly, this type of behavior is consistent with an idea that Einstein once introduced but later disavowed in his general theory of relativity, leading some scientists to suggest that dark energy might successfully be described by a term (known as the "cosmological constant") in Einstein's equations that describe gravity. Nevertheless, even if this idea turns out to be correct, we remain a long way from an actual understanding of dark energy's nature. One way or another, the process-of-science stories of both dark matter and dark energy are likely to provide many more plot twists in coming decades.

Figure 14.17 (pages 248–249) summarizes the evidence we have discussed for dark matter and dark energy.

summary of key concepts

14.1 Evidence for Dark Matter

- **What is the evidence for dark matter?**

Applying Newton's laws of gravitation and motion to the orbits of stars and gas clouds in galaxies leads to the conclusion that the total mass of a galaxy is far larger than the mass of its stars. Because no detectable visible light is coming from this matter, we call it **dark matter**. Similar observations of galaxy clusters indicate that the total mass of dark matter is much more than the total mass in stars and gas.

- **What might dark matter be made of?**

The universe does not appear to contain enough ordinary matter made from protons and neutrons to account for all the dark matter we observe in the universe. We therefore suspect that most of the dark matter consists of **weakly interacting massive particles** of a kind not yet observed here on Earth.

14.2 Gravity versus Expansion

- **How did structures like galaxies form?**

Galaxies can form in regions of the universe where gravity, primarily due to dark matter, is strong enough to stop the expansion and pull gas into protogalactic clouds. The gravity of dark matter also appears to be responsible for arranging galaxies into clusters, superclusters, and **large-scale structures** such as the gigantic chains and sheets that surround great voids.

- **Will the universe continue expanding forever?**

Current evidence indicates that (1) the universe does not contain enough matter for gravity to halt the expansion and (2) the expansion of the universe is accelerating, due to a repulsive force that we attribute to **dark energy**. These facts suggest that the expansion will continue accelerating into the indefinite future.

THE PROCESS OF SCIENCE IN ACTION

14.3 Evidence for Dark Energy

- **What is the evidence for dark energy?**

There are two key lines of evidence for dark energy: (1) Observations of distant supernovae indicate that the expansion of the universe is accelerating, and something must be causing the repulsion that counteracts the attractive force of gravity. (2) The overall flat geometry of the universe indicates that the total density of matter plus energy is equal to the critical density, and because the total density of matter is only about 30% of the critical density, the remainder must be dark energy.

Investigations

For instructor-assigned homework and other learning materials, go to www.MasteringAstronomy.com.

Quick Quiz

Choose the best answer to each of the following. Explain your reasoning.

1. Measurements of how orbital speeds depend on distance from the center of our galaxy tell us that stars in the outskirts of the galaxy (a) orbit the galactic center just as fast as stars closer to the center. (b) rotate rapidly on their axes. (c) travel in straight, flat lines rather than elliptical orbits.
2. Dark matter is inferred to exist because (a) we see lots of dark patches in the sky. (b) it explains how the expansion of the universe can be accelerating. (c) we can observe its gravitational influence on visible matter.
3. Strong evidence for the existence of dark matter comes from observations of (a) our solar system. (b) the center of the Milky Way. (c) clusters of galaxies.
4. A photograph of a cluster of galaxies contains distorted images of galaxies that lie behind it at greater distances. This is an example of what astronomers call (a) dark energy. (b) spiral density waves. (c) gravitational lensing.
5. Given the observational evidence, is it possible that dark matter doesn't really exist? (a) No, the evidence for it is too strong for us to think that our understanding could be in error. (b) Yes, but only if there is something wrong with our current understanding of how gravity works on large scales. (c) Yes, but only if all the observations themselves are in error.
6. Based on current evidence, which of the following is considered the most likely candidate for the majority of the dark matter in galaxies? (a) subatomic particles that we have not yet detected in particle physics experiments (b) clusters of relatively dim, red stars (c) supermassive black holes
7. Which of the following provides evidence that the majority of the dark matter consists of weakly interacting massive particles? (a) measurements of the amount of deuterium in the early universe (b) measurements of the rotation curves of galaxies (c) measurements of the universe's current expansion rate
8. Which region of the early universe was most likely to become a galaxy? (a) a region whose matter density was lower than average (b) a region whose matter density was higher than average (c) a region whose matter density was average
9. Dark energy has been hypothesized to exist in order to explain (a) observations suggesting that the expansion of the universe is accelerating. (b) the high orbital speeds of stars far from the center of our galaxy. (c) explosions that seem to create giant voids between galaxies.
10. The major evidence for the idea that the expansion of the universe is accelerating comes from observations of (a) white dwarf supernovae. (b) the orbital speeds of stars within galaxies. (c) the motions of galaxies within galaxy clusters.
11. In which of these possible models for the universe would the expansion *not* continue forever? (a) a critical universe (b) an accelerating universe (c) a recollapsing universe
12. Current evidence suggests that ordinary matter (such as atoms) makes up about what percentage of the total mass-energy of the universe? (a) 90% (b) 50% (c) 5%

Short-Answer/Essay Questions

Explain all answers clearly, with complete sentences and proper essay structure if needed. An asterisk (*) designates a quantitative problem, for which you should show all your work.

13. *What Is Dark Matter?* Describe at least three possible constituents of dark matter. Explain how we would expect each to interact with light and how we might go about detecting its existence.
14. *Dark Matter and Life.* Explain at least two reasons one might argue that dark matter is (or was) essential for life to exist on Earth.
15. *Rotation Curves.* Draw and label a rotation curve for each of the following three hypothetical situations. Make sure the radius axis has approximate distances labeled.
    a. All the mass of the galaxy is concentrated in the center of the galaxy.
    b. The galaxy has a constant mass density inside 20,000 light-years, and zero density outside that.
    c. The galaxy has a constant mass density inside 20,000 light-years, and its enclosed mass increases proportionally to the radius outside that.
16. *Alternative Gravity.* How would gravity have to be different in order to explain the rotation curves of galaxies without the need for dark matter? Would gravity need to be stronger or weaker than expected at very large distances? Explain.
17. *The Future Universe.* Based on current evidence concerning the growth of structure in the universe, briefly describe what you would expect the universe to look like on large scales about 10 billion years from now.
18. *Dark Energy and Supernova Brightness.* When astronomers began measuring the brightnesses and redshifts of distant white dwarf supernovae, they expected to find that the expansion of the universe was slowing down. Instead, they found that it was speeding up. Were the distant supernovae brighter or fainter than expected? Explain why. (*Hint:* In Figure 14.15, the position of a supernova point on the vertical axis depends on its redshift. Its position on the horizontal axis depends on its brightness—supernovae seen further back in time are not as bright as those seen closer in time.)
19. *Dark Matter.* Overall, how convincing do you consider the case for the existence of dark matter? Write a short essay in which you explain what we mean by dark matter, describe the evidence for its existence, and discuss your opinion about the strength of the evidence.
*20. *Mass from Rotation Curves.* Study the line corresponding to spiral galaxy NGC 7541 in Figure 14.4.
    a. Use the orbital velocity formula to determine the mass (in solar masses) of NGC 7541 enclosed within a radius of 30,000 light-years from its center. (*Hint:* 1 light-year = 9.461×10^{15} m.)
    b. Use the orbital velocity formula to determine the mass of NGC 7541 enclosed within a radius of 60,000 light-years from its center.
    c. Based on your answers to parts a and b, what can you conclude about the distribution of mass in this galaxy?

15 Life in the Universe

LEARNING GOALS

15.1 The Search for Life in the Solar System
- What are the necessities of life?
- Could there be life elsewhere in our solar system?

15.2 The Search for Life Among the Stars
- How can we identify potentially habitable planets?
- Is there intelligent life beyond Earth?

 THE PROCESS OF SCIENCE IN ACTION

15.3 Evolution on Earth and Beyond
- What is the evidence for evolution?

We've discussed many fundamental questions about the universe in this book, but we have not yet discussed one of the most profound questions of all: Are we alone? The universe is filled with worlds beyond imagination, but we do not yet know whether any world besides our own has ever been home to life. In this chapter, we'll discuss the search for life in the universe, which includes the *search for extraterrestrial intelligence*, or SETI, which is often conducted with radio telescopes like those in the photo above.

▶ Life in the Universe

15.1 The Search for Life in the Solar System

The idea of life on other worlds is not new. Not long after he discovered the laws of planetary motion, Kepler wrote a story imagining the Moon to be inhabited. William Herschel, co-discoverer of Uranus, suspected life on virtually all the planets. Most famously, in the late 19th century Percival Lowell claimed to see networks of canals on Mars, which he argued were the mark of an advanced civilization. Lowell's ideas became the basis for H. G. Wells's novel *The War of the Worlds* and led to widespread public belief in Martians.

We now have enough spacecraft images to say confidently that no other world in our solar system has ever been home to a civilization. Nevertheless, we have good reason to suspect that some of these worlds have conditions that could allow more primitive life to survive, and scientists around the world are actively engaged in research that should help us learn whether life exists elsewhere in our solar system.

◆ **What are the necessities of life?**

The first step in the search for life in our solar system is deciding what it is that we are looking for. Defining life turns out to be surprisingly difficult, and science fiction writers have imagined all sorts of exotic life forms, such as crystal life or silicon-based life. Our initial searches, however, have generally focused on looking for life that is at least somewhat Earth-like and depends on living conditions similar to those required by life forms on Earth.

The Nature of Life on Earth We can learn about the requirements of life by studying the nature of life on Earth. All life on Earth shares a few common characteristics. For example, all known life uses DNA as its genetic material and builds proteins from the same set of building blocks, known as amino acids. Beyond such basic similarities, however, life on Earth is remarkably diverse.

For most of human history, people generally assumed that all living organisms belonged to either the *plant kingdom* or the *animal kingdom*. But in the mid-20th century, as scientists began to study microbial life in greater detail, we learned that most microbes do not belong to either of those groups. Today, biologists can identify relationships among living species by comparing the sequences of bases in their DNA. For example, two organisms whose DNA sequences differ in only one place for a particular gene are probably more closely related than two organisms whose gene sequences differ in five places. By making many such DNA comparisons, biologists have begun to assemble a *tree of life* that depicts the relationships among all living species. Figure 15.1 shows key features of the tree of life as it is known today. Notice that life on Earth is divided into three major *domains* (Bacteria, Archaea, and Eukarya), and that plants and animals represent only two small and fairly closely related branches of just one domain.

The tree of life offers several key lessons that can help us search for life beyond Earth. First and foremost, it tells us that the plant and animal life with which we are most familiar is *not* typical of most life on Earth. Instead, most life on Earth is microscopic, and this microscopic life exhibits far greater diversity in its genetic material than we find among plants and animals. This suggests that we are much more likely to find microscopic life on other worlds than to find large life forms similar to plants or animals.

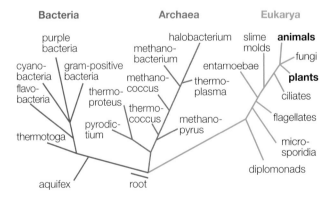

▲ FIGURE 15.1

Key features of the tree of life, showing relationships among species determined by comparison of DNA sequences in different organisms. Just two small branches represent *all* plant and animal species on Earth. (Only a few of the many relationships are shown and labeled.)

253

▲ FIGURE 15.2
This photograph shows a volcanic vent on the ocean floor that spews out hot, mineral-rich water. DNA studies indicate that the microbes living near these vents are evolutionarily older than nearly all other living organisms, suggesting that life may have first arisen in similar environments.

think about it The Copernican revolution brought about a major change in human perspective by teaching us that Earth is not the center of the physical universe. Recent discoveries in biology offer a similar lesson: Despite our apparent dominance over plants and other animals, microbes dominate Earth's overall ecology. In this sense, we are no more the "center" of Earth's biological universe than Earth is the center of the physical universe. Do you think this discovery should change our human perspective? Defend your opinion.

A second important lesson is that, despite the great diversity that we see in the tree of life, all life on Earth appears to be related. This suggests that all life evolved from a common ancestor. Organisms on branches located closer to the "root" of the tree of life must contain DNA that has changed less over time, suggesting that these organisms more closely resemble the organisms that lived early in Earth's history. Many of the modern-day organisms that lie closest to this root are microbes that live in very hot water around seafloor volcanic vents in the deep oceans (Figure 15.2). Unlike most life at Earth's surface, which depends on sunlight, these organisms obtain energy from chemical reactions in water heated volcanically by Earth itself. We conclude that sunlight is *not* a necessity for life, a fact that opens possibilities for life even on worlds where sunlight is very weak.

Other key lessons come from more detailed study of the diverse organisms in the tree of life and the conditions under which they live. Most of these organisms live under conditions that seem "extreme" to us. For example, while we humans require oxygen to breathe, most microbes do not. Moreover, many life forms can survive or even thrive in conditions that would be almost instantly lethal to us. Microbes near the seafloor volcanic vents thrive in water that is hotter than the normal boiling point; the high pressure of the deep sea keeps this very hot water liquid, but it would still kill any plant or animal quickly. In the freezing cold and dry valleys of Antarctica, microbes have been found living *inside* rocks, surviving on tiny droplets of liquid water and energy from sunlight. Other microscopic organisms have been found living in water-infiltrated rock several kilometers underground, in conditions of extreme acidity and saltiness, and in nuclear reactors where they are exposed to high doses of radiation. We have even found organisms, including some tiny animals (Figure 15.3), that can survive for at least some time in the near-vacuum of space, making it conceivable that life might migrate from one world to another by hitching a ride on rocks blasted into space by impacts. Organisms that live in extreme environments are collectively known

▶ FIGURE 15.3
It almost doesn't look real, but this photograph shows a tiny animal called a tardigrade (also called a "water bear") that is about a millimeter long. Tardigrades can survive an incredible range of "extreme" conditions, including at least some time in the near-vacuum of space.

as *extremophiles*, and their discovery shows that life can survive under a much wider range of conditions than we would have imagined just a few decades ago.

Requirements for Habitability We can use our understanding of the nature of life on Earth to determine the kinds of places that qualify as **habitable worlds**—worlds on which at least some type of Earth-like life could survive. Note that a habitable world need not necessarily have life, only conditions that could support life. So what exactly does Earth life need to survive?

If we compare all the different forms of life on Earth, including the extremophiles, we find that all share three basic requirements:

- A source of nutrients (elements and molecules) from which to build living cells
- Energy to fuel the activities of life, either from sunlight, from chemical reactions, or from the heat of Earth itself
- Liquid water, which serves several critical functions, including the transport of nutrients into living cells and of waste products out of them

Interestingly, only the requirement for liquid water poses much of a constraint. Molecules that could serve as building blocks for life are present almost everywhere—even on meteorites and comets. Many worlds are large enough to retain internal heat that could provide energy for life, and all worlds in our solar system have sunlight bathing their surfaces, although the inverse square law for light means that sunlight provides less energy to worlds farther from the Sun [Section 8.2]. Nutrients and energy should therefore be available to at least some degree on almost every planet and moon. In contrast, liquid water seems to be relatively rare. And although it is conceivable that other liquids—such as liquid methane or ethane—could substitute for liquid water on some worlds, these liquids are also rare. The search for habitable worlds and life therefore begins with a search for liquids, especially liquid water.

◆ Could there be life elsewhere in our solar system?

The requirement for liquid water or some other liquid seems to rule out the possibility of life on most worlds in our solar system. Mercury and the Moon lack any liquids at all, as do most small moons, asteroids, and comets. Venus's surface is far too hot for liquid water (though its clouds contain droplets of water laced with acid). The jovian planets may have droplets of liquid water and other liquids in their clouds, but the strong vertical winds on these planets mean that any droplets cannot stay liquid for long, making the planets unlikely homes for life. Nevertheless, several worlds *do* show evidence of liquid water or some other liquid mixture that seems to offer at least some possibility of life.

The Potential for Life on Mars Many scientists consider Mars to be the best candidate in our solar system for past or present life beyond Earth. Extensive evidence indicates that Mars had liquid water on its surface in the distant past [Section 5.2], in which case it would have been habitable at that time. If the period of habitability lasted long enough, it is conceivable that life could have arisen and taken hold on Mars. Even if life did not arise on Mars indigenously, it's possible that meteorites ejected from Earth by impacts might have carried microbes from Earth to Mars. If the microbes survived the journey and landed on Mars at a time when it was habitable, then they might have thrived upon arrival.

It is even possible that Mars remains habitable today, though not on its surface. The low atmospheric pressure ensures that liquid water cannot remain stable on the surface, so we do not expect any life to exist there. However, Mars still has plenty of ice and some internal heat, which means that there could be pockets of liquid water or water-infiltrated rock underground. Indeed, recent evidence suggests that Mars may have a large (20-kilometers-wide), salty lake under its south polar ice cap. If there was life on Mars in the distant past, then

▲ **FIGURE 15.4**
Artist's conception of the *Curiosity* rover's "ChemCam" instrument firing its laser at a rock on Mars. The laser vaporizes bits of rocky material so that it can analyze this material spectroscopically. The inset shows a rock that *Curiosity* first drilled (large hole) and then studied spectroscopically by using its laser to vaporize the dusty debris; the laser made the row of smaller holes.

some microbes might still live in such underground environments, much like the microbes that live in underground rock on Earth.

The first missions to search for life on Mars were the two *Viking* landers that touched down on Mars in 1976. Each lander was equipped with a robotic arm that scooped up soil samples and fed them into on-board, robotically controlled experiments. Although the experiments turned up some very interesting results, most scientists concluded that no life had been found, and that only follow-up missions could help us learn whether Mars ever harbored life.

Recent missions have taught us much more about the potential for life on Mars. The *Spirit* and *Opportunity* rovers, which landed on Mars in 2004, and the more sophisticated *Curiosity* rover, which landed on Mars in 2012, have been equipped with numerous instruments designed to help us learn about the habitability of Mars (Figure 15.4). As we discussed in Chapter 5, these rovers have found mineral evidence lending strong support to the idea that Mars once had lakes or seas on its surface.

Another intriguing discovery is the presence of low and varying levels of methane in the Martian atmosphere. This may be relevant to the search for life because methane cannot last long in the atmosphere unless it is continually replenished, and on Earth the primary source of methane replenishment is life. For Mars, the source is more likely to be volcanic or other geological activity, but such activity would still seem to increase the chance of finding life on Mars, since the same internal heat necessary for methane emission could probably keep some subsurface water in liquid form. Scientists hope to learn more by continuing to study the methane levels and from data about the Martian interior that should be coming soon from the *Mars InSight* mission, scheduled to land on Mars shortly after this book goes to press in 2018.

think about it Find a list of current Mars missions and major results from these missions. What is the latest assessment of the possibility of life on Mars?

Life on Jovian Moons After Mars, the next most likely candidate for life in our solar system is probably Jupiter's moon Europa. Recall that strong evidence suggests that Europa contains a deep ocean of liquid water beneath its icy crust [Section 6.1]. The ice and rock from which Europa formed undoubtedly included the chemical ingredients necessary for life, and Europa's internal heat (due primarily to tidal heating) should be strong enough to power volcanic vents on the seafloor. Given that life thrives around undersea volcanic vents on Earth, it seems reasonable to imagine similar life on Europa. And if life got started near such vents, it could have spread farther in Europa's oceans from there.

The possibility of life on Europa is especially interesting because, unlike potential life on Mars, it would not necessarily have to be microscopic. After all, the several kilometers of surface ice that hide Europa's suspected ocean could also hide large creatures swimming within it. However, the potential energy sources for life on Europa are far more limited than the energy sources for life on Earth, mainly because sunlight could not fuel photosynthesis in the subsurface ocean. As a result, most scientists suspect that any life on Europa would probably be small and primitive.

As we discussed in Chapter 6, Jupiter's moons Ganymede and Callisto may also have subsurface oceans, though these moons would have even less energy for life than Europa. If they have life at all, it is almost certainly small and primitive. Nevertheless, Europa, Ganymede, and Callisto offer at least the possibility that Jupiter alone could be orbited by more worlds with life than we find in all the rest of the solar system.

Saturn's moon Titan offers another enticing place to look for life. Its surface is far too cold for liquid water, but it has lakes and rivers of liquid methane and ethane [Section 6.1]. Although many biologists think it unlikely, it is possible that these liquids could support life as does water on Earth. Titan might also have liquid water or a colder ammonia–water mixture deep underground, so even water-based life is possible.

Saturn's moon Enceladus generates excitement among scientists searching for life because it has ice fountains that spray out from the interior, and detailed studies from the *Cassini* spacecraft indicate that these fountains are almost certainly powered by an underground ocean (Figure 15.5). If there is life in this ocean, the icy fountains might spray out flash-frozen life forms, making it relatively easy to identify the life with future spacecraft. Beyond Saturn, there is at least some possibility that subsurface water—and potentially life—could exist on other solar system bodies, including some of the moons of Uranus, Neptune's moon Triton, and perhaps even Pluto.

Overall, we have found at least six potentially habitable worlds in our solar system: Mars, Europa, Ganymede, Callisto, Titan, and Enceladus. Although we cannot yet know whether any of these worlds will prove to have life, the possibility certainly exists.

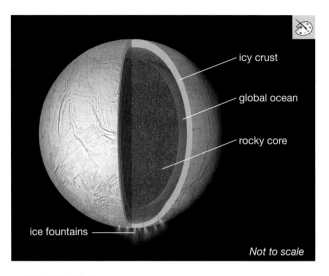

▲ **FIGURE 15.5**
Artist's conception of the interior of Enceladus, based on study of the moon's gravity made by the *Cassini* spacecraft. The ice fountains in the southern hemisphere are presumably driven by energy from the subsurface ocean; the fountains emerge in places where the crust is fractured so water can rise up from below. If life exists in the subsurface ocean, we might find evidence of it spraying into space from the ice fountains. (Recent evidence suggests that Jupiter's moon Europa might also have geysers spraying samples from its subsurface ocean into space.)

15.2 The Search for Life Among the Stars

We already know of thousands of planets orbiting stars besides our Sun, and the statistics indicate that planetary systems orbit most of the more than 100 billion stars in our galaxy. These numbers make prospects for life elsewhere seem quite good, but numbers alone don't tell the whole story. In this section, we'll consider the prospects for life on worlds orbiting other stars.

Before we begin, it's important to distinguish between *surface* life like that on Earth and *subsurface* life like that we envision as a possibility on Mars or Europa. While large telescopes could in principle detect the presence of surface life on

tools of science — Planetary Spacecraft

The search for life in our solar system differs from the search for life beyond it not only because our telescopes can see more detail on nearby worlds but also, more importantly, because we are able to send spacecraft to visit these worlds. In contrast, the stars are so far away that it would take current spacecraft tens of thousands of years to reach even the nearest of them. Our ability to send spacecraft to the planets in our own solar system means we can get close-up images and detailed spectra, and by landing on these worlds we can in principle conduct a direct search for life.

With the exception of the *Apollo* astronaut visits to the Moon between 1969 and 1972, all spacecraft sent to other worlds have been robotic. These spacecraft have computers to control their major components, power sources such as solar cells or nuclear batteries for energy, and scientific instruments such as cameras, spectrometers, and devices to measure magnetic fields. Spacecraft that fly through atmospheres or land on surfaces typically also carry instruments to measure winds, temperatures, and atmospheric or soil composition. The spacecraft carry radio antennas for communication, so that they can receive operating instructions from Earth and radio their data back to Earth.

Some spacecraft simply fly past a world, taking pictures and collecting data during the short time when they are close by. Others go into orbit around a world for longer-term study or land on a surface or probe an atmosphere. To date, we have sent robotic spacecraft to all the terrestrial and jovian planets, to the dwarf planets Ceres and Pluto, and to many moons, asteroids, and comets.

Current spacecraft study of potentially habitable worlds focuses on Mars. Every 26 months, Earth and Mars come into a favorable alignment for spacecraft missions, and scientists hope to continue sending missions to Mars at these launch opportunities. Each mission builds upon previous discoveries, with the eventual goal of learning whether life exists or has ever existed on Mars. A particularly exciting group of missions is scheduled for launch to Mars in 2020, including NASA's *Mars 2020* rover and the European *ExoMars* rover.

Scientists also hope to send new missions to learn about the possibility of life on Jupiter's moon Europa and Saturn's moons Titan and Enceladus. However, funding for such missions has been difficult to come by. As this book was going to press, the European Space Agency was working toward a 2022 launch date for its *Jupiter Icy Moons Explorer (JUICE)* mission to study Europa and other moons of Jupiter, while NASA was in more preliminary stages of planning for detailed study of Europa through a mission called the *Europa Clipper*. Scientists are also developing plans for new missions to Titan and Enceladus and hope funding for such missions will come through soon.

The bottom line is that while spacecraft exploration of the planets remains difficult, we are making great progress. If and when we first discover life on another world, it will almost certainly be thanks to visits by spacecraft.

extrasolar planets, no foreseeable technology could detect life that is hidden deep underground in other star systems (unless it has a noticeable effect on the planet's atmosphere). We therefore will focus on the search for life on planets with habitable surfaces, meaning surfaces with temperatures and pressures that could allow liquid water to exist.

◆ How can we identify potentially habitable planets?

Although we have identified thousands of planets around other stars, current technology remains far too limited to allow us to observe them in enough detail to determine whether any of these planets have habitable surfaces. Nevertheless, we can use our understanding of planetary geology and atmospheres to make educated guesses about whether any particular planet might be habitable.

Four Factors in Surface Habitability If we think back on all the ideas we've discussed in this book, we find four major factors that seem to explain why Earth has been so hospitable to the long-term evolution of life:

1. A distance from the Sun that places our planet within the Sun's **habitable zone** (Figure 15.6), meaning a distance that is great enough to allow water vapor to condense as rain and make oceans, but not so great that all the water freezes

2. Volcanism that released trapped gases from our planet's interior, including water vapor and carbon dioxide, to make the atmosphere and oceans

3. A climate that has remained suitable for liquid water over billions of years, which on Earth has been possible in large part because of the way plate tectonics supports a climate-regulating carbon dioxide cycle

4. A planetary magnetic field that protects our atmosphere from the solar wind

We might therefore expect that other planets that have the same four factors would look very Earth-like and therefore be good candidates for harboring life. So a key question in the search for life is how many planets we should expect to have these factors. While the answer is debatable, there is at least some reason to think that the answer will be a great many planets.

We have already discovered many planets that orbit within habitable zones, and the statistics indicate that our galaxy may have tens of billions of such planets. Finding planets that have the latter three factors might seem more difficult, but many planetary geologists suspect that all three may be present on almost any planet that lies in the habitable zone and has a size and composition similar to those of Earth. For example, we would expect any such planet to have a hot enough interior for volcanism, which would allow for outgassing, and it seems

▶ **FIGURE 15.6**
A star's habitable zone is the region in which a planet of appropriate size and composition could potentially have liquid water on its surface. This figure shows approximate habitable zones around our Sun, a star with one-half the mass of the Sun (spectral type K), and a star with one-tenth the mass of the Sun (spectral type M), shown to scale. The habitable zone becomes increasingly smaller and closer in for stars of lower mass and luminosity.

likely that most such planets would also have a convecting core layer and fast enough rotation to generate a magnetic field. It's less clear whether we should expect Earth-size planets to have plate tectonics. After all, Venus lacks this feature even though it is nearly the same size as Earth. However, recall that the lack of plate tectonics on Venus may be a consequence of its high surface temperature, which comes from the runaway greenhouse effect that occurred because Venus is too close to the Sun to be in the habitable zone [Section 5.2]. If this hypothesis is correct, then Venus might have had plate tectonics and a habitable surface if it had been born a little farther from the Sun. It may also be possible to have climate stability without plate tectonics.

The "recipe" for a habitable planet may therefore be as simple as having a size and composition similar to those of Earth while orbiting within the habitable zone. Because these characteristics seem likely to be common, Earth-like planets with habitable surfaces may be common as well. Some scientists think the range of habitable planets may be even wider, perhaps including some super-Earths and water worlds of the types discussed in Chapter 7. Some scientists even speculate that "orphan planets"—planets that were gravitationally ejected from the star systems in which they were born—could have life, because some models indicate that such planets could have thick atmospheres that would allow for liquid water even without the heat of a nearby star (Figure 15.7).

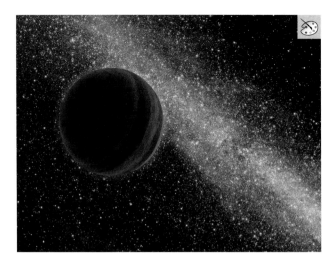

▲ **FIGURE 15.7**
This artist's conception shows an orphan planet—a planet that has escaped from its star system—as it might appear in the infrared. Orphan planets are too far from any star to be observable with reflected visible light and too cool to emit visible light of their own, but they would emit infrared light.

Signatures of Life Actually observing the surfaces of potentially habitable worlds will require far more powerful telescopes than we have today. However, we might still learn whether such worlds harbor life through spectroscopy that may be possible with telescopes not too much more advanced than today's. For example, moderate-resolution infrared spectra could in principle reveal the presence and abundance of atmospheric gases including carbon dioxide, ozone, methane, and water vapor (Figure 15.8), and careful analysis might tell us whether a planet has life. For example, atmospheric oxygen on Earth is a direct result of photosynthetic life. Abundant oxygen in the atmosphere of a distant world might therefore indicate the presence of life, since we know of no nonbiological way to produce an oxygen abundance as high as Earth's. Additional evidence might come from other gases released by life. Scientists are working to improve our understanding of how life influences atmospheric chemistry in the hope that we will be able to recognize particular gas combinations as signatures of life.

◆ Is there intelligent life beyond Earth?

If habitable planets and life turn out to be common, it's natural to wonder if other worlds might also have intelligent life and civilizations. If such civilizations exist, we might be able to find them simply by listening for signals that they are sending into interstellar space, either in deliberate attempts to contact other civilizations or as a means of communicating among themselves. The search for signals from other civilizations is generally known as the **search for extraterrestrial intelligence,** or **SETI** for short. To judge the chances of SETI success, we'd want to know how many civilizations are broadcasting such signals right now.

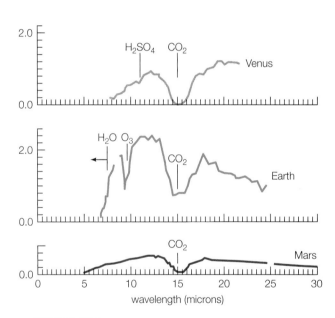

▲ **FIGURE 15.8**
The infrared spectra of Venus, Earth, and Mars as they might be seen from afar, showing absorption features that point to the presence of carbon dioxide (CO_2), ozone (O_3), and sulfuric acid (H_2SO_4) in their atmospheres. While carbon dioxide is present in all three spectra, only Earth has appreciable oxygen (and hence ozone)—a product of photosynthesis. If we could make similar spectral analyses of distant planets, we might detect atmospheric gases that would indicate life.

▶ The Drake Equation

The Drake Equation Given that we do not even know whether microbial life exists anywhere beyond Earth, we certainly can't know how many other civilizations exist. Nevertheless, for the purposes of planning a search for extraterrestrial intelligence, it's helpful to have an organized way of thinking about the number of civilizations that might be out there. In 1961, astronomer Frank Drake wrote a simple equation, now known as the **Drake equation**, designed

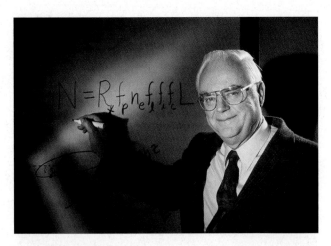

▲ FIGURE 15.9
Astronomer Frank Drake, with the equation he first wrote in 1961. Some of the scientific ideas behind the equation have changed since that time; therefore, with Dr. Drake's approval, we use a modified form of his original equation in this book.

to summarize the factors that would determine the number of civilizations we might contact (Figure 15.9). To keep our discussion simple, we'll consider only the number of potential civilizations in our own galaxy.

In principle, the Drake equation gives us a way to calculate the number of civilizations capable of interstellar communication that are currently sharing the Milky Way Galaxy with us. In a form slightly modified from the original, the Drake equation looks like this:

$$\text{Number of civilizations} = N_{HP} \times f_{life} \times f_{civ} \times f_{now}$$

This equation will make sense once you understand the meaning of each factor:

- N_{HP} is the number of habitable planets in the galaxy—that is, the number of planets that are *potentially* capable of having life.
- f_{life} is the fraction of habitable planets that actually *have* life. For example, $f_{life} = 1$ would mean that all habitable planets have life, and $f_{life} = 1/1,000,000$ would mean that only 1 in a million habitable planets has life. The product $N_{HP} \times f_{life}$ tells us the number of life-bearing planets in the galaxy.
- f_{civ} is the fraction of the life-bearing planets upon which a civilization capable of interstellar communication *has at some time* arisen. For example, $f_{civ} = 1/1000$ would mean that such a civilization has existed on 1 out of 1000 planets with life, while the other 999 out of 1000 have not had a species that learned to build radio transmitters, high-powered lasers, or other devices for interstellar communication. When we multiply this factor by the first two factors to form the product $N_{HP} \times f_{life} \times f_{civ}$, we get the total number of planets upon which intelligent beings have evolved and developed a communicating civilization at some time in the galaxy's history.
- f_{now} is the fraction of these civilization-bearing planets that happen to have a civilization *now*, as opposed to, say, millions or billions of years in the past. This factor is important because we can hope to contact only civilizations that are broadcasting signals we could receive at present. (In estimating f_{now}, we assume that the light-travel time for signals from other stars has been taken into account.)

Because the product of the first three factors tells us the total number of civilizations that have *ever* arisen in the galaxy, multiplying by f_{now} tells us how many civilizations we could potentially make contact with today. In other words, the result of the Drake equation is the number of civilizations that we might hope to contact.

think about it Try the following sample numbers in the Drake equation. Suppose that there are 1000 habitable planets in our galaxy, that 1 in 10 habitable planets has life, that 1 in 4 planets with life has at some point had an intelligent civilization, and that 1 in 5 civilizations that have ever existed is in existence now. How many civilizations would exist at present? Explain.

Estimates with the Drake Equation Unfortunately, we can make a reasonably educated guess only about the first factor in the Drake equation, the number of habitable planets (N_{HP}). As we've discussed, we know that planetary systems are common, and we have at least some good reason to believe that habitable planets could be found in many or most planetary systems. Although there is still some uncertainty, it seems reasonable to hypothesize that our galaxy could have as many as 100 billion habitable planets.

The factor f_{life} presents more difficulty. For the moment, we have no reliable way to estimate the fraction of habitable planets upon which life actually arose. The problem is that we cannot generalize when we have only one example (Earth) to study. However, scientists have come up with plausible ideas for how life arose on Earth, and if these ideas are correct, then it would seem that life should be likely on other planets with similar conditions. Moreover, geological evidence indicates that life arose within a few hundred million years after Earth's formation, and perhaps only tens of millions of years after conditions became conducive for life. This is a short amount of time compared to Earth's history, suggesting that the origin of life was fairly "easy." Together, the biological and geological evidence seem to point to the idea that most or all habitable planets will end up with life, making the fraction f_{life} close to 1. Of course, until we have solid evidence that life really did arise anywhere else, it is also possible that Earth is very lucky to have life and that f_{life} might be so close to zero that life has never arisen on any other planet in our galaxy.

The factor f_{civ}, which represents the fraction of life-bearing planets on which a civilization capable of interstellar communication has arisen, is even more difficult. On Earth, life of some type has flourished for nearly 4 billion years, but our civilization has been capable of sending radio signals to the stars for less than a century. This suggests that even if it was "easy" for life to arise on Earth, it was much harder for that life to evolve into intelligent life and civilization. Some people therefore argue that while life may be common, civilizations will not be. On the other hand, if we assume that Earth is typical, then we might expect that any planet with life would end up with a civilization by the time it reached an age similar to Earth's current age of $4\frac{1}{2}$ billion years. In that case, if habitable planets really are common, as we have guessed, and if life can arise fairly easily, then there might be thousands or millions of civilizations that have arisen at some time in our galaxy (Figure 15.10).

We can now see the importance of the final factor in the Drake equation, f_{now}. For the sake of argument, let's assume that life and intelligence have arisen on thousands or millions of worlds at some time. Then the number of civilizations that exist *now* depends on whether these civilizations still survive. Consider our own example. In the roughly 12 billion years during which our galaxy has existed, we have been capable of interstellar communication via radio for only about 60 years. If we were to destroy ourselves tomorrow, then other civilizations could have received signals from us during only 60 years out of the galaxy's 12-billion-year existence, equivalent to 1 part in 200 million of the galaxy's history. If such a short technological lifetime is typical of civilizations, then f_{now} would be only 1/200,000,000, and some 200 million civilization-bearing planets would need to have existed at one time or another in the Milky Way in order for us to have a decent chance of finding another civilization out there now.

However, we'd expect f_{now} to be so small only if we are on the brink of self-destruction, because the fraction will grow larger for as long as our civilization survives. That is, if civilizations are at all common, survivability is the key factor in whether any are out there now. If most civilizations self-destruct shortly after achieving the technology for interstellar communication, then we are almost certainly alone in the galaxy at present. But if most survive and thrive for thousands or millions of years, the Milky Way may be brimming with civilizations—and most of them would be far more advanced than our own.

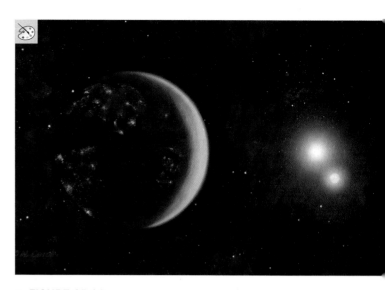

▲ **FIGURE 15.10**
This painting shows a hypothetical planet much like Earth—with the lights of its own civilization visible on the night side—orbiting a close binary star system.

think about it Describe a few reasons why a civilization capable of interstellar communication would also be capable of self-destruction. Overall, do you believe our civilization can survive for thousands or millions of years? Defend your opinion.

▲ **FIGURE 15.11**
The Allen Telescope Array in Hat Creek, California, is used to search for radio signals from extraterrestrial civilizations.

SETI If there are indeed other civilizations out there, then in principle we ought to be able to make contact with them. Based on our current understanding of physics, it seems likely that even advanced civilizations would communicate much as we do—by encoding signals in radio waves or other forms of light. Most SETI researchers use large radio telescopes to search for alien radio signals (Figure 15.11). These telescopes can scan millions of radio frequency bands simultaneously. A few researchers are also beginning to check other parts of the electromagnetic spectrum. For example, some scientists use visible-light telescopes to search for communications encoded as laser pulses. Of course, it's also possible that advanced civilizations have invented communication technologies that we cannot even imagine, let alone detect. In that case, we might be capable of detecting communications only from relatively young civilizations like our own, since more advanced civilizations would be using technologies that we do not yet have.

Alien Visitation and Fermi's Paradox A basic assumption throughout this chapter has been that we do not yet know whether other civilizations exist. However, opinion polls show that many people believe that we have already been visited by aliens in UFOs. Could it be that the evidence for extraterrestrial life is already here?

Claims of UFOs and alien encounters frequently make the news. However, while plenty of evidence has been offered for scientific examination, so far none of it has been enough to convince scientists that the claims of visitation are real. It's possible that more convincing evidence will be found in the future, but for now there is no scientific evidence to back claims of aliens on Earth.

This lack of scientific evidence for alien life—either through SETI contact or through evidence of visitation to Earth—raises questions of its own. Although interstellar travel remains far beyond our current technology, the rate at which our technology is advancing suggests that we should eventually be capable of such travel. And by modeling the way a civilization might spread among the stars, we find that once a civilization was capable of star travel at even a modest fraction of the speed of light, it would take that civilization no more than a few tens of millions of years to colonize most or all of the galaxy. This time is so short compared to the age of our galaxy that it leads us to an astonishing conclusion: If civilizations are common, then someone else should have been capable of creating a galactic civilization by now. In fact, it should have been done a long time ago.

To see why, let's take some sample numbers. Suppose the factors in the Drake equation tell us that the overall odds of a civilization arising around a star are about the same as your odds of winning the lottery: 1 in a million. Taking a low estimate of 100 billion stars in the Milky Way Galaxy, we come up with some 100,000 civilizations in our galaxy alone. Moreover, current evidence suggests that stars and planetary systems like our own could have formed for at least 5 billion years before our solar system was even born, in which case the first of these 100,000 civilizations would have arisen at least 5 billion years ago. Others would have arisen, on average, about every 50,000 years (because 5 billion years ÷ 100,000 civilizations = 50,000 years between each civilization). Under these assumptions, we would expect the youngest civilization besides ours to be some 50,000 years ahead of us technologically, and most would be millions or billions of years ahead of us.

We thereby encounter a strange paradox: Plausible arguments suggest that a galactic civilization should already exist, yet we have so far found no evidence of such a civilization. This paradox is often called *Fermi's paradox*, after the Nobel Prize–winning physicist Enrico Fermi. During a 1950 conversation with other scientists about the possibility of extraterrestrial intelligence, Fermi responded to speculations by asking "So where is everybody?"

This paradox has many possible solutions, but broadly speaking we can group them into three categories:

1. *We are alone.* There is no galactic civilization because civilizations are extremely rare—so rare that we are the first to have arisen on the galactic scene, perhaps even the first in the universe.

2. *Civilizations are common, but no one has colonized the galaxy.* There are at least three possible reasons why this might be the case. Perhaps interstellar travel is much harder or more expensive than we might guess, and civilizations are unable to venture far from their home worlds. Perhaps the desire to explore is unusual, and other societies either never leave their home star systems or stop exploring before they've colonized much of the galaxy. Most ominously, perhaps many civilizations have arisen, but they have all destroyed themselves before achieving the ability to colonize the stars.

3. *There IS a galactic civilization,* but it has not yet revealed its existence to us.

We do not know which, if any, of these explanations is the correct solution to Fermi's paradox. However, each category of solution has astonishing implications for our own species.

Consider the first solution—that we are alone. If this is true, then our civilization is a remarkable achievement. It implies that through all of cosmic evolution, among countless star systems, we are the first piece of our galaxy or the universe ever to know that the rest of the universe exists. Through us, the universe has attained self-awareness. Some philosophers and many religions argue that the ultimate purpose of life is to become truly self-aware. If so, and if we are alone, then the destruction of our civilization and the loss of our scientific knowledge would represent an inglorious end to something that took the universe some 14 billion years to achieve. From this point of view, humanity becomes all the more precious, and the collapse of our civilization would be all the more tragic.

The second category of solutions has much more terrifying implications. If thousands of civilizations before us have all failed to achieve interstellar travel on a large scale, what hope do we have? Unless we somehow think differently than all other civilizations, this solution says that we will never go far in space. Because we have always explored when the opportunity arose, this solution almost inevitably leads to the conclusion that failure will come about because we destroy ourselves. We can only hope that this answer is wrong.

The third solution is perhaps the most intriguing. It says that we are newcomers on the scene of a galactic civilization that has existed for millions or billions of years before us. Perhaps this civilization is deliberately leaving us alone for the time being and will someday decide the time is right to invite us to join it.

No matter what the answer turns out to be, learning it is sure to mark a turning point in the brief history of our species. Moreover, this turning point is likely to be reached within the next few decades or centuries. We already have the ability to destroy our civilization. If we do so, then our fate is sealed. But if we survive long enough to develop technology that can take us to the stars, the possibilities seem almost limitless.

THE PROCESS OF SCIENCE IN ACTION

15.3 Evolution on Earth and Beyond

Up to this point, we have been discussing the prospects for finding habitable worlds and life beyond Earth. However, we have not yet said exactly what life is, or how we would recognize it if we came across it. For example, while aliens that look like plants or animals might be obviously "alive," what about more exotic possibilities, such as life based on an entirely different chemistry from that of life on Earth?

Although we do not yet have a foolproof definition of life, scientists have identified two key characteristics that we assume all life must share. The first is an ability to reproduce, so that the death of one organism does not mean the end of a species. The second is the ability to adapt to environmental changes, because such changes are inevitable on any world, and life would quickly perish if it could not adapt. Today, our understanding of both of these characteristics is incorporated into a single, unifying theory of biology: the *theory of evolution*. Because this theory is widely misunderstood, but so important to understanding the prospects for life in the universe, we devote our final case study in the process of science in action to examining the evidence that makes evolution the underpinning of our understanding of life on Earth and beyond.

▶ **The Theory of Evolution**

◆ What is the evidence for evolution?

The word **evolution** simply means "change with time," and a first step in understanding it is to distinguish between *observations* of evolution and the *theory* of evolution. Observations of evolution come primarily from the fossil record. As we examine layers of sediments deposited over millions of years, such as those in the walls of the Grand Canyon (Figure 15.12), we find that different layers contain fossils of different types. Comparisons of these fossils show that those in older sediment layers are generally more primitive than those in younger layers, and many fossils from the recent geological past represent the extinct ancestors of living species. These observations clearly indicate that living species have evolved over time. The *theory* of evolution, which was first described by Charles Darwin in 1859, is designed to explain how these observed changes occur. In essence, the fossil record provides the evidence that evolution *has* occurred, while Darwin's theory of evolution explains *how* it occurs.

Darwin's theory tells us that evolution proceeds through a process called **natural selection**. The many individuals of any species always differ in small ways. If an individual possesses a trait that gives it an advantage in survival and reproduction, then the trait is likely to be passed on to future generations. We say that nature "selects" the advantageous trait, which is why the process is called natural selection. Over time, natural selection can help individuals of a species become better able to compete for scarce resources. If enough new traits accumulate, natural selection can give rise to an entirely new species.

Accumulating Evidence   Darwin compiled a tremendous body of evidence (by studying both fossils and relationships among living species, such as those among animals of the Galápagos Islands) in support of his idea that natural selection is the primary mechanism by which evolution proceeds. Research since Darwin's time has only given further support to his idea, which is why we call it the *theory* of evolution [Section 3.2]. Some of the strongest evidence in favor of evolution has come from the study of DNA, the genetic material of all life on Earth.

In order for evolution to happen as described by Darwin, there must be some way for organisms to pass their genetic traits on to their offspring and for those traits to diversify over time. DNA is what enables that to happen, and its discovery led scientists to understand how evolution occurs on a molecular level. Living organisms reproduce by copying DNA and passing these copies on to their descendants. A molecule of DNA consists of two long strands—somewhat like the interlocking strands of a zipper—wound together in the spiral shape known as a double helix (Figure 15.13). The instructions for assembling a living organism are written in the precise order of four chemical bases that make up the

▲ **FIGURE 15.12**
The rock layers of the Grand Canyon record more than 500 million years of Earth's history.

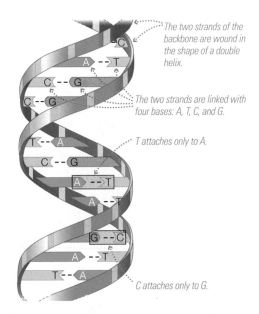

▲ **FIGURE 15.13**
This diagram represents a small piece of a DNA molecule, which looks much like a zipper twisted into a spiral. The important hereditary information is contained in the "teeth" linking the strands. These "teeth" are the DNA bases. Only four DNA bases are used, and they can link the two strands only in specific ways: T attaches only to A, and C attaches only to G. (The color coding is arbitrary and is used only to represent different types of chemical groups; in the backbone, blue and yellow represent sugar and phosphate groups, respectively.)

interlocking portions of the DNA "zipper." (The four chemical bases are represented in Figure 15.13 by the letters A, T, G, and C, which are the first letters of their chemical names.) These bases pair up in a way that ensures that each of the two strands of a DNA molecule contains the same genetic information. By unwinding and allowing new strands to form alongside the original ones (with the new strands made from chemicals floating around inside a cell), a single DNA molecule can give rise to two identical copies of itself. This is how genetic material is copied and passed on to future generations.

Evolution occurs because the passing of genetic information from one generation to the next is not always perfect. An organism's DNA may be altered by occasional copying errors or by external influences, such as ultraviolet light from the Sun or exposure to toxic or radioactive chemicals. Any change in an organism's DNA is called a **mutation**. Many mutations are lethal, killing the cell in which the mutation occurs. Some, however, may improve a cell's ability to survive and reproduce. The cell then passes on this improvement to its offspring.

Our understanding of the molecular mechanism of natural selection has put the theory of evolution on a stronger foundation than ever. While no theory can ever be proved true beyond all doubt, the theory of evolution today is as solid as any theory in science, including the theory of gravity and the theory of atoms. Biologists routinely witness evolution occurring before their eyes among laboratory microorganisms, or over periods of just a few decades among plants and animals subjected to some kind of environmental stress. Moreover, the theory of evolution has become the underpinning of virtually all modern biology, medicine, and agriculture. For example, agricultural scientists apply the idea of natural selection to develop pest control strategies that reduce the populations of harmful insects without harming the populations of beneficial ones; medical researchers test new drugs on animals that are genetically similar to humans, because the theory of evolution tells us that genetically similar species should have somewhat similar physiological responses; and biologists study the relationships between organisms by comparing their DNA.

To summarize, the theory of evolution is backed by an enormous body of evidence and successfully explains both what we observe in the fossil record and what we observe in the laboratory. The fact that we now understand evolution on a molecular level gives us even more confidence in this model of how life changes through time. And perhaps most important to the topic of this chapter, the idea that life evolves through molecular changes that are passed on to offspring could apply even to organisms that use something besides DNA as their hereditary material. We therefore expect that the theory of evolution will apply to life that we might someday discover on other worlds, and that those discoveries will help us further refine and improve the theory.

The Biological Universe Because all life must be able to reproduce and adapt, evolution must hold the key to the diversification of life on any world where it exists. Of course, evolution would not be expected to follow exactly the same course on other worlds, and we do not yet know whether alien life forms might resemble life on Earth, let alone whether they might evolve intelligence and a capacity for space travel and interstellar communication.

Nevertheless, we now understand that the conditions that have made life possible on Earth, summarized in Figure 15.14, seem likely to be common throughout the universe. We may not yet know whether life itself is also common, but we are getting ever closer to learning the answer. Whatever that answer turns out to be, there seems little doubt that life in the universe is intimately connected with the physical universe—which is one reason why a *cosmic perspective* is so important to understanding our own lives.

cosmic context FIGURE 15.14 A Universe of Life

Throughout this book, we have seen that the history of the universe has proceeded in a way that has made our existence on Earth possible. This figure summarizes some of the key ideas, and leads us to ask: If life arose here, shouldn't it also have arisen on many other worlds? We do not yet know the answer, but scientists are actively seeking to learn whether life is rare or common in the universe.

① The protons, neutrons, and electrons in the atoms that make up Earth and life were created out of pure energy during the first few moments after the Big Bang, leaving the universe filled with hydrogen and helium gas.

Matter can be created from energy: $E = mc^2$.

② Ripples in the density of the early universe were necessary for life to form later on. Without those ripples, matter would never have collected into galaxies, stars, and planets.

We observe the seeds of structure formation in the cosmic microwave background.

③ The attractive force of gravity pulls together the matter that makes galaxies, stars, and planets.

Every piece of matter in the universe pulls on every other piece.

④ Our planet and all the life on it is made primarily of elements formed by nuclear fusion in high-mass stars and dispersed into space by supernovae.

^{16}O ^{28}Si
^{31}S (16p, 15n) ^{56}Fe (26p, 30n)
^{16}O ^{28}Si

High-mass stars have cores hot enough to make elements heavier than carbon.

(5) Our galaxy is large enough to retain the elements ejected by supernovae, and it recycles them into new stars and planetary systems.

New elements mix with the interstellar medium, which then forms new stars and planets.

(6) Planets can form in gaseous disks of material around newly formed stars. Earth was built from heavy elements that condensed from the gas as particles of metal and rock, which then gradually accreted to become our planet.

Terrestrial planets formed in warm, inner regions of the solar nebula; jovian planets formed in cooler, outer regions.

(7) Life as we know it requires liquid water, so we define the habitable zone around a star to be the zone in which a suitably large planet can have liquid water on its surface.

The Sun's habitable zone (green) occupies a region from beyond the orbit of Venus to near the orbit of Mars.

(8) Early life has had the time needed to evolve into complex forms—including us—because the stellar thermostat has kept the Sun shining steadily for billions of years.

The stellar thermostat keeps the Sun's fusion rate stable.

summary of key concepts

15.1 The Search for Life in the Solar System

♦ **What are the necessities of life?**

Life on Earth is remarkably diverse, but shares a few commonalities that can help us understand the characteristics that would make a **habitable world**—a world on which life is potentially possible. The three main requirements seem to be a source of nutrients, a source of energy, and a liquid such as liquid water.

♦ **Could there be life elsewhere in our solar system?**

At least six worlds in our solar system offer some possibility of harboring life. Mars shows clear evidence of liquid water on its surface in the distant past, suggesting that it was once habitable; liquid water may still exist underground. Europa probably has a subsurface ocean of liquid water, and undersea volcanoes might provide energy for life. Ganymede and Callisto might have oceans as well. Titan has liquid methane or ethane on its surface and may also have a subsurface ocean of liquid water. Enceladus shows evidence of subsurface water as well, offering yet another possibility for life.

15.2 The Search for Life Among the Stars

♦ **How can we identify potentially habitable planets?**

For a planet to have a surface that is habitable, meaning one with liquid water, it must orbit within its star's **habitable zone**—the region in which temperatures might allow for surface oceans. The planet probably also must have volcanism, plate tectonics, and a magnetic field, but all three requirements may be met by any planet in the habitable zone with size and composition similar to those of Earth. If so, there may be billions of habitable planets in our galaxy, and we may be able to use spectroscopy to learn if they have life.

♦ **Is there intelligent life beyond Earth?**

We don't know, but the **Drake equation** gives us a way to organize our thinking about the question, and scientists are looking for communications from intelligent life through **SETI**, the **search for extraterrestrial intelligence**. Reasonable assumptions make it seem as if civilizations ought to be common, leading to Fermi's paradox, which asks why we do not yet have any clear evidence of other civilizations.

THE PROCESS OF SCIENCE IN ACTION

15.3 Evolution on Earth and Beyond

♦ **What is the evidence for evolution?**

Evidence that species have evolved over time comes from the study of fossils and the fossil record. The **theory of evolution** explains *how* these changes have occurred through the mechanism of **natural selection**. Evidence for evolution by natural selection comes from careful observations of relationships among species, and from our modern understanding of the molecular basis for evolution. We expect the theory of evolution to apply to life elsewhere in the universe as well as to life on Earth.

Investigations

For instructor-assigned homework and other learning materials, go to www.MasteringAstronomy.com.

Quick Quiz

Choose the best answer to each of the following. Explain your reasoning.

1. Plants and animals are (a) the two major forms of life on Earth. (b) the only organisms that have DNA. (c) just two small branches of the tree of life on Earth.
2. Careful study of commonalities among different organisms suggests that all life on Earth (a) shares a common ancestor. (b) evolved at a time when Earth had plentiful oxygen in its atmosphere. (c) depends on sunlight for energy.
3. The modern organisms that appear to be most like the earliest life on Earth are (a) simple animals like worms. (b) green algae. (c) microbes living near undersea volcanic vents.
4. The three key requirements for life are a source of nutrients, a source of energy, and (a) H_2O in some form. (b) a liquid such as liquid water. (c) DNA.
5. A habitable world is a world that (a) has life of some kind. (b) is potentially capable of having life. (c) is potentially capable of having an advanced civilization.
6. Which of the following worlds is *not* considered a candidate for harboring life? (a) Europa (b) Mars (c) the Moon
7. If we someday find life on Mars, we expect to find it (a) near the polar ice caps. (b) in dried-up riverbeds. (c) underground.
8. How does the habitable zone around a star of spectral type G compare to that around a star of spectral type M? (a) It is larger. (b) It is hotter. (c) It is closer to its star.
9. In the Drake equation, what would $f_{life} = 1/2$ mean? (a) Half the stars in the Milky Way Galaxy have a planet with life. (b) Half of all life forms in the universe are intelligent. (c) Half of the habitable worlds in the galaxy have life.
10. According to current scientific understanding, the idea that the Milky Way Galaxy might be home to a civilization millions of years more advanced than ours is (a) a virtual certainty. (b) extremely unlikely. (c) one possible solution to Fermi's paradox.
11. The theory of evolution is (a) a scientific theory backed by extensive evidence. (b) one of several competing scientific models that all seem equally successful in explaining the nature of life on Earth. (c) a speculative hypothesis about how life changes through time.
12. *Natural selection* is the name given to (a) the occasional mutations that occur in DNA. (b) the mechanism by which advantageous traits are passed on from one generation to the next. (c) the idea that species change over time.

Short-Answer/Essay Questions

Explain all answers clearly, with complete sentences and proper essay structure if needed. An asterisk (*) designates a quantitative problem, for which you should show all your work.

13. *Most Likely to Have Life.* Suppose you were asked to vote in a contest to name the world in our solar system (besides Earth) "most likely to have life." Which world would you vote for? Explain and defend your choice in a one-page essay.
14. *Habitable Zones.* Consider the two stars besides the Sun that are shown in Figure 15.6. Briefly describe (a) the size and location of the habitable zone of each star in comparison to the habitable zone of our Sun and (b) the prospects for finding a habitable world around each star. Explain your answers.
15. *Alien Technology.* Suppose that aliens are actually visiting Earth in UFOs. By considering the distances to stars in comparison to the distance to the Moon—which is the farthest distance that humans have so far traveled—describe how the alien technology would have to compare to ours.
16. *Solution to the Fermi Paradox.* Among the various possible solutions to Fermi's question "So where is everybody?" which one do you think is most likely? Write a one- to two-page essay in which you explain why you favor this solution.
17. *What's Wrong with This Picture?* Many science fiction stories have imagined the galaxy divided into a series of empires, each having arisen from a different civilization on a different world, that hold one another at bay because they are all at about the same level of military technology. Is this a realistic scenario? Explain.
18. *Aliens in the Movies.* Choose a science fiction movie or television show that involves an alien species. Do you think aliens like those depicted could really exist? Write a one- to two-page critical review of the movie or show, focusing primarily on the question of whether it portrays the aliens in a scientifically reasonable way.
19. *Extreme Life.* Look for information about a recent discovery of a previously unknown type of extremophile. Describe the organism and the environment in which it lives, and discuss any implications of the finding for the search for life beyond Earth. Summarize your findings in a one-page report.
20. *The Dover Opinion.* Despite the strong scientific evidence for the theory of evolution, the teaching of evolution in public schools has been the subject of a contentious debate that has often landed in the courts. The most recent major decision on this debate was the Dover opinion, issued by a United States District Court in December 2005. Read the opinion, which is available online, and write a short essay describing its implications for the ongoing public controversy about what should be taught in science classes.
21. *The Turning Point.* The end of Section 15.2 presented the idea that humanity is now at a turning point, and that current generations have a greater responsibility for the future than did any previous generation. Do you agree with this assessment? If so, how can we ensure that we deal with this responsibility wisely, in a way that builds a better future for the entire human race? If not, why not?
*22. *Time Between Civilizations.* Suppose that there are 100 billion stars in the Milky Way Galaxy, that 1 in 1 million stars gets a civilization at some point, and that these civilizations have arisen at random times over the past 5 billion years. On average, how much time separates the rise of one civilization in the galaxy from the rise of the next? Based on your answer, should we expect to make radio contact with another civilization that is either at or below our own current level of technological development?
*23. *SETI Search.* Suppose there are 10,000 civilizations broadcasting radio signals in the Milky Way Galaxy right now. On average, how many stars would we have to search before we would expect to hear a signal? Assume there are 400 billion stars in the galaxy. How would the answer change if there were only 100 civilizations instead of 10,000?

Credits

Foldout
Front © ESA and the Planck Collaboration, NASA, The Hubble Heritage Team, STScI, AURA, Stephen J. Mojzsis **Back** Russell Shively/Shutterstock, Inc., Colin Keates, © Dorling Kindersley, Courtesy of the Natural History Museum, London, Dave King, © Dorling Kindersley, Courtesy of the Natural History Museum, London, Linda Whitwam, © Dorling Kindersley, Pius Lee/Shutterstock, NASA

Chapter 1
Opener NASA, ESA, S. Beckwith (STScI) and the HUDF Team **1.2** Jerry Lodriguss/Science Source **1.4** Jeffrey Bennett **1.6** NASA **1.7** (inset) © Fred Espenak, www.AstroPixels.com **1.8** Danny Hooks/Alamy Stock Photo; (Milky Way Galaxy) Michael Carroll **1.9** Jeffrey Bennett **1.11** (second from left to right) Sompol/Shutterstock; NASA; Jon Arnold Images Ltd/Alamy Stock Photo **1.13** NASA/Johns Hopkins University Applied Physics Laboratory/Southwest Research Center/Chandra X-Ray Center; (inset) NASA/JHUAPL/SwRI

Chapter 2
Opener D. Nunuk/Science Source **2.4** Arnulf Husmo/Getty Images **2.5** (inset) NASA, ESA, M. Robberto (Space Telescope Science Institute/ESA) and the Hubble Space Telescope Orion Treasury Project Team **2.10** NASA/Bill Dunford **2.14** (top) Brian Kushner/Alamy Stock Photo; (bottom) Dr Ajay Kumar Singh/Shutterstock **2.15** (top to bottom) JBCorl/Shutterstock; Sarah M. Golonka/Alamy Stock Photo; Phillip Jones/Stocktrek Images/Alamy Stock Photo **2.16** Ben Cooper/LaunchPhotography.com

Chapter 3
Opener Time Life Pictures/NASA/The LIFE Picture Collection/Getty Images **p. 37** World History Archive/Superstock **p. 38** (top) Ian Woolcock/SuperStock; (bottom) New York Public Library/Science Source **p. 40** A. Burkatovski/Fine Art Images/SuperStock **3.6** Mike Hill/Alamy Stock Photo **p. 41** Sir Godfrey Kneller **3.8** (left to right) NASA; Bruce Kluckhohn/Getty Images; NASA **p. 42** (left) Yerkes Observatory Photograph; (right) NOAO/AURA/NSF **3.11** (top left) Tunç Tezel; (bottom left) Cosmographia Petri Apiani/Boston Public Library; (bottom right) The Library of Congress Collection

Chapter 4
Opener NASA Goddard Space Flight Center Image by Reto Stöckli **Table 4.1** NASA Goddard Space Flight Center/APL/SwRI/JPL/Space Science Institute **4.2** NASA/JPL/USGS **4.3a** NASA **4.3b** JAXA; (inset) Institute of Space and Astronomical Science (ISAS) **4.4** (top) NASA/JPL; (bottom, left to right) NASA/JPL-Caltech/SwRI/ASI/INAF/JIRAM; NASA; Lawrence Sromovsky, University of Wisconsin-Madison/W. W. Keck Observatory; NASA/JPL **4.5** Miloslav Druckmuller (Brno University of Technology) **p. 62** (left, clockwise from top) Imagenavi/Getty Images; Creativ Studio Heinemann/Westend61 GmbH/Alamy Stock Photo; Eric Gevaert/Alamy Stock Photo **4.11** NASA/JPL-Caltech/University of Arizona **4.12** Michael Carroll

Chapter 5
Opener NASA/JPL-Caltech/MSSS **5.3** (inset) Doug Duncan **5.4** Brad Snowder/Lunar and Planetary Institute **5.5** Paul Chesley/Getty Images **5.6** NASA **5.7** ShuPhotography/Shutterstock **5.11** Claudio Divizia/Shutterstock; (inset) Malcolm Park/Oxford Scientific/Getty Images **5.12** NASA/Johns Hopkins University Applied Physics Laboratory/Carnegie Institution of Washington; (inset) Johns Hopkins University Applied Physics Laboratory/Carnegie Institution of Washington/NASA **5.13** NASA/JPL **5.14** NASA/JPL-Caltech; (inset) © ESA/DLR/FU Berlin (G. Neukum) **5.15** NASA/JPL-Caltech/UA/USGS **5.16a** EROS Data Center, U.S. Geological Survey **5.16b** R. P. Irwin III and G. A. Franz, National Air and Space Museum, Smithsonian Institution **5.16c** NASA Earth Observing System **5.17** NASA/JPL-Caltech/MSSS **5.18** NASA/JPL-Caltech/MSSS and PSI **5.20** NASA/JPL **5.21** (top) JAXA/ISAS/DARTS/Damia Bouic; (bottom) Russian Academy of Sciences, processing by Ted Stryk **5.22** (left) ARC Science Simulations; (right) NASA **5.29** Map from Goddard Space Flight Center/NASA, accessed on May 5, 2018

Chapter 6
Opener NASA/JPL/Space Science Institute **6.1** NASA/JPL **6.3** NASA/JPL **6.4** NASA/JPL/Space Science Institute; (inset) NASA/JPL-Caltech/SwRI/MSSS/Jason Major **6.5a** NASA **6.5b** NASA/JPL/Space Science Institute **6.5c** NASA/JPL/University of Colorado **6.6** NASA/JPL-Caltech/Space Science Institute **6.9a** NASA/JPL/University of Arizona **6.9b** (left) NASA/Johns Hopkins University Applied Physics Laboratory/Southwest Research Institute; (right) NASA/JPL-Caltech/SwRI/ASI/INAF/JIRAM/Roman Tkachenko **6.11** NASA/JPL/DLR/ASU **6.12** (inset) Britney Schmidt/Dead Pixel VFX/Univ. of Texas at Austin **6.13** NASA/JPL/Ted Stryk; (inset) NASA/JPL/DLR **6.14** NASA/JPL/ASU **6.15** (left to right) NASA/JPL/Space Science Institute; ESA/NASA/JPL/University of Arizona; NASA/JPL/ESA/University of Arizona **6.16** NASA/JPL/USGS **6.17** NASA/JPL/Space Science Institute; (inset) NASA/JPL-Caltech/Space Science Institute **6.18** NASA/JPL/USGS; (inset) NASA/JPL **6.19** NASA/JPL/USGS **6.20a** NASA/JPL-Caltech/UCLA/MPS/DLR/IDA/PSI **6.20b** (left) NASA/JPL-Caltech/UCLA/MPS/DLR/IDA; (right) NASA/JPL-Caltech/JAXA/ESA **6.21** NASA/JPL; (inset) © ESA/Rosetta/MPS for OSIRIS Team MPS/UPD/LAM/IAA/SSO/INTA/UPM/DASP/IDA **6.22** NASA/Johns Hopkins University Applied Physics Laboratory/Southwest Research Institute **6.24** AP Video/Associated Press **6.26** Kirk R. Johnson **6.27** V. L. Sharpton/USRA-Lunar and Planetary Institute

Chapter 7
Opener A.-L. Maire/LBTO **7.6** (bottom right) A.-L. Maire/LBTO **7.9** NASA JPL/STScI/NASA Goddard Space Flight Center **7.12** Frederic Masset

Chapter 8
Opener NASA **8.1** NASA **8.6b** The Institute for Solar Physics **8.8b** NASA **8.12** NASA **8.16** Harvard College Observatory

Chapter 9
Opener NASA, ESA, M. Robberto (Space Telescope Science Institute/ESA) and the Hubble Space Telescope Orion Treasury Project Team **9.1** Stocktrek Images/Getty Images **9.2** i4images rm/Alamy Stock Photo **9.11a** NASA, NOAO, ESA, the Hubble Helix Nebula Team, M. Meixner (STScI), and T. A. Rector (NRAO) **9.11b** NASA, ESA, and the Hubble SM4 ERO Team **9.18** NASA, ESA, J. Hester and A. Loll (Arizona State University) **9.20** Robert Gendler/Stocktrek Images/Science Source **9.21** The Hubble Heritage Team (AURA/STScI/NASA)

Chapter 10
Opener X-ray: NASA/CXC/ASU/J. Hester et al.; optical: NASA/HST/ASU/J. Hester et al. **10.1a** ESA/H. Bond (STScI) and M. Barstow (University of Leicester)/NASA **10.1b** NASA/CXC/SAO **10.6** NASA/CXC/Eureka Scientific/M. Roberts et al. **10.8** European Southern Observatory **10.16** NASA/Dana Berry, Sky Works Digital **10.18** (inset) Handout/MCT/Newscom

Chapter 11
Opener Hubble image: NASA, ESA, K. Kuntz (JHU), F. Bresolin (University of Hawaii), J. Trauger (Jet Propulsion Lab), J. Mould (NOAO), Y. H. Chu (University of Illinois, Urbana), and STScI; CFHT image: Canada-France-Hawaii Telescope/J. C. Cuillandre/Coelum; NOAO image: G. Jacoby, B. Bohannan, M. Hanna/NOAO/AURA/NSF **11.1** Wally Pacholka **p. 184** (top to bottom) Atlas Image courtesy of 2MASS/UMass/IPAC-Caltech/NASA/NSF; Axel Mellinger, A Color All-Sky Panorama Image of the Milky Way, *Publ. Astron. Soc. Pacific* 121, 1180–1187 (2009); ESA/NASA/JPL-Caltech; NASA, Max-Planck-Institut für extraterrestrische Physik (MPE) and S. L. Snowden; ESA/NASA/Planck Collaboration; NASA/DOE/Fermi LAT Collaboration; Mike Carroll **11.5** NASA/CXC/Rutgers/J. Warren & J. Hughes et al. **11.6** NASA, ESA, M. Robberto (Space Telescope Science Institute/ESA) and the Hubble Space Telescope Orion Treasury Project Team **11.7** NASA, ESA, S. Beckwith (STScI), and The Hubble Heritage Team (STScI/AURA) **11.8** UCLA Galactic Center Group–W. M. Keck Observatory Laser Team **11.10** ESA/K. Kuntz (JHU)/F. Bresolin (University of Hawaii)/J. Trauger (Jet Propulsion Lab)/J. Mould (NOAO)/Y. H. Chu (University of Illinois, Urbana)/STScI/NASA **11.11** NASA, ESA, S. Beckwith (STScI), and The Hubble Heritage Team (STScI/AURA) **11.12** NASA, ESA, and The Hubble Heritage Team (STScI/AURA). Acknowledgment: P. Knezek (WIYN) **11.13** NASA/

ESA Hubble Space Telescope **11.14** European Southern Observatory **11.15** (top left) ESO/P. Grosbøl; (left) NASA, ESA, Hubble Space Telescope; (bottom right) NASA/JPL-Caltech/Roma Tre Univ. **11.16** (left) NASA, ESA & Ivo Saviane (European Southern Observatory)/Robert Gendler; (right) ESA/Hubble & NASA **11.18** (clockwise from center) Gemini Observatory—GMOS-S Commissioning Team; Robert Gendler/Stocktrek Images/Science Source; Sloan Digital Sky Survey (SDSS) Collaboration, www.sdss.org; Royal Observatory, Edinburgh/Anglo-Australian Telescope Board/Science Source; ESA/Hubble & NASA. Acknowledgment: Judy Schmidt (Geckzilla) **11.19** J. A. Biretta et al., Hubble Heritage Team (STScI /AURA), NASA **11.20** ESA/Hubble & NASA **11.21** NASA/Hubble Space Telescope

Chapter 12
Opener Byrne Observatory at Sedgwick Reserve and the Palomar Transient Factory/LCOGT.net/BJ Fulton **12.5** (left) NASA, ESA, S. Perlmutter (UC Berkeley, LBNL), A. Koekemoer (STScI), M. Postman (STScI), A. Riess (STScI/JHU), J. Nordin (LBNL, UC Berkeley), D. Rubin (Florida State University), and C. McCully (Rutgers University); (center, right) NASA; ESA; and A. Riess, STScI **12.6** Huntington Library/SuperStock **12.14** NASA/JPL-Caltech/ESA **12.15** Modification of work by CXC/MPE/V. Springel/Max Planck Institute for Astrophysics, Garching

Chapter 13
Opener Planck/CMB/European Space Agency (ESA) **p. 217** ZUMA Press, Inc./Alamy Stock Photo **13.6** Associated Press **13.9** Planck/CMB/European Space Agency (ESA)

Chapter 14
Opener X-ray: NASA/CXC/CfA/M. Markevitch et al.; lensing map: NASA/STScI, ESO WFI, Magellan/University of Arizona/D. Clowe et al.; optical: NASA/STScI, Magellan/University of Arizona/D. Clowe et al. **14.5** NASA, ESA, CXC, C. Ma, H. Ebeling and E. Barrett (University of Hawaii/IfA) et al. and STScI **14.6** NASA, ESA, M. J. Jee and H. Ford (Johns Hopkins University) **14.8** NASA, ESA, J. Richard (CRAL) and J.-P. Kneib (LAM). Acknowledgment: Marc Postman (STScI) **14.10** R. Brent Tully (U. Hawaii) et al., SDvision, DP, CEA/Saclay **14.12** Simulation and Visualization by Andrey Kravtsov (University of Chicago) and Anatoly Klypin (NMSU) **14.13** (Cosmic microwave background) Planck/CMB/European Space Agency (ESA); (NGC 4414) The Hubble Heritage Team, STScI, AURA, NASA; (M87) ESA/Hubble/RELICS/NASA; (Orion Nebula) ESA, M. Robberto (Space Telescope Science Institute/ESA) and the Hubble Space Telescope Orion Treasury Project Team, NASA **14.17** (center left) ESA, H. Lee & H. Ford (Johns Hopkins U.)/NASA; (center right) Simulation and Visualization by Andrey Kravtsov (University of Chicago) and Anatoly Klypin (NMSU)

Chapter 15
Opener Seth Shostak **15.2** © Woods Hole Oceanographic Institution **15.3** Eye of Science/Science Source **15.4** NASA/JPL-Caltech/LANL/J.-L. Lacour, CEA **15.5** NASA/JPL-Caltech **15.9** Seth Shostak **15.10** Michael Carroll **15.11** Seth Shostak **15.12** Darlene Cutshall/Shutterstock **15.14** (Cosmic microwave background) Planck/CMB/European Space Agency (ESA)

Appendixes

A Useful Numbers   A-2

B Useful Formulas   A-3

C A Few Mathematical Skills   A-4

   C.1 Powers of 10   A-4

   C.2 Scientific Notation   A-6

   C.3 Working with Units   A-7

   C.4 Finding a Ratio   A-10

   C.5 The Metric System (SI)   A-11

   C.6 Temperature Scales   A-12

A Useful Numbers

Astronomical Distances

1 AU ≈ 1.496×10^8 km = 1.496×10^{11} m

1 light-year ≈ 9.46×10^{12} km = 9.46×10^{15} m

1 parsec (pc) ≈ 3.09×10^{13} km ≈ 3.26 light-years

1 kiloparsec (kpc) = 1000 pc ≈ 3.26×10^3 light-years

1 megaparsec (Mpc) = 10^6 pc ≈ 3.26×10^6 light-years

Universal Constants

Speed of light: $c = 3.00 \times 10^5$ km/s = 3×10^8 m/s

Gravitational constant: $G = 6.67 \times 10^{-11} \dfrac{m^3}{kg \times s^2}$

Planck's constant: $h = 6.63 \times 10^{-34}$ joule × s

Stefan–Boltzmann constant: $\sigma = 5.67 \times 10^{-8} \dfrac{watt}{m^2 \times Kelvin^4}$

Mass of a proton: $m_p = 1.67 \times 10^{-27}$ kg

Mass of an electron: $m_e = 9.11 \times 10^{-31}$ kg

Useful Sun and Earth Reference Values

Mass of the Sun: $1 M_{Sun} \approx 2 \times 10^{30}$ kg

Radius of the Sun: $1 R_{Sun} \approx 696{,}000$ km

Luminosity of the Sun: $1 L_{Sun} \approx 3.8 \times 10^{26}$ watts

Mass of Earth: $1 M_{Earth} \approx 5.97 \times 10^{24}$ kg

Radius (equatorial) of Earth: $1 R_{Earth} \approx 6378$ km

Acceleration of gravity on Earth: $g = 9.8$ m/s^2

Escape velocity from surface of Earth: $v_{escape} = 11$ km/s = $11{,}000$ m/s

Astronomical Times

1 solar day (average) = 24^h

1 sidereal day ≈ $23^h\ 56^m\ 4.09^s$

1 synodic month (average) ≈ 29.53 solar days

1 sidereal month (average) ≈ 27.32 solar days

1 tropical year ≈ 365.242 solar days

1 sidereal year ≈ 365.256 solar days

Energy and Power Units

Basic unit of energy: 1 joule = $1 \dfrac{kg \times m^2}{s^2}$

Basic unit of power: 1 watt = 1 joule/s

Electron-volt: 1 eV = 1.60×10^{-19} joule

B Useful Formulas

- Universal law of gravitation for the force between objects of mass M_1 and M_2, with distance d between their centers:

$$F = G\frac{M_1 M_2}{d^2}$$

- Newton's version of Kepler's third law (applies to any pair of orbiting objects, such as a star and planet, a planet and moon, or two stars in a binary system; p is the orbital period, a is the distance between the centers of the orbiting objects, and M_1 and M_2 are the object masses):

$$p^2 = \frac{4\pi^2}{G(M_1 + M_2)} a^3$$

- Escape velocity at distance R from center of object of mass M:

$$v_{escape} = \sqrt{\frac{2GM}{R}}$$

- Relationship between a photon's wavelength (λ), frequency (f), and the speed of light (c):

$$\lambda \times f = c$$

- Energy of a photon of wavelength λ or frequency f:

$$E = hf = \frac{hc}{\lambda}$$

- Stefan–Boltzmann law for thermal radiation at temperature T (in Kelvin):

$$\text{emitted power per unit area} = \sigma T^4$$

- Wien's law for the peak wavelength (λ_{max}) thermal radiation at temperature T (in Kelvin):

$$\lambda_{max} = \frac{2{,}900{,}000}{T} \text{ nm}$$

- Doppler shift (radial velocity is positive if the object is moving away from us and negative if it is moving toward us):

$$\frac{\text{radial velocity}}{\text{speed of light}} = \frac{\text{shifted wavelength} - \text{rest wavelength}}{\text{rest wavelength}}$$

- Angular separation (α) of two points with an actual separation s, viewed from a distance d (assuming d is much larger than s):

$$\alpha = \frac{s}{2\pi d} \times 360°$$

- Inverse square law for light (d is the distance to the object):

$$\text{apparent brightness} = \frac{\text{luminosity}}{4\pi d^2}$$

- Parallax formula (distance d to a star with parallax angle p in arcseconds):

$$d \text{ (in parsecs)} = \frac{1}{p \text{ (in arcseconds)}}$$

$$\text{or } d \text{ (in light-years)} = 3.26 \times \frac{1}{p \text{ (in arcseconds)}}$$

- The orbital velocity law, to find the mass M_r contained within the circular orbit of radius r for an object moving at speed v:

$$M_r = \frac{r \times v^2}{G}$$

C A Few Mathematical Skills

This appendix reviews the following mathematical skills: powers of 10, scientific notation, working with units, finding a ratio, the metric system, and temperature scale conversions. You should refer to this appendix as needed while studying the textbook.

C.1 Powers of 10

Powers of 10 simply indicate how many times to multiply 10 by itself. For example:

$$10^2 = 10 \times 10 = 100$$

$$10^6 = 10 \times 10 \times 10 \times 10 \times 10 \times 10 = 1{,}000{,}000$$

Negative powers are the reciprocals of the corresponding positive powers. For example:

$$10^{-2} = \frac{1}{10^2} = \frac{1}{100} = 0.01$$

$$10^{-6} = \frac{1}{10^6} = \frac{1}{1{,}000{,}000} = 0.000001$$

Table C.1 lists powers of 10 from 10^{-12} to 10^{12}. Note that powers of 10 follow two basic rules:

1. A positive exponent tells how many zeros follow the 1. For example, 10^0 is a 1 followed by no zeros, and 10^8 is a 1 followed by eight zeros.

2. A negative exponent tells how many places are to the right of the decimal point, including the 1. For example, $10^{-1} = 0.1$ has one place to the right of the decimal point; $10^{-6} = 0.000001$ has six places to the right of the decimal point.

Multiplying and Dividing Powers of 10

Multiplying powers of 10 simply requires adding exponents, as the following examples show:

$$10^4 \times 10^7 = \underbrace{10{,}000}_{10^4} \times \underbrace{10{,}000{,}000}_{10^7} = \underbrace{100{,}000{,}000{,}000}_{10^{4+7} = 10^{11}} = 10^{11}$$

$$10^5 \times 10^{-3} = \underbrace{100{,}000}_{10^5} \times \underbrace{0.001}_{10^{-3}} = \underbrace{100}_{10^{5+(-3)} = 10^2} = 10^2$$

$$10^{-8} \times 10^{-5} = \underbrace{0.00000001}_{10^{-8}} \times \underbrace{0.00001}_{10^{-5}} = \underbrace{0.0000000000001}_{10^{-8+(-5)} = 10^{-13}} = 10^{-13}$$

Dividing powers of 10 requires subtracting exponents, as in the following examples:

$$\frac{10^5}{10^3} = \underbrace{100{,}000}_{10^5} \div \underbrace{1000}_{10^3} = \underbrace{100}_{10^{5-3} = 10^2} = 10^2$$

TABLE C.1 Powers of 10

Zero and Positive Powers			Negative Powers		
Power	Value	Name	Power	Value	Name
10^0	1	One			
10^1	10	Ten	10^{-1}	0.1	Tenth
10^2	100	Hundred	10^{-2}	0.01	Hundredth
10^3	1000	Thousand	10^{-3}	0.001	Thousandth
10^4	10,000	Ten thousand	10^{-4}	0.0001	Ten-thousandth
10^5	100,000	Hundred thousand	10^{-5}	0.00001	Hundred-thousandth
10^6	1,000,000	Million	10^{-6}	0.000001	Millionth
10^7	10,000,000	Ten million	10^{-7}	0.0000001	Ten-millionth
10^8	100,000,000	Hundred million	10^{-8}	0.00000001	Hundred-millionth
10^9	1,000,000,000	Billion	10^{-9}	0.000000001	Billionth
10^{10}	10,000,000,000	Ten billion	10^{-10}	0.0000000001	Ten-billionth
10^{11}	100,000,000,000	Hundred billion	10^{-11}	0.00000000001	Hundred-billionth
10^{12}	1,000,000,000,000	Trillion	10^{-12}	0.000000000001	Trillionth

$$\frac{10^3}{10^7} = \underbrace{1000}_{10^3} \div \underbrace{10{,}000{,}000}_{10^7} = \underbrace{0.0001}_{10^{3-7} = 10^{-4}} = 10^{-4}$$

$$\frac{10^{-4}}{10^{-6}} = \underbrace{0.0001}_{10^{-4}} \div \underbrace{0.000001}_{10^{-6}} = \underbrace{100}_{10^{-4-(-6)} = 10^2} = 10^2$$

Powers of Powers of 10

We can use the multiplication and division rules to raise powers of 10 to other powers or to take roots. For example:

$$(10^4)^3 = 10^4 \times 10^4 \times 10^4 = 10^{4+4+4} = 10^{12}$$

Note that we can get the same end result by simply multiplying the two powers:

$$(10^4)^3 = 10^{4 \times 3} = 10^{12}$$

Because taking a root is the same as raising to a fractional power (e.g., the square root is the same as the $\frac{1}{2}$ power, the cube root is the same as the $\frac{1}{3}$ power, etc.), we can use the same procedure for roots, as in the following example:

$$\sqrt{10^4} = (10^4)^{1/2} = 10^{4 \times (1/2)} = 10^2$$

Adding and Subtracting Powers of 10

Unlike for multiplying and dividing powers of 10, there is no shortcut for adding or subtracting powers of 10. The values must be written in longhand notation. For example:

$$10^6 + 10^2 = 1{,}000{,}000 + 100 = 1{,}000{,}100$$

$$10^8 + 10^{-3} = 100{,}000{,}000 + 0.001 = 100{,}000{,}000.001$$

$$10^7 - 10^3 = 10{,}000{,}000 - 1000 = 9{,}999{,}000$$

Summary

We can summarize our findings using n and m to represent any numbers:

- To *multiply* powers of 10, *add* exponents: $10^n \times 10^m = 10^{n+m}$

- To *divide* powers of 10, *subtract* exponents: $\dfrac{10^n}{10^m} = 10^{n-m}$

- To *raise* powers of 10 to other powers, multiply exponents: $(10^n)^m = 10^{n \times m}$

C.2 Scientific Notation

When we are dealing with large or small numbers, it's generally easier to write them with powers of 10. For example, it's much easier to write the number 6,000,000,000,000 as 6×10^{12}. This format, in which a number *between* 1 and 10 is multiplied by a power of 10, is called **scientific notation**.

Converting a Number to Scientific Notation

We can convert numbers written in ordinary notation to scientific notation with a simple two-step process:

1. Move the decimal point to come after the *first* nonzero digit.

2. The number of places the decimal point moves tells you the power of 10; the power is *positive* if the decimal point moves to the left and *negative* if it moves to the right.

Examples:

$$3042 \xrightarrow{\text{decimal needs to move 3 places to left}} 3.042 \times 10^3$$

$$0.00012 \xrightarrow{\text{decimal needs to move 4 places to right}} 1.2 \times 10^{-4}$$

$$226 \times 10^2 \xrightarrow{\text{decimal needs to move 2 places to left}} (2.26 \times 10^2) \times 10^2 = 2.26 \times 10^4$$

Converting a Number from Scientific Notation

We can convert numbers written in scientific notation to ordinary notation by the reverse process:

1. The power of 10 indicates how many places to move the decimal point; move it to the *right* if the power of 10 is positive and to the *left* if it is negative.

2. If moving the decimal point creates any open places, fill them with zeros.

Examples:

$$4.01 \times 10^2 \xrightarrow{\text{move decimal 2 places to right}} 401$$

$$3.6 \times 10^6 \xrightarrow{\text{move decimal 6 places to right}} 3{,}600{,}000$$

$$5.7 \times 10^{-3} \xrightarrow{\text{move decimal 3 places to left}} 0.0057$$

Multiplying or Dividing Numbers in Scientific Notation

Multiplying or dividing numbers in scientific notation simply requires operating on the powers of 10 and the other parts of the number separately.

Examples:

$$(6 \times 10^2) \times (4 \times 10^5) = (6 \times 4) \times (10^2 \times 10^5) = 24 \times 10^7 = (2.4 \times 10^1) \times 10^7 = 2.4 \times 10^8$$

$$\frac{4.2 \times 10^{-2}}{8.4 \times 10^{-5}} = \frac{4.2}{8.4} \times \frac{10^{-2}}{10^{-5}} = 0.5 \times 10^{-2-(-5)} = 0.5 \times 10^3 = (5 \times 10^{-1}) \times 10^3 = 5 \times 10^2$$

Note that, in both these examples, we first found an answer in which the number multiplied by a power of 10 was *not* between 1 and 10. We therefore followed the procedure for converting the final answer to scientific notation.

Addition and Subtraction with Scientific Notation

In general, we must write numbers in ordinary notation before adding or subtracting.

Examples:

$$(3 \times 10^6) + (5 \times 10^2) = 3{,}000{,}000 + 500 = 3{,}000{,}500 = 3.0005 \times 10^6$$

$$(4.6 \times 10^9) - (5 \times 10^8) = 4{,}600{,}000{,}000 - 500{,}000{,}000 = 4{,}100{,}000{,}000 = 4.1 \times 10^9$$

When both numbers have the *same* power of 10, we can factor out the power of 10 first.

Examples:

$$(7 \times 10^{10}) + (4 \times 10^{10}) = (7 + 4) \times 10^{10} = 11 \times 10^{10} = 1.1 \times 10^{11}$$

$$(2.3 \times 10^{-22}) - (1.6 \times 10^{-22}) = (2.3 - 1.6) \times 10^{-22} = 0.7 \times 10^{-22} = 7.0 \times 10^{-23}$$

C.3 Working with Units

Showing the units of a problem as you solve it usually makes the work much easier and also provides a useful way of checking your work. If an answer does not come out with the units you expect, you probably did something wrong. In general, working with units is very similar to working with numbers, as the following guidelines and examples show.

Five Guidelines for Working with Units

Before you begin any problem, think ahead and identify the units you expect for the final answer. Then operate on the units along with the numbers as you solve the problem. The following five guidelines may be helpful when you are working with units:

1. Mathematically, it doesn't matter whether a unit is singular (e.g., meter) or plural (e.g., meters); we can use the same abbreviation (e.g., m) for both.

2. You cannot add or subtract numbers unless they have the *same* units. For example, 5 apples + 3 apples = 8 apples, but the expression 5 apples + 3 oranges cannot be simplified further.

3. You *can* multiply units, divide units, or raise units to powers. Look for key words that tell you what to do.

   - *Per* suggests division. For example, we write a speed of 100 kilometers per hour as

$$100 \frac{\text{km}}{\text{hr}} \quad \text{or} \quad \frac{100 \text{ km}}{1 \text{ hr}}$$

- *Of* suggests multiplication. For example, if you launch a 50-kg space probe at a launch cost *of* $10,000 per kilogram, the total cost is

$$50 \text{ kg} \times \frac{\$10,000}{\text{kg}} = \$500,000$$

- *Square* suggests raising to the second power. For example, we write an area of 75 square meters as 75 m^2.

- *Cube* suggests raising to the third power. For example, we write a volume of 12 cubic centimeters as 12 cm^3.

4. Often the number you are given is not in the units you wish to work with. For example, you may be given that the speed of light is 300,000 km/s but need it in units of m/s for a particular problem. To convert the units, simply multiply the given number by a *conversion factor*: a fraction in which the numerator (top of the fraction) and denominator (bottom of the fraction) are equal, so that the value of the fraction is 1; the number in the denominator must have the units that you wish to change. In the case of changing the speed of light from units of km/s to m/s, you need a conversion factor for kilometers to meters. Thus, the conversion factor is

$$\frac{1000 \text{ m}}{1 \text{ km}}$$

Note that this conversion factor is equal to 1, since 1000 meters and 1 kilometer are equal, and that the units to be changed (km) appear in the denominator. We can now convert the speed of light from units of km/s to m/s simply by multiplying by this conversion factor:

$$\underbrace{300,000 \frac{\text{km}}{\text{s}}}_{\text{speed of light in km/s}} \times \underbrace{\frac{1000 \text{ m}}{1 \text{ km}}}_{\text{conversion from km to m}} = \underbrace{3 \times 10^8 \frac{\text{m}}{\text{s}}}_{\text{speed of light in m/s}}$$

Note that the units of km cancel, leaving the answer in units of m/s.

5. It's easier to work with units if you replace division with multiplication by the reciprocal. For example, suppose you want to know how many minutes are represented by 300 seconds. We can find the answer by dividing 300 seconds by 60 seconds per minute:

$$300 \text{ s} \div 60 \frac{\text{s}}{\text{min}}$$

However, it is easier to see the unit cancellations if we rewrite this expression by replacing the division with multiplication by the reciprocal (this process is easy to remember as "invert and multiply"):

$$300 \text{ s} \div 60 \frac{\text{s}}{\text{min}} = 300 \text{ s} \times \underbrace{\frac{1 \text{ min}}{60 \text{ s}}}_{\substack{\text{invert} \\ \text{and multiply}}} = 5 \text{ min}$$

We now see that the units of seconds (s) cancel in the numerator of the first term and the denominator of the second term, leaving the answer in units of minutes.

More Examples of Working with Units

Example 1. How many seconds are there in 1 day?

Solution: We can answer the question by setting up a *chain* of unit conversions in which we start with 1 *day* and end up with *seconds*. We use the facts that there are 24 hours per day (24 hr/day), 60 minutes per hour (60 min/hr), and 60 seconds per minute (60 s/min):

$$1 \text{ day} \times \frac{24 \text{ hr}}{\text{day}} \times \frac{60 \text{ min}}{\text{hr}} \times \frac{60 \text{ s}}{\text{min}} = 86{,}400 \text{ s}$$

(starting value; conversion from day to hr; conversion from hr to min; conversion from min to s)

Note that all the units cancel except *seconds*, which is what we want for the answer. There are 86,400 seconds in 1 day.

Example 2. Convert a distance of 10^8 cm to km.

Solution: The easiest way to make this conversion is in two steps, since we know that there are 100 centimeters per meter (100 cm/m) and 1000 meters per kilometer (1000 m/km):

$$10^8 \text{ cm} \times \frac{1 \text{ m}}{100 \text{ cm}} \times \frac{1 \text{ km}}{1000 \text{ m}} = 10^8 \text{ cm} \times \frac{1 \text{ m}}{10^2 \text{ cm}} \times \frac{1 \text{ km}}{10^3 \text{ m}} = 10^3 \text{ km}$$

Alternatively, if we recognize that the number of kilometers should be smaller than the number of centimeters (because kilometers are larger), we might decide to do this conversion by dividing as follows:

$$10^8 \text{ cm} \div \frac{100 \text{ cm}}{\text{m}} \div \frac{1000 \text{ m}}{\text{km}}$$

In this case, before carrying out the calculation, we replace each division with multiplication by the reciprocal:

$$10^8 \text{ cm} \div \frac{100 \text{ cm}}{\text{m}} \div \frac{1000 \text{ m}}{\text{km}} = 10^8 \text{ cm} \times \frac{1 \text{ m}}{100 \text{ cm}} \times \frac{1 \text{ km}}{1000 \text{ m}}$$

$$= 10^8 \text{ cm} \times \frac{1 \text{ m}}{10^2 \text{ cm}} \times \frac{1 \text{ km}}{10^3 \text{ m}}$$

$$= 10^3 \text{ km}$$

Note that we again get the answer that 10^8 cm is the same as 10^3 km, or 1000 km.

Example 3. Suppose you accelerate at 9.8 m/s² for 4 seconds, starting from rest. How fast will you be going?

Solution: The question asked "how fast?" so we expect to end up with a speed. Therefore, we multiply the acceleration by the amount of time you accelerated:

$$9.8 \frac{\text{m}}{\text{s}^2} \times 4 \text{ s} = (9.8 \times 4) \frac{\text{m} \times \text{s}}{\text{s}^2} = 39.2 \frac{\text{m}}{\text{s}}$$

Note that the units end up as a speed, showing that you will be traveling 39.2 m/s after 4 seconds of acceleration at 9.8 m/s².

Example 4. A reservoir is 2 km long and 3 km wide. Calculate its area, in both square kilometers and square meters.

Solution: We find its area by multiplying its length and width:

$$2 \text{ km} \times 3 \text{ km} = 6 \text{ km}^2$$

Next we need to convert this area of 6 km² to square meters, using the fact that there are 1000 meters per kilometer (1000 m/km). Note that we must square the term 1000 m/km when converting from km² to m²:

$$6 \text{ km}^2 \times \left(1000 \frac{\text{m}}{\text{km}}\right)^2 = 6 \text{ km}^2 \times 1000^2 \frac{\text{m}^2}{\text{km}^2} = 6 \text{ km}^2 \times 1{,}000{,}000 \frac{\text{m}^2}{\text{km}^2}$$

$$= 6{,}000{,}000 \text{ m}^2$$

The reservoir area is 6 km², which is the same as 6 million m².

C.4 Finding a Ratio

Suppose you want to compare two quantities, such as the average density of Earth and the average density of Jupiter. The way we do such a comparison is by dividing, which tells us the *ratio* of the two quantities. In this case, Earth's average density is 5.52 g/cm³ and Jupiter's average density is 1.33 g/cm³, so the ratio is

$$\frac{\text{average density of Earth}}{\text{average density of Jupiter}} = \frac{5.52 \text{ g/cm}^3}{1.33 \text{ g/cm}^3} = 4.15$$

Notice how the units cancel on both the top and the bottom of the fraction. We can state our result in two equivalent ways:

- The ratio of Earth's average density to Jupiter's average density is 4.15.
- Earth's average density is 4.15 times Jupiter's average density.

Sometimes, the quantities that you want to compare may each involve an equation. In such cases, you could, of course, find the ratio by first calculating each of the two quantities individually and then dividing. However, it is much easier if you first express the ratio as a fraction, putting the equation for one quantity on top and the other on the bottom. Some of the terms in the equation may then cancel out, making any calculations much easier.

Example 1. Compare the kinetic energy of a car traveling at 100 km/hr to that of the same car traveling at 50 km/hr.

Solution: We do the comparison by finding the ratio of the two kinetic energies, recalling that the formula for kinetic energy is $\frac{1}{2}mv^2$. Since we are not told the mass of the car, you might at first think that we don't have enough information to find the ratio. However, notice what happens when we put the equations for each kinetic energy into the ratio, calling the two speeds v_1 and v_2:

$$\frac{\text{K.E. car at } v_1}{\text{K.E. car at } v_2} = \frac{\frac{1}{2}m_{\text{car}}v_1^2}{\frac{1}{2}m_{\text{car}}v_2^2} = \frac{v_1^2}{v_2^2} = \left(\frac{v_1}{v_2}\right)^2$$

All the terms cancel except those with the two speeds, leaving us with a very simple formula for the ratio. Now we put in 100 km/hr for v_1 and 50 km/hr for v_2:

$$\frac{\text{K.E. car at 100 km/hr}}{\text{K.E. car at 50 km/hr}} = \left(\frac{100 \text{ km/hr}}{50 \text{ km/hr}}\right)^2 = 2^2 = 4$$

The ratio of the car's kinetic energies at 100 km/hr and 50 km/hr is 4. That is, the car has four times as much kinetic energy at 100 km/hr as it has at 50 km/hr.

Example 2. Compare the strength of gravity between Earth and the Sun to the strength of gravity between Earth and the Moon.

Solution: We do the comparison by taking the ratio of the Earth–Sun gravity to the Earth–Moon gravity. In this case, each quantity is found from the equation of Newton's law of gravity. (See Section 3.3.) Thus, the ratio is

$$\frac{\text{Earth–Sun gravity}}{\text{Earth–Moon gravity}} = \frac{G\frac{M_{\text{Earth}}M_{\text{Sun}}}{(d_{\text{Earth–Sun}})^2}}{G\frac{M_{\text{Earth}}M_{\text{Moon}}}{(d_{\text{Earth–Moon}})^2}} = \frac{M_{\text{Sun}}}{(d_{\text{Earth–Sun}})^2} \times \frac{(d_{\text{Earth–Moon}})^2}{M_{\text{Moon}}}$$

Note how all but four of the terms cancel; the last step comes from replacing the division with multiplication by the reciprocal (the "invert and multiply" rule for division). We can simplify the work further by rearranging the terms so that we have the masses and distances together:

$$\frac{\text{Earth–Sun gravity}}{\text{Earth–Moon gravity}} = \frac{M_{\text{Sun}}}{M_{\text{Moon}}} \times \frac{(d_{\text{Earth–Moon}})^2}{(d_{\text{Earth–Sun}})^2}$$

Now it is just a matter of looking up the numbers and calculating:

$$\frac{\text{Earth–Sun gravity}}{\text{Earth–Moon gravity}} = \frac{1.99 \times 10^{30} \text{ kg}}{7.35 \times 10^{22} \text{ kg}} \times \frac{(384.4 \times 10^3 \text{ km})^2}{(149.6 \times 10^6 \text{ km})^2} = 179$$

In other words, the Earth–Sun gravity is 179 times as strong as the Earth–Moon gravity.

C.5 The Metric System (SI)

The modern version of the metric system, known as *Système Internationale d'Unites* (French for "International System of Units") or **SI**, was formally established in 1960. Today, it is the primary measurement system in nearly every country in the world with the exception of the United States. Even in the United States, it is the system of choice for science and international commerce.

The basic units of length, mass, and time in the SI are

- the **meter** for length, abbreviated m
- the **kilogram** for mass, abbreviated kg
- the **second** for time, abbreviated s

Multiples of metric units are formed by powers of 10, using a prefix to indicate the power. For example, *kilo* means 10^3 (1000), so a kilometer is 1000 meters; a microgram is 0.000001 gram, because *micro* means 10^{-6}, or one millionth. Some of the more common prefixes are listed in Table C.2.

TABLE C.2 SI (Metric) Prefixes

Small Values			Large Values		
Prefix	Abbreviation	Value	Prefix	Abbreviation	Value
Deci	d	10^{-1}	Deca	da	10^1
Centi	c	10^{-2}	Hecto	h	10^2
Milli	m	10^{-3}	Kilo	k	10^3
Micro	μ	10^{-6}	Mega	M	10^6
Nano	n	10^{-9}	Giga	G	10^9
Pico	p	10^{-12}	Tera	T	10^{12}

Metric Conversions

Table C.3 lists conversions between metric units and units used commonly in the United States. Note that the conversions between kilograms and pounds are valid only on Earth, because they depend on the strength of gravity.

Example 1. International athletic competitions generally use metric distances. Compare the length of a 100-meter race to that of a 100-yard race.

Solution: Table C.3 shows that 1 m = 1.094 yd, so 100 m is 109.4 yd. Note that 100 meters is almost 110 yards; a good "rule of thumb" to remember is that distances in meters are about 10% longer than the corresponding number of yards.

TABLE C.3 Metric Conversions

To Metric	From Metric
1 inch = 2.540 cm	1 cm = 0.3937 inch
1 foot = 0.3048 m	1 m = 3.28 feet
1 yard = 0.9144 m	1 m = 1.094 yards
1 mile = 1.6093 km	1 km = 0.6214 mile
1 pound = 0.4536 kg	1 kg = 2.205 pounds

Example 2. How many square kilometers are in 1 square mile?

Solution: We use the square of the miles-to-kilometers conversion factor:

$$(1 \text{ mi}^2) \times \left(\frac{1.6093 \text{ km}}{1 \text{ mi}}\right)^2 = (1 \text{ mi}^2) \times \left(1.6093^2 \ \frac{\text{km}^2}{\text{mi}^2}\right) = 2.5898 \text{ km}^2$$

Therefore, 1 square mile is 2.5898 square kilometers.

C.6 Temperature Scales

In the United States, we usually use the Fahrenheit temperature scale in our daily lives. But most of the rest of the world uses the Celsius scale, and in science we often use the Kelvin scale.

The figure shows the relationship among the three temperature scales. Notice that the Fahrenheit scale is defined so that under ordinary conditions at sea level on Earth, water freezes at 32°F and boils at 212°F. Most of the rest of the world uses the Celsius scale, on which water freezes at 0°C and boils at 100°C. The Kelvin scale is very similar to the Celsius scale, because the boiling point of water is exactly 100 K higher than the freezing point. However, the zero point of the Kelvin scale is defined to be the coldest possible temperature, or *absolute zero*. This puts the freezing point of water at 273.15 K and the boiling point at 373.15 K.

You can use these relationships to convert among the three scales. Because the only difference between the Kelvin scale and the Celsius scale is the starting point, we can convert from Kelvin to Celsius and from Celsius to Fahrenheit with the following approximate relationships:

$$T_{\text{Celsius}} \approx T_{\text{Kelvin}} - 273$$

$$T_{\text{Fahrenheit}} = (1.8 \times T_{\text{Celsius}}) + 32$$

Example 1. Convert a temperature of 300 K to Celsius and Fahrenheit.

Solution: We use the first formula to convert to Celsius:

$$T_{\text{Celsius}} \approx T_{\text{Kelvin}} - 273 = 300 - 273 = 27°C$$

We use the second formula to convert to Fahrenheit:

$$T_{\text{Fahrenheit}} = (1.8 \times T_{\text{Celsius}}) + 32 = (1.8 \times 27) + 32 \approx 81°F$$

Three common temperature scales: Kelvin, Celsius, and Fahrenheit. [The degree symbol (°) is not usually used with the Kelvin scale.]

Glossary

absolute magnitude A measure of an object's luminosity; defined to be the apparent magnitude the object would have if it were located exactly 10 parsecs away.

absolute zero The coldest possible temperature, which is 0 K = −273.15°C.

absorption (of light) The process by which matter absorbs radiative energy.

absorption line A dark band (a "line") on an otherwise bright rainbow of light, occurring when light viewed through a diffraction element such as a prism shows a deficit of photons at or near a specific wavelength.

accelerating universe A universe in which a repulsive force (see cosmological constant) causes the expansion of the universe to accelerate with time. Its galaxies will recede from one another increasingly faster, and it will become cold and dark more quickly than a coasting universe.

acceleration The rate at which an object's velocity changes. Its standard units are m/s^2.

acceleration of gravity The acceleration of a falling object. On Earth, the acceleration of gravity, designated by g, is $9.8 \, m/s^2$.

accretion The process by which small objects gather together to make larger objects.

accretion disk A rapidly rotating disk of material that gradually falls inward as it orbits a starlike object (e.g., white dwarf, neutron star, or black hole).

active galactic nuclei The unusually luminous centers of some galaxies, thought to be powered by accretion onto supermassive black holes. Quasars are the brightest type of active galactic nuclei; radio galaxies also contain active galactic nuclei.

active galaxy A term sometimes used to describe a galaxy that contains an *active galactic nucleus*.

adaptive optics A technique in which telescope mirrors flex rapidly to compensate for the bending of starlight caused by atmospheric turbulence.

Algol paradox A paradox concerning the binary star Algol, which contains a subgiant star that is less massive than its main-sequence companion.

altitude (above horizon) The angular distance between the horizon and an object in the sky.

amino acids The building blocks of proteins.

analemma The figure-8 path traced by the Sun over the course of a year when viewed at the same place and the same time each day; it represents the discrepancies between apparent and mean solar time.

Andromeda Galaxy M31, the Great Galaxy in Andromeda; the nearest large spiral galaxy to the Milky Way.

angular momentum Momentum attributable to rotation or revolution. The angular momentum of an object moving in a circle of radius r is the product $m \times v \times r$.

angular resolution (of a telescope) The smallest angular separation that two pointlike objects can have and still be seen as distinct points of light (rather than as a single point of light).

angular size (or **angular distance**) A measure of the angle formed by extending imaginary lines outward from our eyes to span an object (or the space between two objects).

annihilation *See* matter–antimatter annihilation.

annular solar eclipse A solar eclipse during which the Moon is directly in front of the Sun but its angular size is not large enough to fully block the Sun; thus, a ring (or *annulus*) of sunlight is still visible around the Moon's disk.

Antarctic Circle The circle on Earth with latitude 66.5°S.

antielectron The antimatter equivalent of an electron. It is identical to an electron in virtually all respects, except it has a positive rather than a negative electrical charge.

antimatter Any particle with the same mass as a particle of ordinary matter but whose other basic properties, such as electrical charge, are precisely opposite.

aphelion The point at which an object orbiting the Sun is farthest from the Sun.

apogee The point at which an object orbiting Earth is farthest from Earth.

apparent brightness The amount of light reaching us *per unit area* from a luminous object; often measured in units of $watts/m^2$.

apparent magnitude A measure of the apparent brightness of an object in the sky, based on the ancient system developed by Hipparchus.

apparent retrograde motion The apparent motion of a planet, as viewed from Earth, during the period of a few weeks or months when it moves westward relative to the stars in our sky.

apparent solar time Time measured by the actual position of the Sun in the local sky, defined so that noon is when the Sun is *on* the meridian.

arcminute (or **minute of arc**) 1/60 of 1°.

arcsecond (or **second of arc**) 1/60 of an arcminute, or 1/3600 of 1°.

Arctic Circle The circle on Earth with latitude 66.5°N.

asteroid A relatively small and rocky object that orbits a star; asteroids are officially considered part of a category known as "small solar system bodies."

asteroid belt The region of our solar system between the orbits of Mars and Jupiter in which asteroids are heavily concentrated.

astrobiology The study of life on Earth and beyond; it emphasizes research into questions of the origin of life, the conditions under which life can survive, and the search for life beyond Earth.

astrometric method The detection of extrasolar planets through the side-to-side motion of a star caused by gravitational tugs from the planet.

astronomical unit (AU) The average distance (semimajor axis) of Earth from the Sun, which is about 150 million km.

atmosphere A layer of gas that surrounds a planet or moon, usually very thin compared to the size of the object.

atmospheric pressure The surface pressure resulting from the overlying weight of an atmosphere.

atmospheric structure The layering of a planetary atmosphere due to variations in temperature with altitude. For example, Earth's atmospheric structure from the ground up consists of the troposphere, stratosphere, thermosphere, and exosphere.

atomic hydrogen gas Gas composed mostly of hydrogen atoms, though in space it is generally mixed with helium and small amounts of other elements as well; it is the most common form of interstellar gas.

atomic mass number The combined number of protons and neutrons in an atom.

atomic number The number of protons in an atom.

atoms The basic units of the chemical elements, consisting of a nucleus made from protons and neutrons, surrounded by a cloud of electrons.

aurora Dancing lights in the sky caused by charged particles entering our atmosphere; called the *aurora borealis* in the Northern Hemisphere and the *aurora australis* in the Southern Hemisphere.

axis tilt (of a planet in our solar system) The amount by which a planet's axis is tilted with respect to a line perpendicular to the ecliptic plane.

azimuth (usually called **direction** in this book) Direction around the horizon from due north, measured clockwise in degrees. For example, the azimuth of due north is 0°, due east is 90°, due south is 180°, and due west is 270°.

bar The standard unit of pressure, approximately equal to Earth's atmospheric pressure at sea level.

barred spiral galaxies Spiral galaxies that have a straight bar of stars cutting across their centers.

baryonic matter Ordinary matter made from atoms (so called because the nuclei of atoms contain protons and neutrons, which are both baryons).

baryons Particles, including protons and neutrons, that are made from three quarks.

basalt A type of dark, high-density volcanic rock that is rich in iron and magnesium-based silicate minerals; it forms a runny (easily flowing) lava when molten.

belts (on a jovian planet) Dark bands of sinking air that encircle a jovian planet at a particular set of latitudes.

Big Bang The name given to the event thought to mark the birth of the universe.

Big Bang theory The scientific theory of the universe's earliest moments, stating that all the matter in our observable universe came into being at a single moment in time as an extremely hot, dense mixture of subatomic particles and radiation.

Big Crunch The name given to the event that would presumably end the universe if gravity ever reverses the universal expansion and the universe someday begins to collapse.

binary star system A star system that contains two stars.

biosphere The "layer" of life on Earth.

blackbody radiation *See* thermal radiation.

black hole A bottomless pit in spacetime. Nothing can escape from within a black hole, and we can never again detect or observe an object that falls into a black hole.

black hole merger An event in which two black holes merge to form a single more massive black hole, accompanied by the release of gravitational waves.

blazer A term sometimes used to describe an active galactic nucleus in which a powerful jet of material happens to be aimed nearly in the direction of Earth.

BL Lac objects A class of active galactic nuclei that probably represent the centers of radio galaxies whose jets happen to be pointed directly at us.

blowout Ejection of the hot, gaseous contents of a superbubble when it grows so large that it bursts out of the cooler layer of gas filling the galaxy's disk.

blueshift A Doppler shift in which spectral features are shifted to shorter wavelengths, observed when an object is moving toward the observer.

bosons Particles, such as photons, to which the exclusion principle does not apply.

bound orbits Orbits on which an object travels repeatedly around another object; bound orbits are elliptical in shape.

brown dwarf An object too small to become an ordinary star because electron degeneracy pressure halts its gravitational collapse before fusion becomes self-sustaining; brown dwarfs have masses less than $0.08 M_{Sun}$.

bubble (interstellar) An expanding shell of hot, ionized gas driven by stellar winds or supernovae, with very hot and very low density gas inside.

bulge (of a spiral galaxy) The central portion of a spiral galaxy that is roughly spherical (or football shaped) and bulges above and below the plane of the galactic disk.

Cambrian explosion The dramatic diversification of life on Earth that occurred between about 540 and 500 million years ago.

carbonate rock A carbon-rich rock, such as limestone, that forms underwater from chemical reactions between sediments and carbon dioxide. On Earth, most of the outgassed carbon dioxide currently resides in carbonate rocks.

carbon dioxide cycle (CO_2 cycle) The process that cycles carbon dioxide between Earth's atmosphere and surface rocks.

carbon stars Stars whose atmospheres are especially carbon-rich, thought to be near the ends of their lives; carbon stars are the primary sources of carbon in the universe.

Cassini division A large, dark gap in Saturn's rings, visible through small telescopes on Earth.

CCD (charge-coupled device) A type of electronic light detector that has largely replaced photographic film in astronomical research.

celestial coordinates The coordinates of right ascension and declination that fix an object's position on the celestial sphere.

celestial equator (CE) The extension of Earth's equator onto the celestial sphere.

celestial navigation Navigation on the surface of the Earth accomplished through observations of the Sun and stars.

celestial sphere The imaginary sphere on which objects in the sky appear to reside when observed from Earth.

Celsius (temperature scale) The temperature scale commonly used in daily activity internationally, defined so that, on Earth's surface, water freezes at 0°C and boils at 100°C.

center of mass (of orbiting objects) The point at which two or more orbiting objects would balance if they were somehow connected; it is the point around which the orbiting objects actually orbit.

central dominant galaxy A giant elliptical galaxy found at the center of a dense cluster of galaxies, apparently formed by the merger of several individual galaxies.

Cepheid variable star (or **Cepheid** for short) A particularly luminous type of pulsating variable star that follows *Leavitt's law* (also called the period–luminosity relation) and hence is very useful for measuring cosmic distances.

Chandrasekhar limit *See* white dwarf limit.

charged particle belts Zones in which ions and electrons accumulate and encircle a planet.

chemical element *See* element (chemical).

chemical enrichment The process by which the abundance of heavy elements (heavier than helium) in the interstellar medium gradually increases over time as these elements are produced by stars and released into space.

chemical potential energy Potential energy that can be released through chemical reactions; for example, food contains chemical potential energy that your body can convert to other forms of energy.

chondrites Another name for primitive meteorites. The name comes from the round chondrules within them. *Achondrites*, meaning "without chondrules," is another name for processed meteorites.

chromosphere The layer of the Sun's atmosphere below the corona; most of the Sun's ultraviolet light is emitted from this region, in which the temperature is about 10,000 K.

circulation cells (or **Hadley cells**) Large-scale cells (similar to convection cells) in a planet's atmosphere that transport heat between the equator and the poles.

circumpolar star A star that always remains above the horizon for a particular latitude.

climate The long-term average of weather.

close binary A binary star system in which the two stars are very close together.

closed universe A universe in which spacetime curves back on itself to the point where its overall shape is analogous to that of the surface of a sphere.

cluster of galaxies A collection of a few dozen or more galaxies bound together by gravity; smaller collections of galaxies are simply called *groups*.

cluster of stars A group of anywhere from several hundred to a million or so stars; star clusters come in two types—open clusters and globular clusters.

CNO cycle The cycle of reactions by which intermediate- and high-mass stars fuse hydrogen into helium.

coasting universe A model of the universe in which the universe expands forever with little change in its rate of expansion; in the absence of a repulsive force (*see* cosmological constant), a coasting universe is one in which the actual mass density is *smaller* than the critical density.

color-coded image An image that represents information or forms of light in any way that makes an object appear different than it would appear if we looked at its true, visible-light colors. Sometimes called a *false-color image*.

coma (of a comet) The dusty atmosphere of a comet, created by sublimation of ices in the nucleus when the comet is near the Sun.

comet A relatively small, icy object that orbits a star. Like asteroids, comets are officially considered part of a category known as "small solar system bodies."

comparative planetology The study of the solar system by examining and understanding the similarities and differences among worlds.

compound (chemical) A substance made from molecules consisting of two or more atoms with different atomic numbers.

condensates Solid or liquid particles that condense from a cloud of gas.

condensation The formation of solid or liquid particles from a cloud of gas.

conduction (of energy) The process by which thermal energy is transferred by direct contact from warm material to cooler material.

conjunction (of a planet with the Sun) An event in which a planet and the Sun line up in our sky.

conservation of angular momentum (law of) The principle that, in the absence of net torque (twisting force), the total angular momentum of a system remains constant.

conservation of energy (law of) The principle that energy (including mass-energy) can be neither created nor destroyed, but can only change from one form to another.

conservation of momentum (law of) The principle that, in the absence of net force, the total momentum of a system remains constant.

constellation A region of the sky; 88 official constellations cover the celestial sphere.

continental crust The thicker lower-density crust that makes up Earth's continents. It is made when remelting of seafloor crust allows lower-density rock to separate and erupt to the surface. Continental crust ranges in age from very young to as old as about 4 billion years (or more).

continuous spectrum A spectrum (of light) that spans a broad range of wavelengths without interruption by emission or absorption lines.

convection The energy transport process in which warm material expands and rises while cooler material contracts and falls.

convection cell An individual small region of convecting material.

convection zone (of a star) A region in which energy is transported outward by convection.

Copernican revolution The dramatic change, initiated by Copernicus, that occurred when we learned that Earth is a planet orbiting the Sun rather than the center of the universe.

core of a planet The dense central region of a planet that has undergone differentiation.

core of a star The central region of a star, in which nuclear fusion can occur.

Coriolis effect The effect due to rotation that causes air or objects on a rotating surface or planet to deviate from straight-line trajectories.

corona (solar) The tenuous uppermost layer of the Sun's atmosphere; most of the Sun's x-rays are emitted from this region, in which the temperature is about 1 million K.

coronal holes Regions of the corona that barely show up in x-ray images because they are nearly devoid of hot coronal gas.

coronal mass ejections Bursts of charged particles from the Sun's corona that travel outward into space.

cosmic microwave background The remnant radiation from the Big Bang, which we detect using radio telescopes sensitive to microwaves (which are short-wavelength radio waves).

cosmic rays Particles such as electrons, protons, and atomic nuclei that zip through interstellar space at close to the speed of light.

cosmological constant The name given to a term in Einstein's equations of general relativity. If it is not zero, then it represents a repulsive force or a type of energy (sometimes called *dark energy* or *quintessence*) that might cause the expansion of the universe to accelerate with time.

cosmological horizon The boundary of our observable universe, which is where the lookback time is equal to the age of the universe. Beyond this boundary in spacetime, we cannot see anything at all.

Cosmological Principle The idea that matter is distributed uniformly throughout the universe on very large scales, meaning that the universe has neither a center nor an edge.

cosmological redshift The redshift we see from distant galaxies, caused by the fact that expansion of the universe stretches all the photons within it to longer, redder wavelengths.

cosmology The study of the overall structure and evolution of the universe.

cosmos An alternative name for the universe.

crescent phase The phase of the Moon (or of a planet) in which just a small portion (less than half) of the visible face is illuminated by sunlight.

critical density The precise average density for the entire universe that marks the dividing line between a recollapsing universe and one that will expand forever.

critical universe A model of the universe in which the universe expands more and more slowly as time progresses; in the absence of a repulsive force (*see* cosmological constant), a critical universe is one in which the average mass density *equals* the critical density.

crust (of a planet) The low-density surface layer of a planet that has undergone differentiation.

curvature of spacetime A change in the geometry of space that is produced in the vicinity of a massive object and is responsible for the force we call gravity. The overall geometry of the universe may also be curved, depending on its overall mass-energy content.

cycles per second Units of frequency for a wave; describes the number of peaks (or troughs) of a wave that pass by a given point each second. Equivalent to *hertz*.

dark energy Name sometimes given to energy that could be causing the expansion of the universe to accelerate. See cosmological constant.

dark matter Matter that we infer to exist from its gravitational effects but from which we have not detected any light; dark matter apparently dominates the total mass of the universe.

daylight saving time Standard time plus 1 hour, so that the Sun appears on the meridian around 1 p.m. rather than around noon.

decay (radioactive) *See* radioactive decay.

December solstice Both the point on the celestial sphere where the ecliptic is farthest south of the celestial equator and the moment in time when the Sun appears at that point each year (around December 21).

declination (dec) The angular north-south distance between the celestial equator and a location on the celestial sphere; analogous to latitude, but on the celestial sphere.

deferent The large circle upon which a planet follows its circle-upon-circle path around Earth in the (Earth-centered) Ptolemaic model of the universe. *See also* epicycle.

degeneracy pressure A type of pressure unrelated to an object's temperature, which arises when electrons (electron degeneracy pressure) or neutrons (neutron degeneracy pressure) are packed so tightly that the exclusion and uncertainty principles come into play.

degenerate object An object, such as a brown dwarf, white dwarf, or neutron star, in which degeneracy pressure is the primary pressure pushing back against gravity.

density (mass) The amount of mass per unit volume of an object. The average density of any object can be found by dividing its mass by its volume. Standard metric units are kilograms per cubic meter, but density is more commonly stated in units of grams per cubic centimeter.

deuterium A form of hydrogen in which the nucleus contains a proton and a neutron, rather than only a proton (as is the case for most hydrogen nuclei).

differential rotation Rotation in which the equator of an object rotates at a different rate than the poles.

differentiation The process by which gravity separates materials according to density, with high-density materials sinking and low-density materials rising.

diffraction grating A finely etched surface that can split light into a spectrum.

diffraction limit The angular resolution that a telescope could achieve if it were limited only by the interference of light waves; it is smaller (i.e., better angular resolution) for larger telescopes.

dimension (mathematical) Measure of the number of independent directions in which movement is possible; for example, the surface of Earth is two-dimensional because only two independent directions of motion are possible (north-south and east-west).

direction (in local sky) One of the two coordinates (the other is altitude) needed to pinpoint an object in the local sky. It is the direction, such as north, south, east, or west, in which you must face to see the object. *See also* azimuth.

disk (of a galaxy) The portion of a spiral galaxy that looks like a disk and contains an interstellar medium with cool gas and dust; stars of many ages are found in the disk.

disk population The stars that orbit within the disk of a spiral galaxy; sometimes called *Population I*.

DNA (deoxyribonucleic acid) The molecule that constitutes the genetic material of life on Earth.

Doppler effect (or **Doppler shift**) The effect that shifts the wavelengths of spectral features in objects that are moving toward or away from the observer.

Doppler method The detection of extrasolar planets through the motion of a star toward and away from the observer caused by gravitational tugs from the planet.

double shell–fusion star A star that is fusing helium into carbon in a shell around an inert carbon core and is fusing hydrogen into helium in a shell at the top of the helium layer.

down quark One of the two quark types (the other is the up quark) found in ordinary protons and neutrons. It has a charge of $-\frac{1}{3}$.

Drake equation An equation that lays out the factors that play a role in determining the number of communicating civilizations in our galaxy.

dust (or **dust grains**) Tiny solid flecks of material; in astronomy, we often discuss interplanetary dust (found within a star system) or interstellar dust (found between the stars in a galaxy).

dust tail (of a comet) One of two tails seen when a comet passes near the Sun (the other is the *plasma tail*). It is composed of small solid particles pushed away from the Sun by the radiation pressure of sunlight.

dwarf elliptical galaxy A small elliptical galaxy with less than about a billion stars.

dwarf galaxies Relatively small galaxies, consisting of less than about 10 billion stars.

dwarf planet An object that orbits the Sun and is massive enough for its gravity to have made it nearly round in shape, but that does not qualify as an official planet because it has not cleared its orbital neighborhood. The dwarf planets of our solar system include the asteroid Ceres and the Kuiper belt objects Pluto, Eris, Haumea, and Makemake.

dwarf spheroidal galaxy A subclass of dwarf elliptical galaxies that are round and diskless like other elliptical galaxies but are particularly small and much less bright. They are the most numerous subtype of galaxy in the Local Group and possibly in the universe.

Earth-orbiters (spacecraft) Spacecraft designed to study Earth or the universe from Earth orbit.

eccentricity A measure of how much an ellipse deviates from a perfect circle; defined as the center-to-focus distance divided by the length of the semimajor axis.

eclipse An event in which one astronomical object casts a shadow on another or crosses our line of sight to the other object.

eclipse seasons Periods during which lunar and solar eclipses can occur because the nodes of the Moon's orbit are aligned with Earth and the Sun.

eclipsing binary A binary star system in which the two stars happen to be orbiting in the plane of our line of sight, so that each star will periodically eclipse the other.

ecliptic The Sun's apparent annual path among the constellations.

ecliptic plane The plane of Earth's orbit around the Sun.

ejecta (from an impact) Debris ejected by the blast of an impact.

electrical charge A fundamental property of matter that is described by its amount and as either positive or negative; more technically, a measure of how a particle responds to the electromagnetic force.

electromagnetic field An abstract concept used to describe how a charged particle would affect other charged particles at a distance.

electromagnetic radiation Another name for light of all types, from radio waves through gamma rays.

electromagnetic spectrum The complete spectrum of light, including radio waves, infrared light, visible light, ultraviolet light, x-rays, and gamma rays.

electromagnetic wave A synonym for *light*, which consists of waves of electric and magnetic fields.

electromagnetism (or **electromagnetic force**) One of the four fundamental forces; it is the force that dominates atomic and molecular interactions.

electron degeneracy pressure Degeneracy pressure exerted by electrons, as in brown dwarfs and white dwarfs.

electrons Fundamental particles with negative electric charge; the distribution of electrons in an atom gives the atom its size.

electron-volt (eV) A unit of energy equivalent to 1.60×10^{-19} joule.

electroweak era The era of the universe during which only three forces operated (gravity, strong force, and electroweak force), lasting from 10^{-38} second to 10^{-10} second after the Big Bang.

electroweak force The force that exists at high energies when the electromagnetic force and the weak force exist as a single force.

element (chemical) A substance made from individual atoms of a particular atomic number.

ellipse A type of oval that happens to be the shape of bound orbits. An ellipse can be drawn by moving a pencil along a string whose ends are tied to two tacks; the locations of the tacks are the *foci* (singular: *focus*) of the ellipse.

elliptical galaxies Galaxies that appear rounded in shape, often longer in one direction, like a football. They have no disks and contain little cool gas and dust compared to spiral galaxies, though they often contain hot, ionized gas.

elongation (greatest) For Mercury or Venus, the point at which it appears farthest from the Sun in our sky.

emission (of light) The process by which matter emits energy in the form of light.

emission line A bright band (a "line") of single color, superimposed on a fainter or completely absent rainbow of light, occurring when light viewed through a diffraction element such as a prism shows an excess of photons at or near a specific wavelength.

emission nebula *See* ionization nebula.

energy Broadly speaking, what can make matter move. The three basic types of energy are kinetic, potential, and radiative.

energy balance (in a star) The balance between the rate at which fusion releases energy in the star's core and the rate at which the star's surface radiates this energy into space.

epicycle The small circle upon which a planet moves while simultaneously going around a larger circle (the *deferent*) around Earth in the (Earth-centered) Ptolemaic model of the universe.

equation of time An equation describing the discrepancies between apparent and mean solar time.

equinox *See* March equinox *and* September equinox.

equivalence principle The fundamental starting point for general relativity, which states that the effects of gravity are exactly equivalent to the effects of acceleration.

era of atoms The era of the universe lasting from about 500,000 years to about 1 billion years after the Big Bang, during which it was cool enough for neutral atoms to form.

era of galaxies The present era of the universe, which began with the formation of galaxies when the universe was about 1 billion years old.

era of nuclei The era of the universe lasting from about 3 minutes to about 380,000 years after the Big Bang, during which matter in the universe was fully ionized and opaque to light. The cosmic background radiation was released at the end of this era.

era of nucleosynthesis The era of the universe lasting from about 0.001 second to about 3 minutes after the Big Bang, by the end of which virtually all of the neutrons and about one-seventh of the protons in the universe had fused into helium.

erosion The wearing down or building up of geological features by wind, water, ice, and other phenomena of planetary weather.

eruption The process of releasing hot lava onto a planet's surface.

escape velocity The speed necessary for an object to completely escape the gravity of a large body such as a moon, planet, or star.

evaporation The process by which atoms or molecules escape into the gas phase from a liquid.

event Any particular point along a worldline; all observers will agree on the reality of an event but may disagree about its time and location.

event horizon The boundary that marks the "point of no return" between a black hole and the outside universe; events that occur within the event horizon can have no influence on our observable universe.

evolution (biological) The gradual change in populations of living organisms responsible for transforming life on Earth from its primitive origins to its great diversity today.

exchange particle A type of subatomic particle that transmits one of the four fundamental forces; according to the standard model of physics, these particles are always exchanged whenever two objects interact through a force.

excited state (of an atom) Any arrangement of electrons in an atom that has more energy than the ground state.

exclusion principle The law of quantum mechanics that states that two fermions cannot occupy the same quantum state at the same time.

exosphere The hot, outer layer of an atmosphere, where the atmosphere "fades away" to space.

expansion (of the universe) The idea that the space between galaxies or clusters of galaxies is growing with time.

exposure time The amount of time during which light is collected to make a single image.

extrasolar planet A planet orbiting a star other than our Sun.

extremophiles Living organisms that are adapted to conditions that are "extreme" by human standards, such as very high or low temperature or a high level of salinity or radiation.

Fahrenheit (temperature scale) The temperature scale commonly used in daily activity in the United States; defined so that, on Earth's surface, water freezes at 32°F and boils at 212°F.

fall equinox *See* September equinox, which is commonly called the fall equinox by people living in the Northern Hemisphere.

false-color image *See* color-coded image.

fault (geological) A place where lithospheric plates slip sideways relative to one another.

feedback processes Processes in which a small change in some property (such as temperature) leads to changes in other properties that either amplify or diminish the original small change.

fermions Particles, such as electrons, neutrons, and protons, that obey the exclusion principle.

Fermi's paradox The question posed by Enrico Fermi about extraterrestrial intelligence—"So where is everybody?"—which asks why we have not observed other civilizations even though simple arguments would suggest that some ought to have spread throughout the galaxy by now.

field An abstract concept used to describe how a particle would interact with a force. For example, the idea of a gravitational field describes how a particle would react to the local strength of gravity, and the idea of an electromagnetic field describes how a charged particle would respond to forces from other charged particles.

filter (for light) A material that transmits only particular wavelengths of light.

fireball A particularly bright meteor.

first-quarter phase The phase of the Moon that occurs one-quarter of the way through each cycle of phases, in which precisely half of the visible face is illuminated by sunlight.

fission *See* nuclear fission.

flare star A small, spectral type M star that displays particularly strong flares on its surface.

flat geometry (or **Euclidean geometry**) The type of geometry in which the rules of geometry for a flat plane hold, such as that the shortest distance between two points is a straight line and that the sum of the angles in a triangle is 180°.

flat universe A universe in which the overall geometry of spacetime is flat (Euclidean), as would be the case if the density of the universe was equal to the critical density.

flybys Spacecraft that fly past a target object (such as a planet), usually just once, as opposed to entering a bound orbit of the object.

focal plane The place where an image created by a lens or mirror is in focus.

foci Plural of *focus*.

focus (of a lens or mirror) The point at which rays of light that were initially parallel (such as those from a distant star) converge.

focus (of an ellipse) One of two special points within an ellipse that lie along the major axis; these are the points around which we could stretch a pencil and string to draw an ellipse. When one object orbits a second object, the second object lies at one focus of the orbit.

force Anything that can cause a change in momentum.

formation properties (of planets) In this book, for the purpose of understanding geological processes, four formation properties that planets are defined to be born with: size (mass and radius), distance from the Sun, composition, and rotation rate.

fossil Any relic of an organism that lived and died long ago.

frame of reference *See* reference frame.

free-fall The condition in which an object is falling without resistance; objects are weightless when in free-fall.

free-float frame A frame of reference in which all objects are weightless and hence float freely.

frequency The rate at which peaks of a wave pass by a point, measured in units of 1/s, often called *cycles per second* or *hertz*.

frost line The boundary in the solar nebula beyond which ices could condense; only metals and rocks could condense within the frost line.

fundamental forces The four forces that appear to govern all interactions that occur in the universe today: gravity, electromagnetism, the strong force, and the weak force.

fundamental particles Subatomic particles that cannot be divided into anything smaller.

fusion *See* nuclear fusion.

galactic cannibalism The term sometimes used to describe the process by which large galaxies merge with other galaxies in collisions. *Central dominant galaxies* are products of galactic cannibalism.

galactic fountain A model for the cycling of gas in the Milky Way Galaxy in which fountains of hot, ionized gas rise from the disk into the halo and then cool and form clouds as they sink back into the disk.

galactic wind A wind of low-density but extremely hot gas flowing out from a starburst galaxy, created by the combined energy of many supernovae.

galaxy A great island of stars in space, containing millions, billions, or even trillions of stars, all held together by gravity and orbiting a common center.

galaxy cluster *See* cluster of galaxies.

galaxy evolution The formation and development of galaxies.

Galilean moons The four moons of Jupiter that were discovered by Galileo: Io, Europa, Ganymede, and Callisto.

gamma-ray burst A sudden burst of gamma rays from deep space. At least some relatively long (many seconds to minutes) gamma-ray bursts come from powerful supernovae (hypernovae), while at least some very short bursts (less than a few seconds) come from neutron star mergers.

gamma rays Light with very short wavelengths (and hence high frequencies)—shorter than those of x-rays.

gap moons Tiny moons located within a gap in a planet's ring system. The gravity of a gap moon helps clear the gap.

gas phase The phase of matter in which atoms or molecules can move essentially independently of one another.

gas pressure The force (per unit area) pushing on any object due to surrounding gas. *See also* pressure.

general theory of relativity Einstein's generalization of his special theory of relativity so that the theory also applies when we consider effects of gravity or acceleration.

genetic code The "language" that living cells use to read the instructions chemically encoded in DNA.

geocentric model Any of the ancient Greek models that were used to predict planetary positions under the assumption that Earth lay in the center of the universe.

geocentric universe The ancient belief that Earth is the center of the entire universe.

geological activity Processes that change a planet's surface long after formation, such as volcanism, tectonics, and erosion.

geological processes The processes responsible for a planet's surface features; the four basic geological processes are impact cratering, volcanism, tectonics, and erosion.

geological time scale The time scale used by scientists to describe major eras in Earth's past.

geology The study of surface features (on a moon, planet, or asteroid) and the processes that create them.

geostationary satellite A satellite that appears to stay stationary in the sky as viewed from Earth's surface, because it orbits in the same time it takes Earth to rotate and orbits in Earth's equatorial plane.

geosynchronous satellite A satellite that orbits Earth in the same time it takes Earth to rotate (one sidereal day).

giant galaxies Galaxies that are unusually large, typically containing a trillion or more stars. Most giant galaxies are elliptical, and many contain multiple nuclei near their centers.

giant impact A collision between a forming planet and a very large planetesimal, such as is thought to have formed our Moon.

giant molecular cloud A very large cloud of cold, dense interstellar gas, typically containing up to a million solar masses worth of material. *See also* molecular clouds.

giants (luminosity class III) Stars that appear just below the supergiants on the H-R diagram because they are somewhat smaller in radius and lower in luminosity.

gibbous phase The phase of the Moon (or of a planet) in which more than half but less than all of the visible face is illuminated by sunlight.

global positioning system (GPS) A system of navigation by satellites orbiting Earth.

global warming An expected increase in Earth's global average temperature caused by human input of carbon dioxide and other greenhouse gases into the atmosphere.

global wind patterns (or **global circulation**) Wind patterns that remain fixed on a global scale, determined by the combination of surface heating and the planet's rotation.

globular cluster A spherically shaped cluster of up to a million or more stars; globular clusters are found primarily in the halos of galaxies and contain only very old stars.

gluons The exchange particles for the strong force.

grand unified theory (GUT) A theory that unifies three of the four fundamental forces—the strong force, the weak force, and the electromagnetic force (but not gravity)—in a single model.

granulation (on the Sun) The bubbling pattern visible in the photosphere, produced by the underlying convection.

gravitation (law of) *See* universal law of gravitation.

gravitational constant The experimentally measured constant G that appears in the law of universal gravitation:

$$G = 6.67 \times 10^{-11} \frac{m^3}{kg \times s^2}$$

gravitational contraction The process in which gravity causes an object to contract, thereby converting gravitational potential energy into thermal energy.

gravitational encounter An encounter in which two (or more) objects pass near enough so that each can feel the effects of the other's gravity and they can therefore exchange energy.

gravitational equilibrium A state of balance in which the force of gravity pulling inward is precisely counteracted by pressure pushing outward; also referred to as *hydrostatic equilibrium*.

gravitational lensing The magnification or distortion (into arcs, rings, or multiple images) of an image caused by light bending through a gravitational field, as predicted by Einstein's general theory of relativity.

gravitationally bound system Any system of objects, such as a star system or a galaxy, that is held together by gravity.

gravitational potential energy Energy that an object has by virtue of its position in a gravitational field; an object has more gravitational potential energy when it has a greater distance that it can potentially fall.

gravitational redshift A redshift caused by the fact that time runs slowly in gravitational fields.

gravitational time dilation The slowing of time that occurs in a gravitational field, as predicted by Einstein's general theory of relativity.

gravitational waves Waves, predicted by Einstein's general theory of relativity, that are created by changes in a local gravitational field and cause distortions of spacetime as they propagate outward through the universe at the speed of light.

gravitons The exchange particles for the force of gravity.

gravity One of the four fundamental forces; it is the force that dominates on large scales.

grazing incidence (in telescopes) Reflections in which light grazes a mirror surface and is deflected at a small angle; commonly used to focus high-energy ultraviolet light and x-rays.

great circle A circle on the surface of a sphere whose center is at the center of the sphere.

greatest elongation *See* elongation (greatest).

Great Red Spot A large, high-pressure storm on Jupiter.

greenhouse effect The process by which greenhouse gases in an atmosphere make a planet's surface temperature warmer than it would be in the absence of an atmosphere.

greenhouse gases Gases, such as carbon dioxide, water vapor, and methane, that are particularly good absorbers of infrared light but are transparent to visible light.

Gregorian calendar Our modern calendar, introduced by Pope Gregory in 1582.

ground state (of an atom) The lowest possible energy state of the electrons in an atom.

group of galaxies A few to a few dozen galaxies bound together by gravity. *See also* cluster of galaxies.

GUT era The era of the universe during which only two forces operated (gravity and the grand-unified-theory, or GUT, force), lasting from 10^{-43} second to 10^{-38} second after the Big Bang.

GUT force The proposed force that exists at very high energies when the strong force, the weak force, and the electromagnetic force (but not gravity) all act as one.

H II region *See* ionization nebula.

habitable world A world with environmental conditions under which life could *potentially* arise or survive.

habitable zone The region around a star in which planets could potentially have surface temperatures at which liquid water could exist.

Hadley cells *See* circulation cells.

half-life The time it takes for half of the nuclei in a given quantity of a radioactive substance to decay.

halo (of a galaxy) The spherical region surrounding the disk of a spiral galaxy.

halo component (of a galaxy) The portion of any galaxy that is spherical (or football-like) in shape and contains very little cool gas; it generally contains only very old stars. For spiral galaxies, the halo component includes both the halo and the bulge (but not the disk); elliptical galaxies have only a halo component.

halo population Stars that orbit within the halo component of a galaxy; sometimes called *Population II*. Elliptical galaxies have only a halo population (they lack a disk population), while spiral galaxies have halo population stars in their bulges and halos.

Hawking radiation Radiation predicted to arise from the evaporation of black holes.

heavy bombardment The period in the first few hundred million years after the solar system formed during which the tail end of planetary accretion created most of the craters found on ancient planetary surfaces.

heavy elements In astronomy, generally all elements *except* hydrogen and helium.

helium-capture reactions Fusion reactions that fuse a helium nucleus into some other nucleus; such reactions can fuse carbon into oxygen, oxygen into neon, neon into magnesium, and so on.

helium flash The event that marks the sudden onset of helium fusion in the previously inert helium core of a low-mass star.

helium-fusing star (or **helium core-fusion star**) A star that is currently fusing helium into carbon in its core.

helium fusion The fusion of three helium nuclei to form one carbon nucleus; also called the *triple-alpha reaction*.

hertz (Hz) The standard unit of frequency for light waves; equivalent to units of 1/s.

Hertzsprung-Russell (H-R) diagram A graph plotting individual stars as points, with stellar luminosity on the vertical axis and spectral type (or surface temperature) on the horizontal axis.

high-mass stars Stars born with masses above about $8M_{Sun}$; these stars will end their lives by exploding as supernovae.

horizon A boundary that divides what we can see from what we cannot see.

horizontal branch The horizontal line of stars that represents helium-fusing stars on an H-R diagram for a cluster of stars.

horoscope A predictive chart made by an astrologer; in scientific studies, horoscopes have never been found to have any validity as predictive tools.

hot Jupiter A class of planet that is Jupiter-like in size but orbits very close to its star, causing it to have a very high surface temperature.

hot spot (geological) A place within a plate of the lithosphere where a localized plume of hot mantle material rises.

hour angle (HA) The angle or time (measured in hours) since an object was last on the meridian in the local sky; defined to be 0 hours for objects that are on the meridian.

Hubble's constant A number that expresses the current rate of expansion of the universe; designated H_0, it is usually stated in units of km/s/Mpc. The reciprocal of Hubble's constant is the age the universe would have *if* the expansion rate had never changed.

Hubble's law Mathematical expression of the idea that more distant galaxies move away from us faster: $v = H_0 \times d$, where v is a galaxy's speed away from us, d is its distance, and H_0 is Hubble's constant.

hydrogen compounds Compounds that contain hydrogen and were common in the solar nebula, such as water (H_2O), ammonia (NH_3), and methane (CH_4).

hydrogen shell fusion Hydrogen fusion that occurs in a shell surrounding a stellar core.

hydrosphere The "layer" of water on Earth consisting of oceans, lakes, rivers, ice caps, and other liquid water and ice.

hydrostatic equilibrium *See* gravitational equilibrium.

hyperbola The precise mathematical shape of one type of unbound orbit (the other is a parabola) allowed under the force of gravity; at great distances from the attracting object, a hyperbolic path looks like a straight line.

hypernova A term sometimes used to describe a supernova (explosion) of a star so massive that it leaves a black hole behind.

hyperspace Any space with more than three dimensions.

hypothesis A tentative model proposed to explain some set of observed facts, but which has not yet been rigorously tested and confirmed.

ice ages Periods of global cooling during which the polar caps, glaciers, and snow cover extend closer to the equator.

ices (in solar system theory) Materials that are solid only at low temperatures, such as the hydrogen compounds water, ammonia, and methane.

ideal gas law The law relating the pressure, temperature, and number density of particles in an ideal gas.

image A picture of an object made by focusing light.

imaging (in astronomical research) The process of obtaining pictures of astronomical objects.

impact The collision of a small body (such as an asteroid or comet) with a larger object (such as a planet or moon).

impact basin A very large impact crater, often filled by a lava flow.

impact crater A bowl-shaped depression left by the impact of an object that strikes a planetary surface (as opposed to burning up in the atmosphere).

impact cratering The excavation of bowl-shaped depressions (*impact craters*) by asteroids or comets striking a planet's surface.

impactor The object responsible for an impact.

inflation (of the universe) A sudden and dramatic expansion of the universe thought to have occurred at the end of the GUT era.

infrared light Light with wavelengths that fall in the portion of the electromagnetic spectrum between radio waves and visible light.

inner solar system Generally, the region of our solar system out to about the orbit of Mars.

intensity (of light) A measure of the amount of energy coming from light of specific wavelength in the spectrum of an object.

interferometry A telescopic technique in which two or more telescopes are used in tandem to produce much better angular resolution than the telescopes could achieve individually.

intermediate-mass stars Stars born with masses between about $2M_{Sun}$ and $8M_{Sun}$; these stars end their lives by ejecting a planetary nebula and becoming a white dwarf.

interstellar cloud A cloud of gas and dust between the stars.

interstellar dust grains Tiny solid flecks of carbon and silicon minerals found in cool interstellar clouds; they resemble particles of smoke and form in the winds of red giant stars.

interstellar medium The gas and dust that fills the space between stars in a galaxy.

interstellar ramjet A hypothesized type of spaceship that uses a giant scoop to sweep up interstellar gas for use in a nuclear fusion engine.

interstellar reddening The change in the color of starlight as it passes through dusty gas. The light appears redder because dust grains absorb and scatter blue light more effectively than red light.

intracluster medium Hot, x-ray-emitting gas found between the galaxies within a cluster of galaxies.

inverse square law A law followed by any quantity that decreases with the square of the distance between two objects.

inverse square law for light The law stating that an object's apparent brightness depends on its actual luminosity and the inverse square of its distance from the observer:

$$\text{apparent brightness} = \frac{\text{luminosity}}{4\pi \times (\text{distance})^2}$$

inversion (atmospheric) A local weather condition in which air is colder near the surface than higher up in the troposphere—the opposite of the usual condition, in which the troposphere is warmer at the bottom.

ionization The process of stripping one or more electrons from an atom.

ionization nebula A colorful, wispy cloud of gas that glows because neighboring hot stars irradiate it with ultraviolet photons that can ionize hydrogen atoms; also called an *emission nebula* or *H II region*.

ionosphere A portion of the thermosphere in which ions are particularly common (because of ionization by x-rays from the Sun).

ions Atoms with a positive or negative electrical charge.

Io torus A donut-shaped charged-particle belt around Jupiter that approximately traces Io's orbit.

irregular galaxies Galaxies that look neither spiral nor elliptical.

isotopes Forms of an element that have the same number of protons but different numbers of neutrons.

jets High-speed streams of gas ejected from an object into space.

joule The international unit of energy, equivalent to about 1/4000 of a Calorie.

jovian nebulae The clouds of gas that swirled around the jovian planets, from which the moons formed.

jovian planets Giant gaseous planets similar in overall composition to Jupiter.

Julian calendar The calendar introduced in 46 B.C. by Julius Caesar and used until the Gregorian calendar replaced it.

June solstice Both the point on the celestial sphere where the ecliptic is farthest north of the celestial equator and the moment in time when the Sun appears at that point each year (around June 21).

Kelvin (temperature scale) The temperature scale most commonly used in science, defined such that absolute zero is 0 K and water freezes at 273.15 K.

Kepler's first law Law stating that the orbit of each planet about the Sun is an ellipse with the Sun at one focus.

Kepler's laws of planetary motion Three laws discovered by Kepler that describe the motion of the planets around the Sun.

Kepler's second law The principle that, as a planet moves around its orbit, it sweeps out equal areas in equal times. This tells us that a planet moves faster when it is closer to the Sun (near perihelion) than when it is farther from the Sun (near aphelion) in its orbit.

Kepler's third law The principle that the square of a planet's orbital period is proportional to the cube of its average distance from the Sun (semimajor axis), which tells us that more distant planets move more slowly in their orbits; in its original form, written $p^2 = a^3$. See also Newton's version of Kepler's third law.

kilonova A name occasionally applied to the observed afterglow of events thought to be due to the merger of two neutron stars.

kinetic energy Energy of motion, given by the formula $\frac{1}{2} mv^2$.

Kirchhoff's laws A set of rules that summarizes the conditions under which objects produce thermal, absorption line, or emission line spectra. In brief: (1) An opaque object produces thermal radiation. (2) An absorption line spectrum occurs when thermal radiation passes through a thin gas that is cooler than the object emitting the thermal radiation. (3) An emission line spectrum occurs when we view a cloud of gas that is warmer than any background source of light.

Kirkwood gaps On a plot of asteroid semimajor axes, regions with few asteroids as a result of orbital resonances with Jupiter.

K-T event (or **K-T impact**) The collision of an asteroid or comet 65 million years ago that caused the mass extinction best known for wiping out the dinosaurs. *K* and *T* stand for the geological layers above and below the event.

Kuiper belt The comet-rich region of our solar system that resides between about 30 and 100 AU from the Sun. Kuiper belt comets have orbits that lie fairly close to the plane of planetary orbits and travel around the Sun in the same direction as the planets.

Kuiper belt object Any object orbiting the Sun within the region of the Kuiper belt, although the term is most often used for relatively large objects. For example, Pluto and Eris are considered large Kuiper belt objects.

Large Magellanic Cloud One of two small, irregular galaxies (the other is the Small Magellanic Cloud) located about 150,000 light-years away; it probably orbits the Milky Way Galaxy.

large-scale structure (of the universe) Generally, the structure of the universe on size scales larger than that of clusters of galaxies.

latitude The angular north-south distance between Earth's equator and a location on Earth's surface.

leap year A calendar year with 366 rather than 365 days. Our current calendar (the Gregorian calendar) incorporates a leap year every 4 years (by adding February 29) except in century years that are not divisible by 400.

Leavitt's law The relation, discovered by Henrietta Leavitt, that describes how the luminosity of a Cepheid variable star is related to the period between peaks in its brightness: the longer the Cepheid's period, the more luminous the star. Also called the *Cepheid period–luminosity relation*.

length contraction The effect in which you observe lengths to be shortened in reference frames moving relative to you.

lens (gravitational) See gravitational lensing.

lenticular galaxies Galaxies that look lens-shaped when seen edge-on, resembling spiral galaxies without arms. They tend to have less cool gas than normal spiral galaxies but more cool gas than elliptical galaxies.

leptons Fermions *not* made from quarks, such as electrons and neutrinos.

life track A track drawn on an H-R diagram to represent the changes in a star's surface temperature and luminosity during its life; also called an *evolutionary track*.

light-collecting area (of a telescope) The area of the primary mirror or lens that collects light in a telescope.

light curve A graph of an object's intensity against time.

light gases (in solar system theory) Hydrogen and helium, which never condense under solar nebula conditions.

light pollution Human-made light that hinders astronomical observations.

light-year (ly) The distance that light can travel in 1 year, which is 9.46 trillion km.

liquid phase The phase of matter in which atoms or molecules are held together but move relatively freely.

lithosphere The relatively rigid outer layer of a planet; it generally encompasses the crust and the upper portion of the mantle.

Local Bubble (interstellar) The bubble of hot gas in which our Sun and other nearby stars apparently reside. See also bubble (interstellar).

Local Group The group of about 40 galaxies to which the Milky Way Galaxy belongs.

local sidereal time (LST) Sidereal time for a particular location, defined according to the position of the March equinox in the local sky. More formally, the local sidereal time at any moment is defined to be the hour angle of the March equinox.

local sky The sky as viewed from a particular location on Earth (or another solid object). Objects in the local sky are pinpointed by the coordinates of *altitude* and *direction* (or *azimuth*).

local solar neighborhood The portion of the Milky Way Galaxy that is located relatively close (within a few hundred to a couple thousand light-years) to our Sun.

Local Supercluster The supercluster of galaxies to which the Local Group belongs.

longitude The angular east-west distance between the prime meridian (which passes through Greenwich, England) and a location on Earth's surface.

lookback time The amount of time since the light we see from a distant object was emitted. If an object has a lookback time of 400 million years, we are seeing it as it looked 400 million years ago.

low-mass stars Stars born with masses less than about $2M_{Sun}$; these stars end their lives by ejecting a planetary nebula and becoming a white dwarf.

luminosity The total power output of an object, usually measured in watts or in units of solar luminosities ($L_{Sun} = 3.8 \times 10^{26}$ watts).

luminosity class A category describing the region of the H-R diagram in which a star falls. Luminosity class I represents supergiants, III represents giants, and V represents main-sequence stars; luminosity classes II and IV are intermediate to the others.

lunar eclipse An event that occurs when the Moon passes through Earth's shadow, which can happen only at full moon. A lunar eclipse may be total, partial, or penumbral.

lunar maria The regions of the Moon that look smooth from Earth and actually are impact basins.

lunar month *See* synodic month.

lunar phase *See* phase (of the Moon or a planet).

magma Underground molten rock.

magnetic braking The process by which a star's rotation slows as its magnetic field transfers its angular momentum to the surrounding nebula.

magnetic field The region surrounding a magnet in which it can affect other magnets or charged particles.

magnetic field lines Lines that represent how the needles on a series of compasses would point if they were laid out in a magnetic field.

magnetosphere The region surrounding a planet in which charged particles are trapped by the planet's magnetic field.

magnitude system A system for describing stellar brightness by using numbers, called *magnitudes*, based on an ancient Greek way of describing the brightnesses of stars in the sky. This system uses *apparent magnitude* to describe a star's apparent brightness and *absolute magnitude* to describe a star's luminosity.

main sequence The prominent line of points (representing *main-sequence stars*) running from the upper left to the lower right on an H-R diagram.

main-sequence lifetime The length of time for which a star of a particular mass can shine by fusing hydrogen into helium in its core.

main-sequence stars (luminosity class V) Stars whose temperature and luminosity place them on the main sequence of the H-R diagram. Main-sequence stars release energy by fusing hydrogen into helium in their cores.

main-sequence turnoff point The point on a cluster's H-R diagram where its stars turn off from the main sequence; the age of the cluster is equal to the main-sequence lifetime of stars at the main-sequence turnoff point.

mantle (of a planet) The rocky layer that lies between a planet's core and crust.

March equinox Both the point in Pisces on the celestial sphere where the ecliptic crosses the celestial equator and the moment in time when the Sun appears at that point each year (around March 21).

Martian meteorites Meteorites found on Earth that are thought to have originated on Mars.

mass A measure of the amount of matter in an object.

mass-energy The potential energy of mass, which has an amount $E = mc^2$.

mass exchange (in close binary star systems) The process in which tidal forces cause matter to spill from one star to a companion star in a close binary system.

mass extinction An event in which a large fraction of the species living on Earth go extinct, such as the event in which the dinosaurs died out about 65 million years ago.

mass increase (in relativity) The effect in which an object moving past you seems to have a mass greater than its rest mass.

massive star supernova A supernova that occurs when a massive star dies, initiated by the catastrophic collapse of its iron core; often called a *Type II supernova*.

mass-to-light ratio The mass of an object divided by its luminosity, usually stated in units of solar masses per solar luminosity. Objects with high mass-to-light ratios must contain substantial quantities of dark matter.

matter–antimatter annihilation An event that occurs when a particle of matter and a particle of antimatter meet and convert all of their mass-energy to photons.

mean solar time Time measured by the average position of the Sun in the local sky over the course of the year.

meridian A half-circle extending from your horizon (altitude 0°) due south, through your zenith, to your horizon due north.

metallic hydrogen Hydrogen that is so compressed that the hydrogen atoms all share electrons and thereby take on properties of metals, such as conducting electricity. It occurs only under very high-pressure conditions, such as those found deep within Jupiter.

metals (in solar system theory) Elements, such as nickel, iron, and aluminum, that condense at fairly high temperatures.

meteor A flash of light caused when a particle from space burns up in our atmosphere.

meteorite A rock from space that lands on Earth.

meteor shower A period during which many more meteors than usual can be seen.

Metonic cycle The 19-year period, discovered by the Babylonian astronomer Meton, over which the lunar phases occur on the same dates.

microwaves Light with wavelengths in the range of micrometers to millimeters. Microwaves are generally considered to be a subset of the radio-wave portion of the electromagnetic spectrum.

mid-ocean ridges Long ridges of undersea volcanoes on Earth, along which mantle material erupts onto the ocean floor and pushes apart the existing seafloor on either side. These ridges are essentially the source of new seafloor crust, which then makes its way along the ocean bottom for millions of years before returning to the mantle at a subduction zone.

Milankovitch cycles The cyclical changes in Earth's axis tilt and orbit that can change the climate and cause ice ages.

Milky Way The name used both for our galaxy and for the band of light we see in the sky when we look into the plane of our galaxy.

millisecond pulsars Pulsars with rotation periods of a few thousandths of a second.

minor planets An alternative name for *asteroids*.

model (scientific) A representation of some aspect of nature that can be used to explain and predict real phenomena without invoking myth, magic, or the supernatural.

molecular bands The tightly bunched lines in an object's spectrum that are produced by molecules.

molecular cloud fragments (or **molecular cloud cores**) The densest regions of molecular clouds, which usually go on to form stars.

molecular clouds Cool, dense interstellar clouds in which the low temperatures allow hydrogen atoms to pair up into hydrogen molecules (H_2).

molecular dissociation The process by which a molecule splits into its component atoms.

molecule Technically, the smallest unit of a chemical element or compound; in this text, the term refers only to combinations of two or more atoms held together by chemical bonds.

momentum The product of an object's mass and velocity.

moon An object that orbits a planet.

moonlets Very small moons that orbit within the ring systems of jovian planets.

mutations Errors in the copying process when a living cell replicates itself.

natural selection The process by which mutations that make an organism better able to survive get passed on to future generations.

neap tides The lower-than-average tides on Earth that occur at first- and third-quarter moon, when the tidal forces from the Sun and Moon oppose each other.

nebula A cloud of gas in space, usually one that is glowing.

nebular capture The process by which icy planetesimals capture hydrogen and helium gas to form jovian planets.

nebular theory The scientific theory that describes how our solar system formed from a cloud of interstellar gas and dust.

net force The overall force to which an object responds; the net force is equal to the rate of change in the object's momentum, or equivalently to the object's mass × acceleration.

neutrino A type of fundamental particle that has extremely low mass and responds only to the weak force; neutrinos are leptons and come in three types—electron neutrinos, muon neutrinos, and tau neutrinos.

neutron degeneracy pressure Degeneracy pressure exerted by neutrons, as in neutron stars.

neutrons Particles with no electrical charge found in atomic nuclei, built from three quarks.

neutron star The compact corpse of a high-mass star left over after a supernova; it typically has a mass comparable to the mass of the Sun in a volume just a few kilometers in radius.

neutron star merger An event in which two neutron stars that were orbiting in a binary system merge together, accompanied by the release of a tremendous amount of energy, some of which is in the form of gravitational waves and some in the form of a gamma-ray burst.

newton The standard unit of force in the metric system:

$$1 \text{ newton} = 1 \frac{\text{kg} \times \text{m}}{\text{s}^2}$$

Newton's first law of motion Principle that, in the absence of a net force, an object moves with constant velocity.

Newton's laws of motion Three basic laws that describe how objects respond to forces.

Newton's second law of motion Law stating how a net force affects an object's motion. Specifically, force = rate of change in momentum, or force = mass × acceleration.

Newton's third law of motion Principle that, for any force, there is always an equal and opposite reaction force.

Newton's universal law of gravitation See universal law of gravitation.

Newton's version of Kepler's third law A generalization of Kepler's third law used to calculate the masses of orbiting objects from measurements of orbital period and distance; usually written as

$$p^2 = \frac{4\pi^2}{G(M_1 + M_2)} a^3$$

nodes (of Moon's orbit) The two points in the Moon's orbit where it crosses the ecliptic plane.

nonbaryonic matter Matter that is not part of the normal composition of atoms, such as neutrinos or the hypothetical WIMPs; more technically, particles that are not made from three quarks.

nonscience As used in this book, any way of searching for knowledge that makes no claim to follow the scientific method, such as seeking knowledge through intuition, tradition, or faith.

north celestial pole (NCP) The point on the celestial sphere directly above Earth's North Pole.

nova The dramatic brightening of a star that lasts for a few weeks and then subsides; it occurs when a burst of hydrogen fusion ignites in a shell on the surface of an accreting white dwarf in a binary star system.

nuclear fission The process in which a larger nucleus splits into two (or more) smaller particles.

nuclear fusion The process in which two (or more) smaller nuclei slam together and make one larger nucleus.

nucleus (of a comet) The solid portion of a comet—the only portion that exists when the comet is far from the Sun.

nucleus (of an atom) The compact center of an atom made from protons and neutrons.

observable universe The portion of the entire universe that, at least in principle, can be seen from Earth.

Occam's razor A principle often used in science, holding that scientists should prefer the simpler of two models that agree equally well with observations; named after the medieval scholar William of Occam (1285–1349).

Olbers' paradox A paradox pointing out that if the universe were infinite in both age and size (with stars found throughout the universe), then the sky would not be dark at night.

Oort cloud A huge, spherical region centered on the Sun, extending perhaps halfway to the nearest stars, in which trillions of comets orbit the Sun with random inclinations, orbital directions, and eccentricities.

opacity A measure of how much light a material absorbs compared to how much it transmits; materials with higher opacity absorb more light.

opaque A word used to describe a material that absorbs light.

open cluster A cluster of up to several thousand stars; open clusters are found only in the disks of galaxies and often contain young stars.

open universe A universe in which spacetime has an overall shape analogous to the surface of a saddle.

opposition The point at which a planet appears opposite the Sun in our sky.

optical quality The ability of a lens, mirror, or telescope to obtain clear and properly focused images.

orbit The path followed by a celestial body because of gravity; an orbit may be *bound* (elliptical) or *unbound* (parabolic or hyperbolic).

orbital energy The sum of an orbiting object's kinetic and gravitational potential energies.

orbital resonance A situation in which one object's orbital period is a simple ratio of another object's period, such as 1/2, 1/4, or 5/3. In such cases, the two objects periodically line up with each other, and the extra gravitational attractions at these times can affect the objects' orbits.

orbital velocity law A variation on Newton's version of Kepler's third law that allows us to use a star's orbital speed and distance from the galactic center to determine the total mass of the galaxy contained *within* the star's orbit; mathematically,

$$M_r = \frac{r \times v^2}{G}$$

where M_r is the mass contained within the star's orbit, r is the star's distance from the galactic center, v is the star's orbital velocity, and G is the gravitational constant.

orbiters (of other worlds) Spacecraft that go into orbit of another world for long-term study.

organic molecule Generally, any molecule containing carbon and of a type associated with carbon-based life, though not necessarily a product of life. Note that we do not usually consider molecules such as carbon dioxide (CO_2) and carbonate minerals to be organic, since they are more commonly found independent of life.

outer solar system Generally, the region of our solar system beginning at about the orbit of Jupiter.

outgassing The process of releasing gases from a planetary interior, usually through volcanic eruptions.

oxidation Chemical reactions, often with rocks on the surface of a planet, that remove oxygen from the atmosphere.

ozone The molecule O_3, which is a particularly good absorber of ultraviolet light.

ozone depletion The decline in levels of atmospheric ozone found worldwide on Earth, especially in Antarctica, in recent years.

ozone hole A place where the concentration of ozone in the stratosphere is dramatically lower than is the norm.

pair production The process in which a concentration of energy spontaneously turns into a particle and its antiparticle.

parabola The precise mathematical shape of a special type of unbound orbit allowed under the force of gravity. If an object in a parabolic orbit loses only a tiny amount of energy, it will become bound.

paradigm (in science) A general pattern of thought that tends to shape scientific study during a particular time period.

paradox A situation that, at least at first, seems to violate common sense or contradict itself. Resolving paradoxes often leads to deeper understanding.

parallax The apparent shifting of an object against the background, due to viewing it from different positions. *See also* stellar parallax.

parallax angle Half of a star's annual back-and-forth shift due to stellar parallax; it is related to the star's distance according to the formula

$$\text{distance in parsecs} = \frac{1}{p}$$

where p is the parallax angle in arcseconds.

parsec (pc) The distance to an object with a parallax angle of 1 arcsecond; approximately equal to 3.26 light-years.

partial lunar eclipse A lunar eclipse during which the Moon becomes only partially covered by Earth's umbral shadow.

partial solar eclipse A solar eclipse during which the Sun becomes only partially blocked by the disk of the Moon.

particle accelerator A machine designed to accelerate subatomic particles to high speeds in order to create new particles or to test fundamental theories of physics.

particle era The era of the universe lasting from 10^{-10} to 0.001 second after the Big Bang, during which subatomic particles were continually created and destroyed, and ending when matter annihilated antimatter.

peculiar velocity (of a galaxy) The component of a galaxy's velocity relative to the Milky Way that deviates from the velocity expected by Hubble's law.

penumbra The lighter, outlying regions of a shadow.

penumbral lunar eclipse A lunar eclipse during which the Moon passes only within Earth's penumbral shadow and does not fall within the umbra.

perigee The point at which an object orbiting Earth is nearest to Earth.

perihelion The point at which an object orbiting the Sun is closest to the Sun.

period–luminosity relation *See* Leavitt's law.

phase (of matter) The state determined by the way in which atoms or molecules are held together; the common phases are solid, liquid, and gas.

phase (of the Moon or a planet) The state determined by the portion of the visible face of the Moon (or of a planet) that is illuminated by sunlight. For the Moon, the phases cycle through new, waxing crescent, first-quarter, waxing gibbous, full, waning gibbous, third-quarter, waning crescent, and back to new.

photon An individual particle of light, characterized by a wavelength and a frequency.

photosphere The visible surface of the Sun, where the temperature averages just under 6000 K.

pixel An individual "picture element" in a digital image.

Planck era The era of the universe prior to the Planck time.

Planck's constant A universal constant, abbreviated h, with a value of $h = 6.626 \times 10^{-34}$ joule $\times$ s.

Planck time The time when the universe was 10^{-43} second old, before which random energy fluctuations were so large that our current theories are powerless to describe what might have been happening.

planet A moderately large object that orbits a star and shines primarily by reflecting light from its star. More precisely, according to a definition approved in 2006, a planet is an object that (1) orbits a star (but is itself neither a star nor a moon); (2) is massive enough for its own gravity to give it a nearly round shape; and (3) has cleared the neighborhood around its orbit. Objects that meet the first two criteria but not the third, including Ceres, Pluto, and Eris, are designated *dwarf planets*.

planetary geology The extension of the study of Earth's surface and interior to apply to other solid bodies in the solar system, such as terrestrial planets and jovian planet moons.

planetary migration A process through which a planet can move from the orbit on which it is born to a different orbit that is closer to or farther from its star.

planetary nebula The glowing cloud of gas ejected from a low-mass star at the end of its life.

planetesimals The building blocks of planets, formed by accretion in the solar nebula.

plasma A gas consisting of ions and electrons.

plasma tail (of a comet) One of two tails seen when a comet passes near the Sun (the other is the *dust tail*). It is composed of ionized gas blown away from the Sun by the solar wind.

plates (on a planet) Pieces of a lithosphere that apparently float upon the denser mantle below.

plate tectonics The geological process in which plates are moved around by stresses in a planet's mantle.

polarization (of light) The property of light describing how the electric and magnetic fields of light waves are aligned; light is said to be *polarized* when all of the photons have their electric and magnetic fields aligned in some particular way.

Population I *See* disk population.

Population II *See* halo population.

positron *See* antielectron.

potential energy Energy stored for later conversion into kinetic energy; includes gravitational potential energy, electrical potential energy, and chemical potential energy.

power The rate of energy usage, usually measured in watts (1 watt = 1 joule/s).

precession The gradual wobble of the axis of a rotating object around a vertical line.

precipitation Condensed atmospheric gases that fall to the surface in the form of rain, snow, or hail.

pressure The force (per unit area) pushing on an object. In astronomy, we are generally interested in pressure applied by surrounding gas (or plasma). Ordinarily, such pressure is related to the temperature of the gas (*see* thermal pressure). In objects such as white dwarfs and neutron stars, pressure may arise from a quantum effect (*see* degeneracy pressure). Light can also exert pressure (*see* radiation pressure).

primary mirror The large, light-collecting mirror of a reflecting telescope.

prime focus (of a reflecting telescope) The first point at which light focuses after bouncing off the primary mirror; located in front of the primary mirror.

prime meridian The meridian of longitude that passes through Greenwich, England; defined to be longitude 0°.

primitive meteorites Meteorites that formed at the same time as the solar system itself, about 4.6 billion years ago. Primitive meteorites from the inner asteroid belt are usually stony, and those from the outer belt are usually carbon-rich.

processed meteorites Meteorites that apparently once were part of a larger object that "processed" the original material of the solar nebula into another form. Processed meteorites can be rocky if chipped from the surface or mantle, or metallic if blasted from the core.

proper motion The motion of an object in the plane of the sky, perpendicular to our line of sight.

protogalactic cloud A huge, collapsing cloud of intergalactic gas from which an individual galaxy formed.

proton–proton chain The chain of reactions by which low-mass stars (including the Sun) fuse hydrogen into helium.

protons Particles with positive electrical charge found in atomic nuclei, built from three quarks.

protoplanetary disk A disk of material surrounding a young star (or protostar) that may eventually form planets.

protostar A forming star that has not yet reached the point where sustained fusion can occur in its core.

protostellar disk A disk of material surrounding a protostar; essentially the same as a protoplanetary disk, but may not necessarily lead to planet formation.

protostellar wind The relatively strong wind from a protostar.

protosun The central object in the forming solar system that eventually became the Sun.

pseudoscience Something that purports to be science or may appear to be scientific but that does not adhere to the testing and verification requirements of the scientific method.

Ptolemaic model The geocentric model of the universe developed by Ptolemy in about 150 A.D.

pulsar A neutron star from which we observe rapid pulses of radiation as it rotates.

pulsating variable stars Stars that grow alternately brighter and dimmer as their outer layers expand and contract in size.

quantum laws The laws that describe the behavior of particles on a very small scale; *see also* quantum mechanics.

quantum mechanics The branch of physics that deals with the very small, including molecules, atoms, and fundamental particles.

quantum state The complete description of the state of a subatomic particle, including its location, momentum, orbital angular momentum, and spin, to the extent allowed by the uncertainty principle.

quantum tunneling The process in which, thanks to the uncertainty principle, an electron or other subatomic particle appears on the other side of a barrier that it does not have the energy to overcome in a normal way.

quarks The building blocks of protons and neutrons; quarks are one of the two basic types of fermions (leptons are the other).

quasar The brightest type of active galactic nucleus.

radar mapping Imaging of a planet by bouncing radar waves off its surface, especially important for Venus and Titan, where thick clouds mask the surface.

radar ranging A method of measuring distances within the solar system by bouncing radio waves off planets.

radial motion The component of an object's motion directed toward or away from us.

radial velocity The portion of any object's total velocity that is directed toward or away from us. This part of the velocity is the only part that we can measure with the Doppler effect.

radiation pressure Pressure exerted by photons of light.

radiation zone (of a star) A region of the interior in which energy is transported primarily by radiative diffusion.

radiative diffusion The process by which photons gradually migrate from a hot region (such as the solar core) to a cooler region (such as the solar surface).

radiative energy Energy carried by light; the energy of a photon is Planck's constant times its frequency, or $h \times f$.

radioactive decay The spontaneous change of an atom into a different element, in which its nucleus breaks apart or a proton turns into an electron. This decay releases heat in a planet's interior.

radioactive element (or **radioactive isotope**) A substance whose nucleus tends to fall apart spontaneously.

radio galaxy A galaxy that emits unusually large quantities of radio waves; thought to contain an active galactic nucleus powered by a supermassive black hole.

radio lobes The huge regions of radio emission found on either side of radio galaxies. The lobes apparently contain plasma ejected by powerful jets from the galactic center.

radiometric dating The process of determining the age of a rock (i.e., the time since it solidified) by comparing the present amount of a radioactive substance to the amount of its decay product.

radio waves Light with very long wavelengths (and hence low frequencies)—longer than those of infrared light.

random walk A type of haphazard movement in which a particle or photon moves through a series of bounces, with each bounce sending it in a random direction.

recession velocity (of a galaxy) The speed at which a distant galaxy is moving away from us because of the expansion of the universe.

recollapsing universe A model of the universe in which the collective gravity of all its matter eventually halts and reverses the expansion, causing the galaxies to come crashing back together and the universe to end in a fiery Big Crunch.

red giant A giant star that is red in color.

red-giant winds The relatively dense but slow winds from red giant stars.

redshift (Doppler) A Doppler shift in which spectral features are shifted to longer wavelengths, observed when an object is moving away from the observer.

reference frame (or **frame of reference**) In the theory of relativity, what two people (or objects) share if they are *not* moving relative to each other.

reflecting telescope A telescope that uses mirrors to focus light.

reflection (of light) The process by which matter changes the direction of light.

reflection nebula A nebula that we see as a result of starlight reflected from interstellar dust grains. Reflection nebulae tend to have blue and black tints.

refracting telescope A telescope that uses lenses to focus light.

resonance *See* orbital resonance.

rest wavelength The wavelength of a spectral feature in the absence of any Doppler shift or gravitational redshift.

retrograde motion Motion that is backward compared to the norm. For example, we see Mars in apparent retrograde motion during the periods of time when it moves westward, rather than the more common eastward, relative to the stars.

revolution The orbital motion of one object around another.

right ascension (RA) The angular east-west distance between the March equinox and a location on the celestial sphere; analogous to longitude, but on the celestial sphere.

rings (planetary) The collections of numerous small particles orbiting a planet within its Roche tidal zone.

Roche tidal zone The region within two to three planetary radii (of any planet) in which the tidal forces tugging an object apart become comparable to the gravitational forces holding it together; planetary rings are always found within the Roche tidal zone.

rocks (in solar system theory) Materials common on the surface of Earth, such as silicon-based minerals, that are solid at temperatures and pressures found on Earth but typically melt or vaporize at temperatures of 500–1300 K.

rotation The spinning of an object around its axis.

rotation curve A graph that plots rotational (or orbital) velocity against distance from the center for any object or set of objects.

runaway greenhouse effect A positive feedback cycle in which heating caused by the greenhouse effect causes more greenhouse gases to enter the atmosphere, which further enhances the greenhouse effect.

saddle-shaped geometry (or **hyperbolic geometry**) The type of geometry in which the rules—such as that two lines that begin parallel eventually diverge—are most easily visualized on a saddle-shaped surface.

Sagittarius Dwarf A small dwarf elliptical galaxy that is currently passing through the disk of the Milky Way Galaxy.

saros cycle The period over which the basic pattern of eclipses repeats, which is about 18 years $11\frac{1}{3}$ days.

satellite Any object orbiting another object.

scattered light Light that is reflected into random directions.

Schwarzschild radius A measure of the size of the event horizon of a black hole.

science The search for knowledge that can be used to explain or predict natural phenomena in a way that can be confirmed by rigorous observations or experiments.

scientific method An organized approach to explaining observed facts through science.

scientific theory A model of some aspect of nature that has been rigorously tested and has passed all tests to date.

seafloor crust On Earth, the thin, dense crust of basalt created by seafloor spreading.

seafloor spreading On Earth, the creation of new seafloor crust at mid-ocean ridges.

search for extraterrestrial intelligence (SETI) The name given to observing projects designed to search for signs of intelligent life beyond Earth.

secondary mirror A small mirror in a reflecting telescope, used to reflect light gathered by the primary mirror toward an eyepiece or instrument.

sedimentary rock A rock that formed from sediments created and deposited by erosional processes. The sediments tend to build up in distinct layers, or *strata*.

seismic waves Earthquake-induced vibrations that propagate through a planet.

selection effect (or **selection bias**) A type of bias that arises from the way in which objects of study are selected and that can lead to incorrect conclusions. For example, when you are counting animals in a jungle it is easiest to see brightly colored animals, which could mislead you into thinking that these animals are the most common.

semimajor axis Half the distance across the long axis of an ellipse; in this text, it is usually referred to as the *average* distance of an orbiting object, abbreviated *a* in the formula for Kepler's third law.

September equinox Both the point in Virgo on the celestial sphere where the ecliptic crosses the celestial equator and the moment in time when the Sun appears at that point each year (around September 21).

Seyfert galaxies The name given to a class of galaxies that are found relatively nearby and that have nuclei much like those of quasars, except that they are less luminous.

shepherd moons Tiny moons within a planet's ring system that help force particles into a narrow ring; a variation on *gap moons*.

shield volcano A shallow-sloped volcano made from the flow of low-viscosity basaltic lava.

shock wave A wave of pressure generated by gas moving faster than the speed of sound.

sidereal day The time of 23 hours 56 minutes 4.09 seconds between successive appearances of any particular star on the meridian; essentially, the true rotation period of Earth.

sidereal month The time required for the Moon to orbit Earth once (as measured against the stars); about $27\frac{1}{4}$ days.

sidereal period (of a planet) A planet's actual orbital period around the Sun.

sidereal time Time measured according to the position of stars in the sky rather than the position of the Sun in the sky. *See also* local sidereal time.

sidereal year The time required for Earth to complete exactly one orbit as measured against the stars; about 20 minutes longer than the tropical year on which our calendar is based.

silicate rock A silicon-rich rock.

singularity The place at the center of a black hole where, in principle, gravity crushes all matter to an infinitely tiny and dense point.

Small Magellanic Cloud One of two small, irregular galaxies (the other is the Large Magellanic Cloud) located about 150,000 light-years away; it probably orbits the Milky Way Galaxy.

small solar system body An asteroid, comet, or other object that orbits a star but is too small to qualify as a planet or dwarf planet.

snowball Earth Name given to a hypothesis suggesting that, some 600–700 million years ago, Earth experienced a period in which it became cold enough for glaciers to exist worldwide, even in equatorial regions.

solar activity Short-lived phenomena on the Sun, including the emergence and disappearance of individual sunspots, prominences, and flares; sometimes called *solar weather*.

solar circle The Sun's orbital path around the galaxy, which has a radius of about 28,000 light-years.

solar day 24 hours, which is the average time between appearances of the Sun on the meridian.

solar eclipse An event that occurs when the Moon's shadow falls on Earth, which can happen only at new moon. A solar eclipse may be total, partial, or annular.

solar flares Huge and sudden releases of energy on the solar surface, probably caused when energy stored in magnetic fields is suddenly released.

solar luminosity The luminosity of the Sun, which is approximately 4×10^{26} watts.

solar maximum The time during each sunspot cycle at which the number of sunspots is the greatest.

solar minimum The time during each sunspot cycle at which the number of sunspots is the smallest.

solar nebula The piece of interstellar cloud from which our own solar system formed.

solar neutrino problem A now-solved problem in which, during the latter decades of the 20th century, there appeared to be disagreement between the predicted and observed number of neutrinos coming from the Sun.

solar prominences Vaulted loops of hot gas that rise above the Sun's surface and follow magnetic field lines.

solar sail A large, highly reflective (and thin, to minimize mass) piece of material that can "sail" through space using pressure exerted by sunlight.

solar system (or **star system**) A star (or sometimes more than one star) and all the objects that orbit it.

solar thermostat *See* stellar thermostat; the solar thermostat is the same idea applied to the Sun.

solar wind A stream of charged particles ejected from the Sun.

solid phase The phase of matter in which atoms or molecules are held rigidly in place.

solstice *See* December solstice *and* June solstice.

sound wave A wave of alternately rising and falling pressure.

south celestial pole (SCP) The point on the celestial sphere directly above Earth's South Pole.

spacetime The inseparable, four-dimensional combination of space and time.

spacetime diagram A graph that plots a spatial dimension on one axis and time on another axis.

special theory of relativity Einstein's theory that describes the relativity of time and space based on the fact that the laws of nature are the same for everyone and that everyone always measures the same speed of light.

spectral lines Bright or dark lines that appear in an object's spectrum, which we can see when we pass the object's light through a prismlike device that spreads out the light like a rainbow.

spectral resolution The degree of detail that can be seen in a spectrum; the higher the spectral resolution, the more detail we can see.

spectral type A way of classifying a star by the lines that appear in its spectrum; it is related to surface temperature. The basic spectral types are designated by letters (OBAFGKM, with O for the hottest stars and M for the coolest) and are subdivided with numbers from 0 through 9.

spectrograph An instrument used to record spectra.

spectroscopic binary A binary star system whose binary nature is revealed because we detect the spectral lines of one or both stars alternately becoming blueshifted and redshifted as the stars orbit each other.

spectroscopy (in astronomical research) The process of obtaining spectra from astronomical objects.

spectrum (of light) *See* electromagnetic spectrum.

speed The rate at which an object moves. Its units are distance divided by time, such as m/s or km/hr.

speed of light The speed at which light travels, which is about 300,000 km/s.

spherical geometry The type of geometry in which the rules—such as that lines that begin parallel eventually meet—are those that hold on the surface of a sphere.

spheroidal component (of a galaxy) *See* halo component.

spheroidal galaxy Another name for an *elliptical galaxy*. *See also* dwarf spheroidal galaxy.

spheroidal population *See* halo population.

spin *See* spin angular momentum.

spin angular momentum The inherent angular momentum of a fundamental particle; often simply called *spin*.

spiral arms The bright, prominent arms, usually in a spiral pattern, found in most spiral galaxies.

spiral density waves Gravitationally driven waves of enhanced density that move through a spiral galaxy and are responsible for maintaining its spiral arms.

spiral galaxies Galaxies that look like flat white disks with yellowish bulges at their centers. The disks are filled with cool gas and dust, interspersed with hotter ionized gas, and usually display beautiful spiral arms.

spreading centers (geological) Places where hot mantle material rises upward between plates and then spreads sideways, creating new seafloor crust.

spring equinox *See* March equinox, which is commonly called the spring equinox by people living in the Northern Hemisphere.

spring tides The higher-than-average tides on Earth that occur at new and full moon, when the tidal forces from the Sun and Moon both act along the same line.

standard candle An object for which we have some means of knowing its true luminosity, so that we can use its apparent brightness to determine its distance with the luminosity–distance formula.

standard model (of physics) The current theoretical model that describes the fundamental particles and forces in nature.

standard time Time measured according to the internationally recognized time zones.

star A large, glowing ball of gas that generates energy through nuclear fusion in its core. The term *star* is sometimes applied to objects that are in the process of becoming true stars (e.g., protostars) and to the remains of stars that have died (e.g., neutron stars).

starburst galaxy A galaxy in which stars are forming at an unusually high rate.

star cluster *See* cluster of stars.

star–gas–star cycle The process of galactic recycling in which stars expel gas into space, where it mixes with the interstellar medium and eventually forms new stars.

star system *See* solar system.

state (quantum) *See* quantum state.

steady state theory A now-discredited hypothesis that held that the universe had no beginning and looks about the same at all times.

Stefan–Boltzmann constant A constant that appears in the laws of thermal radiation, with value

$$\sigma = 5.67 \times 10^{-8} \frac{\text{watt}}{\text{m}^2 \times \text{K}^4}$$

stellar evolution The formation and development of stars.

stellar parallax The apparent shift in the position of a nearby star (relative to distant objects) that occurs as we view the star from different positions in Earth's orbit of the Sun each year.

stellar thermostat The regulation of a star's core temperature that comes about when a star is in both energy balance (the rate at which fusion releases energy in the star's core is balanced with the rate at which the star's surface radiates energy into space) and gravitational equilibrium.

stellar wind A stream of charged particles ejected from the surface of a star.

stratosphere An intermediate-altitude layer of Earth's atmosphere that is warmed by the absorption of ultraviolet light from the Sun.

stratovolcano A steep-sided volcano made from viscous lavas that cannot flow very far before solidifying.

string theory New ideas, not yet well-tested, that attempt to explain all of physics in a much simpler way than current theories.

stromatolites Rocks thought to be fossils of ancient microbial colonies.

strong force One of the four fundamental forces; it is the force that holds atomic nuclei together.

subduction (of tectonic plates) The process in which one plate slides under another.

subduction zones Places where one plate slides under another.

subgiant (luminosity class IV) A star that is between being a main-sequence star and being a giant; subgiants have inert helium cores and hydrogen-fusing shells.

sublimation The process by which atoms or molecules escape into the gas phase from the solid phase.

summer solstice *See* June solstice, which is commonly called the summer solstice by people living in the Northern Hemisphere.

sunspot cycle The period of about 11 years over which the number of sunspots on the Sun rises and falls.

sunspots Blotches on the surface of the Sun that appear darker than surrounding regions.

superbubble Essentially a giant interstellar bubble, formed when the shock waves of many individual bubbles merge to form a single giant shock wave.

superclusters The largest known structures in the universe, consisting of many clusters of galaxies, groups of galaxies, and individual galaxies.

super-Earth An extrasolar planet whose size and density suggest it has an Earth-like composition, made up largely of rock and metal.

supergiants (luminosity class I) The very large and very bright stars that appear at the top of an H-R diagram.

supermassive black holes Giant black holes, with masses millions to billions of times that of our Sun, thought to reside in the centers of many galaxies and to power active galactic nuclei.

supernova The explosion of a star.

Supernova 1987A A supernova witnessed on Earth in 1987; it was the nearest supernova seen in nearly 400 years and helped astronomers refine theories of supernovae.

supernova remnant A glowing, expanding cloud of debris from a supernova explosion.

surface area–to–volume ratio The ratio defined by an object's surface area divided by its volume; this ratio is larger for smaller objects (and vice versa).

synchronous rotation The rotation of an object that always shows the same face to an object that it is orbiting because its rotation period and orbital period are equal.

synchrotron radiation A type of radio emission that occurs when electrons moving at nearly the speed of light spiral around magnetic field lines.

synodic month (or **lunar month**) The time required for a complete cycle of lunar phases, which averages about $29\frac{1}{2}$ days.

synodic period (of a planet) The time between successive alignments of a planet and the Sun in our sky; measured from opposition to opposition for a planet beyond Earth's orbit, or from superior conjunction to superior conjunction for Mercury and Venus.

tangential motion The component of an object's motion directed across our line of sight.

tangential velocity The portion of any object's total velocity that is directed across (perpendicular to) our line of sight. This part of the velocity cannot be measured with the Doppler effect. It can be measured only by observing the object's gradual motion across our sky.

tectonics The disruption of a planet's surface by internal stresses.

temperature A measure of the average kinetic energy of particles in a substance.

terrestrial planets Rocky planets similar in overall composition to Earth.

theories of relativity (special and general) Einstein's theories that describe the nature of space, time, and gravity.

theory (in science) *See* scientific theory.

theory of evolution The scientific theory, first advanced by Charles Darwin, that explains how evolution occurs through the process of natural selection.

thermal emitter An object that produces a thermal radiation spectrum; sometimes called a *blackbody*.

thermal energy The collective kinetic energy, as measured by temperature, of the many individual particles moving within a substance.

thermal escape The process in which atoms or molecules in a planet's exosphere move fast enough to escape into space.

thermal pressure The ordinary pressure in a gas arising from motions of particles that can be attributed to the object's temperature.

thermal pulses The predicted upward spikes in the rate of helium fusion, occurring every few thousand years, that occur near the end of a low-mass star's life.

thermal radiation The spectrum of radiation produced by an opaque object that depends only on the object's temperature; sometimes called *blackbody radiation*.

thermosphere A high, hot, x-ray-absorbing layer of an atmosphere, just below the exosphere.

third-quarter phase The phase of the Moon that occurs three-quarters of the way through each cycle of phases, in which precisely half of the visible face is illuminated by sunlight.

tidal force A force that occurs when the gravity pulling on one side of an object is larger than that on the other side, causing the object to stretch.

tidal friction Friction within an object that is caused by a tidal force.

tidal heating A source of internal heating created by tidal friction. It is particularly important for satellites with eccentric orbits, such as Io and Europa.

time dilation The effect in which you observe time running more slowly in reference frames moving relative to you.

time monitoring (in astronomical research) The process of tracking how the light intensity from an astronomical object varies with time.

torque A twisting force that can cause a change in an object's angular momentum.

total apparent brightness *See* apparent brightness. The word "total" is sometimes added to make clear that we are talking about light across all wavelengths, not just visible light.

totality (eclipse) The portion of a total lunar eclipse during which the Moon is fully within Earth's umbral shadow or a total solar eclipse during which the Sun's disk is fully blocked by the Moon.

total luminosity *See* luminosity. The word "total" is sometimes added to make clear that we are talking about light across all wavelengths, not just visible light.

total lunar eclipse A lunar eclipse in which the Moon becomes fully covered by Earth's umbral shadow.

total solar eclipse A solar eclipse during which the Sun becomes fully blocked by the disk of the Moon.

transit An event in which a planet passes in front of a star (or the Sun) as seen from Earth. Only Mercury and Venus can be seen in transit of our Sun. The search for transits of extrasolar planets is an important planet detection strategy.

transmission (of light) The process in which light passes through matter without being absorbed.

transparent A word used to describe a material that transmits light.

tree of life (evolutionary) A diagram that shows relationships between different species as inferred from genetic comparisons.

triple-alpha reaction *See* helium fusion.

Trojan asteroids Asteroids found within two stable zones that share Jupiter's orbit but lie 60° ahead of and behind Jupiter.

tropical year The time from one March equinox to the next, on which our calendar is based.

Tropic of Cancer The circle on Earth with latitude 23.5°N, which marks the northernmost latitude at which the Sun ever passes directly overhead (which it does at noon on the June solstice).

Tropic of Capricorn The circle on Earth with latitude 23.5°S, which marks the southernmost latitude at which the Sun ever passes directly overhead (which it does at noon on the December solstice).

tropics The region on Earth surrounding the equator and extending from the Tropic of Capricorn (latitude 23.5°S) to the Tropic of Cancer (latitude 23.5°N).

troposphere The lowest atmospheric layer, in which convection and weather occur.

turbulence Rapid and random motion.

21-cm line A spectral line from atomic hydrogen with wavelength 21 cm (in the radio portion of the spectrum).

ultraviolet light Light with wavelengths that fall in the portion of the electromagnetic spectrum between visible light and x-rays.

umbra The dark central region of a shadow.

unbound orbits Orbits on which an object comes in toward a large body only once, never to return; unbound orbits may be parabolic or hyperbolic in shape.

uncertainty principle The law of quantum mechanics that states that we can never know both a particle's position and its momentum, or both its energy and the time it has the energy, with absolute precision.

universal law of gravitation The law expressing the force of gravity (F_g) between two objects, given by the formula

$$F_g = G \frac{M_1 M_2}{d^2}$$

$$\left(\text{where } G = 6.67 \times 10^{-11} \frac{\text{m}^3}{\text{kg} \times \text{s}^2}\right)$$

universal time (UT) Standard time in Greenwich, England (or anywhere on the prime meridian).

universe The sum total of all matter and energy.

up quark One of the two quark types (the other is the down quark) found in ordinary protons and neutrons. It has a charge of $+\frac{2}{3}$.

vaporization The process by which atoms or molecules escape into the gas phase from the liquid or solid phase. More technically, vaporization from a liquid is called *evaporation* and vaporization from a solid is called *sublimation*.

velocity The combination of speed and direction of motion; it can be stated as a speed in a particular direction, such as 100 km/hr due north.

virtual particles Particles that "pop" in and out of existence so rapidly that, according to the uncertainty principle, they cannot be directly detected.

viscosity The thickness of a liquid described in terms of how rapidly it flows; low-viscosity liquids flow quickly (e.g., water), while high-viscosity liquids flow slowly (e.g., molasses).

visible light The light our eyes can see, ranging in wavelength from about 400 to 700 nm.

visual binary A binary star system in which both stars can be resolved through a telescope.

voids Huge volumes of space between superclusters that appear to contain very little matter.

volatiles Substances, such as water, carbon dioxide, and methane, that are usually found as gases, liquids, or surface ices on the terrestrial worlds.

volcanic plains Vast, relatively smooth areas created by the eruption of very runny lava.

volcanism The eruption of molten rock, or lava, from a planet's interior onto its surface.

waning (phases) The set of phases in which less and less of the visible face of the Moon is illuminated; the phases that come after full moon but before new moon.

water world An extrasolar planet whose size and density suggest it is made largely of water, without a substantial amount of hydrogen or helium gas.

watt The standard unit of power in science; defined as 1 watt = 1 joule/s.

wavelength The distance between adjacent peaks (or troughs) of a wave.

waxing (phases) The set of phases in which more and more of the visible face of the Moon is becoming illuminated; the phases that come after new moon but before full moon.

weak bosons The exchange particles for the weak force.

weak force One of the four fundamental forces; it is the force that mediates nuclear reactions, and it is the only force besides gravity felt by weakly interacting particles.

weakly interacting particles Particles, such as neutrinos and WIMPs, that respond only to the weak force and gravity; that is, they do not feel the strong force or the electromagnetic force.

weather The ever-varying combination of winds, clouds, temperature, and pressure in a planet's troposphere.

weight The net force that an object applies to its surroundings; in the case of a stationary body on the surface of Earth, it equals mass × acceleration of gravity.

weightlessness A weight of zero, as occurs during free-fall.

white dwarf limit (or **Chandrasekhar limit**) The maximum possible mass for a white dwarf, which is about $1.4 M_{Sun}$.

white dwarfs The hot, compact corpses of low-mass stars, typically with a mass similar to that of the Sun compressed to a volume the size of Earth.

white dwarf supernova A supernova that occurs when an accreting white dwarf reaches the white dwarf limit, ignites runaway carbon fusion, and explodes like a bomb; often called a *Type Ia supernova*.

WIMPs A possible form of dark matter consisting of subatomic particles that are dark because they do not respond to the electromagnetic force; stands for *weakly interacting massive particles*.

winter solstice *See* December solstice, which is commonly called the winter solstice by people living in the Northern Hemisphere.

worldline A line that represents an object on a spacetime diagram.

wormholes The name given to hypothetical tunnels through hyperspace that might connect two distant places in the universe.

x-ray binary A binary star system that emits substantial amounts of x-rays, thought to be from an accretion disk around a neutron star or black hole.

x-ray burster An object that emits a burst of x-rays every few hours to every few days; each burst lasts a few seconds and is thought to be caused by helium fusion on the surface of an accreting neutron star in a binary system.

x-ray bursts Bursts of x-rays coming from sudden ignition of fusion on the surface of an accreting neutron star in an x-ray binary system.

x-rays Light with wavelengths that fall in the portion of the electromagnetic spectrum between ultraviolet light and gamma rays.

Zeeman effect The splitting of spectral lines by a magnetic field.

zenith The point directly overhead, which has an altitude of 90°.

zodiac The constellations on the celestial sphere through which the ecliptic passes.

zones (on a jovian planet) Bright bands of rising air that encircle a jovian planet at a particular set of latitudes.

Index

Page references preceded by a "t" refer to tables.
Page references followed by an "n" refer to notes.

A

Abell 383 galaxy cluster, 237
absolute zero, A-12
absorption line, 137, 152
absorption line spectrum, 132
accelerating universe, 244–245, 246–247
acceleration of gravity (*g*), A-2
accretion, 65, 69
    and gas capture in jovian planets, 66
    in neutron star close binary systems, 171
    in supermassive black holes, 188, 196
    of terrestrial planets, 65–66, 70, 267
    in white dwarf binary systems, 168
accretion disk, 168
    around neutron star, 171
    around supermassive black hole, 175, 196
    around white dwarf, 168
active galactic nuclei, 195
    luminosity of, 195–196
    power source for, 196–197
age
    radiometric, 71–72
    of star clusters, 159–160
    of universe, 4, 5, 12–13, 208–209
Akatsuki mission, 86
Aldebaran, 138, 139, 143
Aldrin, Buzz, 7
aliens, 262–263
Allen Telescope Array, 262
Almagest (Ptolemy), 37
Alpha Centauri, 7, 8, 135, t137
Alphonso X, king of Spain, 37
altitude, 18
    and jovian cloud layers, 98–99
aluminum, in solar nebula, 64
Alvarez, Luis and Walter, 109, 110, 111
amino acids, 253
ammonia
    on jovian planets, 58, 96, 98
    in solar nebula, 45
ammonium hydrosulfide, 98
ancient societies, and definition of planets, 13
Andromeda Galaxy (M31), 4, 193, 204
angular distance, 26
angular momentum
    in galaxy formation, 189, 192
    law of conservation of, 62, 63, 168, 170, 189
    and neutron star, 170
    in solar system formation, 63
angular resolution, of telescope, 42
angular separation, A-3
angular size, 26
annihilation, 216, 219, 220
annular eclipse, 28, 29
Antarctica, life in, 254
Antennae galaxies (NGC 4038/4039), 193
antielectron, 216, 223
antimatter, 216, 217, 219, 220
antineutron, 216

antiproton, 216, 220, 223
aphelion, 39, 47
Apollo missions, 7, 71, 257
apparent brightness, 134, 135, 136, 201, 202
apparent retrograde motion, 31–32, 36–37, 44, 46–47
Archaea, 253
Arches National Park, 17
arcminute, 26
arcsecond, 26
Arctic Circle, 22
Arcturus, t137, 143
argon, 102
argon-40, 71
Ariel (moon of Uranus), 100
Aristarchus, 32, 33, 36, 37, 46
Aristotle, 36, 38, 39, 40, 49
Armstrong, Neil, 7
asteroid belt, 57, 58, 60, 105–106, 107, 108
asteroids, 2, 13–14, 15, 57, 58, 60, 69, 255. *See also names of specific asteroids*
    definition of, 5
    formation of, 66–67, 69
    impacts on Earth of, 11, 77, 109–111
    orbits of, 106
    regions of, 105–106
    risk of impacts from, 108–109
astrometric method, and extrasolar planets, 114–115, 118, 120, t121
astronomical distance, measurement of, 2–5
astronomical motions, 5
astronomical objects, 5
astronomical unit (AU), 2–3, 5, 201, A-2
astronomy, 118, 128, 152, 184
    ancient Greek, 31–33, 36–37, 46
    Copernican revolution in, 37–38, 39, 43, 44, 46–47, 113
atmosphere
    and conditions for life, 78–81, 87–89
    of Earth, 78–81
    effects of on light, 81
    of Jupiter, 97–98
    of Mars, 78, 80
    of Neptune, 98
    of Saturn, 98
    of Sun, 131–133
    of Titan, 102–103
    of Uranus, 98
    of Venus, 78, 80
atomic hydrogen gas
    emission lines of, 184
    formation and distribution of in Milky Way, 185–186
atoms
    and composition of universe, 219
    energy levels in, 152
    era of, 219, 221, 223
aurora, on Earth, 77, 133
autumnal equinox, 19
axis (plural: axes)
    of Earth, 5
    of ellipse, 39

axis tilt
    of Earth, 18–22
    of planets in solar system, t55, 56–57
    seasons and, 18–22
    of Uranus, 57
azimuth, 18

B

bacteria, 253
balloon
    and expansion of universe, 207, 208, 209
    and flattening of space with inflation, 229
bar, 97
barred spiral galaxies, 190
baseball, and Newton's first law, 41
Bell, Jocelyn, 170
Betelgeuse, 4, 134, 135, 136, t137, 138–139, 140, 156
    on Hertzsprung-Russell diagram, 143
bias, t48
Big Bang, 5, 9, 10, 11, 208, 215, 216, 217, 222–223, 239, 242, 266. *See also* cosmic microwave background; critical density
    and abundances of light elements, 225–226
    and electroweak era, 219, 222
    and era of atoms, 219, 221, 223
    and era of galaxies, 219, 221, 222
    and era of nuclei, 219, 220–221, 223, 224, 225, 228, 229
    and era of nucleosynthesis, 219, 220, 221, 223, 225, 226
    evidence for, 221–226
    and fundamental forces, 217–218
    and galaxy formation, 225, 227
    and GUT era, 219, 222, 227
    and helium formation, 225–226
    inflation following, 227–220
    and particle creation and annihilation, 216
    and particle era, 219–220, 223
    and Planck era, 219, 222
    and uniformity of early universe, 228
"Big Crunch," 244
Big Dipper, pointer stars of, 26
binary star systems, 136–137, 138
    black holes in, 175
    and gravitational waves, 178
    neutron stars in, 171
    white dwarfs in, 168
    x-ray, 167, 171, 175, 184
black hole, 51, 163, 171–175, 177
    active galactic nuclei and, 195, 196–197
    at center of Milky Way galaxy, 187–188, 196
    event horizon of, 173, 174
    evidence for, 175
    falling into, 174–175
    formation of, 172–173
    galaxy formation and, 195, 197
    singularity and, 173
    spacetime curvature and, 173–174
    supermassive, 175, 195, 196–197
    from supernova, 158, 163
blue cloud, 194, 195

I-1

blueshift, 115, 116, 118
bombardment. *See* heavy bombardment
boson, Higgs, 217
Brahe, Tycho, 38, 40, 42, 44, 47
brightness, apparent, 134, 135, 136, 201, 202
Brown, Michael, 14
brown dwarf, t137, 150–151, 152, 167, 176, 177
 as dark matter, 237
bulge
 of galaxy, 190, 197
 of Milky Way, 183, 185
bulge stars, 185
Butterfly Nebula, 155

C

Callisto (moon of Jupiter), 100, 102
 possibility of life on, 256, 257
Canis Major dwarf galaxy, 189
Cannon, Annie Jump, 140–141
"cantaloupe terrain," 104
captured moons, 67, 104
carbon, 11, 153, 156, 177, 226
 fusion of, 154, 156, 157, 168
 isotopes of, 70
carbonate rocks, 86, 87, 88, 89
carbon dioxide, 80
 on Earth, 86
 global warming and, 90–92
 on Mars, 81, 84
 on Venus, 54, 81, 86
 from volcanic outgassing, 77
carbon dioxide cycle (CO_2 cycle), 83
carbon monoxide (CO), 186
Cassini division, 98, 99
Cassini spacecraft, 59, 95, 98, 99, 103, 104, 257
Catholic Church, Galileo and, 40
celestial equator, 23
celestial poles, 23, 24
celestial sphere, constellations on, 23–24
Celsius scale, A-12
Centaurus constellation, 7
center of mass, 114, 115, 118, 120
Cepheid period-luminosity relation, 202
Cepheid variable stars, 202–203, 204, 205, 206
Ceres, 13–14, 105, 106, 257
Chandra X-Ray Observatory, 167, 169, 186, 236
Chandrasekhar limit, 168
charged particle belts, around Earth, 77
Charon (moon of Pluto), 14
Chelyabinsk, Russia, asteroid explosion over, 108, 109
chemical potential energy, 62
CHEOPS mission, 117
chromosphere, 132, 133
climate
 on Earth, 70, 87, 88, 89, 90–92
 on Mars, 82–85
close binary systems. *See* binary star systems
closed universe, 229
clouds. *See also* Magellanic Clouds; molecular clouds; protogalactic clouds
 on jovian planets, 97–98
clusters. *See* galaxy clusters; globular clusters; star clusters
coasting universe, 244, 245, 246
COBE (Cosmic Background Explorer) spacecraft, 224, 225
collisions. *See also* heavy bombardment; impact craters
 of galaxies, 193, 194, 212, 232, 243
 mass extinctions from, 109–111
 in planet formation, 64, 65–66, 67, 68, 69
 planetary rings from, 99
color
 of Earth's sky, 80, 81
 of galaxies, 190, 191, 194–195
 of jovian planets, 98
 of stars, 136, 137–138, 142
Columbus, Christopher, 36
comets, 2, 5, 57, 59–60, 66–67, 69, 106, 255. *See also names of specific comets*
 current risk of impacts from, 108–109, 111
 influence of jovian planets on, 108
 Kuiper belt objects as, 106
 observation of by Tycho, 38, 40
 regions of, 105–107
condensation, in solar system formation, 64–65, 69
conservation laws
 of angular momentum, 62, 63, 168, 170, 189
 of energy, 62, 63
constellations, 22–25
 Sun's apparent position in, 24–25
continuous spectrum, 132
convection, 76, 78, 84, 85, 89, 131
convection zone, in Sun, 131, 133
Copernican revolution, 37–38, 39, 43, 44, 46–47, 113
Copernicus, Nicholas, 37–38, 39, 43, 44, 47
core
 of Europa, 101, 102
 of jovian planet, 97
 of low-mass star, 162–163, 167, 177
 of Sun, 62, 129–130, 131, 133, 148, 152–154
 of terrestrial planet, 76
corona, of Sun, 29, 132, 133
coronal mass ejection, 128, 133
COROT 7b, 123
COROT 14b, 123
cosmic address, 2, 3
Cosmic Background Explorer (COBE) spacecraft, 224, 225
cosmic calendar, 11–13, 162–163
cosmic microwave background, 215, 221, 223, 224–225, 240, 242, 266
 characteristics of, 224–225
 inflation and, 227–228, 229, 230
 origin of, 224
 photons from era of nuclei in, 221, 223, 224
 temperature of, 224, 225, 228, 229, 230, 238, 247
 thermal radiation spectrum of, 224, 225
cosmic rays, 184
cosmological horizon, 210
Cosmological Principle, 207, 240
cosmological redshift, 209–210
cosmos. *See* universe
Crab Nebula, 159, 170
crescent moon, 25, 27
critical, as scientific term, t48
critical density, 228, 229, 237, 244, 245, 246
 and total density of universe, 247
critical universe, 244, 245, 246
crust, planetary, 76

Crux (Southern Cross), 7
Curiosity rover, 74, 83–84, 256
curvature of spacetime. *See* spacetime
Cygnus (constellation), 175
Cygnus X-1 system, 175

D

Daphnis (moon of Saturn), 99
dark energy, 232, 233, 239, 244, 248–249, 250
 evidence for, 245–247, 250
dark matter, 185, 232, 233, 243, 246, 248–249, 250
 alternatives to, 236–237
 composition of, 237–238
 distribution of in early universe, 239
 and distribution of mass in Milky Way Galaxy, 233–234
 in elliptical galaxies, 235
 evidence for, 233–238
 evidence against neutrinos as, 238
 and expansion of universe, 232, 239–241
 in galaxy clusters, 232, 236–237, 239, 248
 and galaxy formation, 239, 243
 gravitational lensing and, 235, 236, 237, 248
 in spiral galaxies, 234–235
 and temperature of hot gas in clusters, 235–236
 and understanding of gravity, 236–237
 WIMPs as, 238, 247
"dark side" of Moon, 27
Darwin, Charles, 45, 264
Dawn spacecraft, 106
days
 length of, A-2
 names of, 13
December solstice, 19, 21
deep-sea vents, 102, 254, 256
deferent, 37
degeneracy pressure, 154, 166, 175, 176, 177
 in brown dwarfs, 150–151, 152, 177
 electron, 151, 152, 158, 167, 168, 172, 177
 in helium-fusing stage of star life cycle, 154
 in high-mass star, 156, 157, 158, 177
 neutron, 169, 173, 177
 supernova explosions and, 158, 163
 in white dwarfs, 152, 154, 167, 177
Deimos (moon of Mars), 67
density. *See also* critical density
 in early universe, 220
 of extrasolar planets, 120, t121
 of jovian planets, 61, 95, 97
 of planets in solar system, t55, 65
 of protogalactic cloud, 192–193
 of white dwarfs, 139, 167
deoxyribonucleic acid. *See* DNA
De Revolutionibus Orbium Coelestium (Copernicus), 37, 47
deuterium, 237–238, 247
 in early universe, 220, 226
deviation, t48
differentiation, 76, 97
dinosaurs, 11–12, 109–111
Dione (moon of Saturn), 100, 103
disk
 formation of from solar nebula, 63, 69
 formation of stars in, 9
 of galaxy, 190
 around jovian planets, 66
 of Milky Way, 2, 183, 184, 185, 189

disk population, 188
disk stars, 183, 185, 188–189
distance, 2–4, 7
   angular, 26
   to galaxies, 201–203, 205–206
   in light-years, 2, 3, 4, 5, 10, A-2
   orbital speed and, 39, 118
   in parsecs, 201, A-2
   of planets from Sun, t55, 88, 89, 95
   units of, A-2
distance chain, 201–203, 205–206
distance measurements, and age of universe, 208–210
DNA, 253, 254, 264–265
Doppler effect, 117
   in calculating distances to galaxies, 209
   in measuring motions of stars, 115
Doppler method, for detection of extrasolar planets, 115, 118, 120, t121
Doppler shifts, 132, 197, 234, 235
   equation for, A-3
   in gas orbiting galactic centers, 197
   and interpretation of spectrum, 116
double shell–fusion star, 154, 163, 177
Drake, Frank, 259, 260
Drake equation, 259–261, 262
dust grains, interstellar, in Milky Way, 184
dwarf elliptical galaxies, 190
dwarf galaxies, 190
dwarf planet(s), 100, 257
   Ceres as, 14, 105
   classification of, 5
   Eris as, 6, 14, 15, 59, 106
   Pluto as, 6, 14, 15, 59, 106–107
dwarf spheroidal galaxies, 190

E

Earth, 13, 53, 56, 58, 94, 123.
   See also life on Earth; terrestrial planets
   age of, 71, 72
   age of oldest rocks on, 71
   atmosphere of, 78–81, 87
   axis tilt of, 18–22
   circumference of, 36
   climate of, 88
   composition of, 9, 11
   cosmic address of, 2, 3
   data on, t55, A-2
   differentiation of interior of, 75–76
   distance of from Sun, 19
   erosion on, 77, 78
   features important for life on, 87–89
   formation of, 11, 12
   formation of Moon of by giant impact, 67–68, 72
   geological activity of, 75–78
   greenhouse effect on, 80–81, 86–87, 88, 90–92
   habitability of, 87–89, 254–255
   impact cratering on, 77
   infrared spectrum of, 259
   interior structure of, 75–76, 97
   magnetic field of, 76, 77
   motion of, 2, 56, 60
   orbit of, 2, 5, 7, 10, 18, 19, 20, 24–25, t55
   origin of water on, 67, 87
   plate tectonics on, 78, 88
   rotation of, 5, 17, 18, 23, 25, 101
   shape of, 36
   size of, 15, 36, A-2
   surface of, 58
   threat of impacts on, 108–109
   tides on, 100–101
   view of universe from, 4
   volcanism on, 75, 77–78
Earth-centered universe. *See* geocentric model of universe
earthquakes, 75
eclipses
   conditions for, 29–30
   detection of extrasolar planets from, 115–117, 119, t121
   lunar, 28, 29
   predicting, 30
   solar, 28, 29
   variations on, 28–29
eclipse seasons, 29–30
eclipsing binary star systems, 136–137
ecliptic, 23, 24–25
ecliptic plane, 29–30
egg balancing, and spring equinox, 45
Einstein, Albert, 50–51, 125
   equation of, 62, 130, 173, 196, 216, 217
   general theory of relativity of, 10, 44, 45, t48, 50, 51, 172, 173, 174, 178, 208, 227, 228, 236, 237, 245, 247
   special theory of relativity of, 172
electromagnetic force, 130, 217–218
electromagnetic radiation. *See* light
electromagnetic spectrum, 79, 129, 184, 262
electromagnetic wave, 79
electron–antielectron pairs, 216, 223
electron degeneracy pressure, 151, 152, 158, 167, 168, 172, 177
electron-volt (eV), A-2
electrons, 216, 217, 219, 220, 221, 223, 224, 266
   degeneracy pressure and, 150–152, 158, 167, 168, 172, 177
   in early universe, 216–217, 219, 220, 221, 223, 224
   energy levels of, 151, 152, 186
   mass of, A-2
electrostatic forces, 65
electroweak era, 219, 222
electroweak force, 218, 219
elements. *See also* names of specific elements
   abundances of in solar system, 158–159
   and Big Bang theory, 225–226
   detection of by emission and absorption lines, 132
   formation of, 9, 11
   galactic recycling of, 9, 11, 185–186
   from high-mass stars, 158–159
ellipse, 39
elliptical galaxies, 190–191, 192, 193, 194, 211, 235, 243
   from collisions of spiral galaxies, 193, 194
   evidence of dark matter in, 235
   in Hubble's galaxy classes, 191, 192
elliptical orbits, 39, 47, 67, 101, 104, 120, 171
emission line spectrum, 132
emission lines, 132, 152
emission nebulae, 186
Enceladus (moon of Saturn), 100
   geological activity on, 103–104
   possibility of life on, 257
energy. *See also* dark energy
   chemical potential, 62
   conservation of, 62, 63
   gravitational potential, 62, 63, 75, 148, 150, 168, 176, 196
   kinetic, 62, 63, 171, 196
   law of conservation of, 62
   mass-, 62, 216, 237, 247
   of photons, A-3
   potential, 62
   radiative, 62
   thermal, 62, 63, 75, 97, 148, 150, 152, 168, 176, 177, 196
   units of, A-2
energy balance, 148–150
energy-level transitions, 152, 186
enhance/enrich, as scientific terms, t48
epicycle, 37
equations, useful, A-3
equinoxes, 19, 21, 45
eras of universe, 218–221, 222–223
Eratosthenes, 36
Eris, 6, 14, 15, 59, 106
   data on, t55
erosion, 77, 78, 89, 100
   on Earth, 75, 78
   lack of on Venus, 78, 85
   on Mars, 78, 82–84
   on Titan, 104
error, t48
escape velocity
   from black hole, 172, 173
   from Earth, 171, A-2
   equation for, A-3
ethane, on Titan, 59, 102, 103, 257
Eukarya, 253
Europa (moon of Jupiter), 58, 100, 101–102
   possibility of life on, 256, 257
Europa Clipper, 257
European Southern Observatory, 170
European Space Agency (ESA), 103, 115, 117, 257
event horizon, 173, 174
evolution, theory of, 45, 263–265
evolutionary track, of star. *See* life track
exclusion principle, 152
ExoMars rover, 257
exotic matter, 237, 238, 247. *See also* dark matter
expansion of universe. *See* universe, expansion of
exponents, A-4–A-7
extrasolar planets
   atmospheres of, 120, t121
   characteristics of, 121–124
   detection of, 114–121
   habitability of, 255–257, 258–259
   measurable properties of, 116–121
   and nebular theory, 124–126
   orbital properties of, 117, 118, 120, 121
   physical properties of, 120, 121–123
   probability of life on, 257–263
extremophiles, 254–255
eyewitness testimony, 44–45

F

Fahrenheit scale, A-12
fall equinox, 19, 21

feedback, t48
Fermi, Enrico, 262
Fermi's paradox, 262–263
51 Pegasi, 115
first law of motion, Newton's, 41
first law of planetary motion, Kepler's, 39, 40, 42, 47, 51
fission, 129, 157, 158
flat universe, 227, 228, 229, 237, 247
Fleming, Williamina, 140
focus, of ellipse, 39
force(s)
  electromagnetic, 130, 217–218, 219
  electroweak, 218, 219
  fundamental, 217–218
  gravitational, 50, 130, 217, 218, 219, 233
  GUT, 218, 219, 227
  in Newton's laws, 41
  strong, 130, 217, 218, 219, 227
  "super-," 218, 219
  tidal, 68, 100–101, 175, 193
  weak, 217, 218, 219, 238
formulas, A-3
fossil fuels, and global warming, 90, 91
fossils, 89, 109, 264–265
frequency, 79
frost line, 64, 65
full moon, 25, 26, 27, 28
full shadow, of Moon, 28
fundamental forces, 217–218
fusion. *See* helium fusion; hydrogen fusion; nuclear fusion

G

G11.2-03 supernova remnant, 169
GAIA, 115, 118, 201
galactic recycling
  in formation of solar system, 9, 11
  in star–gas–star cycle, 185–186, 189, 243, 267
galaxies. *See also* Milky Way Galaxy; spiral galaxies
  barred spiral, 190
  bulge of, 190
  center of, 195
  collisions of, 193, 194, 212, 232, 243
  colors of, 190, 191, 194–195
  dark matter in, 232, 234–235, 248
  definition of, 2, 5, 182
  differences among, 192–195
  distances to, 201–203, 205–206, 211, 212
  distribution of in universe, 239–241
  dwarf, 190
  elliptical, 190–191, 192, 193, 194, 211, 235, 243
  era of, 219, 221, 222
  evolution of, 192–195, 210–212, 223, 242–243
  family albums of, 211
  formation of, 188–189, 192, 197, 212, 223, 242–243
  giant, 190
  halo of, 190
  Hubble's classes of, 191–192
  Hubble's law for velocity of, 204–205, 206, 209, 239, 240
  irregular, 190, 191, 194, 211
  lenticular, 190
  motions of, 240
  Seyfert, 195
  vs. solar system, 10
  star-forming regions of, 186–187
  types of, 190–191
  in universe expansion, 10, 204–205
galaxy clusters, 2, 5, 190, 193, 194, 240, 244
  dark matter in, 232, 235–236, 239, 248
  gravitational lensing by, 236, 237, 248
  hot gas in, 235, 236
Gale Crater (Mars), 83–84
Galilean moons, 100–102. *See also* jovian moons; *names of specific moons*
Galileo Galilei, 13, 39–40, 41, 42, 43, 47, 49
Galileo spacecraft, 95, 97, 100, 101
Galle, Johann, 50
gamma rays, 216, 220, 223, 226
  in electromagnetic spectrum, 79
  from Milky Way, 184
Ganymede (moon of Jupiter), 100, 101, 102, possibility of life on, 256, 257
gap moons, 99
gas cloud, collapse of, 10, 61, 62, 63, 69, 148, 162
gas giants, 97
gases
  from dying stars, 153, 154, 155, 158, 185–186
  in formation of jovian planets, 66–69, 96–107
  greenhouse, 80–81, 90
  in Milky Way, 185–186, 188, 234
  in molecular clouds, 147–148, 185, 186
  vs. plasma, 129
  in protostars, 148, 162
  sulfur-bearing, from volcanic outgassing, 78
Gemini North telescope, 42
general theory of relativity, 10, 44, 45, t48, 50, 51, 172, 173, 174, 178, 208, 227, 228, 236, 237, 245, 247
  and black holes, 172–175
  and curvature of spacetime, 172, 173, 227, 228, 236
  and gravitational lensing, 236
genetic code. *See* DNA
geocentric model of universe, 36–37, 44, 46–47
geological activity
  of Earth, 75, 76, 77, 78
  of jovian moons, 99–105
  of Mars, 75, 77, 78, 82
  of Mercury, 75, 77, 78, 82
  of Moon, 75, 77, 78, 82
  of moons of Neptune, 104–105
  of moons of Saturn, 103–104
  of moons of Uranus, 99–100, 104–105
  and planetary size, 75–78, 81
  processes in, 76–79, 82
  of terrestrial planets, 75–78
  of Venus, 75, 77, 78
giant galaxies, 190
giant impact, formation of Moon by, 68
giants, 138–139, 140, 141, 142, 143. *See also* red giant
gibbous moon, 25, 27
GJ 1214b, 123
global warming, 89–92
globular cluster, 160, 161, 189, 190, 235
  in Milky Way Galaxy, 183
Grand Canyon, 78, 82, 264
grand unified theories (GUTs), 218, 219, 227
gravitation, universal law of, 41, 49–50, 96, A-3
gravitational constant, 50, 96, A-2
gravitational contraction, 148, 150, 152, 153, 154, 156, 157
gravitational encounter, 65, 67, 107, 108
gravitational equilibrium
  of gas in galaxy clusters, 236
  in Sun, 148, 149
gravitational lenses, 235, 236, 237, 248
gravitational lensing
  detection of dark matter from, 236, 248
  by galaxy clusters, 235, 236, 237, 248
gravitational potential energy, 62, 75, 168, 176
  in solar system formation, 63, 148, 150
gravitational redshift, 174, 175
gravitational tugs, detection of extrasolar planets from, 114–116, 118, t121
gravitational waves, 175, 178–179
gravity. *See also* gravitation, universal law of
  acceleration of, A-2
  after Big Bang, 10, 217, 218, 219, 266
  of black hole, 173–174
  and Einstein's theory of relativity, 50–51
  escape velocity and, 171, 172
  and expansion of universe, 239–245, 250
  within galaxies, 5
  inverse square law for, 49, 50, 61
  in neutron stars, 158, 166, 169, 173, 177
  Newton's theory of, 10, 45, 49–50, 51, 126
  in planetary rings, 98–99
  of pure energy, 173
  in solar nebula collapse, 61–63
  and spacetime curvature, 172, 173
  in star formation, 147, 148
  and star life cycle, 162, 176–177
  strength of, 49, 50
  and structure in universe, 240, 242–243
  theory vs. fact of, 49–51
  tides and, 100–101
Great Red Spot (Jupiter), 6, 98
Greek science, ancient, 31–33, 36–37, 46
greenhouse effect, 80–81, 86–87, 88, 90–92
  on Mars, 81, 84, 85
  runaway, 86–87, 88
  on Titan, 102–103
  on Venus, 81, 85, 86–87
greenhouse gases, 80–81
groups of galaxies, 5, 190. *See also* galaxy clusters
GUT era, 219, 222, 227
GUT force, 218, 219, 227
Guth, Alan, 227
GUTs (grand unified theories), 218, 219, 227

H

H II regions, 186
habitable world(s), 87–89, 254–259. *See also* life
  distance from Sun and, 89
  in Drake equation, 260–261
  planetary size and, 89
  volcanic/tectonic activity and, 89
habitable zone, 121, 258, 267
Hale–Bopp, Comet, 106
half-life, 70
Halley, Edmund, 50
Halley's Comet, 50
hallmarks of science, 43–48, 248–249
halo, of Milky Way, 160, 183, 188, 189, 190, 234
halo population, 188
halo stars, 188, 189
  dark matter and, 234–235, 237
  orbits of, 185

Harvard College Observatory, 140
HAT-P-32b, 123
Hayabusa, 58
HD 189733, 117
HD 189733b, 123
HD 209458b, 123
heavenly perfection, ancient belief in, 36, 40
heavenly spheres, Greek model of, 36, 37
heavy bombardment, 67, 77, 104
    as enabler of life on Earth , 67
    impact craters from, 67, 81, 82, 83, 84, 85, 89
    lunar maria and, 81
helium, 64, 106, 153, 154, 157, 185, 186, 233, 237, 239, 243. *See also* helium fusion
    abundance of, 158, 159, 223, 225, 226
    condensation temperature of, 64
    in early universe, 9, 11, 212, 220, 221, 223, 225–226
    in jovian planets, 58, t60, 66, 69, 96, 97
    in planet formation, 65–66
    in solar nebula, 61, 64, 69
    in Sun, 130, 150, 153
helium-capture reactions, 157
helium-fusing star, 153–154, 155, 162, 163. *See also* star life cycles
    thermal pressure and gravity in, 177
helium fusion, 149–150, 153–154, 155, 156, 157, 161, 163, 167, 177. *See also* nuclear fusion
Helix Nebula, 155
Herschel, William, 253
Hertzsprung, Ejnar, 141
Hertzsprung-Russell (H-R) diagrams, 141–144, 149–150, 154–156, 160
    main-sequence turnoff point on, 160, 161, 200
    Pleiades on, 160, 161
    star clusters on, 160, 161
    Sun's life track on, 154–156
    supergiants on, 141, 142, 143, 156, 160
    white dwarfs on, 141, 142, 143, 161
Higgs boson, 217
high-mass stars, 143, 146, 150, 175
    advanced nuclear fusion in, 156–157
    death of, 152, 156–159, 163
    life stages of, 162–163, 176–177
    supernova explosions of, 266
"high noon," 22
Himalayas, 78
horizon, 4, 17, 18, 22, 23, 24, 25, 27
hot Jupiters, 115, 123, 125
HR8799, 113, 119
H-R diagrams. *See* Hertzsprung-Russell (H-R) diagrams
Hubble, Edwin, 204, 221
    galaxy classes of, 191, 192
Hubble eXtreme Deep Field, 136, 211
Hubble Space Telescope, 1, 4, 42, 136, 155, 159, 160, 167, 186, 187, 193, 196, 197, 203, 205, 236, 237
    angular resolution of, 42
    and value of Hubble's constant, 205
Hubble's constant, 204, 205, 208–209, 247
Hubble's law, 204–205, 206, 209, 239, 240
Hulse-Taylor system, 178
human activity, and global warming, 89, 90, 91–92
Huygens probe, 103
Hyades Cluster, 161
hydrogen, 64, 84, 85, 86, 102, 106, 125, 147, 153, 158, 159, 160, 162, 168, 186, 219, 220, 221, 225, 226, 233. *See also* atomic hydrogen gas; hydrogen fusion
    condensation temperature of, 64–65
    in early universe, 9, 11, 67, 212, 219, 220, 221, 226
    energy levels in, 152
    fusion of in Sun, 62, 129–130
    in jovian planets, 57, 58, t60, 67, 69, 95, 96, 97, 98
    metallic, 97
    in molecular clouds, 147, 186
    in planet formation, 66–67
    in solar nebula, 61, 64
    spectral lines of, 117, t137, 140, 169
hydrogen compounds, 67, 102, 122, 123
    on jovian planets, 58
    in solar nebula, 64
hydrogen fusion, 138, 139, 140, 141, 142, 144, 149, 150, 151, 153, 154, 155, 156, 157, 160, 161, 162, 163, 168, 176. *See also* nuclear fusion
hydrogen shell fusion, 153, 154, 156, 161, 163, 176
hypothesis (plural: hypotheses), 43, 48

I

Iapetus (moon of Saturn), 100
IAU (International Astronomical Union), 14
ice, 64, 65, 69
    on Earth, 90, 91
    on Enceladus, 104, 257
    on Europa, 101, 102
    on Ganymede, 102
    on Mars, 82, 83
    in planetary rings, 98, 99
    on Pluto, 107
    on Titan, 103
ice geology, 100, 105
impact
    and extinction of dinosaurs, 109–111
    threat of, 108–109
impact cratering, 77, 100
impact craters, 67, 77, 107, 108. *See also* collisions
    on Ceres, 106
    on Mars, 58, 77
    on Mercury, 82
    on Moon, 67, 77, 81
    on Triton, 104
    on Venus, 85
    on Vesta, 105, 106
inflation, 219, 222, 242
    evidence for, 229
    and gaps in Big Bang model, 226–230
    GUT era and, 219, 222, 227
    and total universe density vs. critical density, 228–229
    and uniformity of early universe, 228
infrared light, 79, 80–81
    from Milky Way galaxy, 184
inner planets. *See* terrestrial planets
intelligence, extraterrestrial. *See* SETI
intensity, of light, 79, 132
interferometer, 179
interior structures
    of Earth, 76, 98
    of Europa, 102, 105
    of jovian planets and moons, 97, 100, 101, 105
    of Mars, 76
    of Mercury, 76
    of Moon, 76
    of Sun, 130–131, 149
    of Venus, 76
internal heat
    of Earth, 75, 76, 77, 78, 89
    of jovian planets and moons, 97, 100–102, 103, 104, 105
    of Mars, 75, 77, 78
    of Mercury, 75
    of Venus, 77, 78
International Astronomical Union (IAU), 14
interstellar clouds, 61, 62. *See also* gases; molecular clouds; solar nebula
    formation of stars from, 23, 146, 147–148, 158, 162
interstellar medium, 183, 184, 185–186
inverse square law
    for gravity, 49, 50, 61, 64
    for light, 135, 196, 201, 203, 255, A-3
Io (moon of Jupiter), 58, 100, 105
    tidal heating on, 101
    volcanism on, 100
ionization nebulae, 186, 187
ions, 221
iridium, 110
iron, 11, 64, 76, 122, 123, 157, 158, 159, 163, 169, 170, 172, 175, 177
irregular galaxies, 190, 191, 194, 211
isotopes, 70
Itokawa (asteroid), 58

J

James Webb Space Telescope, 212
joule, A-2
jovian moons, 40, 66, 67. *See also names of specific moons*
    geological activity of, 99–105
    habitability of, 256–257
jovian planets, 15, 57, 58–59. *See also names of specific planets*
    clouds on, 97–98
    colors of, 98
    formation of, 66, 69, 96, 107
    influence of on Earth, 108
    interiors of, 97
    moons of, 40, 66, 67, 99–105, 256–257
    rings of, 98–99
    size and composition of, 95–96
    vs. terrestrial planets, 57, 60
    water on, 255
    weather on, 97–98
June solstice, 19, 20, 22
Juno mission, 59, 95, 98
Jupiter, 13, 56, 59, 66, 95, 123. *See also* jovian planets
    atmosphere of, 97–98
    axis tilt of, 19, t55, 56
    and center of mass of solar system, 114, 118
    color of, 98
    composition of, 95–96
    data on, t55, 95
    gravitational force of, 101, 114
    Great Red Spot of, 6, 98
    influence of on small bodies, 106, 108
    interior of, 95–96, 97

Jupiter (*Continued*)
  moons of, 40, t55, 58, 99, 100–102, 256, 257. *See also* names of specific moons
  size of, 15, 95
  total force of, 101
  weather on, 97–98
Jupiter Icy Moons Explorer (*JUICE*), 257
Jupiters, hot, 115, 123, 125

K

Kant, Immanuel, 61
Keck Telescope, 59, 188
Kelvin scale, A-12
Kepler, Johannes, 13, 38–39, 40, 43, 44, 47, 253
Kepler 10b, 120, 123
Kepler 11f, 123
Kepler 16, 123
Kepler mission, 115, 117, 122, 124
Kepler's laws of planetary motion, 39, 40, 42, 44, 47, 51, 201
Kilauea, 77
kilogram, A-11
kiloparsec, 201, A-2
kinetic energy, 62, 63, 171, 196
Kuiper belt, 14, 57, 59–60, 105, 106–107, 108

L

Laniakea, 2, 240
Large Binocular Telescope, 113, 119
Large Hadron Collider, 216, 217, 238
Large Magellanic Cloud, 59, 191, 202
large-scale structures, 239–241, 249
Laser Interferometer Gravitational-Wave Observatory (LIGO), 179
latitude, 22, 23–24, 26, 98, 129, 134
lava, 71, 76, 77, 81–82, 100, 104, 105
law of conservation of angular momentum, 62, 63, 168, 170, 189
law of conservation of energy, 62
law of thermal radiation, A-3
laws of motion, Newton's, 41, 43, 171, 235, 236, 248
laws of planetary motion, Kepler's, 38–39, 40, 41, 44, 47, 50, 201, 253
leap years, 19
Leavitt, Henrietta, 202
Levitt's law, 202, 203, 204, 205
lenticular galaxies, 190
Leverrier, Urbain, 50
life. *See also* habitable world(s); life on Earth
  necessities of, 253–255
  search for extraterrestrial intelligence and, 252, 259–263
  in solar system, 253–257
  surface vs. subsurface, 257–258
  tree of, 253
  in universe, 252, 257–259, 266–267
life on Earth, 11, 74, 87–89, 253–255
  atmosphere and, 78–81, 88, 259
  climate change and, 89–92
  on cosmic calendar, 12–13
  and habitability, 255, 258–259
  heavy bombardment and, 67
  and mass extinctions, 109–111
  necessities of, 253–255
  plate tectonics and, 88
  at seafloor volcanic vents, 102, 254, 256
  and theory of evolution, 263–265

life track, of star, 154–156, 161
Ligeia Mare (Titan), 103
light, 62, 129, 222. *See also specific forms of light*
  atmospheric effects on, 80–81
  basic properties of, 79
  and Doppler effect, 117
  forms of, 79, 184
  and gravitational lensing, 236
  intensity of, 79, 132
  inverse square law for, 135, 196, 201, 203, 255, A-3
  spectra of, 132
  speed of, 3, 10, t48, 79, 130, 131, 167–168, 169, 171–172, 173, 178, 193, 195, 262, A-2
  as thermal radiation, 79, 81
light-collecting area, of telescope, 42
light-year, 2, 4, 5, 10, A-2
LIGO, 179
lithium, 11, 220, 226
lithosphere, 76, 77, 78, 82, 85, 88
Local Group, 190, 205, 207
  Milky Way Galaxy in, 2, 3
local sky, 18, 23, 24
Local Supercluster, 2, 3, 240
lookback time, 4n, 209, 210, 211, 246
Lowell, Percival, 45, 253
low-mass stars, 144, 146, 150
  brown dwarfs as, 150
  death of, 152–156
  habitable zone around, 258
  life stages of, 152–156, 162–163, 176–177
luminosity, 129, 134, 135, 137–144, 146, 147, 149, 176, 179. *See also* Hertzsprung-Russell (H-R) diagrams
  of Cepheid variable stars, 202–203, 205
  and dark matter in galaxy, 234
  and inverse square law for light, 135, 196, 201, 203, 255, A-3
  measuring, 134–135
  patterns in, 140–144
  of quasars, 195–196, 197
  range of, 135
  relation of to period, 203
  of standard candles, 201–202
  of Sun, 129, 135, 147, 152–155, A-2
  of supernovae, 158, 169, 203
lunar eclipses, 28, 29
lunar maria, 81–82
lunar phases, 25–28

M

M4 globular cluster, 161
M31. *See* Andromeda Galaxy
M51 galaxy, 187
M80 globular cluster, 160
M87 elliptical galaxy, 191, 195, 196, 197, 243
M101 spiral galaxy, 190
Magellan spacecraft, 85
Magellanic Clouds, 59, 191
magma, 76
magma chamber, 77
magnesium, fusion of, 157
magnetic field, 76, 79, 89
  of Earth, 76, 77
  of Mars, 84, 85
  of neutron star, 170

  of pulsars, 170
  of Sun, 133
magnetosphere, 77, 85, 89
main sequence, 138, 139, 140, 141, 142, 143, 144, 150
main-sequence lifetime, 138, 139, 143, 144, 162–163
main-sequence stage, 161
main-sequence stars
  as dark matter, 237
  on H-R diagrams, 141, 142, 143
  lifetimes of, 138, 139, 152–159, 162–163, 176
  luminosity of, 138, 140–144, 150, 153, 155, 202
  mass and properties of, 149–151
main-sequence turnoff point, 160, 161
mantle, 76
March equinox, 19, 21
Mars, 13, 56, 58, 61, 67, 82–85. *See also* terrestrial planets
  apparent retrograde motion of, 31, 32, 46, 47
  atmosphere of, 78, 80, 84–85
  "canals" on, 45, 253
  color of, 85
  data on, t55
  distance from Earth of, 7
  geological activity of, 75, 77, 78, 82
  and greenhouse effect, 81, 84, 85
  infrared spectrum of, 259
  internal heat of, 77
  internal structures of, 77
  magnetic field of, 84, 85
  moons of, 60, 67
  possibility of life on, 255–256, 257
  size of, 15
  surface of, 58, 74, 83, 84
  water on, 82–85, 285
Mars Express, 83
Mars InSight mission, 256
Mars Reconnaissance Orbiter, 67, 83
Mars 2020 rover, 257
mass
  of black holes, 173
  center of, 114, 118, 120
  of Earth, A-2
  of electron, A-2
  of extrasolar planets, 120, 121, 122, 123
  on H-R diagram, 143, 144
  of Milky Way Galaxy, 231–234
  in Newton's second law, 41
  in Newton's version of Kepler's third law, 96, 136, 235
  of planets in solar system, t55
  of proton, A-2
  of spiral galaxies, 234–235
  stellar, 134, 136–137, 143, 144, 149–151
  of Sun, 6, t129, A-2
  and velocity, 62
  of white dwarfs, 154, 155, 163
mass-energy, 62, 216, 237, 247
mass extinction, 109–111
mass per nuclear particle, 157, 158
massive star supernova, 169
mathematics
  and science, 10
  skills for, A-4–A-12
Mathilde (asteroid), 58
matter. *See also* dark matter
  and antimatter, 216, 217, 219, 220

density of, 237, 238, 239, 240, 242, 244, 247
exotic, 237, 238, 247
ordinary, 237–238, 247
structure of, 217
matter-antimatter particles, 216, 219
Maury, Antonia, 140
MAVEN orbiter, 85
McNaught, Comet, 59
megaparsec, 201, A-2
Mercury, 13, 54, t55, 56, 58, 61, 68, 100, 234, 255. *See also* terrestrial planets
apparent retrograde motion of, 31
data on, t55
geological activity of, 76, 77, 78, 82, 93
giant impact on, 68
impact craters on, 77, 82
internal heat of, 77
internal structure of, 77
orbit of, 50, t55, 56, 68, 139, 172
outgassing on, 78
size of, 15
surface of, 58, 82
meridian, 18
merry-go-round, rotation curve for, 233, 234
MESSENGER spacecraft, 82
metals
in meteorites, 110
in solar nebula, 64, 69
meteor, 13
Meteor Crater, 77, 110
meteorites, 108, 110, 255
age of solar system from, 72
meter, A-11
methane, 58, 80
on Mars, 256
in solar nebula, 64
on Titan, 59, 102, 103, 257
on Uranus and Neptune, t55, 58, 96, 98
metric system, A-11–A-12
microbes, 253, 254, 255
microwaves, 79. *See also* cosmic microwave background
midsummer, 22
migration, of planets, 125
Milky Way Galaxy, 2, 3, 183–189, 201, 202, 203, 204, 205, 209, 211, 212, 240
bulge of, 190
center of, 187–188
civilizations in, 260, 261, 262
dark matter in, 233–234
disk vs. spheroidal populations of, 190
formation of, 10, 11, 12, 188–189, 190
gas in, 185–186, 234
mass of, 233, 234
motion of stars within, 183, 185, 233–234
number of stars in, 8
planetary systems in, 122–124
recycling of gases in, 185–186
rotation curve for, 233, 234
size of, 7–8
spiral arms of, 183, 186–187, 190
star-forming regions of, 186
structure of, 183, 187–188
views of, 59, 183, 184
Mimas (moon of Saturn), 100
impact of on Saturn's rings, 99
Miranda (moon of Uranus), 100, 104

model
scientific, 36, 43, 44, 45, 46, 47, t48
stellar, testing, 159–163
molecular clouds
in Milky Way, 186, 188
and spiral density waves, 187
and star formation, 147–148
in star–gas–star cycle, 185, 186
momentum, angular, law of conservation of, 62, 63, 168, 170, 189
month, 27
synodic, A-2
Moon (of Earth), 13, 57, 61, 100, 255, 257
age of, 71–72
angular size of, 26
"dark side" of, 27
in daytime, 27
distance of from Earth, 7
and Earth tides, 28, 101
eclipses of, 28–30
as fifth terrestrial world, 75
formation of, 67–68, 72
Galileo's observations of, 41
geological activity of, 75, 78, 81–82
impact craters on, 67
interior structure of, 76
movement of in night sky, 25–27
orbit of, 29–30
outgassing on, 78
phases of, 25–28
rocks from, 71–72, 81
surface of, 81–82
synchronous rotation of, 27, 101
tidal force of Earth on, 101
moonlet, 99
moons, 5. *See also* jovian moons; *names of specific moons*
captured, 67, 100, 104
features of, 66, 67–68
gap, 99
orbits of, 67
of planets in solar system, t55, 61, 66
tidal forces and, 100–105
motion. *See also* planetary motion
Newton's laws of, 41, 50, 235, 236–237, 248
patterns of, in solar system, 56, 60, 61–64
Mount Sharp (Mars), 84
Mount Wilson Observatory, 204
multiple shell–fusion star, 163, 177
mutation, 265

N

NASA, 74, 103, 109, 115, 116, 117, 133, 224, 225, 257
natural selection, 264
Near-Earth Asteroid Rendezvous (NEAR) spacecraft, 58
nebula, 61. *See also* planetary nebula; solar nebula
nebular theory of solar system formation, 61–64, 67, 68, 69, 96, 124–126
neon, fusion of, 157
Neptune, 13, 56, 59, 95, 123. *See also* jovian planets
atmosphere of, 98
color of, 98
composition of, 96
data on, t55, 95

discovery of, 13, 50
formation of, 96
interior structure of, 97
moons of, 100, 104–105, 257
orbit of, 14
and orbits of comets, 106
size of, 6, 15, 95, 96
Nereid (moon of Neptune), 100
neutrinos
in particle era, 220
solar, 131
from supernovae, 158
weak force and, 238
neutron degeneracy pressure, 169, 173, 177
neutrons, 70, 130, 152, 157, 158, 163, 216, 266
in era of nucleosynthesis, 219, 220, 225–226
in particle era, 219, 220
neutron stars, 158, 166, 169–171, 175, 177
in binary systems, 171
black holes from, 158, 167, 171, 173, 175, 177
discovery of, 170
mass of, 169, 173
merger of, 178, 179
from supernovae, 163, 169, 172
New Horizons spacecraft, 7, 14, 100, 107
new moon, 25, 27
Newton, Isaac, 10, 40–41, 45, 49
Newton's laws of motion, 41, 50, 235, 236–237, 248
Newton's universal law of gravitation, 41, 49–50, 96, A-3
Newton's version of Kepler's third law, 96, 116, 121, 129, 136, 188, 235, A-3
NGC 188 cluster, 161
NGC 801 spiral galaxy, 235
NGC 1300 barred spiral galaxy, 190
NGC 2998 spiral galaxy, 235
NGC 4038/4039 spiral galaxies, 193
NGC 4414 spiral galaxy, 243
NGC 7541 spiral galaxy, 235
nickel, 64, 76
night sky
angular size and distance in, 26
constellations in, 22–25
Milky Way in, 183
Moon movement in, 25–27
planets in, 31–32
nitrogen, 11, 77, 102
nodes, of Moon's orbit, 29, 30
nonscience, vs. science, 43–45
north celestial pole, 23, 24
North Star (Polaris), 18, t137
nova (plural: novae), 168
nuclear fission, 129, 157, 218
nuclear fusion, 5, 9, 11, 62, 95–96, 129, 147, 148, 149, 218, 266. *See also* helium fusion; hydrogen fusion
in brown dwarfs, t137, 150
of carbon, 154, 156, 157, 168
in era of nucleosynthesis, 219, 220
in high-mass stars, 150, 156–157, 162
iron and, 157
neutrinos from, 131
vs. nuclear fission, 129, 157
in red giant stage, 152–153, 156, 161, 162, 163
in star formation, 147, 148
in Sun, 129–130, 149, 150

nuclear reactions, in death of high-mass star, 156–157
nuclei, era of, 219, 220–221, 223, 224, 225, 228, 229
nucleosynthesis, era of, 219, 220, 223, 237–238

O

OBAFGKM sequence, of spectral types, 136, t137, 140, 141
Oberon (moon of Uranus), 100
objectivity, in science, 45
observable universe, 4, 5, 8, 207
observations, in science, 42, 43, 44–45
observatories. *See also* telescope(s)
 European Southern, 170
 Mount Wilson, 204
 Virgo, 179
 Yerkes, 42
Occam's razor, 44
Olympus Mons (Mars), 82
Oort cloud, 57, 60, 68, 105, 106, 107, 108
open cluster, 160
open universe, 229
Opportunity rover, 256
orbit(s), 5
 of asteroids, 94, 105, 106
 around center of mass, 114, 115, 118, 120
 circular, belief in, 38, 47
 of Earth, 2, 5, 7, 10, 18, 19, 20, 24–25, t55
 elliptical, 39, 47, 67, 101, 104, 120, 171
 of Eris, t55
 and escape velocity, 171
 of extrasolar planets, 117, 118, 120, 121
 of Io, 101
 of Jupiter, t55
 in Kepler's laws of planetary motion, 39, 40, 47
 of Mars, t55
 of Mercury, 50, t55, 56, 68, 139, 172
 of Moon, 25–27, 29–30
 of Neptune, t55
 and orbital velocity formula, 235, A-3
 of planets in solar system, t55, 56–57
 of Pluto, 14, t55
 of Saturn, t55
 solar system formation and, 63–64
 of stars, 114, 118, 183–185
 of Sun, 114, 118, 235
 of Triton, 104
 of Uranus, 50, t55
 of Venus, t55
orbital distance, 96
orbital period, 39, t55, 96, 116
orbital resonance, 99
 of asteroids and Jupiter, 105–106, 108
 of jovian moons, 101
 of Kuiper belt objects and Neptune, 106, 108
 in planetary rings, 99
orbital speed
 dark matter and, 233–234, 248
 in Kepler's third law, 39
 rotation curves for, 233, 234, 235
orbital velocity, 235, A-3. *See also* orbital speed
ordinary matter, 237–238, 247
Orion, 4, 22–23, 135, t137, 139, 156
Orion Nebula, 23, 146, 186, 243
orphan planet, 259
outer planets. *See* jovian planets

outgassing, 77
oxygen, 11, 157, 254, 259
 life on Earth and, 87
ozone, 87, 259
 absorption of ultraviolet light by, 80
ozone layer, 74, 80, 87

P

paradigm, in science, 45
parallax. *See* stellar parallax
parallax angle, 201
parallax formula, A-3
parsec (pc), 201, A-2
partial lunar eclipse, 28, 29
partial shadow, 28
partial solar eclipse, 28, 29
particle accelerator, 217, 238
particle era, 219–220, 223
particles, creation and annihilation of in early universe, 216–217
penumbra, 28
penumbral lunar eclipse, 28, 29
Penzias, Arno, 221, 224
perihelion, 39
period–luminosity relation, 202
Persei cluster, 161
phases of Moon, 25–28
Philosophiae Naturalis Principia Mathematica (Newton), 41, 49
Phobos (moon of Mars), 67
photons, 79
 after Big Bang, 216, 219, 220, 221, 222, 223
 in cosmic microwave background, 221, 223, 224
 from energy-level transitions, 152
 energy of, A-3
 in expanding universe, 209
 and gravitational lensing, 236
 and greenhouse effect, 81
 from solar fusion reactions, 130–131
photosphere, 131–132, 133
photosynthesis, atmospheric oxygen from, 87, 259
physics, quantum, 173
Pickering, Edward, 140, 141
Planck, Max, 219
Planck era, 219, 222
Planck satellite, 225, 229
Planck's constant, A-2
planet(s). *See also* dwarf planets; extrasolar planets; jovian planets; terrestrial planets; *names of specific planets*
 apparent retrograde motion of, 31–32, 36–37, 44, 46–47
 definition of, 5, 12–15
 Earth-like, 122–124, 258, 259
 features of, 65–66
 formation of, 64–66, 267
 habitable, 260
 light from, 80
 migration of, 125
 names of, 13
 orbits of, t55, 56–57, 125
 orphan, 259
 relative sizes of, 15
 types of in solar system, 60, 65–66
planetary migration, 125
planetary motion, 46–47. *See also* orbit(s)

 ancient Greeks' explanation of, 31–33, 36–37
 apparent retrograde, 31–32, 36–37, 44, 46–47
 Kepler's laws of, 39–40, 42, 44, 47, 51, 201
 Newton's version of Kepler's third law of, 96, 116, 129, 136, 188, 235, A-3
 Ptolemaic model of, 36–37, 38, 41, 44
planetary nebula, 154, 155, 156, 163, 167, 185
planetary size, and geological history, 89
planetary types, 125
planetesimals, 65–68, 69, 106, 107.
 See also asteroids; comets; meteorites
"planet nine," 15
plasma, 129, 130, 131, 133, 219, 220
plate tectonics, 78. *See also* tectonics
 absence of on Venus, 86, 259
 carbon dioxide cycle and, 88
 life on Earth and, 88
Plato, 36
Pleiades, 160–161
Pluto, 100, 256
 data on, t55
 distance from Sun of, 7
 as dwarf planet, 5, 6, 14, 15, 60, 106–107
 location of, 59–60
 orbit of, 14
 moon of, 14
 properties of, t55
 relative size of, 15
Polaris, t137
 axis tilt of Earth and, 18
 on H-R diagram, 143
Populations I and II, 188
positron, 216
potassium-40, 71
potential energy, 62
power, units of, A-2
powers of 10, A-4–A-6
pressure. *See also* degeneracy pressure
 in brown dwarf, 150–151
 and star life cycle, 147, 148, 149, 176–177
 thermal, 150
 in white dwarf, 167
primary mirror, of telescope, 42
Principia (Newton), 41, 49
prism, 132
Procyon, 134, 143
Procyon B, 139, 143
protogalactic clouds, 212, 221, 225, 239, 243
 dark matter and, 239
 in early universe, 212, 221
 in formation of galaxies, 188–189, 192–193
protons, 70, 130, 149, 152, 154, 157, 158, 159, 216, 217, 222–223, 266
 in era of nucleosynthesis, 219, 220, 225–226
 mass of, A-2
 in particle era, 219, 220
protostar, 148, 150, 152, 162, 176
Proxima Centauri, 135, t137, 139, 143
Ptolemaic model, 36–37, 38, 41, 44
Ptolemy, Claudius, 36, 37, 39
pulsars, 170–171. *See also* neutron stars
Pythagoras, 36

Q

quantum fluctuations, 227, 242
quantum laws, 152
quantum mechanics, 51, 132, 151, 152, 167, 227

quantum physics, singularity and, 173
quarks, 217, 220
quasars, 195, 196, 197

R

radar ranging, 201, 205, 206
radiation. *See also* light
 from Big Bang, 215, 221, 224–225
 thermal, 79, 81, 131, 136
radiation zone, in Sun, 131, 133
radiative energy, 62
radio telescopes, 174, 252, 262
radio waves, 79, 170, 184, 201, 205, 233
 from Milky Way galaxy, 184
 from neutron stars, 170
radioactive decay, 70
radioactive isotopes, dating by, 71, 72
radiometric dating, 70–72
raisin cake analogy, for expansion of universe, 206–207
random walk, 131
ratios, finding, A-10–A-11
recession velocity, 204, 205
recollapsing universe, 244, 245, 246
recycling, of stellar material, 9, 11, 185–186, 189, 243, 267
red giant, 152–153, 154–156, 162, 163
Red Sea, 78
red sequence, 194, 195
red supergiant, 156, 162
redshift, 115, 116, 117, 118, 204, 205, 209, 210, 239, 246
 cosmological, 209–210
 gravitational, 174, 175
 of spectra of galaxies, 204, 205
reflecting telescope, 42
refracting telescope, 42
relativity. *See* general theory of relativity; special theory of relativity
repulsive force, 244, 246
resonance. *See* orbital resonance
retrograde motion, apparent. *See* apparent retrograde motion
revolution, 5
Rhea (moon of Saturn), 100
Rigel, t137, 143
rings, of jovian planets, t55, 59, 98–99
rock
 carbonate, 86, 87, 88, 89
 condensation temperatures of, 64
 radiometric dating of, 71–72
 sedimentary, 78
 in solar nebula, 64
rock droplets, 110
rocket, and Newton's first law, 41
Rosetta spacecraft, 106
rotation, 5
 and conservation of angular momentum, 62, 63
 of Earth, 5, 17, 18, 23, 25, 101
 and movement of stars in night sky, 23–25
 of planets in solar system, t55
 of solar nebula, 63
 of Sun, t55, 56, 129
 synchronous, 27, 101
 of Uranus, t55, 57, 60
 of Venus, t55, 60, 68, 85
rotation curve, 233, 234, 235

runaway greenhouse effect, 86–87, 88
Russell, Henry Norris, 141

S

saddle-shaped universe, 228, 229, 247
Sagan, Carl, 11
 cosmic calendar of, 12–13
Sagittarius A* (Sgr A*), 188
Sagittarius dwarf galaxy, 189
saros cycle, 30
satellite, natural, 5. *See also* moons
Saturn, 13, 57, 59, 94, 95, 123. *See also* jovian planets
 atmosphere of, 98, 99
 axis tilt of, 19
 composition of, 95, 96
 data on, t55, 95
 interior of, 97
 moons of, 60, 99, 100, 102–104, 256–257. *See also* names of specific moons
 possibility of life on moons of, 256–257
 rings of, 98–99
 size of, 15
Schwarzschild radius, 173, 174
science, 12, 35, 36, 37, 39
 hallmarks of, 43–48, 248–249
 vs. nonscience, 43–45
 objectivity in, 45
 Occam's razor in, 44
 theory in, 45, t48
 verifiable observations in, 42, 44–45
science fiction movies, 105, 187
scientific classification, 12–15
scientific method, 43
scientific models, 36, 43, 44, 45, 46, 47, t48
scientific notation, A-6–A-7
scientific theory, 45–48
scientific usage of everyday terms, t48
Scorpius, 147
sea level, increase in, 91
seafloor volcanic vents, 102, 254, 256
search for extraterrestrial intelligence (SETI), 252, 259–263
seasons, 18–22
 eclipse, 29–30
second, A-11
second law of motion, Newton's, 41
second law of planetary motion, Kepler's, 39, 40, 47
secondary mirror, of telescope, 42, 136
sedimentary rock, 78
"seeds," of planets, 64–65, 69
semimajor axis, of ellipse, 39
September equinox, 19, 21
SETI, 252, 259–263
Seven Sisters (Pleiades), 160, 161
Seyfert galaxies, 195
Sgr A*, 188
shocked quartz, 110
SI units, A-11–A-12
sidereal day, A-2
sidereal month, A-2
sidereal year, A-2
silicon, fusion of, 157
singularity, 173
Sirius, 4, 136, t137, 139, 143
 as binary system, 167
67P/Churyumov-Gerasimenko (comet), 106

sky. *See also* night sky
 color of, 80, 81
 local, 18, 23, 24
Sloan Digital Sky Survey, 194, 240, 241, 242
Sloan Great Wall, 240, 241
Small Magellanic Cloud, 59
small solar system bodies, 5, 105, 108–109. *See also* asteroids; comets
solar activity, 133
solar day, A-2
solar eclipses, 28, 29
solar flares, 133
solar maximum, 133, 134
solar minimum, 133, 134
solar nebula, 61–63, 69, 148, 189
 materials in, 64–65
 processes affecting, 63
solar neutrinos, 131
solar system, 2, 3, 5, 6–7
 abundance of elements in, 158–159
 age of, 70–72
 birth of, 61–69
 exceptions to general rules of, 57, 60–61, 67–68
 vs. galaxy, 10
 layout of, 56–57
 life in, 253–257
 location of within Milky Way Galaxy, 2, 3, 233
 major features of, 60, 64–68
 nebular theory of formation of, 61–64, 67, 68, 69, 96, 124–126
 patterns of motion in, 56, 60, 61–64
 Ptolemaic model of, 36, 37, 38, 41, 44
 regions of, 54–60
 rotation curve for, 234
 Voyage scale model of, 6–7
solar weather, 133
solar wind, 54, 66, 69, 77, 84, 85, 89, 96, 106, 132, 133
 Earth's magnetosphere and, 76, 77
solstices, 19, 20, 21
sound
 Doppler effect on, 117
 of space, in science fiction movies, 187
south celestial pole, 23, 24
space, sound of, 187
spacecraft and missions, 257
 Akatsuki, 86
 Apollo, 7, 71, 257
 Cassini, 59, 95, 98, 99, 103, 104, 257
 CHEOPS, 117
 Cosmic Background Explorer (COBE), 224, 225
 Curiosity rover, 74, 83–84, 256
 Dawn, 106
 Europa Clipper, 257
 ExoMars rover, 257
 GAIA, 115, 118, 201
 Galileo, 95, 97, 100, 101
 Hayabusa, 58
 Huygens probe, 103
 Juno, 59, 95, 98
 Jupiter Icy Moons Explorer (JUICE), 257
 Kepler, 115, 117, 122, 124
 Magellan, 85
 Mars Express, 83
 Mars InSight, 256
 Mars Reconnaissance Orbiter, 67, 83

spacecraft and missions (*Continued*)
   Mars 2020, 257
   MAVEN, 85
   MESSENGER, 82
   NEAR (Near-Earth Asteroid Rendezvous), 58
   New Horizons, 7, 14, 100, 107
   Opportunity rover, 25
   Planck satellite, 225, 229
   Rosetta, 106
   Spirit rover, 83, 256
   TESS, 117
   TRACE, 133
   Venera, 86
   Viking, 83, 256
   Voyager, 7, 59, 104
   Wilkinson Microwave Anisotropy Probe (WMAP), 225, 229
spaceship, and Newton's first law, 41
spacetime, 172, 173, 227, 228, 229, 236
special theory of relativity, 172
spectra. *See* spectrum
spectral lines
   as chemical "fingerprints," 132
   and measurement of Doppler shifts, 116
   of stars, 136, t137
spectral type, 136, t137, 138, 140–141, 142, 160, 161, 258
spectroscopy, 132, 152, 259
spectrum (plural: spectra), 79
   absorption line, 132
   continuous, 132
   Doppler shifts in, 115, 117, 118, 120, 132, 136, 209
   electromagnetic, 79, 129, 184, 262
   emission line, 132
   infrared, 259
   thermal radiation, 79, 132, 136, 224, 225
speed. *See also* velocity
   of Earth orbit, 10
   of light, 3, 10, t48, 79, 130, 131, 167–168, 169, 171–172, 173, 178, 195, 262, A-2
   of moving object, formula for, 117
   orbital, 39, 233, 234, 235, 248
spherical universe, 228, 229, 247
Spica, 139, 143
spiral arms, 183, 186–187
spiral density waves, 187
spiral galaxies, 183, 190, 192, 193, 194, 204, 211, 243. *See also* Milky Way Galaxy
   barred, 190
   collisions of, 193, 194
   dark matter in, 234–235
   formation of, 188, 189, 192
   in Hubble's galaxy classes, 191, 192
   rotation curve for, 234, 235
Spirit rover, 83, 256
spring equinox, 19, 21, 45
standard candle, 201–202, 203, 205, 206
star(s), 5, 13. *See also* high-mass stars; low-mass stars; luminosity; neutron stars; protostars; star clusters
   birth of, 146, 147–148, 158, 162
   bulge, 185
   classification of, 140
   colors of, 136, 137–138, 186, 187
   death of, 151–159
   disk, 183–185, 187, 188, 189

   distances to, 7, 201–203
   dying, gas from, 153, 154, 155, 185–186
   element formation in, 9, 11
   gravitational potential energy of, 168, 176
   habitable zones around, 121, 258, 267
   halo, 185, 188–189, 190
   life cycle of. *See* star life cycles
   lifetimes of, 138, 139, 143, 144
   light from, 79
   luminosity of, 134–135
   main-sequence, 138, 140–144, 149–151, 162–163, 176, 202, 237
   masses of, 134, 136–137, 143, 144, 150, 176
   in Milky Way, orbits of, 183, 185, 187, 189
   movement of in night sky, 23–25
   orbital movement of, 114, 115, 117, 118
   patterns among, 137–144
   photographs of, 136
   vs. planet, 13
   properties of, 134–140
   sizes of, 136–137, 141, 143
   spectral type of, 136, t137, 140, 144, 258
   surface temperature of, 134, 136, t137, 138, 139, 140, 141, 142–143
   twinkling of, 42
star clusters, 148, 159–161
   formation of, 148
   in Milky Way Galaxy, 183–185
star–gas–star cycle, 185–186, 189, 243, 267
star life cycles, 9, 11, 147–159
   See also star–gas–star cycle
   and balance between pressure and gravity, 147–148, 166, 176–177
   and degeneracy pressure, 150, 151, 154–158, 163, 177
   of high-mass stars, 162–163
   importance of mass in, 144, 149–151
   and limits on stellar masses, 150
   of low-mass stars, 162–163
   and molecular clouds, 147–148, 186
   protostars in, 148, 150, 152, 162, 176
   in star-forming regions of Milky Way, 186
"star stuff," 9–11, 146, 158–159
star system, 5
Star Trek, 7
Star Wars, 7
state, as scientific term, t48
Stefan–Boltzmann constant, A-2
Stefan–Boltzmann law, A-3
stellar lives. *See* star life cycles
stellar material, recycling of, 9, 11, 185–186, 189, 243, 267
stellar models, testing, 159–163
stellar parallax, 32–33, 38, 40, 135
   distances measured by, 201, 205, 206
stellar thermostat, 149, 176, 267
storms
   on jovian planets, 59, 97, 98
stratosphere, 80
strong force, 130, 217, 218, 219, 227
Subaru (Pleiades), 160–161
sulfur, 157
sulfur-bearing gases, 77
summer solstice, 19, 22
Sun, 2, 13, 54
   age of, 71

   angular size of, 26
   atmosphere of, 131–134
   and center of mass of solar system, 114, 115
   color of, 136, t137
   composition of, t129
   core temperature of, t129, 149
   daily path of, 18
   death of, 152–156
   distances of planets from, t55, 89, 95
   eclipses of, 28, 29
   gravitational contraction in, 148, 152, 153
   gravitational equilibrium in, 148, 149
   gravitational tug of planets on, 114–115, 118
   gravity of, 96
   habitable zone of, 258, 267
   at "high noon," 22
   on H-R diagram, 142, 143, 154–156
   interior of, 130–131, 149
   life stages of, 152–156, 162–163
   life track of, 154–156
   lifetime of, 138, 139
   light from, 79, 80, 86, 87, 129–130, 131
   location of in Milky Way Galaxy, 183, 233
   luminosity of, 129, 147, 152, 153, 154, 155, A-2
   magnetic field of, 133
   mass of, 6, 96, t129, A-2
   misconception regarding burning of, 130
   neutrinos from, 131
   nuclear fusion in, 129–130, 149
   orbit of, 114, 118, 233
   position of in constellations, 24–25
   properties of, 129–130
   red giant stage of, 152–154
   release of energy from, 62, 130–131
   rotation of, 129
   size of, 6, 7, 10, t129, A-2
   spectral type of, 136, t137, 138, 139, 140, 142
   structure of, 133
   sunspots on, 40, 129, 133, 134
   surface temperature of, 129
   thermal radiation from, 130–131, 132
   thermostat of, 149, 267
sunrise, 80
sunset, 80
sunspot cycle, 133, 134
sunspots, 40, 129, 133, 134
superclusters, 2, 5, 239, 240, 244
super-Earths, 122, 123, 125
"super force," 218, 219
supergiants, 138–139, 140, 141, 142, 156, 160, 162, 163
supergravity, 218
supermassive black holes, 175, 195, 196–197
   as energy source for active galactic nuclei, 196–197
supernova (plural: supernovae), 11, 38, 61, 159, 162, 167, 170, 172, 173, 177, 186, 187, 188, 193
   black holes formed from, 158, 163
   from death of high-mass star, 152, 156, 158, 163, 167
   distribution of elements by, 185, 266
   hot gas bubbles from, 185
   luminosities of, 158, 169, 203
   observations of, 38, 41
   spectra of, 169

Type I vs. Type II, 169n
white dwarf, 168–169, 200, 203, 205, 245–246
supernova remnant, 159, 169, 170, 186
superstrings, 218
supersymmetry, 218
surface temperature, of star, 134, 136, t137, 138, 139, 140, 141, 142–143
synchronous rotation, 27, 101
synodic month, A-2
Système Internationale d'Unites (SI), A-11–A-12

T

tail, of comet, 59
tardigrade, 254
Tatooine, 123
Taurus, 160
tectonics, 77, 78, 100
  on Earth, 78, 87, 88
  on Mars, 82, 88
  on Mercury, 82
  on Miranda, 104
  on Venus, 85, 88, 259
telescope(s), 40, 107, 115, 124, 184. *See also* Hubble Space Telescope
  Allen Array, 262
  angular resolution of, 42
  basic design of, 42
  Chandra X-Ray, 167, 169, 186, 236
  Gemini North, 42
  James Webb Space, 212
  Keck, 59, 188
  Large Binocular, 113, 119
  light-collecting area of, 42
  Mount Wilson Observatory, 204
  radio, 174, 252, 262
  reflecting, 42
  refracting, 42
  use of by Galileo, 40, 47
  Very Large, 170
  WMAP, 225, 229
temperature
  average global, 89, 90, 91
  and carbon dioxide cycle, 88
  and color of star, 136, 137–138
  of cosmic microwave background, 224, 225, 228, 229, 230, 238, 247
  global warming and, 90–92
  greenhouse effect and, 80–81, 85–87
  of hot gas in galaxy clusters, 235, 236
  in H-R diagrams, 141, 142
  of molecular clouds, 147, 148, 185, 186
  nuclear fusion rate in Sun and, 130, 149
  of planets in solar system, t55
  scales for, A-12
  in solar nebula, 64, 65
  spectral type and, 136, 140
  of star surface, 134, 136, t137, 138, 139, 140, 141, 142–143
  and thermal energy, 63, 97, 148, 150
  of universe after Big Bang, 215–217, 218, 219, 220, 223
terrestrial planets, 15, 54–58, 61, 122. *See also* names of specific planets
  atmospheres of, 78–81
  differences among, 81–87

formation of, 65–66, 69
geological activity and features of, 75–78
internal heat of, 75–76, 89
internal structures of, 76
vs. jovian planets, 57, 60
magnetic fields of, 76, 77, 89
metal-rich, 123
TESS mission, 117
Tethys (moon of Saturn), 100
theory
  of evolution, 45, 263–265
  of gravity, 45, 49–51
  of relativity, 45
  scientific, 45–48, 52
thermal energy, 62, 63, 75, 97, 148, 150, 152, 168, 176, 177, 196
thermal pressure, 150, 176–177
thermal radiation, 79, 81, 131, 136
  Stefan–Boltzmann law for, A-3
  Wien's law for, A-3
thermal radiation spectrum, 79, 132, 136, 224, 225
thermostat, stellar, 149, 176, 267
third law of motion, Newton's, 41
third law of planetary motion, Kepler's, 39, 41, 42, 47, 96
3C 273 quasar, 196
tidal bulges, 101
tidal force, 101
  near black hole, 175
  in colliding galaxies, 193
  on Earth, 100–101
  on Io, 100, 101
  near jovian planets, 101
  on Moon, 68, 100–101
tidal friction, 101
tidal heating, on jovian moons, 100, 101–102, 105, 256
tidal tails, 193, 194
tides, on Earth, 100–101
time
  astronomical units of, A-2
  near black holes, 173–175
  and distance light travels, 4
  lookback, 4n, 209, 210–211, 246
Titan (moon of Saturn), 58–59, 94, 100
  atmosphere of, 102–103
  geological activity on, 103
  possibility of life on, 256–257
Titania (moon of Uranus), 100
Tombaugh Regio (Pluto), 107
total lunar eclipse, 28, 29
total solar eclipse, 28, 29
totality, of eclipses, 28, 29
TRACE mission, 133
transit method, detection of extrasolar planets by, 115–117, 119, 120, t121
transits, 115, 117, 119, 120, t121
TRAPPIST-1 system, 121
tree of life, 253
tritium, in early universe, 226
Triton (moon of Neptune), 100, 104
  possibility of life on, 257
tropical year, A-2
tropics, 18, 22
Tunguska, Siberia, asteroid explosion over, 109
twinkling, of stars, 42
Tycho Brahe, 38, 40, 42, 44, 47

U

UFOs, 44, 45, 262
UGC 2885 spiral galaxy, 235
ultraviolet light, 74, 79, 80, 84, 86
  loss of water from Mars and, 84
  loss of water from Venus and, 86
ultraviolet radiation, 80, 154
umbra, 28
Umbriel (moon of Uranus), 100
uncertainty, as scientific term, t48
uncertainty principle, 152
units, A-7–A-10
universal law of gravitation. *See* gravitation, universal law of
universe, 2, 3, 4, 5. *See also* Big Bang; cosmic microwave background; inflation; life, in universe
  accelerating, 244–245, 246–247
  age of, 4, 10, 11–12, 208–210
  ancient views of, 2
  birth of, 9, 215
  coasting, 244, 245, 246
  composition of, 220, 225–226, 233
  and cosmological horizon, 210
  critical, 244, 245, 246
  critical density of, 228, 229, 237, 244, 245, 246
  early conditions in, 216–221, 222–223
  Earth's place in, 2–8
  eras in, 218–221, 222–223
  expansion of, 5, 10, 200, 206–207, 208, 216, 219, 221, 222–223, 232, 239, 241, 244–245
  geocentric model of, 36–37, 44, 46–47
  geometry of, 227, 228–229, 237, 247
  history of, 8–12
  observable, 4–5, 8, 209–210
  recollapsing, 244, 245, 246
  scale of, 2–8
  size of, 6–8
  structure in, 226–227, 239–241, 242–243, 249
  temperature of, 216
Uranus, 13, 57, 59, 68, 123. *See also* jovian planets
  atmosphere of, 98, 99
  color of, 98
  composition of, 96
  data on, t55, 95
  formation of, 96
  interior structure of, 97
  moons of, 100, 104–105, 257
  orbit of, 50
  rings of, t55, 98
  rotation of, t55, 57, 60
  size of, 15
  weather on, 97, 98

V

Valles Marineris (Mars), 82, 83
values, as scientific term, t48
Vela Nebula, 170
velocity. *See also* speed
  escape, 171–172, 173, A-2, A-3
  and mass, 62
  and orbital velocity formula, 235, A-3
  recession, 204, 205

Venera lander, 86
Venus, 13, 54, 56, 58, 60, 85–87, 123
    See also terrestrial planets
  atmosphere of, 75, 78, 80, 86
  data on, t55
  geological activity of, 75, 77
  infrared spectrum of, 259
  internal heat of, 77
  internal structures of, 76
  lack of plate tectonics on, 259
  phases of, 40, 41
  rotation of, t55, 60, 68, 85
  and runaway greenhouse effect, 81, 85–87, 88, 259
  size of, 15
  surface of, 58, 78, 85, 86, 255
  temperature of, 86–87
  water on, 86, 255
vernal equinox, 19
Very Large Telescope, 170
Vesta (asteroid), 105, 106
Viking missions, 83, 256
Virgo Cluster of galaxies, 191
Virgo observatory, 179
visible light, 54, 79, 80, 81, 116, 117, 119, 129, 130, 131–132, 167, 183, 184. *See also* light
voids, 2, 208, 240, 241
volcanic vents, 102, 254, 256
volcanism, 75, 77–78
  in carbon dioxide cycle, 88
  on Earth, 75, 77–78
  on Io, 100
  on Mars, 85
  on Mercury, 78
  on Moon, 78
  on moons in outer solar system, 100, 105
  on Venus, 58, 75, 77, 78
volcanoes. *See* volcanism

Voyager missions, 7, 59, 104
Voyage scale model of solar system, 6–7

W

waning moon, 27
War of the Worlds, The (Wells), 253
water, 58, 67, 80
  density of, 97
  on Enceladus, 103–104, 257
  on Europa, 58, 101–102
  on extrasolar planets, 255
  on Mars, 82–85, 255
  as requirement for life, 87, 255
  in solar nebula, 64
  on Venus, 86
  from volcanic outgassing, 77
water bear, 254
water vapor. *See* water
"water worlds," 122, 123, 125
watt, A-2
wave nature of light, 79
wavelength(s), of light, 79, 184
  Doppler effect on, 116
  and emission or absorption of energy, 152
waves
  radio, 79, 170, 184, 201, 205, 233
  spiral density, 187
waxing moon, 27
weak force, 217, 218, 219, 238
weakly interacting massive particles (WIMPs), 238, 247
weakly interacting particles, 238
weather
  and global warming, 91
  and seasons, 22
Wells, H. G., 253
white dwarf, 139–140, 141, 151, 152, 161, 167–169, 171, 175, 177
  from death of low-mass star, 152–156, 163
  on H-R diagram, 142, 143, 155
white dwarf limit, 167–168, 172
white dwarf supernova, 200, 244
  in distance measurements, 203, 205
  formation of, 168–169
  and models of universe expansion, 245–246, 247
Wien's law, A-3
Wilkinson Microwave Anisotropy Probe (WMAP), 225, 229
William of Occam, 44
Wilson, Robert, 221, 224
WIMPs (weakly interacting massive particles), 238, 247
winter solstice, 19, 21

X

x-ray binary, 167, 171, 175, 184
x-rays
  absorption of by Earth's atmosphere, 80
  from black hole, 175
  distribution of in Milky Way, 184
  in electromagnetic spectrum, 79
  from hot gas in galaxy clusters, 235, 236
  from Sun, 80
  from supernova remnants, 186

Y

year, tropical vs. sidereal, A-2
Yerkes Observatory, 42
Yucatán Peninsula, impact crater on, 110

Z

zenith, of local sky, 18, 22, 24
zero, absolute, A-12
zodiac, 24, 25